SIXTH EDITION

Fundamentals of Physics

PART 3

SIXTH EDITION

Fundamentals of Physics

PART 3

David Halliday

University of Pittsburgh

Robert Resnick

Rensselaer Polytechnic Institute

Jearl Walker

Cleveland State University

John Wiley & Sons, Inc.

*New York / Chichester / Weinheim / Brisbane
Singapore / Toronto*

ACQUISITIONS EDITOR   Stuart Johnson
DEVELOPMENTAL EDITOR   Ellen Ford
MARKETING MANAGER   Sue Lyons and Bob Smith
ASSOCIATE PRODUCTION DIRECTOR   Lucille Buonocore
SENIOR PRODUCTION EDITOR   Monique Calello
TEXT/COVER DESIGNER   Madelyn Lesure
COVER PHOTO   Tsuyoshi Nishiinoue/Orion Press
PHOTO MANAGER   Hilary Newman
PHOTO RESEARCHER   Jennifer Atkins
DUMMY DESIGNER   Lee Goldstein
ILLUSTRATION EDITORS   Edward Starr and Anna Melhorn
ILLUSTRATION   Radiant/Precision Graphics
COPYEDITOR   Helen Walden
PROOFREADER   Lilian Brady
TECHNICAL PROOFREADER   Georgia Kamvosoulis Mederer
INDEXER   Dorothy M. Jahoda

This book was set in 10/12 Times Roman by Progressive Information Technologies and printed and bound by Von Hoffmann Press, Inc. The cover was printed by Brady Palmer Printing Company.

This book is printed on acid-free paper.   ∞

The paper in this book was manufactured by a mill whose forest management programs include sustained yield harvesting of its timberlands. Sustained yield harvesting principles ensure that the number of trees cut each year does not exceed the amount of new growth.

The Library of Congress has already cataloged as follows:

Library of Congress Cataloging-in-Publication Data

Halliday, David
    Fundamentals of physics/David Halliday, Robert Resnick, Jearl Walker.—6th ed.
        p. cm
    Includes index.
    ISBN 0-471-33235-6 (v. 1 : cloth : alk. paper)
    1. Physics.   I. Resnick, Robert.   II. Walker, Jearl.   III. Title.

    QC21,2.H35 2001
    530–dc21   00-027365

To order books or for customer service call 1-800-CALL-WILEY (225-5945).

ISBN 0-471-36040-6

Printed in the United States of America

10   9   8   7   6   5   4   3   2   1

BRIEF CONTENTS

TABLES

CONTENTS

CHAPTER 27

Current and Resistance   611

What caused the flaming crash of the Zeppelin Hindenburg?

CHAPTER 28

Circuits   633

How does an electric eel produce a large current?

CHAPTER 29

Magnetic Fields   658

Why is an aurora so thin yet so tall and wide?

CHAPTER 30

Magnetic Fields Due to Currents   686

How can cargo be shot into space?

CHAPTER 31

Induction and Inductance   710

How did the electric guitar revolutionize rock?

CHAPTER 32

Magnetism of Matter; Maxwell's Equations 744

How can a frog be levitated by a magnetic field?

CHAPTER 33

Electromagnetic Oscillations and Alternating Current 768

Why do electrical transmission lines have high potential and not high current?

APPENDICES

ANSWERS TO CHECKPOINTS AND ODD-NUMBERED QUESTIONS, EXERCISES, AND PROBLEMS   AN-1

INDEX   I-1

PREFACE

This sixth edition of *Fundamentals of Physics* contains a redesign and major rewrites of the widely used fifth edition, while maintaining many elements of the classic text first written by David Halliday and Robert Resnick. Nearly all the changes are based on suggestions from instructors and students using the fifth edition, from reviewers of the manuscripts for the sixth edition, and from research done on the process of learning. You can send suggestions, corrections, and positive or negative comments to John Wiley & Sons (http://www.wiley.com/college/hrw) or Jearl Walker (mail address: Physics Department, Cleveland State University, Cleveland OH 44115 USA; fax number: (USA) (216) 687-2424; or email address: physics@wiley.com). We may not be able to respond to all suggestions, but we keep and study each of them.

Design Changes

➤ *More open format.* Previous editions have been printed in a double-column format, which many students and instructors have found cluttered and distracting. In this edition, the narrative is presented in a single-column format with a wide margin for note-taking.

➤ *Streamlined presentation.* It is a common complaint of all texts that they cover too much material. As a response to this criticism, the sixth edition has been shortened in two ways.

    1. Material regarding special relativity and quantum physics has been moved from the early chapters to the later chapters devoted to those subjects.

    2. The essential sample problems have been retained in this book, but the more specialized sample problems have been shifted to the Problem Supplement that is automatically provided with this book. (The Problem Supplement is described on the next page.)

➤ *Vector notation.* Vectors are now presented with an overhead arrow (such as $\vec{F}$) instead of as a bold symbol (such as **F**).

➤ *Emphasis on metric units.* Except in Chapter 1 (in which various systems of units are employed) and certain problems involving baseball (in which English units are traditional), metric units are used almost exclusively.

➤ *Structured versus unstructured order of problems.* The homework problems in this book are still ordered approximately according to their difficulty and grouped under section titles corresponding to the narrative of the chapter. However, many of the homework problems of the fifth edition have been shifted, without order or grouping, to the Problem Supplement. (The total number of problems in this book and in the Problem Supplement exceeds what was available in the fifth edition.)

➤ *Icons for additional help.* When worked-out solutions are provided either in print or electronically for certain of the odd-numbered homework problems, the statements for those problems include a trailing icon to alert both student and instructor as to where the solutions are located. An icon guide is provided here and at the beginning of each set of homework problems:

> ssm   Solution is in the Student Solutions Manual.
> www Solution is available on the World Wide Web at:
>   http://www.wiley.com/college/hrw
> ilw   Solution is available on the Interactive LearningWare.

These resources are described later in this preface.

Pedagogy Changes

➤ *Reasoning versus plug-and-chug.* The primary goal of this book is to teach students to reason through challenging situations, from basic principles to a solution. Although some plug-and-chug homework problems remain in this book, most homework problems emphasize reasoning.

➤ *Key Ideas in the sample problems.* The solutions to all 360 sample problems in this book and in the Problem Supplement have been rewritten to begin with one or more Key Ideas based on basic principles.

➤ *Lengthened solutions to sample problems.* Most of the solutions to the sample problems are now longer because they build step by step from the beginning Key Ideas to an answer, often repeating some of the important reasoning of the narrative preceding the sample problems. For example, see Sample Problem 8-3 on page 148 and Sample Problem 10-2 on pages 200–201.

➤ *Use of vector-capable calculators.* When vector calculations in a sample problem can be performed directly on-screen with a vector-capable calculator, the solution of the sample problem indicates that fact but still carries through the traditional component analysis. When vector calculations cannot be performed directly on-screen, the solution explains why.

➤ *Problems with applied physics,* based on published research, have been added in many places, either as sample problems or homework problems. For example, see Sample Problem 11 6 on page 229, homework problem 64 on page 71, and homework problem 56 on page 214. For an example of homework problems that build with a continuing story, see problems 4, 32, and 48 (on pages 112, 114, and 115, respectively) in Chapter 6.

Content Changes

➤ *Chapter 5 on force and motion* contains clearer explanations of the gravitational force, weight, and normal force (pages 80–82).

➤ *Chapter 7 on kinetic energy and work* begins with a rough definition of energy. It then defines kinetic energy, work, and the work–kinetic energy theorem in ways that are more closely tied to Newton's second law than in the fifth edition, while keeping those definitions consistent with thermodynamics (pages 117–120).

➤ *Chapter 8 on the conservation of energy* avoids the much criticized definition of work done by a nonconservative force by explaining, instead, the energy transfers that occur due to a nonconservative force (page 153). (The wording still allows an instructor to superimpose a definition of work done by a nonconservative force.)

➤ *Chapter 10 on collisions* now presents the general situation of inelastic one-dimensional collisions (pages 198–200) before the special situation of elastic one-dimensional collisions (pages 202–204).

➤ *Chapters 16, 17, and 18 on SHM and waves* have been rewritten to better ease a student into these difficult subjects.

➤ *Chapter 21 on entropy* now presents a Carnot engine as the ideal heat engine with the greatest efficiency.

Chapter Features

➤ *Opening puzzlers.* A curious puzzling situation opens each chapter and is explained somewhere within the chapter, to entice a student to read the chapter.

➤ *Checkpoints* are stopping points that effectively ask the student, "Can you answer this question with some reasoning based on the narrative or sample problem that you just read?" If not, then the student should go back over that previous material before traveling deeper into the chapter. For ex-

ample, see Checkpoint 3 on page 78 and Checkpoint 1 on page 101. **Answers to all checkpoints are in the back of the book.**

➤ *Sample problems* have been chosen to help the student organize the basic concepts of the narrative and to develop problem solving skills. Each sample problem builds step by step from one or more Key Ideas to a solution.

➤ *Problem-solving tactics* contain helpful instructions to guide the beginning physics student as to how to solve problems and to avoid common errors.

➤ *Review & Summary* is a brief outline of the chapter contents that contains the essential concepts but which is not a substitute for reading the chapter.

➤ *Questions* are like the checkpoints and require reasoning and understanding rather than calculations. **Answers to the odd-number questions are in the back of the book.**

➤ *Exercises & Problems* are ordered approximately according to difficulty and grouped under section titles. **The odd-numbered ones are answered in the back of the book.** Worked-out solutions to the odd-numbered problems with trailing icons are available either in print or electronically. (See the icon guide at the beginning of the Exercises & Problems.) A problem number with a star indicates an especially challenging problem.

➤ *Additional Problems* appear at the end of the Exercises & Problems in certain chapters. They are not sorted according to section titles and many involve applied physics.

Problem Supplement

A problem supplement is automatically provided with this book. The *Problem Supplement #1* (green book) will be provided until May 15, 2002. Thereafter, the *Problem Supplement #2* (blue book) will be provided. The blue book will have a different set of questions and homework problems and will contain more sample problems. The features of both versions of the problem supplement are the following:

➤ *Additional sample problems* that were shifted from the main book, plus many new ones. All begin with the basic *Key Ideas* and then build step by step to a solution.

➤ *Questions* include:
   1. *Checkpoint-style questions,* as in the main book.
   2. *Organizing questions,* which request that equations be set up for common situations, as a warm-up for the homework problems.

3. *Discussion questions* from the fourth and earlier editions of the book (back by request).

➤ *Exercises & Problems.* More homework problems, including many shifted from the main book. These are *not* ordered according to difficulty, section titles, or appearance of the associated physics in the chapter. Some of the new problems involve applied physics. In some chapters the homework problems end with *Clustered Problems,* in which similar problems are grouped together. In the other chapters, the homework problems end with *Tutorial Problems,* in which solutions are worked out.

Versions of the Text

The sixth edition of *Fundamentals of Physics* is available in a number of different versions, to accommodate the individual needs of instructors and students. The Regular Edition consists of Chapters 1 through 38 (ISBN 0-471-32000-5). The Extended Edition contains seven additional chapters on quantum physics and cosmology (Chapters 1–45) (ISBN 0-471-33236-4). Both editions are available as single, hardcover books, or in the following alternative versions:

➤ *Volume 1—Chapters 1–21 (Mechanics/Thermodynamics), hardcover, 0-471-33235-6*

➤ *Volume 2—Chapters 22–45 (E&M and Modern Physics), hardcover, 0-471-36037-6*

➤ *Part 1—Chapters 1–12, paperback, 0-471-33234-8*

➤ *Part 2—Chapters 13–21, paperback, 0-471-36041-4*

➤ *Part 3—Chapters 22–33, paperback, 0-471-36040-6*

➤ *Part 4—Chapters 34–38, paperback, 0-471-36039-2*

➤ *Part 5—Chapters 39–45, paperback, 0-471-36038-4*

Supplements

The sixth edition of *Fundamentals of Physics* is supplemented by a comprehensive ancillary package carefully developed to help teachers teach and students learn.

Instructor's Supplements

➤ *Instructor's Manual* by J. RICHARD CHRISTMAN, U.S. Coast Guard Academy. This manual contains lecture notes outlining the most important topics of each chapter, demonstration experiments, laboratory and computer projects, film and video sources, answers to all Questions, Exercises & Problems, and Checkpoints, and a correlation guide to the Questions and Exercises & Problems in the previous edition.

➤ *Instructor's Solutions Manual* by JAMES WHITENTON, Southern Polytechnic University. This manual provides worked-out solutions for all the exercises and problems found at the end of each chapter within the text and in the Problem Supplement #1. *This supplement is available only to instructors.*

➤ *Test Bank* by J. RICHARD CHRISTMAN, U.S. Coast Guard Academy. More than 2200 multiple-choice questions are included in this manual. These items are also available in the Computerized Test Bank (see below).

➤ *Instructor's Resource CD.* This CD contains:
- All of the Instructor's Solutions Manual in both LaTex and PDF files.
- Computerized Test Bank in both IBM and Macintosh versions, with full editing features to help instructors customize tests.
- All text illustrations suitable for both classroom presentation and printing.

➤ *Transparencies.* More than 200 four-color illustrations from the text are provided in a form suitable for projection in the classroom.

➤ *On-line Course Management.*
- WebAssign, CAPA, and Wiley eGrade are on-line homework and quizzing programs that give instructors the ability to deliver and grade homework and quizzes over the Internet.
- Instructors will also have access to WebCT course materials. WebCT is a powerful Web site program that allows instructors to set up complete on-line courses with chat rooms, bulletin boards, quizzing, student tracking, etc. Please contact your local Wiley representative for more information.

Student's Supplements

➤ *A Student Companion* by J. RICHARD CHRISTMAN, U.S. Coast Guard Academy. This student study guide consists of a traditional print component and an accompanying Web site, which together provide a rich, interactive environment for review and study. The Student Companion Web site includes self-quizzes, simulation exercises, hints for solving end-of-chapter problems, the *Interactive LearningWare* program (see the next page), and links to other Web sites that offer physics tutorial help.

➤ *Student Solutions Manual* by J. RICHARD CHRISTMAN, U.S. Coast Guard Academy and EDWARD DERRINGH, Wentworth Institute. This manual provides students with complete worked-out solutions to 30 percent of the exercises and problems found at the end of each chapter within the text. These problems are indicated with an `ssm` icon in the text.

➤ *Interactive LearningWare.* This software guides students through solutions to 200 of the end-of-chapter problems. The solutions process is developed interactively, with appropriate feedback and access to error-specific help for the most common mistakes. These problems are indicated with an `ilw` icon in the text.

➤ *CD-Physics, 3.0.* This CD-ROM based version of *Fundamentals of Physics,* Sixth Edition, contains the complete, extended version of the text, *A Student's Companion,* the *Student's Solutions Manual,* the *Interactive LearningWare,* and numerous simulations all connected with extensive hyperlinking.

➤ *Take Note!* This bound notebook lets students take notes directly onto large, black-and-white versions of textbook illustrations. All of the illustrations from the transparency set are included. In-class time spent copying illustrations is substantially reduced by this supplement.

➤ *Physics Web Site.* This Web site, **http://www.wiley.com/college/hrw**, was developed specifically for *Fundamentals of Physics,* Sixth Edition, and is designed to further assist students in the study of physics and offers additional physics resources. The site also includes solutions to selected end-of-chapter problems. These problems are identified with a `www` icon in the text.

ACKNOWLEDGMENTS

A textbook contains far more contributions to the elucidation of a subject than those made by the authors alone. J. Richard Christman, of the U.S. Coast Guard Academy, has once again created many fine supplements for us; his knowledge of our book and his recommendations to students and faculty are invaluable. James Tanner, of Georgia Institute of Technology, and Gary Lewis, of Kennesaw State College, have provided us with innovative software, closely tied to the text exercises and problems. James Whitenton, of Southern Polytechnic State University, and Jerry Shi, of Pasadena City College, performed the Herculean task of working out solutions for every one of the Exercises & Problems in the text. We thank John Merrill, of Brigham Young University, and Edward Derringh, of the Wentworth Institute of Technology, for their many contributions in the past. We also thank George W. Hukle of Oxnard, California, and Frank G. Jacobs of Evanston, Illinois, for their check of the answers for the problems in the book.

At John Wiley publishers, we have been fortunate to receive strong coordination and support from our former editor, Cliff Mills. Cliff guided our efforts and encouraged us along the way. When Cliff moved on to other responsibilities at Wiley, we were ably guided to completion by his successor, Stuart Johnson. Ellen Ford has coordinated the developmental editing and multilayered preproduction process. Sue Lyons, our marketing manager, has been tireless in her efforts on behalf of this edition. Joan Kalkut has built a fine supporting package of ancillary materials. Thomas Hempstead managed the reviews of manuscript and the multiple administrative duties admirably.

We thank Lucille Buonocore, our production director and Monique Calello, our production editor, for pulling all the pieces together and guiding us through the complex production process. We also thank Maddy Lesure, for her design; Helen Walden for her copyediting; Edward Starr and Anna Melhorn, for managing the illustration program; Georgia Kamvosoulis Mederer, Katrina Avery, and Lilian Brady, for their proofreading; and all other members of the production team.

Hilary Newman and her team of photo researchers were inspired in their search for unusual and interesting photographs that communicate physics principles beautifully. We also owe a debt of gratitude for the line art to the late John Balbalis, whose careful hand and understanding of physics can still be seen in every diagram.

We especially thank Edward Millman for his developmental work on the manuscript. With us, he has read every word, asking many questions from the point of view of a student. Many of his questions and suggested changes have added to the clarity of this volume.

We owe a particular debt of gratitude to the numerous students who used the previous editions of *Fundamentals of Physics* and took the time to fill out the response cards and return them to us. As the ultimate consumers of this text, students are extremely important to us. By sharing their opinions with us, your students help us ensure that we are providing the best possible product and the most value for their textbook dollars. We encourage the users of this book to contact us with their thoughts and concerns so that we can continue to improve this text in the years to come.

Finally, our external reviewers have been outstanding and we acknowledge here our debt to each member of that team:

Edward Adelson
Ohio State University

Mark Arnett
Kirkwood Community College

Arun Bansil
Northeastern University

J. Richard Christman
U.S. Coast Guard Academy

Robert N. Davie, Jr.
St. Petersburg Junior College

Cheryl K. Dellai
Glendale Community College

Eric R. Dietz
California State University at Chico

N. John DiNardo
Drexel University

Harold B. Hart
Western Illinois University

Rebecca Hartzler
Edmonds Community College

Joey Huston
Michigan State University

Hector Jimenez
University of Puerto Rico

Sudhakar B. Joshi
York University

Leonard M. Kahn
University of Rhode Island

Yuichi Kubota
Cornell University

Priscilla Laws
Dickinson College

Edbertho Leal
Polytechnic University of Puerto Rico

Dale Long
Virginia Tech

Andreas Mandelis
University of Toronto

Paul Marquard
Caspar College

James Napolitano
Rensselaer Polytechnic Institute

Des Penny
Southern Utah University

Joe Redish
University of Maryland

Timothy M. Ritter
University of North Carolina at Pembroke

Gerardo A. Rodriguez
Skidmore College

John Rosendahl
University of California at Irvine

Michael Schatz
Georgia Institute of Technology

Michael G. Strauss
University of Oklahoma

Dan Styer
Oberlin College

Marshall Thomsen
Eastern Michigan University

Fred F. Tomblin
New Jersey Institute of Technology

B. R. Weinberger
Trinity College

William M. Whelan
Ryerson Polytechnic University

William Zimmerman, Jr.
University of Minnesota

Reviewers of the Fifth and Previous Editions

Maris A. Abolins
Michigan State University

Barbara Andereck
Ohio Wesleyan University

Albert Bartlett
University of Colorado

Michael E. Browne
University of Idaho

Timothy J. Burns
Leeward Community College

Joseph Buschi
Manhattan College

Philip A. Casabella
Rensselaer Polytechnic Institute

Randall Caton
Christopher Newport College

J. Richard Christman
U.S. Coast Guard Academy

Roger Clapp
University of South Florida

W. R. Conkie
Queen's University

Peter Crooker
University of Hawaii at Manoa

William P. Crummett
Montana College of Mineral Science and Technology

Eugene Dunnam
University of Florida

Robert Endorf
University of Cincinnati

F. Paul Esposito
University of Cincinnati

Jerry Finkelstein
San Jose State University

Alexander Firestone
Iowa State University

Alexander Gardner
Howard University

Andrew L. Gardner
Brigham Young University

John Gieniec
Central Missouri State University

John B. Gruber
San Jose State University

Ann Hanks
American River College

Samuel Harris
Purdue University

Emily Haught
Georgia Institute of Technology

Laurent Hodges
Iowa State University

John Hubisz
North Carolina State University

Joey Huston
Michigan State University

Darrell Huwe
Ohio University

Claude Kacser
University of Maryland

Leonard Kleinman
University of Texas at Austin

Earl Koller
Stevens Institute of Technology

Arthur Z. Kovacs
Rochester Institute of Technology

Kenneth Krane
Oregon State University

Sol Krasner
University of Illinois at Chicago

Peter Loly
University of Manitoba

Robert R. Marchini
Memphis State University

David Markowitz
University of Connecticut

Howard C. McAllister
University of Hawaii at Manoa

W. Scott McCullough
Oklahoma State University

James H. McGuire
Tulane University

David M. McKinstry
Eastern Washington University

Joe P. Meyer
Georgia Institute of Technology

Roy Middleton
University of Pennsylvania

Irvin A. Miller
Drexel University

Eugene Mosca
United States Naval Academy

Michael O'Shea
Kansas State University

Patrick Papin
San Diego State University

George Parker
North Carolina State University

Robert Pelcovits
Brown University

Oren P. Quist
South Dakota State University

Jonathan Reichart
SUNY—Buffalo

Manuel Schwartz
University of Louisville

Darrell Seeley
Milwaukee School of Engineering

Bruce Arne Sherwood
Carnegie Mellon University

John Spangler
St. Norbert College

Ross L. Spencer
Brigham Young University

Harold Stokes
Brigham Young University

Jay D. Strieb
Villanova University

David Toot
Alfred University

J. S. Turner
University of Texas at Austin

T. S. Venkataraman
Drexel University

Gianfranco Vidali
Syracuse University

Fred Wang
Prairie View A & M

Robert C. Webb
Texas A & M University

George Williams
University of Utah

David Wolfe
University of New Mexico

22 Electric Charge

If you adapt your eyes to darkness for about 15 minutes and then have a friend chew a wintergreen LifeSaver, you will see a faint flash of blue light from your friend's mouth with each chomp. (To avoid wear on the teeth, you might crush the candy with pliers, as in the photograph.)

What causes this display of light, commonly called "sparking"?

The answer is in this chapter.

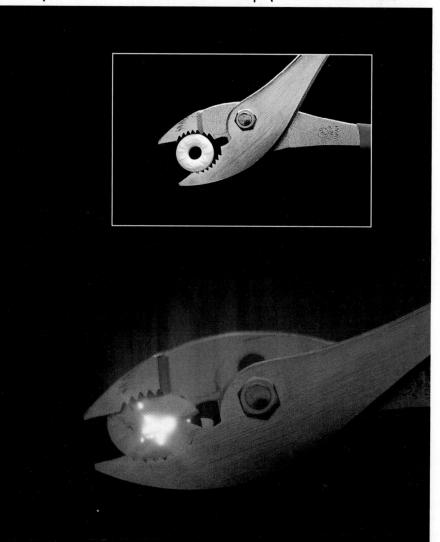

22-1   Electromagnetism

The early Greek philosophers knew that if you rubbed a piece of amber, it would attract bits of straw. This ancient observation is a direct ancestor of the electronic age in which we live. (The strength of the connection is indicated by our word *electron*, which is derived from the Greek word for amber.) The Greeks also recorded the observation that some naturally occurring "stones," known today as the mineral magnetite, would attract iron.

From these modest origins, the sciences of electricity and magnetism developed separately for centuries—until 1820, in fact, when Hans Christian Oersted found a connection between them: an electric current in a wire can deflect a magnetic compass needle. Interestingly enough, Oersted made this discovery while preparing a lecture demonstration for his physics students.

The new science of *electromagnetism* (the combination of electrical and magnetic phenomena) was developed further by workers in many countries. One of the best was Michael Faraday, a truly gifted experimenter with a talent for physical intuition and visualization. That talent is attested to by the fact that his collected laboratory notebooks do not contain a single equation. In the mid-19th century, James Clerk Maxwell put Faraday's ideas into mathematical form, introduced many new ideas of his own, and put electromagnetism on a sound theoretical basis.

Table 32-1 shows the basic laws of electromagnetism, now called Maxwell's equations. We plan to work our way through them in the chapters between here and there, but you might want to glance at them now, to see our goal.

22-2   Electric Charge

If you walk across a carpet in dry weather, you can produce a spark by bringing your finger close to a metal doorknob. Television advertising has alerted us to the problem of "static cling" in clothing (Fig. 22-1). On a grander scale, lightning is familiar to everyone. Each of these phenomena represents a tiny glimpse of the vast amount of *electric charge* that is stored in the familiar objects that surround us and—indeed—in our own bodies. **Electric charge** is an intrinsic characteristic of the fundamental particles making up those objects; that is, it is a characteristic that automatically accompanies those particles wherever they exist.

The vast amount of charge in an everyday object is usually hidden because the object contains equal amounts of the two kinds of charge: *positive charge* and *negative charge*. With such an equality—or *balance*—of charge, the object is said to be *electrically neutral*; that is, it contains no *net* charge. If the two types of charge are not in balance, then there *is* a net charge. We say that an object is *charged* to indicate that it has a charge imbalance, or net charge. The imbalance is always very small compared to the total amounts of positive charge and negative charge contained in the object.

Charged objects interact by exerting forces on one another. To show this, we first charge a glass rod by rubbing one end with silk. At points of contact between the rod and the silk, tiny amounts of charge are transferred from one to the other, slightly upsetting the electrical neutrality of each. (We *rub* the silk over the rod to increase the number of contact points and thus the amount, still tiny, of transferred charge.)

Suppose we now suspend the charged rod from a thread to *electrically isolate* it from its surroundings so that its charge cannot change. If we bring a second, similarly charged, glass rod nearby (Fig. 22-2a), the two rods *repel* each other; that is, each rod experiences a force directed away from the other rod. However, if we rub a plastic rod with fur and bring it near the suspended glass rod (Fig. 22-2b), the

Fig. 22-1 Static cling, an electrical phenomenon that accompanies dry weather, causes these pieces of paper to stick to one another and to the plastic comb, and your clothing to stick to your body.

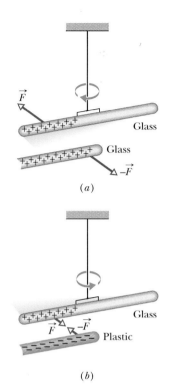

Fig. 22-2 (*a*) Two charged rods of the same signs repel each other. (*b*) Two charged rods of opposite signs attract each other. Plus signs indicate a positive net charge, and minus signs a negative net charge.

two rods *attract* each other; that is, each rod experiences a force directed toward the other rod.

We can understand these two demonstrations in terms of positive and negative charges. When a glass rod is rubbed with silk, the glass loses some of its negative charge and then has a small unbalanced positive charge (represented by the plus signs in Fig. 22-2*a*). When the plastic rod is rubbed with fur, the plastic gains a small unbalanced negative charge (represented by the minus signs in Fig. 22-2*b*). Our two demonstrations reveal the following:

> Charges with the same electrical sign repel each other, and charges with opposite electrical signs attract each other.

In Section 22-4, we shall put this rule into quantitative form as Coulomb's law of *electrostatic force* (or *electric force*) between charges. The term *electrostatic* is used to emphasize that, relative to each other, the charges are either stationary or moving only very slowly.

The "positive" and "negative" labels and signs for electric charge were chosen arbitrarily by Benjamin Franklin. He could easily have interchanged the labels or used some other pair of opposites to distinguish the two kinds of charge. (Franklin was a scientist of international reputation. It has even been said that Franklin's triumphs in diplomacy in France during the American War of Independence were facilitated, and perhaps even made possible, because he was so highly regarded as a scientist.)

The attraction and repulsion between charged bodies have many industrial applications, including electrostatic paint spraying and powder coating, fly-ash collection in chimneys, nonimpact ink-jet printing, and photocopying. Figure 22-3 shows a tiny carrier bead in a Xerox copying machine, covered with particles of black powder called *toner*, which stick to it by means of electrostatic forces. The negatively charged toner particles are eventually attracted from the carrier bead to a rotating drum, where a positively charged image of the document being copied has formed. A charged sheet of paper then attracts the toner particles from the drum to itself, after which they are heat-fused in place to produce the copy.

Fig. 22-3 A carrier bead from a Xerox copying machine; it is covered with toner particles that cling to it by electrostatic attraction. The diameter of the bead is about 0.3 mm.

22-3 Conductors and Insulators

In some materials, such as metals, tap water, and the human body, some of the negative charge can move rather freely. We call such materials **conductors.** In other materials, such as glass, chemically pure water, and plastic, none of the charge can move freely. We call these materials **nonconductors** or **insulators.**

If you rub a copper rod with wool while holding the rod in your hand, you will not be able to charge the rod, because both you and the rod are conductors. The rubbing will cause a charge imbalance on the rod, but the excess charge will immediately move from the rod through you to the floor (which is connected to Earth's surface), and the rod will quickly be neutralized.

In thus setting up a pathway of conductors between an object and Earth's surface, we are said to *ground* the object, and in neutralizing the object (by eliminating an unbalanced positive or negative charge), we are said to *discharge* the object. If instead of holding the copper rod in your hand, you hold it by an insulating handle, you eliminate the conducting path to Earth, and the rod can then be charged by rubbing, as long as you do not touch it directly with your hand.

The properties of conductors and insulators are due to the structure and electrical nature of atoms. Atoms consist of positively charged *protons*, negatively charged

Fig. 22-4 A neutral copper rod is electrically isolated from its surroundings by being suspended on a nonconducting thread. Either end of the copper rod will be attracted by a charged rod. Here, conduction electrons in the copper rod are repelled to the far end of that rod by the negative charge on the plastic rod. Then that negative charge attracts the remaining positive charge on the near end of the copper rod, rotating the copper rod to bring that near end closer to the plastic rod.

Fig. 22-5 This is not a parlor stunt but a serious experiment carried out in 1774 to prove that the human body is a conductor of electricity. The etching shows a person suspended by nonconducting ropes while being charged by a charged rod (which probably touched flesh instead of the trousers). When the person brought his face, left hand, or the conducting ball and rod in his right hand near bits of paper on the plates, charge was induced on the paper, which then flew through the intermediate air to him.

electrons, and electrically neutral *neutrons*. The protons and neutrons are packed tightly together in a central *nucleus*.

The charge of a single electron and that of a single proton have the same magnitude but are opposite in sign. Hence, an electrically neutral atom contains equal numbers of electrons and protons. Electrons are held near the nucleus because they have the electrical sign opposite that of the protons in the nucleus and thus are attracted to the nucleus.

When atoms of a conductor like copper come together to form the solid, some of their outermost (and so most loosely held) electrons do not remain attached to the individual atoms but become free to wander about within the solid, leaving behind positively charged atoms (*positive ions*). We call the mobile electrons *conduction electrons*. There are few (if any) free electrons in a nonconductor.

The experiment of Fig. 22-4 demonstrates the mobility of charge in a conductor. A negatively charged plastic rod will attract either end of an isolated neutral copper rod. What happens is that many of the conduction electrons in the closer end of the copper rod are repelled by the negative charge on the plastic rod. They move to the far end of the copper rod, leaving the near end depleted in electrons and thus with an unbalanced positive charge. This positive charge is attracted to the negative charge in the plastic rod. Although the copper rod is still neutral, it is said to have an *induced charge*, which means that some of its positive and negative charges have been separated owing to the presence of a nearby charge.

Similarly, if a positively charged glass rod is brought near one end of a neutral copper rod, conduction electrons in the copper rod are attracted to that end. That end becomes negatively charged and the other end positively charged, so again an induced charge is set up in the copper rod. Although the copper rod is still neutral, it and the glass rod attract each other. (Figure 22-5 shows another demonstration of induced charge.)

Note that only conduction electrons, with their negative charges, can move; positive ions are fixed in place. Thus, an object becomes positively charged only through the *removal of negative charges*.

Semiconductors, such as silicon and germanium, are materials that are intermediate between conductors and insulators. The microelectronic revolution that has transformed our lives in so many ways is due to devices constructed of semiconducting materials.

Finally, there are **superconductors,** so called because they present no resistance to the movement of electric charge through them. When charge moves through a material, we say that an **electric current** exists in the material. Ordinary materials, even good conductors, tend to resist the flow of charge through them. In a superconductor, however, the resistance is not just small; it is precisely zero. If you set up a current in a superconducting ring, it lasts "forever," with no battery or other source of energy needed to maintain it.

✔**CHECKPOINT 1:**   The figure shows five pairs of plates: *A*, *B*, and *D* are charged plastic plates and *C* is an electrically neutral copper plate. The electrostatic forces between the pairs of plates are shown for three of the pairs. For the remaining two pairs, do the plates repel or attract each other?

22-4 Coulomb's Law

Fig. 22-6 Two charged particles, separated by distance r, repel each other if their charges are (a) both positive and (b) both negative. (c) They attract each other if their charges are of opposite signs. In each of the three situations, the force acting on one particle is equal in magnitude to the force acting on the other particle but has the opposite direction.

Let two charged particles (also called *point charges*) have charge magnitudes q_1 and q_2 and be separated by a distance r. The **electrostatic force** of attraction or repulsion between them has the magnitude

$$F = k \frac{|q_1||q_2|}{r^2} \quad \text{(Coulomb's law)}, \tag{22-1}$$

in which k is a constant. Each particle exerts a force of this magnitude on the other particle; the two forces form a third-law force pair. If the particles *repel* each other, the force on each particle is directed *away from* the other particle (as in Figs. 22-6a and b). If the particles *attract* each other, the force on each particle is directed *toward* the other particle (as in Fig. 22-6c).

Equation 22-1 is called **Coulomb's law** after Charles Augustin Coulomb, whose experiments in 1785 led him to it. Curiously, the form of Eq. 22-1 is the same as that of Newton's equation for the gravitational force between two particles with masses m_1 and m_2 that are separated by a distance r:

$$F = G \frac{m_1 m_2}{r^2}, \tag{22-2}$$

in which G is the gravitational constant.

The constant k in Eq. 22-1, by analogy with the gravitational constant G in Eq. 22-2, may be called the *electrostatic constant*. Both equations describe inverse square laws that involve a property of the interacting particles—the mass in one case and the charge in the other. The laws differ in that gravitational forces are always attractive but electrostatic forces may be either attractive or repulsive, depending on the signs of the two charges. This difference arises from the fact that, although there is only one kind of mass, there are two kinds of charge (and that is why absolute value signs are needed in Eq. 22-1 but not in Eq. 22-2).

Coulomb's law has survived every experimental test; no exceptions to it have ever been found. It holds even within the atom, correctly describing the force between the positively charged nucleus and each of the negatively charged electrons, even though classical Newtonian mechanics fails in that realm and is replaced there by quantum physics. This simple law also correctly accounts for the forces that bind atoms together to form molecules, and for the forces that bind atoms and molecules together to form solids and liquids.

For practical reasons having to do with the accuracy of measurements, the SI unit of charge is derived from the SI unit of electric current, the ampere (A). The SI unit of charge is the **coulomb** (C): *One coulomb is the amount of charge that is transferred through the cross section of a wire in 1 second when there is a current of 1 ampere in the wire.* In Section 30-2 we shall describe how the ampere is defined experimentally. In general, we can write

$$dq = i \, dt, \tag{22-3}$$

in which dq (in coulombs) is the charge transferred by a current i (in amperes) during the time interval dt (in seconds).

For historical reasons (and because doing so simplifies many other formulas), the electrostatic constant k of Eq. 22-1 is usually written $1/4\pi\varepsilon_0$. Then Coulomb's law becomes

$$F = \frac{1}{4\pi\varepsilon_0} \frac{|q_1||q_2|}{r^2} \quad \text{(Coulomb's law)}. \tag{22-4}$$

The constants in Eqs. 22-1 and 22-4 have the value

$$k = \frac{1}{4\pi\varepsilon_0} = 8.99 \times 10^9 \text{ N} \cdot \text{m}^2/\text{C}^2. \qquad (22\text{-}5)$$

The quantity ε_0, called the **permittivity constant,** sometimes appears separately in equations and is

$$\varepsilon_0 = 8.85 \times 10^{-12} \text{ C}^2/\text{N} \cdot \text{m}^2. \qquad (22\text{-}6)$$

Still another parallel between the gravitational force and the electrostatic force is that both obey the principle of superposition. If we have n charged particles, they interact independently in pairs, and the force on any one of them, let us say particle 1, is given by the vector sum

$$\vec{F}_{1,\text{net}} = \vec{F}_{12} + \vec{F}_{13} + \vec{F}_{14} + \vec{F}_{15} + \cdots + \vec{F}_{1n}, \qquad (22\text{-}7)$$

in which, for example, $\vec{F}_{14}$ is the force acting on particle 1 owing to the presence of particle 4. An identical formula holds for the gravitational force.

Finally, the two shell theorems that we found so useful in our study of gravitation have analogs in electrostatics:

> ► A shell of uniform charge attracts or repels a charged particle that is outside the shell as if all the shell's charge were concentrated at its center.

> ► If a charged particle is located inside a shell of uniform charge, there is no net electrostatic force on the particle from the shell.

(In the first theorem, we assume that the charge on the shell is much greater than that of the particle. Then any redistribution of the charge on the shell due to the presence of the particle's charge can be neglected.)

Spherical Conductors

If excess charge is placed on a spherical shell that is made of conducting material, the excess charge spreads uniformly over the (external) surface. For example, if we place excess electrons on a spherical metal shell, those electrons repel one another and tend to move apart, spreading over the available surface until they are uniformly distributed. That arrangement maximizes the distances between all pairs of the excess electrons. According to the first shell theorem, the shell then will attract or repel an external charge as if all the excess charge on the shell were concentrated at its center.

If we remove negative charge from a spherical metal shell, the resulting positive charge of the shell is also spread uniformly over the surface of the shell. For example, if we remove n electrons, there are then n sites of positive charge (sites missing an electron) that are spread uniformly over the shell. According to the first shell theorem, the shell will again attract or repel an external charge as if all the shell's excess charge were concentrated at its center.

✔**CHECKPOINT 2:**   The figure shows two protons (symbol p) and one electron (symbol e) on an axis. What are the directions of (a) the electrostatic force on the central proton due to the electron, (b) the electrostatic force on the central proton due to the other proton, and (c) the net electrostatic force on the central proton?

Sample Problem 22-1

(a) Figure 22-7a shows two positively charged particles fixed in place on an x axis. The charges are $q_1 = 1.60 \times 10^{-19}$ C and $q_2 = 3.20 \times 10^{-19}$ C, and the particle separation is $R = 0.0200$ m. What are the magnitude and direction of the electrostatic force $\vec{F}_{12}$ on particle 1 from particle 2?

SOLUTION: The Key Idea here is that, because both particles are positively charged, particle 1 is repelled by particle 2, with a force magnitude given by Eq. 22-4. Thus, the direction of force $\vec{F}_{12}$ on particle 1 is *away from* particle 2, in the negative direction of the x axis, as indicated in the free-body diagram of Fig. 22-7b. Using Eq. 22-4 with separation R substituted for r, we can write the magnitude F_{12} of this force as

$$F_{12} = \frac{1}{4\pi\varepsilon_0} \frac{|q_1||q_2|}{R^2}$$
$$= (8.99 \times 10^9 \text{ N} \cdot \text{m}^2/\text{C}^2)$$
$$\times \frac{(1.60 \times 10^{-19} \text{ C})(3.20 \times 10^{-19} \text{ C})}{(0.0200 \text{ m})^2}$$
$$= 1.15 \times 10^{-24} \text{ N}.$$

Thus, force $\vec{F}_{12}$ has the following magnitude and direction (relative to the positive direction of the x axis):

$$1.15 \times 10^{-24} \text{ N} \quad \text{and} \quad 180°. \qquad \text{(Answer)}$$

We can also write $\vec{F}_{12}$ in unit-vector notation as

$$\vec{F}_{12} = -(1.15 \times 10^{-24} \text{ N})\hat{\text{i}}. \qquad \text{(Answer)}$$

(b) Figure 22-7c is identical to Fig. 22-7a except that particle 3 now lies on the x axis between particles 1 and 2. Particle 3 has charge $q_3 = -3.20 \times 10^{-19}$ C and is at a distance $\frac{3}{4}R$ from particle 1. What is the net electrostatic force $\vec{F}_{1,\text{net}}$ on particle 1 due to particles 2 and 3?

SOLUTION: One Key Idea here is that the presence of particle 3 does not alter the electrostatic force on particle 1 from particle 2. Thus, force $\vec{F}_{12}$ still acts on particle 1. Similarly, the force $\vec{F}_{13}$ that acts on particle 1 due to particle 3 is not affected by the presence of particle 2. Because particles 1 and 3 have charge of opposite sign, particle 1 is attracted to particle 3. Thus, force $\vec{F}_{13}$ is directed *toward* particle 3, as indicated in the free-body diagram of Fig. 22-7d.

To find the magnitude of $\vec{F}_{13}$, we can rewrite Eq. 22-4 as

$$F_{13} = \frac{1}{4\pi\varepsilon_0} \frac{|q_1||q_3|}{(\frac{3}{4}R)^2}$$
$$= (8.99 \times 10^9 \text{ N} \cdot \text{m}^2/\text{C}^2)$$
$$\times \frac{(1.60 \times 10^{-19} \text{ C})(3.20 \times 10^{-19} \text{ C})}{(\frac{3}{4})^2(0.0200 \text{ m})^2}$$
$$= 2.05 \times 10^{-24} \text{ N}.$$

We can also write $\vec{F}_{13}$ in unit-vector notation:

$$\vec{F}_{13} = (2.05 \times 10^{-24} \text{ N})\hat{\text{i}}.$$

A second Key Idea here is that the net force $\vec{F}_{1,\text{net}}$ on particle 1 is the vector sum of $\vec{F}_{12}$ and $\vec{F}_{13}$; that is, from Eq. 22-7, we can

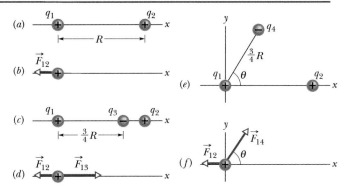

Fig. 22-7 Sample Problem 22-1. (a) Two charged particles of charges q_1 and q_2 are fixed in place on an x axis, with separation R. (b) The free-body diagram for particle 1, showing the electrostatic force on it from particle 2. (c) Particle 3 is now fixed in place on the x axis, along with particles 1 and 2. (d) The free-body diagram for particle 1. (e) Particle 4 is fixed in place on a line at angle θ to the x axis, again with particles 1 and 2. (f) The free-body diagram for particle 1.

write the net force $\vec{F}_{1,\text{net}}$ on particle 1 in unit-vector notation as

$$\vec{F}_{1,\text{net}} = \vec{F}_{12} + \vec{F}_{13}$$
$$= -(1.15 \times 10^{-24} \text{ N})\hat{\text{i}} + (2.05 \times 10^{-24} \text{ N})\hat{\text{i}}$$
$$= (9.00 \times 10^{-25} \text{ N})\hat{\text{i}}. \qquad \text{(Answer)}$$

Thus, $\vec{F}_{1,\text{net}}$ has the following magnitude and direction (relative to the positive direction of the x axis):

$$9.00 \times 10^{-25} \text{ N} \quad \text{and} \quad 0°. \qquad \text{(Answer)}$$

(c) Figure 22-7e is identical to Fig. 22-7a except that particle 4 is now positioned as shown. Particle 4 has charge $q_4 = -3.20 \times 10^{-19}$ C, is at a distance $\frac{3}{4}R$ from particle 1, and lies on a line that makes an angle $\theta = 60°$ with the x axis. What is the net electrostatic force $\vec{F}_{1,\text{net}}$ on particle 1 due to particles 2 and 4?

SOLUTION: The Key Idea here is that the net force $\vec{F}_{1,\text{net}}$ is the vector sum of $\vec{F}_{12}$ and a new force $\vec{F}_{14}$ acting on particle 1 due to particle 4. Because particles 1 and 4 have charges of opposite sign, particle 1 is attracted to particle 4. Thus, force $\vec{F}_{14}$ on particle 1 is directed *toward* particle 4, at angle $\theta = 60°$, as indicated in the free-body diagram of Fig. 22-7f.

To find the magnitude of $\vec{F}_{14}$, we can rewrite Eq. 22-4 as

$$F_{14} = \frac{1}{4\pi\varepsilon_0} \frac{|q_1||q_4|}{(\frac{3}{4}R)^2}$$
$$= (8.99 \times 10^9 \text{ N} \cdot \text{m}^2/\text{C}^2)$$
$$\times \frac{(1.60 \times 10^{-19} \text{ C})(3.20 \times 10^{-19} \text{ C})}{(\frac{3}{4})^2(0.0200 \text{ m})^2}$$
$$= 2.05 \times 10^{-24} \text{ N}.$$

Then from Eq. 22-7, we can write the net force $\vec{F}_{1,\text{net}}$ on particle 1 as

$$\vec{F}_{1,\text{net}} = \vec{F}_{12} + \vec{F}_{14}.$$

To evaluate the right side of this equation, we need another Key Idea: Because the forces $\vec{F}_{12}$ and $\vec{F}_{14}$ are not directed along the same axis, we *cannot* sum by simply combining their magnitudes. Instead, we must add them as vectors, using one of the following methods.

Method 1. *Summing directly on a vector-capable calculator.* For $\vec{F}_{12}$, we enter the magnitude 1.15×10^{-24} and the angle $180°$. For $\vec{F}_{14}$, we enter the magnitude 2.05×10^{-24} and the angle $60°$. Then we add the vectors.

Method 2. *Summing in unit-vector notation.* First we rewrite $\vec{F}_{14}$ as

$$\vec{F}_{14} = (F_{14} \cos \theta)\hat{i} + (F_{14} \sin \theta)\hat{j}.$$

Substituting 2.05×10^{-24} N for F_{14} and $60°$ for θ, this becomes

$$\vec{F}_{14} = (1.025 \times 10^{-24} \text{ N})\hat{i} + (1.775 \times 10^{-24} \text{ N})\hat{j}.$$

Then we sum:

$$\vec{F}_{1,\text{net}} = \vec{F}_{12} + \vec{F}_{14}$$
$$= -(1.15 \times 10^{-24} \text{ N})\hat{i}$$
$$+ (1.025 \times 10^{-24} \text{ N})\hat{i} + (1.775 \times 10^{-24} \text{ N})\hat{j}$$
$$\approx (-1.25 \times 10^{-25} \text{ N})\hat{i} + (1.78 \times 10^{-24} \text{ N})\hat{j}.$$
(Answer)

Method 3. *Summing components axis by axis.* The sum of the x components gives us

$$F_{1,\text{net},x} = F_{12,x} + F_{14,x} = F_{12} + F_{14} \cos 60°$$
$$= -1.15 \times 10^{-24} \text{ N} + (2.05 \times 10^{-24} \text{ N})(\cos 60°)$$
$$= -1.25 \times 10^{-25} \text{ N}.$$

The sum of the y components gives us

$$F_{1,\text{net},y} = F_{12,y} + F_{14,y} = 0 + F_{14} \sin 60°$$
$$= (2.05 \times 10^{-24} \text{ N})(\sin 60°)$$
$$= 1.78 \times 10^{-24} \text{ N}.$$

The net force $\vec{F}_{1,\text{net}}$ has the magnitude

$$F_{1,\text{net}} = \sqrt{F_{1,\text{net},x}^2 + F_{1,\text{net},y}^2} = 1.78 \times 10^{-24} \text{ N}. \quad \text{(Answer)}$$

To find the direction of $\vec{F}_{1,\text{net}}$, we take

$$\theta = \tan^{-1} \frac{F_{1,\text{net},y}}{F_{1,\text{net},x}} = -86.0°.$$

However, this is an unreasonable result because $\vec{F}_{1,\text{net}}$ must have a direction between the directions of $\vec{F}_{12}$ and $\vec{F}_{14}$. To correct θ, we add $180°$, obtaining

$$-86.0° + 180° = 94.0°. \quad \text{(Answer)}$$

✔**CHECKPOINT 3:**   The figure here shows three arrangements of an electron e and two protons p. (a) Rank the arrangements according to the magnitude of the net electrostatic force on the electron due to the protons, largest first. (b) In situation c, is the angle between the net force on the electron and the line labeled d less than or more than $45°$?

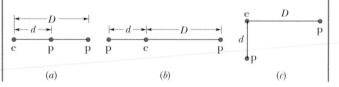

(a) (b) (c)

PROBLEM-SOLVING TACTICS

Tactic 1: *Symbols Representing Charge*
Here is a general guide to the symbols representing charge. If the symbol q, with or without a subscript, is used in a sentence when no electrical sign has been specified, the charge can be either positive or negative. Sometimes the sign is explicitly shown, as in the notation $+q$ or $-q$.

When more than one charged object is being considered, their charges might be given as multiples of a charge magnitude. As examples, the notation $+2q$ means a positive charge with magnitude twice that of some reference charge magnitude q, and $-3q$ means a negative charge with magnitude three times that of the reference charge magnitude q.

Sample Problem 22-2

Figure 22-8a shows two particles fixed in place: a particle of charge $q_1 = +8q$ at the origin and a particle of charge $q_2 = -2q$ at $x = L$. At what point (other than infinitely far away) can a proton be placed so that it is in *equilibrium* (meaning that the net force on it is zero)? Is that equilibrium *stable* or *unstable*?

SOLUTION: The Key Idea here is that, if $\vec{F}_1$ is the force on the proton due to charge q_1 and $\vec{F}_2$ is the force on the proton due to charge q_2, then the point we seek is where $\vec{F}_1 + \vec{F}_2 = 0$. This condition requires that

$$\vec{F}_1 = -\vec{F}_2. \quad (22\text{-}8)$$

This tells us that at the point we seek, the forces acting on the

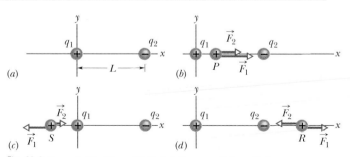

Fig. 22-8 Sample Problem 22-2. (a) Two particles of charges q_1 and q_2 are fixed in place on an x axis, with separation L. (b)–(d) Three possible locations P, S, and R for a proton. At each location, $\vec{F}_1$ is the force on the proton from particle 1 and $\vec{F}_2$ is the force on the proton from particle 2.

proton due to the other two particles must be of equal magnitudes,

$$F_1 = F_2, \qquad (22\text{-}9)$$

and that the forces must have opposite directions.

A proton has a positive charge. Thus, the proton and the particle of charge q_1 are of the same sign, and force $\vec{F}_1$ on the proton must point away from q_1. Also, the proton and the particle of charge q_2 are of opposite signs, so force $\vec{F}_2$ on the proton must point toward q_2. "Away from q_1" and "toward q_2" can be in opposite directions only if the proton is located on the x axis.

If the proton is on the x axis at any point between q_1 and q_2, such as P in Fig. 22-8b, then $\vec{F}_1$ and $\vec{F}_2$ are in the same direction and not in opposite directions as required. If the proton is at any point on the x axis to the left of q_1, such as point S in Fig. 22-8c, then $\vec{F}_1$ and $\vec{F}_2$ are in opposite directions. However, Eq. 22-4 tells us that $\vec{F}_1$ and $\vec{F}_2$ cannot have equal magnitudes there: F_1 must be greater than F_2, because F_1 is produced by a closer charge (with lesser r) of greater magnitude ($8q$ versus $2q$).

Finally, if the proton is at any point on the x axis to the right of q_2, such as point R in Fig. 22-8d, then $\vec{F}_1$ and $\vec{F}_2$ are again in opposite directions. However, because now the charge of greater magnitude (q_1) is *farther* away from the proton than the charge of lesser magnitude (q_2), there is a point at which F_1 is equal to F_2. Let x be the coordinate of this point, and let q_p be the charge of the proton. Then with the aid of Eq. 22-4, we can rewrite Eq. 22-9 as

$$\frac{1}{4\pi\varepsilon_0}\frac{8qq_p}{x^2} = \frac{1}{4\pi\varepsilon_0}\frac{2qq_p}{(x-L)^2}. \qquad (22\text{-}10)$$

(Note that only the charge magnitudes appear in Eq. 22-10.) Rearranging Eq. 22-10 gives us

$$\left(\frac{x-L}{x}\right)^2 = \frac{1}{4}.$$

After taking the square roots of both sides, we have

$$\frac{x-L}{x} = \frac{1}{2},$$

which gives us

$$x = 2L. \qquad \text{(Answer)}$$

The equilibrium at $x = 2L$ is unstable; that is, if the proton is displaced leftward from point R, then F_1 and F_2 both increase but F_2 increases more (because q_2 is closer than q_1), and a net force will drive the proton farther leftward. If the proton is displaced rightward, both F_1 and F_2 decrease but F_2 decreases more, and a net force will then drive the proton farther rightward. In a stable equilibrium, each time the proton is displaced slightly, it would return to the equilibrium position.

PROBLEM-SOLVING TACTICS

Tactic 2: *Drawing Electrostatic Force Vectors*
When you are given a diagram of charged particles, such as Fig. 22-7a, and are asked to find the net electrostatic force on one of them, you should usually draw a free-body diagram showing only the particle of concern and the forces *it* experiences, as in Fig. 22-7b. If, instead, you choose to superimpose those forces on the given diagram showing all the particles, be sure to draw the force vectors with either their tails (preferably) or their heads on the particle of concern. If you draw the vectors elsewhere in the diagram, you invite confusion—and confusion is guaranteed if you draw the vectors on the particles *causing* the forces on the particle of concern.

Sample Problem 22-3

In Fig. 22-9a, two identical, electrically isolated conducting spheres A and B are separated by a (center-to-center) distance a that is large compared to the spheres. Sphere A has a positive charge of $+Q$, and sphere B is electrically neutral. Initially, there is no electrostatic force between the spheres. (Assume that there is no induced charge on the spheres because of their large separation.)

(a) Suppose the spheres are connected for a moment by a conducting wire. The wire is thin enough so that any net charge on it is negligible. What is the electrostatic force between the spheres after the wire is removed?

SOLUTION: A Key Idea here is that when the spheres are wired together, the (negative) conduction electrons on sphere B, which always repel one another, have a way to move farther away from one another (along the wire to positively charged sphere A, which attracts them). (See Fig. 22-9b.) As sphere B loses negative charge, it becomes positively charged, and as A gains negative charge, it becomes *less* positively charged. A second Key Idea is that the spheres must end up with the same charge because they are identical. Thus, the transfer of charge stops when the excess charge on B has increased to $+Q/2$ and the excess charge on A has decreased to $+Q/2$. This condition occurs when a charge of $-Q/2$ has been transferred.

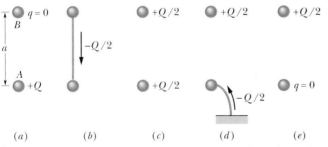

(a) (b) (c) (d) (e)

Fig. 22-9 Sample Problem 22-3. Two small conducting spheres A and B. (a) To start, sphere A is charged positively. (b) Negative charge is transferred between the spheres through a connecting wire. (c) Both spheres are then charged positively. (d) Negative charge is transferred through a grounding wire to sphere A. (e) Sphere A is then neutral.

After the wire has been removed (Fig. 22-9c), we can assume that the charge on either sphere does not disturb the uniformity of the charge distribution on the other sphere, because the spheres are small relative to their separation. Thus, we can apply the first shell theorem to each sphere. By Eq. 22-4 with $q_1 = q_2 = Q/2$ and $r = a$, the electrostatic force between the spheres has a magnitude of

$$F = \frac{1}{4\pi\varepsilon_0}\frac{(Q/2)(Q/2)}{a^2} = \frac{1}{16\pi\varepsilon_0}\left(\frac{Q}{a}\right)^2. \quad \text{(Answer)}$$

The spheres, now positively charged, repel each other.

(b) Next, suppose sphere A is grounded momentarily, and then the ground connection is removed. What now is the electrostatic force between the spheres?

SOLUTION: The Key Idea here is that the ground connection allows electrons, with a total charge of $-Q/2$, to move from the ground to sphere A (Fig. 22-9d), neutralizing that sphere (Fig. 22-9e). With no charge on sphere A, there is no electrostatic force between the two spheres (just as initially, in Fig. 22-9a).

22-5   Charge Is Quantized

In Benjamin Franklin's day, electric charge was thought to be a continuous fluid — an idea that was useful for many purposes. However, we now know that fluids themselves, such as air and water, are not continuous but are made up of atoms and molecules; matter is discrete. Experiment shows that "electrical fluid" is also not continuous but is made up of multiples of a certain elementary charge. Any positive or negative charge q that can be detected can be written as

$$q = ne, \qquad n = \pm 1, \pm 2, \pm 3, \ldots, \qquad (22\text{-}11)$$

in which e, the **elementary charge,** has the value

$$e = 1.60 \times 10^{-19} \text{ C}. \qquad (22\text{-}12)$$

The elementary charge e is one of the important constants of nature. The electron and proton both have a charge of magnitude e (Table 22-1). (Quarks, the constituent particles of protons and neutrons, have charges of $\pm e/3$ or $\pm 2e/3$, but they apparently cannot be detected individually. For this and for historical reasons, we do not take their charges to be the elementary charge.)

You often see phrases — such as "the charge on a sphere," "the amount of charge transferred," and "the charge carried by the electron" — that suggest that charge is a substance. (Indeed, such statements have already appeared in this chapter.) You should, however, keep in mind what is intended: *Particles* are the substance and charge happens to be one of their properties, just as mass is.

When a physical quantity such as charge can have only discrete values rather than any value, we say that the quantity is **quantized.** It is possible, for example, to find a particle that has no charge at all or a charge of $+10e$ or $-6e$, but not a particle with a charge of, say, $3.57e$.

The quantum of charge is small. In an ordinary 100 W lightbulb, for example, about 10^{19} elementary charges enter the bulb every second and just as many leave. However, the graininess of electricity does not show up in such large-scale phenomena (the bulb does not flicker with each electron), just as you cannot feel the individual molecules of water with your hand.

The graininess of electricity is responsible for the blue glow that is emitted by a wintergreen LifeSaver while it is being crushed. When the sugar (sucrose) crystals in the candy rupture, one part of each ruptured crystal has excess electrons while the other part has excess positive ions. Almost immediately, electrons and ions jump across the gap of the rupture to neutralize the two sides. During the jumps, the electrons and positive ions collide with nitrogen molecules in the air that is then flowing into the gap.

The collisions cause the nitrogen to emit ultraviolet light that you cannot see, as well as blue light (from the visible region of the spectrum) that is, however, too

TABLE 22-1   **The Charges of Three Particles**

Particle	Symbol	Charge
Electron	e or e⁻	$-e$
Proton	p	$+e$
Neutron	n	0

dim to see. Oil of wintergreen in the crystals absorbs the ultraviolet light and immediately emits enough blue light to light up a mouth or a pair of pliers. However, if the candy is wet with saliva, the demonstration fails, because the conducting saliva neutralizes the two parts of a fractured crystal before sparking can occur.

✔**CHECKPOINT 4:**   Initially, sphere A has a charge of $-50e$ and sphere B has a charge of $+20e$. The spheres are made of conducting material and are identical in size. If the spheres then touch, what is the resulting charge on sphere A?

Sample Problem 22-4

The nucleus in an iron atom has a radius of about 4.0×10^{-15} m and contains 26 protons.

(a) What is the magnitude of the repulsive electrostatic force between two of the protons that are separated by 4.0×10^{-15} m?

SOLUTION: The Key Idea here is that the protons can be treated as charged particles, so the magnitude of the electrostatic force on one from the other is given by Coulomb's law. Table 22-1 tells us that their charge is $+e$. Thus, Eq. 22-4 gives us

$$F = \frac{1}{4\pi\varepsilon_0} \frac{e^2}{r^2}$$
$$= \frac{(8.99 \times 10^9 \text{ N} \cdot \text{m}^2/\text{C}^2)(1.60 \times 10^{-19} \text{ C})^2}{(4.0 \times 10^{-15} \text{ m})^2}$$
$$= 14 \text{ N.} \qquad \text{(Answer)}$$

This is a small force to be acting on a macroscopic object like a cantaloupe, but an enormous force to be acting on a proton. Such forces should blow apart the nucleus of any element but hydrogen (which has only one proton in its nucleus). However, they don't, not even in nuclei with a great many protons. Therefore, there must be some enormous attractive force to counter this enormous repulsive electrostatic force.

(b) What is the magnitude of the gravitational force between those same two protons?

SOLUTION: The Key Idea here is like that in part (a): Because the protons are particles, the magnitude of the gravitational force on one from the other is given by Newton's equation for the gravitational force (Eq. 22-2). With m_p ($= 1.67 \times 10^{-27}$ kg) representing the mass of a proton, Eq. 22-2 gives us

$$F = G \frac{m_p^2}{r^2}$$
$$= \frac{(6.67 \times 10^{-11} \text{ N} \cdot \text{m}^2/\text{kg}^2)(1.67 \times 10^{-27} \text{ kg})^2}{(4.0 \times 10^{-15} \text{ m})^2}$$
$$= 1.2 \times 10^{-35} \text{ N.} \qquad \text{(Answer)}$$

This result tells us that the (attractive) gravitational force is far too weak to counter the repulsive electrostatic forces between protons in a nucleus. Instead, the protons are bound together by an enormous force called (aptly) the *strong nuclear force*—a force that acts between protons (and neutrons) when they are close together, as in a nucleus.

Although the gravitational force is many times weaker than the electrostatic force, it is more important in large-scale situations because it is always attractive. This means that it can collect many small bodies into huge bodies with huge masses, such as planets and stars, that then exert large gravitational forces. The electrostatic force, on the other hand, is repulsive for charges of the same sign, so it is unable to collect either positive charge or negative charge into large concentrations that would then exert large electrostatic forces.

22-6 Charge Is Conserved

If you rub a glass rod with silk, a positive charge appears on the rod. Measurement shows that a negative charge of equal magnitude appears on the silk. This suggests that rubbing does not create charge but only transfers it from one body to another, upsetting the electrical neutrality of each body during the process. This hypothesis of **conservation of charge,** first put forward by Benjamin Franklin, has stood up under close examination, both for large-scale charged bodies and for atoms, nuclei, and elementary particles. No exceptions have ever been found. Thus, we add electric charge to our list of quantities—including energy and both linear and angular momentum—that obey a conservation law.

Radioactive decay of nuclei, in which a nucleus spontaneously transforms into a different type of nucleus, gives us many instances of charge conservation at the nuclear level. For example, uranium-238, or ^{238}U, which is found in common uranium ore, can decay by emitting an alpha particle (which is a helium nucleus, ^{4}He)

Fig. 22-10 A photograph of trails of bubbles left in a bubble chamber by an electron and a positron. The pair of particles was produced by a gamma ray that entered the chamber from the bottom. Being electrically neutral, the gamma ray did not generate a telltale trail of bubbles along its path, as the electron and positron did.

and transforming to thorium, ^{234}Th:

$$^{238}\text{U} \rightarrow {}^{234}\text{Th} + {}^{4}\text{He} \qquad \text{(radioactive decay).} \qquad (22\text{-}13)$$

The atomic number Z of the radioactive *parent* nucleus ^{238}U is 92, which tells us that this nucleus contains 92 protons and has a charge of 92e. The emitted alpha particle has $Z = 2$, and the *daughter* nucleus ^{234}Th has $Z = 90$. Thus, the amount of charge present before the decay, 92e, is equal to the total amount present after the decay, 90e + 2e. Charge is conserved.

Another example of charge conservation occurs when an electron e^{-} (whose charge is $-e$) and its antiparticle, the *positron* e^{+} (whose charge is $+e$), undergo an *annihilation process* in which they transform into two *gamma rays* (high-energy light):

$$\text{e}^{-} + \text{e}^{+} \rightarrow \gamma + \gamma \qquad \text{(annihilation).} \qquad (22\text{-}14)$$

In applying the conservation-of-charge principle, we must add the charges algebraically, with due regard for their signs. In the annihilation process of Eq. 22-14 then, the net charge of the system is zero both before and after the event. Charge is conserved.

In *pair production*, the converse of annihilation, charge is also conserved. In this process a gamma ray transforms into an electron and a positron:

$$\gamma \rightarrow \text{e}^{-} + \text{e}^{+} \qquad \text{(pair production).} \qquad (22\text{-}15)$$

Figure 22-10 shows such a pair-production event that occurred in a bubble chamber. A gamma ray entered the chamber from the bottom and at one point transformed into an electron and a positron. Because those new particles were charged and moving, each left a trail of tiny bubbles. (The trails were curved because a magnetic field had been set up in the chamber.) The gamma ray, being electrically neutral, left no trail. Still, you can tell exactly where it underwent pair production—at the tip of the curved **V**, where the trails of the electron and positron begin.

REVIEW & SUMMARY

Electric Charge The strength of a particle's electric interaction with objects around it depends on its **electric charge,** which can be either positive or negative. Charges with the same sign repel each other and charges with opposite signs attract each other. An object with equal amounts of the two kinds of charge is electrically neutral, whereas one with an imbalance is electrically charged.

Conductors are materials in which a significant number of charged particles (electrons in metals) are free to move. The charged particles in **nonconductors,** or **insulators,** are not free to move. When charge moves through a material, we say that an **electric current** exists in the material.

The Coulomb and Ampere The SI unit of charge is the **coulomb** (C). It is defined in terms of the unit of current, the ampere (A), as the charge passing a particular point in 1 second when there is a current of 1 ampere at that point.

Coulomb's Law Coulomb's law describes the **electrostatic force** between small (point) electric charges q_1 and q_2 at rest (or nearly at rest) and separated by a distance r:

$$F = \frac{1}{4\pi\varepsilon_0} \frac{|q_1||q_2|}{r^2} \qquad \text{(Coulomb's law).} \qquad (22\text{-}4)$$

Here $\varepsilon_0 = 8.85 \times 10^{-12} \text{ C}^2/\text{N} \cdot \text{m}^2$ is the **permittivity constant,** and $1/4\pi\varepsilon_0 = k = 8.99 \times 10^9 \text{ N} \cdot \text{m}^2/\text{C}^2$.

The force of attraction or repulsion between point charges at rest acts along the line joining the two charges. If more than two charges are present, Eq. 22-4 holds for each pair of charges. The net force on each charge is then found, using the superposition principle, as the vector sum of the forces exerted on the charge by all the others.

The two shell theorems for electrostatics are

A shell of uniform charge attracts or repels a charged particle that is outside the shell as if all the shell's charge were concentrated at its center.

If a charged particle is located inside a shell of uniform charge, there is no net electrostatic force on the particle from the shell.

The Elementary Charge Electric charge is **quantized:** any charge can be written as ne, where n is a positive or negative integer, and e is a constant of nature called the **elementary charge** (approximately 1.60×10^{-19} C). Electric charge is **conserved:** the (algebraic) net charge of any isolated system cannot change.

QUESTIONS

1. Does Coulomb's law hold for all charged objects?

2. A particle of charge q is to be placed, in turn, outside four metal objects, each of uniform charge Q: (1) a large solid sphere, (2) a large spherical shell, (3) a small solid sphere, and (4) a small spherical shell. The distance between the particle and the center of the object is the same, and q is small enough not to alter significantly the uniform distribution of Q. Rank the objects according to the electrostatic force they exert on the particle, greatest first.

3. Figure 22-11 shows four situations in which charged particles are fixed in place on an axis. In which situations is there a point to the left of the particles where an electron will be in equilibrium?

Fig. 22-11 Question 3.

4. Figure 22-12 shows two charged particles on an axis. The charges are free to move. At one point, however, a third charged particle can be placed such that all three particles are in equilibrium. (a) Is that point to the left of the first two particles, to their right, or between them? (b) Should the third particle be positively or negatively charged? (c) Is the equilibrium stable or unstable?

Fig. 22-12 Question 4.

5. In Fig. 22-13, a central particle of charge $-q$ is surrounded by two circular rings of charged particles, of radii r and R, with $R > r$. What are the magnitude and direction of the net electrostatic force on the central particle due to the other particles?

6. Figure 22-14 shows four arrangements of charged particles. Rank the arrangements according to the magnitude of the net electrostatic force on the particle with charge $+Q$, greatest first.

Fig. 22-13 Question 5.

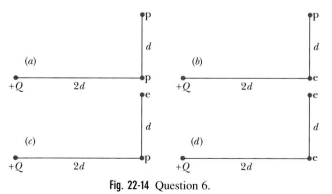

Fig. 22-14 Question 6.

7. Figure 22-15 shows four situations in which particles of charge $+q$ or $-q$ are fixed in place. In each, the particles on the x axis are equidistant from the y axis. First, consider the middle particle in situation 1; the middle particle experiences an electrostatic force from each of the other two particles. (a) Are the magnitudes F of those forces the same or different? (b) Is the magnitude of the net force on the middle particle equal to, greater than, or less than $2F$? (c) Do the x components of the two forces add or cancel? (d) Do their y components add or cancel? (e) Is the direction of the net force on the middle particle that of the canceling components or the adding components? (f) What is the direction of that net force? Now consider the remaining situations: What is the direction of the net force on the middle particle in (g) situation 2, (h) situation 3, and (i) situation 4? (In each, consider the symmetry of the charge distribution and determine the canceling components and the adding components.)

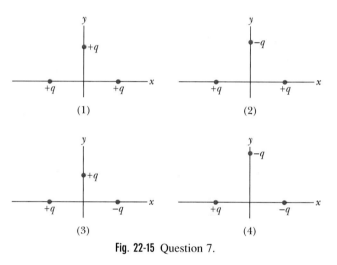

Fig. 22-15 Question 7.

8. A positively charged ball is brought close to a neutral isolated conductor. The conductor is then grounded while the ball is kept close. Is the conductor charged positively or negatively, or is it neutral, if (a) the ball is first taken away and then the ground connection is removed and (b) the ground connection is first removed and then the ball is taken away?

9. (a) A positively charged glass rod attracts an object suspended by a nonconducting thread. Is the object definitely negatively charged or only possibly negatively charged? (b) A positively charged glass rod repels a similarly suspended object. Is the object definitely positively charged or only possibly?

10. In Fig. 22-4, the nearby (negatively charged) plastic rod causes some of the conduction electrons in the copper rod to move to the far end of the copper rod. Why does the flow of the conduction electrons quickly cease? After all, a huge number of them are free to move to that far end.

11. A person standing on an electrically insulated platform touches a charged, electrically isolated conductor. Does this discharge the conductor completely?

EXERCISES & PROBLEMS

ssm　Solution is in the Student Solutions Manual.
www　Solution is available on the World Wide Web at:
　　　　http://www.wiley.com/college/hrw
ilw　Solution is available on the Interactive LearningWare.

SEC. 22-4　Coulomb's Law

1E. What must be the distance between point charge $q_1 = 26.0$ μC and point charge $q_2 = -47.0$ μC for the electrostatic force between them to have a magnitude of 5.70 N?　ssm

2E. A point charge of $+3.00 \times 10^{-6}$ C is 12.0 cm distant from a second point charge of -1.50×10^{-6} C. Calculate the magnitude of the force on each charge.

3E. Two equally charged particles, held 3.2×10^{-3} m apart, are released from rest. The initial acceleration of the first particle is observed to be 7.0 m/s^2 and that of the second to be 9.0 m/s^2. If the mass of the first particle is 6.3×10^{-7} kg, what are (a) the mass of the second particle and (b) the magnitude of the charge of each particle?　ilw

4E. Identical isolated conducting spheres 1 and 2 have equal charges and are separated by a distance that is large compared with their diameters (Fig. 22-16a). The electrostatic force acting on sphere 2 due to sphere 1 is $\vec{F}$. Suppose now that a third identical sphere 3, having an insulating handle and initially neutral, is touched first to sphere 1 (Fig. 22-16b), then to sphere 2 (Fig. 22-16c), and finally removed (Fig. 22-16d). In terms of magnitude F, what is the magnitude of the electrostatic force $\vec{F}'$ that now acts on sphere 2?

(a)

(b)

(c)

(d)

Fig. 22-16　Exercise 4.

5P. In Fig. 22-17, what are the (a) horizontal and (b) vertical components of the net electrostatic force on the charged particle in the lower left corner of the square if $q = 1.0 \times 10^{-7}$ C and $a = 5.0$ cm?　ilw

6P. Point charges q_1 and q_2 lie on the x axis at points $x = -a$ and $x = +a$, respectively. (a) How must q_1 and q_2 be related for the

Fig. 22-17　Problem 5.

net electrostatic force on point charge $+Q$, placed at $x = +a/2$, to be zero? (b) Repeat (a) but with point charge $+Q$ now placed at $x = +3a/2$.

7P. Two identical conducting spheres, fixed in place, attract each other with an electrostatic force of 0.108 N when separated by 50.0 cm, center-to-center. The spheres are then connected by a thin conducting wire. When the wire is removed, the spheres repel each other with an electrostatic force of 0.0360 N. What were the initial charges on the spheres?　ssm

8P. In Fig. 22-18, three charged particles lie on a straight line and are separated by distances d. Charges q_1 and q_2 are held fixed. Charge q_3 is free to move but happens to be in equilibrium (no net electrostatic force acts on it). Find q_1 in terms of q_2.

Fig. 22-18　Problem 8.

9P. Two *free* particles (that is, free to move) with charges $+q$ and $+4q$ are a distance L apart. A third charge is placed so that the entire system is in equilibrium. (a) Find the location, magnitude, and sign of the third charge. (b) Show that the equilibrium is unstable.　ssm　www

10P. Two fixed particles, of charges $q_1 = +1.0$ μC and $q_2 = -3.0$ μC, are 10 cm apart. How far from each should a third charge be located so that no net electrostatic force acts on it?

11P. (a) What equal positive charges would have to be placed on Earth and on the Moon to neutralize their gravitational attraction? Do you need to know the lunar distance to solve this problem? Why or why not? (b) How many kilograms of hydrogen would be needed to provide the positive charge calculated in (a)?　ssm

12P. The charges and coordinates of two charged particles held fixed in the xy plane are $q_1 = +3.0$ μC, $x_1 = 3.5$ cm, $y_1 = 0.50$ cm, and $q_2 = -4.0$ μC, $x_2 = -2.0$ cm, $y_2 = 1.5$ cm. (a) Find the magnitude and direction of the electrostatic force on q_2. (b) Where could you locate a third charge $q_3 = +4.0$ μC such that the net electrostatic force on q_2 is zero?

13P. A certain charge Q is divided into two parts q and $Q - q$, which are then separated by a certain distance. What must q be in terms of Q to maximize the electrostatic repulsion between the two charges?　ssm　ilw

14P. A particle with charge Q is fixed at each of two opposite corners of a square, and a particle with charge q is placed at each of the other two corners. (a) If the net electrostatic force on each particle with charge Q is zero, what is Q in terms of q? (b) Is there any value of q that makes the net electrostatic force on each of the four particles zero? Explain.

15P. In Fig. 22-19, two tiny conducting balls of identical mass m and identical charge q hang from nonconducting threads of length L. Assume that θ is so small that

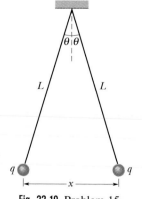

Fig. 22-19　Problem 15.

tan θ can be replaced by its approximate equal, sin θ. (a) Show that, for equilibrium,

$$x = \left(\frac{q^2 L}{2\pi\varepsilon_0 mg} \right)^{1/3},$$

where x is the separation between the balls. (b) If $L = 120$ cm, $m = 10$ g, and $x = 5.0$ cm, what is q? ssm

16P. Explain what happens to the balls of Problem 15b if one of them is discharged (loses it charge q to, say, the ground), and find the new equilibrium separation x, using the given values of L and m and the computed value of q.

17P. Figure 22-20 shows a long, nonconducting, massless rod of length L, pivoted at its center and balanced with a block of weight W at a distance x from the left end. At the left and right ends of the rod are attached small conducting spheres with positive charges q and $2q$, respectively. A distance h directly beneath each of these spheres is a fixed sphere with positive charge Q. (a) Find the distance x when the rod is horizontal and balanced. (b) What value should h have so that the rod exerts no vertical force on the bearing when the rod is horizontal and balanced? ssm

Fig. 22-20 Problem 17.

SEC. 22-5 Charge Is Quantized

18E. What is the magnitude of the electrostatic force between a singly charged sodium ion (Na^+, of charge $+e$) and an adjacent singly charged chlorine ion (Cl^-, of charge $-e$) in a salt crystal if their separation is 2.82×10^{-10} m?

19E. What is the total charge in coulombs of 75.0 kg of electrons? ssm

20E. How many megacoulombs of positive (or negative) charge are in 1.00 mol of neutral molecular-hydrogen gas (H_2)?

21E. The magnitude of the electrostatic force between two identical ions that are separated by a distance of 5.0×10^{-10} m is 3.7×10^{-9} N. (a) What is the charge of each ion? (b) How many electrons are "missing" from each ion (thus giving the ion its charge imbalance)? ssm

22E. Two tiny, spherical water drops, with identical charges of -1.00×10^{-16} C, have a center-to-center separation of 1.00 cm. (a) What is the magnitude of the electrostatic force acting between them? (b) How many excess electrons are on each drop, giving it its charge imbalance?

23E. How many electrons would have to be removed from a coin to leave it with a charge of $+1.0 \times 10^{-7}$ C? ilw

24E. An electron is in a vacuum near the surface of Earth. Where should a second electron be placed so that the electrostatic force it exerts on the first electron balances the gravitational force on the first electron due to Earth?

25P. Earth's atmosphere is constantly bombarded by *cosmic ray protons* that originate somewhere in space. If the protons all passed through the atmosphere, each square meter of Earth's surface would intercept protons at the average rate of 1500 protons per second. What would be the corresponding electric current intercepted by the total surface area of the planet? ilw

26P. Calculate the number of coulombs of positive charge in 250 cm^3 of (neutral) water (about a glassful).

27P. In the basic CsCl (cesium chloride) crystal structure, Cs^+ ions form the corners of a cube and a Cl^- ion is at the cube's center (Fig. 22-21). The edge length of the cube is 0.40 nm. The Cs^+ ions are each deficient by one electron (and thus each has a charge of $+e$), and the Cl^- ion has one excess electron (and thus has a charge of $-e$). (a) What is the magnitude of the net electrostatic force exerted on the Cl^- ion by the eight Cs^+ ions at the corners of the cube? (b) If one of the Cs^+ ions is missing, the crystal is said to have a *defect*; what is the magnitude of the net electrostatic force exerted on the Cl^- ion by the seven remaining Cs^+ ions? ssm www

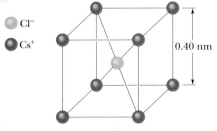

Fig. 22-21 Problem 27.

28P. We know that the negative charge on the electron and the positive charge on the proton are equal. Suppose, however, that these magnitudes differ from each other by 0.00010%. With what force would two copper coins, placed 1.0 m apart, repel each other? Assume that each coin contains 3×10^{22} copper atoms. (*Hint:* A neutral copper atom contains 29 protons and 29 electrons.) What do you conclude?

SEC. 22-6 Charge Is Conserved

29E. Identify X in the following nuclear reactions (in the first, n represents a neutron): (a) $^1H + {}^9Be \rightarrow X + n$; (b) $^{12}C + {}^1H \rightarrow X$; (c) $^{15}N + {}^1H \rightarrow {}^4He + X$. Appendix F will help. ssm

Additional Problem

30. In Problem 13, let $q = \alpha Q$. (a) Write an expression for the magnitude F of the force between the charges in terms of α, Q, and the charge separation d. (b) Graph F as a function of α. Graphically find the values of α that give (c) the maximum value of F and (d) half the maximum value of F.

23 Electric Fields

During the frequent eruptions of the Sakurajima volcano in Japan, multiple electrical discharges (sparks) flash over the volcano's crater, lighting up the sky and sending out sound waves that resemble thunder. However, this is not a lightning display in a thunderstorm, with electrified clouds of water drops discharging to the ground. This is something different.

How does the region above the volcano become electrified, and is there any way to tell whether the sparks travel up from the crater or down to it?

The answer is in this chapter.

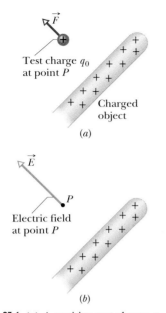

Fig. 23-1 (*a*) A positive test charge q_0 placed at point P near a charged object. An electrostatic force $\vec{F}$ acts on the test charge. (*b*) The electric field $\vec{E}$ at point P produced by the charged object.

23-1 Charges and Forces: A Closer Look

Suppose we fix a positive point charge q_1 in place and then put a second positive point charge q_2 near it. From Coulomb's law we know that q_1 exerts a repulsive electrostatic force on q_2 and, given enough data, we could determine the magnitude and direction of that force. Still, a nagging question remains: How does q_1 "know" of the presence of q_2? Since the charges do not touch, how can q_1 exert a force on q_2?

This question about *action at a distance* can be answered by saying that q_1 sets up an **electric field** in the space surrounding it. At any given point P in that space, the field has both magnitude and direction. The magnitude depends on the magnitude of q_1 and the distance between P and q_1. The direction depends on the direction from q_1 to P and the electrical sign of q_1. Thus when we place q_2 at P, q_1 interacts with q_2 through the electric field at P. The magnitude and direction of that electric field determine the magnitude and direction of the force acting on q_2.

Another action-at-a-distance problem arises if we move q_1, say, toward q_2. Coulomb's law tells us that when q_1 is closer to q_2, the repulsive electrostatic force acting on q_2 must be greater—and it is. However, here the nagging question is: Does the electric field at q_2, and thus the force acting on q_2, change immediately?

The answer is no. Instead, the information about the move by q_1 travels outward from q_1 (in all directions) as an electromagnetic wave at the speed of light c. The change in the electric field at q_2, and thus the change in the force acting on q_2, occurs when the wave finally reaches q_2.

23-2 The Electric Field

The temperature at every point in a room has a definite value. You can measure the temperature at any given point or combination of points by putting a thermometer there. We call the resulting distribution of temperatures a *temperature field.* In much the same way, you can imagine a *pressure field* in the atmosphere; it consists of the distribution of air pressure values, one for each point in the atmosphere. These two examples are of *scalar fields*, because temperature and air pressure are scalar quantities.

The electric field is a *vector field*; it consists of a distribution of *vectors,* one for each point in the region around a charged object, such as a charged rod. In principle, we can define the electric field at some point near the charged object, such as point P in Fig. 23-1*a*, as follows: We first place a *positive* charge q_0, called a *test charge*, at the point. We then measure the electrostatic force $\vec{F}$ that acts on the test charge. Finally, we define the electric field $\vec{E}$ at point P due to the charged object as

$$\vec{E} = \frac{\vec{F}}{q_0} \qquad \text{(electric field).} \qquad (23\text{-}1)$$

Thus, the magnitude of the electric field $\vec{E}$ at point P is $E = F/q_0$, and the direction of $\vec{E}$ is that of the force $\vec{F}$ that acts on the *positive* test charge. As shown in Fig. 23-1*b*, we represent the electric field at P with a vector whose tail is at P. To define the electric field within some region, we must similarly define it at all points in the region.

The SI unit for the electric field is the newton per coulomb (N/C). Table 23-1 shows the electric fields that occur in a few physical situations.

Although we use a positive test charge to define the electric field of a charged object, that field exists independently of the test charge. The field at point P in Figure 23-1*b* existed both before and after the test charge of Fig. 23-1*a* was put there. (We

TABLE 23-1 **Some Electric Fields**

Field Location or Situation	Value (N/C)
At the surface of a uranium nucleus	3×10^{21}
Within a hydrogen atom, at a radius of 5.29×10^{-11} m	5×10^{11}
Electric breakdown occurs in air	3×10^{6}
Near the charged drum of a photocopier	10^{5}
Near a charged comb	10^{3}
In the lower atmosphere	10^{2}
Inside the copper wire of household circuits	10^{-2}

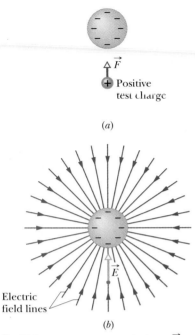

Fig. 23-2 (*a*) The electrostatic force $\vec{F}$ acting on a positive test charge near a sphere of uniform negative charge. (*b*) The electric field vector $\vec{E}$ at the location of the test charge, and the electric field lines in the space near the sphere. The field lines extend *toward* the negatively charged sphere. (They originate on distant positive charges.)

assume that in our defining procedure, the presence of the test charge does not affect the charge distribution on the charged object, and thus does not alter the electric field we are defining.)

To examine the role of an electric field in the interaction between charged objects, we have two tasks: (1) calculating the electric field produced by a given distribution of charge, and (2) calculating the force that a given field exerts on a charge placed in it. We perform the first task in Sections 23-4 through 23-7 for several charge distributions. We perform the second task in Sections 23-8 and 23-9 by considering a point charge and a pair of point charges in an electric field. First, however, we discuss a way to visualize electric fields.

23-3 Electric Field Lines

Michael Faraday, who introduced the idea of electric fields in the 19th century, thought of the space around a charged body as filled with *lines of force*. Although we no longer attach much reality to these lines, now usually called **electric field lines,** they still provide a nice way to visualize patterns in electric fields.

The relation between the field lines and electric field vectors is this: (1) At any point, the direction of a straight field line or the direction of the tangent to a curved field line gives the direction of $\vec{E}$ at that point, and (2) the field lines are drawn so that the number of lines per unit area, measured in a plane that is perpendicular to the lines, is proportional to the *magnitude* of $\vec{E}$. This second relation means that where the field lines are close together, E is large; and where they are far apart, E is small.

Figure 23-2*a* shows a sphere of uniform negative charge. If we place a *positive* test charge anywhere near the sphere, an electrostatic force pointing *toward* the center of the sphere will act on the test charge as shown. In other words, the electric field vectors at all points near the sphere are directed radially toward the sphere. This pattern of vectors is neatly displayed by the field lines in Fig. 23-2*b*, which point in the same directions as the force and field vectors. Moreover, the spreading of the field lines with distance from the sphere tells us that the magnitude of the electric field decreases with distance from the sphere.

If the sphere of Fig. 23-2 were of uniform *positive* charge, the electric field vectors at all points near the sphere would be directed radially *away from* the sphere. Thus, the electric field lines would also extend radially away from the sphere. We then have the following rule:

▶ Electric field lines extend away from positive charge (where they originate) and toward negative charge (where they terminate).

Figure 23-3*a* shows part of an infinitely large, nonconducting *sheet* (or plane) with a uniform distribution of positive charge on one side. If we were to place a positive test charge at any point near the sheet of Fig. 23-3*a*, the net electrostatic force acting on the test charge would be perpendicular to the sheet, because forces acting in all other directions would cancel one another as a result of the symmetry. Moreover, the net force on the test charge would point away from the sheet as shown. Thus, the electric field vector at any point in the space on either side of the sheet is also perpendicular to the sheet and directed away from it (Figs. 23-3*b* and *c*). Since the charge is uniformly distributed along the sheet, all the field vectors have the same magnitude. Such an electric field, with the same magnitude and direction at every point, is a *uniform electric field.*

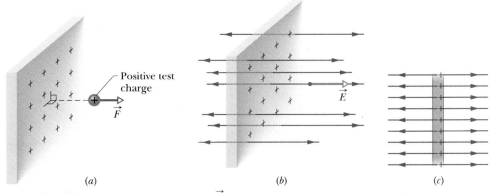

(a)   (b)   (c)

Fig. 23-3 (a) The electrostatic force $\vec{F}$ on a positive test charge near a very large, nonconducting sheet with uniformly distributed positive charge on one side. (b) The electric field vector $\vec{E}$ at the location of the test charge, and the electric field lines in the space near the sheet. The field lines extend *away from* the positively charged sheet. (c) Side view of (b).

Of course, no real nonconducting sheet (such as a flat expanse of plastic) is infinitely large, but if we consider a region that is near the middle of a real sheet and not near its edges, the field lines through that region are arranged as in Figs. 23-3b and c.

Figure 23-4 shows the field lines for two equal positive charges. Figure 23-5 shows the pattern for two charges that are equal in magnitude but of opposite sign, a configuration that we call an **electric dipole.** Although we do not often use field lines quantitatively, they are very useful to visualize what is going on. Can you not almost "see" the charges being pushed apart in Fig. 23-4 and pulled together in Fig. 23-5?

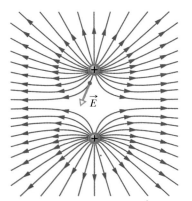

Fig. 23-4 Field lines for two equal positive point charges. The charges repel each other. (The lines terminate on distant negative charges.) To "see" the actual three-dimensional pattern of field lines, mentally rotate the pattern shown here about an axis passing through both charges in the plane of the page. The three-dimensional pattern and the electric field it represents are said to have *rotational symmetry* about that axis. The electric field vector at one point is shown; note that it is tangent to the field line through that point.

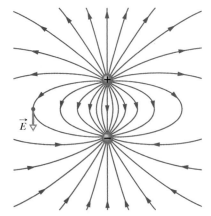

Fig. 23-5 Field lines for a positive and a nearby negative point charge that are equal in magnitude. The charges attract each other. The pattern of field lines and the electric field it represents have rotational symmetry about an axis passing through both charges in the plane of the page. The electric field vector at one point is shown; the vector is tangent to the field line through the point.

Sample Problem 23-1

How does the magnitude of the electric field vary with distance from the center of the uniformly charged sphere in Fig. 23-2? Use an argument based on the electric field lines.

SOLUTION: One Key Idea here is that the field lines are uniformly distributed around the sphere and extend outward from it without interruption. Thus, if we place a concentric spherical shell of radius r around the charged sphere, all the field lines terminating on the charged sphere must pass through the concentric shell. Let the num-ber of field lines be N. Then, because the shell has surface area $4\pi r^2$, the number of field lines per unit area passing through the shell is $N/4\pi r^2$.

A second Key Idea is that the magnitude E of the electric field is proportional to the number of lines per unit area perpendicular to the lines. Since the shell is perpendicular to the field lines, E is proportional to $N/4\pi r^2$. Because r is the only variable in that term, E varies as the inverse square of the distance from the center of the charged sphere.

23-4 The Electric Field Due to a Point Charge

To find the electric field due to a point charge q (or charged particle) at any point a distance r from the point charge, we put a positive test charge q_0 at that point. From Coulomb's law (Eq. 22-4), the magnitude of the electrostatic force acting on q_0 is

$$F = \frac{1}{4\pi\varepsilon_0} \frac{|q||q_0|}{r^2}.$$ (23-2)

The direction of $\vec{F}$ is directly away from the point charge if q is positive, and directly toward the point charge if q is negative. The magnitude of the electric field vector is, from Eq. 23-1,

$$E = \frac{F}{q_0} = \frac{1}{4\pi\varepsilon_0} \frac{|q|}{r^2} \qquad \text{(point charge).}$$ (23-3)

The direction of $\vec{E}$ is the same as that of the force on the positive test charge: directly away from the point charge if q is positive, and toward it if q is negative.

Because there is nothing special about the point we chose for q_0, Eq. 23-3 gives the field at every point around the point charge q. The field for a positive point charge is shown in Fig. 23-6 in vector form (not as field lines).

We can quickly find the net, or resultant, electric field due to more than one point charge. If we place a positive test charge q_0 near n point charges $q_1, q_2, \ldots, q_n$, then, from Eq. 22-7, the net force $\vec{F}_0$ from the n point charges acting on the test charge is

$$\vec{F}_0 = \vec{F}_{01} + \vec{F}_{02} + \cdots + \vec{F}_{0n}.$$

Therefore, from Eq. 23-1, the net electric field at the position of the test charge is

$$\vec{E} = \frac{\vec{F}_0}{q_0} = \frac{\vec{F}_{01}}{q_0} + \frac{\vec{F}_{02}}{q_0} + \cdots + \frac{\vec{F}_{0n}}{q_0}$$
$$= \vec{E}_1 + \vec{E}_2 + \cdots + \vec{E}_n.$$ (23-4)

Here $\vec{E}_i$ is the electric field that would be set up by point charge i acting alone. Equation 23-4 shows us that the principle of superposition applies to electric fields as well as to electrostatic forces.

Fig. 23-6 The electric field vectors at several points around a positive point charge.

✓**CHECKPOINT 1:**   The figure here shows a proton p and an electron e on an x axis. What is the direction of the electric field due to the electron at (a) point S and (b) point R? What is the direction of the net electric field at (c) point R and (d) point S?

Sample Problem 23-2

Figure 23-7a shows three particles with charges $q_1 = +2Q$, $q_2 = -2Q$, and $q_3 = -4Q$, each a distance d from the origin. What net electric field $\vec{E}$ is produced at the origin?

SOLUTION: The Key Idea is that charges q_1, q_2, and q_3 produce electric field vectors $\vec{E}_1$, $\vec{E}_2$, and $\vec{E}_3$, respectively, at the origin, and the net electric field is the vector sum $\vec{E} = \vec{E}_1 + \vec{E}_2 + \vec{E}_3$. To find this sum, we first must find the magnitudes and orientations of the three field vectors. To find the magnitude of $\vec{E}_1$, which is due to q_1, we use Eq. 23-3, substituting d for r and $2Q$ for $|q|$ and obtaining

$$E_1 = \frac{1}{4\pi\varepsilon_0}\frac{2Q}{d^2}.$$

Similarly, we find the magnitudes of the fields $\vec{E}_2$ and $\vec{E}_3$ to be

$$E_2 = \frac{1}{4\pi\varepsilon_0}\frac{2Q}{d^2} \quad \text{and} \quad E_3 = \frac{1}{4\pi\varepsilon_0}\frac{4Q}{d^2}.$$

We next must find the orientations of the three electric field vectors at the origin. Because q_1 is a positive charge, the field vector it produces points directly *away* from it, and because q_2 and q_3 are both negative, the field vectors they produce point directly *toward* each of them. Thus, the three electric fields produced at the origin by the three charged particles are oriented as in Fig. 23-7b. (*Caution:* Note that we have placed the tails of the vectors at the point

where the fields are to be evaluated; doing so decreases the chance of error.)

We can now add the fields vectorially as outlined for forces in Sample Problem 22-1c. However, here we can use symmetry to simplify the procedure. From Fig. 23-7b, we see that $\vec{E}_1$ and $\vec{E}_2$ have the same direction. Hence, their vector sum has that direction and has the magnitude

$$E_1 + E_2 = \frac{1}{4\pi\varepsilon_0}\frac{2Q}{d^2} + \frac{1}{4\pi\varepsilon_0}\frac{2Q}{d^2}$$

$$= \frac{1}{4\pi\varepsilon_0}\frac{4Q}{d^2},$$

which happens to equal the magnitude of field $\vec{E}_3$.

We must now combine two vectors, $\vec{E}_3$ and the vector sum $\vec{E}_1 + \vec{E}_2$, that have the same magnitude and that are oriented symmetrically about the x axis, as shown in Fig. 23-7c. From the symmetry of Fig. 23-7c, we realize that the equal y components of our two vectors cancel and the equal x components add. Thus, the net electric field $\vec{E}$ at the origin is in the positive direction of the x axis and has the magnitude

$$E = 2E_{3x} = 2E_3 \cos 30°$$

$$= (2)\frac{1}{4\pi\varepsilon_0}\frac{4Q}{d^2}(0.866) = \frac{6.93Q}{4\pi\varepsilon_0 d^2}. \quad \text{(Answer)}$$

(a)

(b)   *(c)*

Fig. 23-7 Sample Problem 23-2. (*a*) Three particles with charges q_1, q_2, and q_3 are at the same distance d from the origin. (*b*) The electric field vectors $\vec{E}_1$, $\vec{E}_2$, and $\vec{E}_3$ at the origin due to the three particles. (*c*) The electric field vector $\vec{E}_3$ and the vector sum $\vec{E}_1 + \vec{E}_2$ at the origin.

✓**CHECKPOINT 2:** The figure here shows four situations in which charged particles are at equal distances from the origin. Rank the situations according to the magnitude of the net electric field at the origin, greatest first.

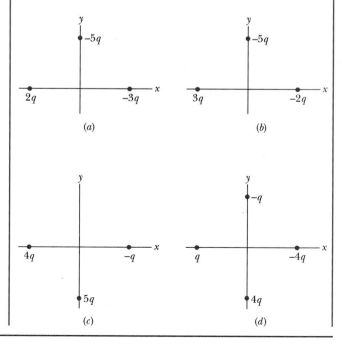

(a)   *(b)*

(c)   *(d)*

23-5 The Electric Field Due to an Electric Dipole

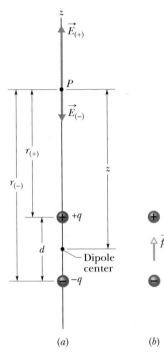

Fig. 23-8 (a) An electric dipole. The electric field vectors $\vec{E}_{(+)}$ and $\vec{E}_{(-)}$ at point P on the dipole axis result from the dipole's two charges. P is at distances $r_{(+)}$ and $r_{(-)}$ from the individual charges that make up the dipole. (b) The dipole moment $\vec{p}$ of the dipole points from the negative charge to the positive charge.

Figure 23-8a shows two charged particles of magnitude q but of opposite sign, separated by a distance d. As was noted in connection with Fig. 23-5, we call this configuration an *electric dipole*. Let us find the electric field due to the dipole of Fig. 23-8a at a point P, a distance z from the midpoint of the dipole and on the axis through the particles, which is called the *dipole axis*.

From symmetry, the electric field $\vec{E}$ at point P—and also the fields $\vec{E}_{(+)}$ and $\vec{E}_{(-)}$ due to the separate charges that make up the dipole—must lie along the dipole axis, which we have taken to be a z axis. Applying the superposition principle for electric fields, we find that the magnitude E of the electric field at P is

$$E = E_{(+)} - E_{(-)}$$
$$= \frac{1}{4\pi\varepsilon_0}\frac{q}{r_{(+)}^2} - \frac{1}{4\pi\varepsilon_0}\frac{q}{r_{(-)}^2}$$
$$= \frac{q}{4\pi\varepsilon_0(z - \frac{1}{2}d)^2} - \frac{q}{4\pi\varepsilon_0(z + \frac{1}{2}d)^2}. \qquad (23\text{-}5)$$

After a little algebra, we can rewrite this equation as

$$E = \frac{q}{4\pi\varepsilon_0 z^2}\left[\left(1 - \frac{d}{2z}\right)^{-2} - \left(1 + \frac{d}{2z}\right)^{-2}\right]. \qquad (23\text{-}6)$$

We are usually interested in the electrical effect of a dipole only at distances that are large compared with the dimensions of the dipole—that is, at distances such that $z \gg d$. At such large distances, we have $d/2z \ll 1$ in Eq. 23-6. We can then expand the two quantities in the brackets in that equation by the binomial theorem (Appendix E), obtaining for those quantities

$$\left[\left(1 + \frac{2d}{2z(1!)} + \cdots\right) - \left(1 - \frac{2d}{2z(1!)} + \cdots\right)\right].$$

Thus,
$$E = \frac{q}{4\pi\varepsilon_0 z^2}\left[\left(1 + \frac{d}{z} + \cdots\right) - \left(1 - \frac{d}{z} + \cdots\right)\right]. \qquad (23\text{-}7)$$

The unwritten terms in the two expansions in Eq. 23-7 involve d/z raised to progressively higher powers. Since $d/z \ll 1$, the contributions of those terms are progressively less, and to approximate E at large distances, we can neglect them. Then, in our approximation, we can rewrite Eq. 23-7 as

$$E = \frac{q}{4\pi\varepsilon_0 z^2}\frac{2d}{z} = \frac{1}{2\pi\varepsilon_0}\frac{qd}{z^3}. \qquad (23\text{-}8)$$

The product qd, which involves the two intrinsic properties q and d of the dipole, is the magnitude p of a vector quantity known as the **electric dipole moment** $\vec{p}$ of the dipole. (The unit of $\vec{p}$ is the Coulomb–meter.) Thus, we can write Eq. 23-8 as

$$E = \frac{1}{2\pi\varepsilon_0}\frac{p}{z^3} \qquad \text{(electric dipole).} \qquad (23\text{-}9)$$

The direction of $\vec{p}$ is taken to be from the negative to the positive end of the dipole, as indicated in Fig. 23-8b. We can use $\vec{p}$ to specify the orientation of a dipole.

Equation 23-9 shows that, if we measure the electric field of a dipole only at distant points, we can never find q and d separately, only their product. The field at

TABLE 23-2 Some Measures of Electric Charge

Name	Symbol	SI Unit
Charge	q	C
Linear charge density	λ	C/m
Surface charge density	σ	C/m^2
Volume charge density	ρ	C/m^3

distant points would be unchanged if, for example, q were doubled and d simultaneously halved. Thus, the dipole moment is a basic property of a dipole.

Although Eq. 23-9 holds only for distant points along the dipole axis, it turns out that E for a dipole varies as $1/r^3$ for *all* distant points, regardless of whether they lie on the dipole axis; here r is the distance between the point in question and the dipole center.

Inspection of Fig. 23-8 and of the field lines in Fig. 23-5 shows that the direction of $\vec{E}$ for distant points on the dipole axis is always the direction of the dipole moment vector $\vec{p}$. This is true whether point P in Fig. 23-8a is on the upper or the lower part of the dipole axis.

Inspection of Eq. 23-9 shows that if you double the distance of a point from a dipole, the electric field at the point drops by a factor of 8. If you double the distance from a single point charge, however (see Eq. 23-3), the electric field drops only by a factor of 4. Thus the electric field of a dipole decreases more rapidly with distance than does the electric field of a single charge. The physical reason for this rapid decrease in electric field for a dipole is that from distant points a dipole looks like two equal but opposite charges that almost—but not quite—coincide. Thus, their electric fields at distant points almost—but not quite—cancel each other.

23-6 The Electric Field Due to a Line of Charge

So far we have considered the electric field that is produced by one or, at most, a few point charges. We now consider charge distributions that consist of a great many closely spaced point charges (perhaps billions) that are spread along a line, over a surface, or within a volume. Such distributions are said to be **continuous** rather than discrete. Since these distributions can include an enormous number of point charges, we find the electric fields that they produce by means of calculus rather than by considering the point charges one by one. In this section we discuss the electric field caused by a line of charge. We consider a charged surface in the next section. In the next chapter, we shall find the field inside a uniformly charged sphere.

When we deal with continuous charge distributions, it is most convenient to express the charge on an object as a *charge density* rather than as a total charge. For a line of charge, for example, we would report the linear charge density (or charge per unit length) λ, whose SI unit is the coulomb per meter. Table 23-2 shows the other charge densities we shall be using.

Figure 23-9 shows a thin ring of radius R with a uniform positive linear charge density λ around its circumference. We may imagine the ring to be made of plastic or some other insulator, so that the charges can be regarded as fixed in place. What is the electric field $\vec{E}$ at point P, a distance z from the plane of the ring along its central axis?

To answer, we cannot just apply Eq. 23-3, which gives the electric field set up by a point charge, because the ring is obviously not a point charge. However, we can mentally divide the ring into differential elements of charge that are so small that they are like point charges, and then we can apply Eq. 23-3 to each of them. Next, we can add the electric fields set up at P by all the differential elements. The vector sum of all those fields gives us the field set up at P by the ring.

Let ds be the (arc) length of any differential element of the ring. Since λ is the charge per unit length, the element has a charge of magnitude

$$dq = \lambda \, ds. \tag{23-10}$$

Fig. 23-9 A ring of uniform positive charge. A differential element of charge occupies a length ds (greatly exaggerated for clarity). This element sets up an electric field $d\vec{E}$ at point P. The component of $d\vec{E}$ along the central axis of the ring is $dE \cos \theta$.

This differential charge sets up a differential electric field $d\vec{E}$ at point P, which is a distance r from the element. Treating the element as a point charge, and using Eq. 23-10, we can rewrite Eq. 23-3 to express the magnitude of $d\vec{E}$ as

$$dE = \frac{1}{4\pi\varepsilon_0}\frac{dq}{r^2} = \frac{1}{4\pi\varepsilon_0}\frac{\lambda\,ds}{r^2}.\qquad(23\text{-}11)$$

From Fig. 23-9, we can rewrite Eq. 23-11 as

$$dE = \frac{1}{4\pi\varepsilon_0}\frac{\lambda\,ds}{(z^2 + R^2)}.\qquad(23\text{-}12)$$

Figure 23-9 shows us that $d\vec{E}$ is at an angle θ to the central axis (which we have taken to be a z axis) and has components perpendicular to and parallel to that axis.

Every charge element in the ring sets up a differential field $d\vec{E}$ at P, with magnitude given by Eq. 23-12. All the $d\vec{E}$ vectors have identical components parallel to the central axis, in both magnitude and direction. All these $d\vec{E}$ vectors have components perpendicular to the central axis as well; these perpendicular components are identical in magnitude but point in different directions. In fact, for any perpendicular component that points in a given direction, there is another one that points in the opposite direction. The sum of this pair of components, like the sum of all other pairs of oppositely directed components, is zero.

Thus, the perpendicular components cancel and we need not consider them further. This leaves the parallel components; they all have the same direction, so the net electric field at P is their sum.

The parallel component of $d\vec{E}$ shown in Fig. 23-9 has magnitude $dE\cos\theta$. The figure also shows us that

$$\cos\theta = \frac{z}{r} = \frac{z}{(z^2 + R^2)^{1/2}}.\qquad(23\text{-}13)$$

Then Eqs. 23-13 and 23-12 give us, for the parallel component of $d\vec{E}$,

$$dE\cos\theta = \frac{z\lambda}{4\pi\varepsilon_0(z^2 + R^2)^{3/2}}\,ds.\qquad(23\text{-}14)$$

To add the parallel components $dE\cos\theta$ produced by all the elements, we integrate Eq. 23-14 around the circumference of the ring, from $s = 0$ to $s = 2\pi R$. Since the only quantity in Eq. 23-14 that varies during the integration is s, the other quantities can be moved outside the integral sign. The integration then gives us

$$E = \int dE\cos\theta = \frac{z\lambda}{4\pi\varepsilon_0(z^2 + R^2)^{3/2}}\int_0^{2\pi R} ds$$

$$= \frac{z\lambda(2\pi R)}{4\pi\varepsilon_0(z^2 + R^2)^{3/2}}.\qquad(23\text{-}15)$$

Since λ is the charge per length of the ring, the term $\lambda(2\pi R)$ in Eq. 23-15 is q, the total charge on the ring. We then can rewrite Eq. 23-15 as

$$E = \frac{qz}{4\pi\varepsilon_0(z^2 + R^2)^{3/2}}\qquad\text{(charged ring).}\qquad(23\text{-}16)$$

If the charge on the ring is negative, instead of positive as we have assumed, the magnitude of the field at P is still given by Eq. 23-16. However, the electric field vector then points toward the ring instead of away from it.

Let us check Eq. 23-16 for a point on the central axis that is so far away that $z \gg R$. For such a point, the expression $z^2 + R^2$ in Eq. 23-16 can be approximated as z^2, and Eq. 23-16 becomes

$$E = \frac{1}{4\pi\varepsilon_0} \frac{q}{z^2} \qquad \text{(charged ring at large distance).} \qquad (23\text{-}17)$$

This is a reasonable result, because from a large distance, the ring "looks like" a point charge. If we replace z with r in Eq. 23-17, we indeed do have Eq. 23-3, the magnitude of the electric field due to a point charge.

Let us next check Eq. 23-16 for a point at the center of the ring—that is, for $z = 0$. At that point, Eq. 23-16 tells us that $E = 0$. This is a reasonable result, because if we were to place a test charge at the center of the ring, there would be no net electrostatic force acting on it; the force due to any element of the ring would be canceled by the force due to the element on the opposite side of the ring. By Eq. 23-1, if the force at the center of the ring were zero, the electric field there would also have to be zero.

Sample Problem 23-3

Figure 23-10a shows a plastic rod having a uniformly distributed charge $-Q$. The rod has been bent in a 120° circular arc of radius r. We place coordinate axes such that the axis of symmetry of the rod lies along the x axis and the origin is at the center of curvature P of the rod. In terms of Q and r, what is the electric field $\vec{E}$ due to the rod at point P?

SOLUTION: The Key Idea here is that, because the rod has a continuous charge distribution, we must find an expression for the electric fields due to differential elements of the rod and then sum those fields via calculus. Consider a differential element having arc length ds and located at an angle θ above the x axis (Fig. 23-10b). If we let λ represent the linear charge density of the rod, our element ds has a differential charge of magnitude

$$dq = \lambda \, ds. \qquad (23\text{-}18)$$

Our element produces a differential electric field $d\vec{E}$ at point P, which is a distance r from the element. Treating the element as a point charge, we can rewrite Eq. 23-3 to express the magnitude of $d\vec{E}$ as

$$dE = \frac{1}{4\pi\varepsilon_0} \frac{dq}{r^2} = \frac{1}{4\pi\varepsilon_0} \frac{\lambda \, ds}{r^2}. \qquad (23\text{-}19)$$

The direction of $d\vec{E}$ is toward ds, because charge dq is negative.

Our element has a symmetrically located (mirror image) element ds' in the bottom half of the rod. The electric field $d\vec{E}'$ set up at P by ds' also has the magnitude given by Eq. 23-19, but the field vector points toward ds' as shown in Fig. 23-10b. If we resolve the electric field vectors of ds and ds' into x and y components as shown in Fig. 23-10b, we see that their y components cancel (because they have equal magnitudes and are in opposite directions). We also see that their x components have equal magnitudes and are in the same direction.

Thus, to find the electric field set up by the rod, we need sum (via integration) only the x components of the differential electric fields set up by all the differential elements of the rod. From

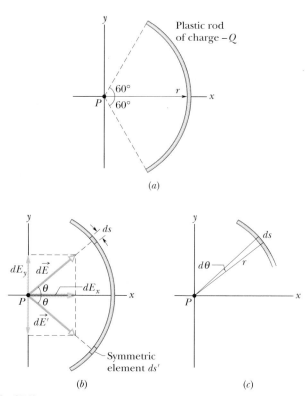

(a)

(b)   (c)

Fig. 23-10 Sample Problem 23-3. (a) A plastic rod of charge $-Q$ is a circular section of radius r and central angle 120°; point P is the center of curvature of the rod. (b) A differential element in the top half of the rod, at an angle θ to the x axis and of arc length ds, sets up a differential electric field $d\vec{E}$ at P. An element ds', symmetric to ds about the x axis, sets up a field $d\vec{E}'$ at P with the same magnitude. (c) Arc length ds makes an angle $d\theta$ about point P.

Fig. 23-10b and Eq. 23-19, we can write the component dE_x set up by ds as

$$dE_x = dE \cos \theta = \frac{1}{4\pi\varepsilon_0} \frac{\lambda}{r^2} \cos \theta \, ds. \quad (23\text{-}20)$$

Equation 23-20 has two variables, θ and s. Before we can integrate it, we must eliminate one variable. We do so by replacing ds, using the relation

$$ds = r \, d\theta,$$

in which $d\theta$ is the angle at P that includes arc length ds (Fig. 23-10c). With this replacement, we can integrate Eq. 23-20 over the angle made by the rod at P, from $\theta = -60°$ to $\theta = 60°$; that will give us the magnitude of the electric field at P due to the rod:

$$E = \int dE_x = \int_{-60°}^{60°} \frac{1}{4\pi\varepsilon_0} \frac{\lambda}{r^2} \cos \theta \, r \, d\theta$$

$$= \frac{\lambda}{4\pi\varepsilon_0 r} \int_{-60°}^{60°} \cos \theta \, d\theta = \frac{\lambda}{4\pi\varepsilon_0 r} \left[\sin \theta \right]_{-60°}^{60°}$$

$$= \frac{\lambda}{4\pi\varepsilon_0 r} [\sin 60° - \sin(-60°)]$$

$$= \frac{1.73\lambda}{4\pi\varepsilon_0 r}. \quad (23\text{-}21)$$

(If we had reversed the limits on the integration, we would have gotten the same result but with a minus sign. Since the integration gives only the magnitude of $\vec{E}$, we would then have discarded the minus sign.)

To evaluate λ, we note that the rod has an angle of 120° and so is one-third of a full circle. Its arc length is then $2\pi r/3$, and its linear charge density must be

$$\lambda = \frac{\text{charge}}{\text{length}} = \frac{Q}{2\pi r/3} = \frac{0.477Q}{r}.$$

Substituting this into Eq. 23-21 and simplifying give us

$$E = \frac{(1.73)(0.477Q)}{4\pi\varepsilon_0 r^2}$$

$$= \frac{0.83Q}{4\pi\varepsilon_0 r^2}. \quad \text{(Answer)}$$

The direction of $\vec{E}$ is toward the rod, along the axis of symmetry of the charge distribution. We can write $\vec{E}$ in unit-vector notation as

$$\vec{E} = \frac{0.83Q}{4\pi\varepsilon_0 r^2} \hat{i}.$$

PROBLEM-SOLVING TACTICS

Tactic 1: *A Field Guide for Lines of Charge*
Here is a generic guide for finding the electric field $\vec{E}$ produced at a point P by a line of uniform charge, either circular or straight. The general strategy is to pick out an element dq of the charge, find $d\vec{E}$ due to that element, and integrate $d\vec{E}$ over the entire line of charge.

Step 1. If the line of charge is circular, let ds be the arc length of an element of the distribution. If the line is straight, run an x axis along it and let dx be the length of an element. Mark the element on a sketch.

Step 2. Relate the charge dq of the element to the length of the element with either $dq = \lambda \, ds$ or $dq = \lambda \, dx$. Consider dq and λ to be positive, even if the charge is actually negative. (The sign of the charge is used in the next step.)

Step 3. Express the field $d\vec{E}$ produced at P by dq with Eq. 23-3, replacing q in that equation with either $\lambda \, ds$ or $\lambda \, dx$. If the charge on the line is positive, then at P draw a vector $d\vec{E}$ that points directly away from dq. If the charge is negative, draw the vector pointing directly toward dq.

Step 4. Always look for any symmetry in the situation. If P is on an axis of symmetry of the charge distribution, resolve the field $d\vec{E}$ produced by dq into components that are perpendicular and parallel to the axis of symmetry. Then consider a second element dq' that is located symmetrically to dq about the line of symmetry. At P draw the vector $d\vec{E}'$ that this symmetrical element produces, and resolve it into components. One of the components produced by dq is a *canceling component*; it is canceled by the corresponding component produced by dq' and needs no further attention. The other com-

ponent produced by dq is an *adding component*; it adds to the corresponding component produced by dq'. Add the adding components of all the elements via integration.

Step 5. Here are four general types of uniform charge distributions, with strategies for simplifying the integral of step 4.

Ring, with point P on (central) axis of symmetry, as in Fig. 23-9. In the expression for dE, replace r^2 with $z^2 + R^2$, as in Eq. 23-12. Express the adding component of $d\vec{E}$ in terms of θ. That introduces $\cos \theta$, but θ is identical for all elements and thus is not a variable. Replace $\cos \theta$ as in Eq. 23-13. Integrate over s, around the circumference of the ring.

Circular arc, with point P at the center of curvature, as in Fig. 23-10. Express the adding component of $d\vec{E}$ in terms of θ. That introduces either $\sin \theta$ or $\cos \theta$. Reduce the resulting two variables s and θ to one, θ, by replacing ds with $r \, d\theta$. Integrate over θ, as in Sample Problem 23-3, from one end of the arc to the other end.

Straight line, with point P on an extension of the line, as in Fig. 23-11a. In the expression for dE, replace r with x. Integrate over x, from end to end of the line of charge.

Straight line, with point P at perpendicular distance y from the line of charge, as in Fig. 23-11b. In the expression for dE, replace r with an expression involving x and y. If P is on the perpendicular bisector of the line of charge, find an expression for the adding component of $d\vec{E}$. That will introduce either $\sin \theta$ or $\cos \theta$. Reduce the resulting two variables x and θ to one, x, by replacing the trigonometric function with an expression (its definition) involving x and y. Integrate over x from end to end of the line of charge. If P is not on a line

Fig. 23-11 (*a*) Point *P* is on an extension of the line of charge. (*b*) *P* is on a line of symmetry of the line of charge, at perpendicular distance *y* from that line. (*c*) Same as (*b*) except that *P* is not on a line of symmetry.

Step 6.   One arrangement of the integration limits gives a positive result. The reverse arrangement gives the same result with a minus sign; discard the minus sign. If the result is to be stated in terms of the total charge *Q* of the distribution, replace λ with *Q/L*, in which *L* is the length of the distribution. For a ring, *L* is the ring's circumference.

✓**CHECKPOINT 3:**   The figure here shows three nonconducting rods, one circular and two straight. Each has a uniform charge of magnitude *Q* along its top half and another along its bottom half. For each rod, what is the direction of the net electric field at point *P*?

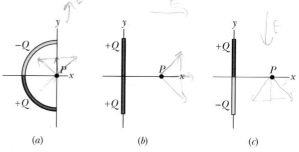

(*a*)   (*b*)   (*c*)

of symmetry, as in Fig. 23-11*c*, set up an integral to sum the components dE_x, and integrate over *x* to find E_x. Also set up an integral to sum the components dE_y, and integrate over *x* again to find E_y. Use the components E_x and E_y in the usual way to find the magnitude *E* and the orientation of $\vec{E}$.

23-7   The Electric Field Due to a Charged Disk

Figure 23-12 shows a circular plastic disk of radius *R* that has a positive surface charge of uniform density σ on its upper surface (see Table 23-2). What is the electric field at point *P*, a distance *z* from the disk along its central axis?

Our plan is to divide the disk into concentric flat rings and then to calculate the electric field at point *P* by adding up (that is, by integrating) the contributions of all the rings. Figure 23-12 shows one such ring, with radius *r* and radial width *dr*. Since σ is the charge per unit area, the charge on the ring is

$$dq = \sigma \, dA = \sigma(2\pi r \, dr), \qquad (23\text{-}22)$$

where *dA* is the differential area of the ring.

We have already solved the problem of the electric field due to a ring of charge. Substituting *dq* from Eq. 23-22 for *q* in Eq. 23-16, and replacing *R* in Eq. 23-16 with *r*, we obtain an expression for the electric field *dE* at *P* due to our flat ring:

$$dE = \frac{z\sigma 2\pi r \, dr}{4\pi\varepsilon_0(z^2 + r^2)^{3/2}},$$

which we may write as

$$dE = \frac{\sigma z}{4\varepsilon_0} \frac{2r \, dr}{(z^2 + r^2)^{3/2}}. \qquad (23\text{-}23)$$

We can now find *E* by integrating Eq. 23-23 over the surface of the disk—that is, by integrating with respect to the variable *r* from $r = 0$ to $r = R$. Note that *z* remains constant during this process. We get

$$E = \int dE = \frac{\sigma z}{4\varepsilon_0} \int_0^R (z^2 + r^2)^{-3/2}(2r) \, dr. \qquad (23\text{-}24)$$

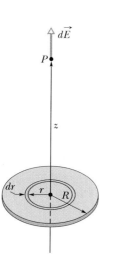

Fig. 23-12 A disk of radius *R* and uniform positive charge. The ring shown has radius *r* and radial width *dr*. It sets up a differential electric field $d\vec{E}$ at point *P* on its central axis.

To solve this integral, we cast it in the form $\int X^m \, dX$ by setting $X = (z^2 + r^2)$, $m = -\frac{3}{2}$, and $dX = (2r) \, dr$. For the recast integral we have

$$\int X^m \, dX = \frac{X^{m+1}}{m+1},$$

so Eq. 23-24 becomes

$$E = \frac{\sigma z}{4\varepsilon_0} \left[\frac{(z^2 + r^2)^{-1/2}}{-\frac{1}{2}} \right]_0^R. \qquad (23\text{-}25)$$

Taking the limits in Eq. 23-25 and rearranging, we find

$$E = \frac{\sigma}{2\varepsilon_0} \left(1 - \frac{z}{\sqrt{z^2 + R^2}} \right) \qquad \text{(charged disk)} \qquad (23\text{-}26)$$

as the magnitude of the electric field produced by a flat, circular, charged disk at points on its central axis. (In carrying out the integration, we assumed that $z \geq 0$.)

If we let $R \rightarrow \infty$ while keeping z finite, the second term in the parentheses in Eq. 23-26 approaches zero, and this equation reduces to

$$E = \frac{\sigma}{2\varepsilon_0} \qquad \text{(infinite sheet)}. \qquad (23\text{-}27)$$

This is the electric field produced by an infinite sheet of uniform charge located on one side of a nonconductor such as plastic. The electric field lines for such a situation are shown in Fig. 23-3.

We also get Eq. 23-27 if we let $z \rightarrow 0$ in Eq. 23-26 while keeping R finite. This shows that at points very close to the disk, the electric field set up by the disk is the same as if the disk were infinite in extent.

23-8   A Point Charge in an Electric Field

In the preceding four sections we worked at the first of our two tasks: given a charge distribution, to find the electric field it produces in the surrounding space. Here we begin the second task: to determine what happens to a charged particle when it is in an electric field that is produced by other stationary or slowly moving charges.

What happens is that an electrostatic force acts on the particle, as given by

$$\vec{F} = q\vec{E}, \qquad (23\text{-}28)$$

in which q is the charge of the particle (including its sign) and $\vec{E}$ is the electric field that other charges have produced at the location of the particle. (The field is *not* the field set up by the particle itself; to distinguish the two fields, the field acting on the particle in Eq. 23-28 is often called the *external field*. A charged particle (or object) is not affected by its own electric field.) Equation 23-28 tells us

> ► The electrostatic force $\vec{F}$ acting on a charged particle located in an external electric field $\vec{E}$ has the direction of $\vec{E}$ if the charge q of the particle is positive and has the opposite direction if q is negative.

✔CHECKPOINT 4:   (a) In the figure, what is the direction of the electrostatic force on the electron due to the electric field shown? (b) In which direction will the electron accelerate if it is moving parallel to the y axis before it encounters the electric field? (c) If, instead, the electron is initially moving rightward, will its speed increase, decrease, or remain constant?

Fig. 23-13 The Millikan oil-drop apparatus for measuring the elementary charge *e*. When a charged oil drop drifted into chamber C through the hole in plate P_1, its motion could be controlled by closing and opening switch S and thereby setting up or eliminating an electric field in chamber C. The microscope was used to view the drop, to permit timing of its motion.

Measuring the Elementary Charge

Equation 23-28 played a role in the measurement of the elementary charge *e* by American physicist Robert A. Millikan in 1910–1913. Figure 23-13 is a representation of his apparatus. When tiny oil drops are sprayed into chamber A, some of them become charged, either positively or negatively, in the process. Consider a drop that drifts downward through the small hole in plate P_1 and into chamber C. Let us assume that this drop has a negative charge *q*.

If switch S in Fig. 23-13 is open as shown, battery B has no electrical effect on chamber C. If the switch is closed (the connection between chamber C and the positive terminal of the battery is then complete), the battery causes an excess positive charge on conducting plate P_1 and an excess negative charge on conducting plate P_2. The charged plates set up a downward-directed electric field $\vec{E}$ in chamber C. According to Eq. 23-28, this field exerts an electrostatic force on any charged drop that happens to be in the chamber and affects its motion. In particular, our negatively charged drop will tend to drift upward.

By timing the motion of oil drops with the switch opened and with it closed and thus determining the effect of the charge *q*, Millikan discovered that the values of *q* were always given by

$$q = ne, \qquad \text{for } n = 0, \pm1, \pm2, \pm3, \dots, \qquad (23\text{-}29)$$

in which *e* turned out to be the fundamental constant we call the *elementary charge*, 1.60×10^{-19} C. Millikan's experiment is convincing proof that charge is quantized, and he earned the 1923 Nobel prize in physics in part for this work. Modern measurements of the elementary charge rely on a variety of interlocking experiments, all more precise than the pioneering experiment of Millikan.

Ink-Jet Printing

The need for high-quality, high-speed printing has caused a search for an alternative to impact printing, such as occurs in a standard typewriter. Building up letters by squirting tiny drops of ink at the paper is one such alternative.

Figure 23-14 shows a negatively charged drop moving between two conducting deflecting plates, between which a uniform, downward-directed electric field $\vec{E}$ has been set up. The drop is deflected upward according to Eq. 23-28 and then strikes the paper at a position that is determined by the magnitudes of $\vec{E}$ and the charge *q* of the drop.

In practice, *E* is held constant and the position of the drop is determined by the charge *q* delivered to the drop in the charging unit, through which the drop must pass before entering the deflecting system. The charging unit, in turn, is activated by electronic signals that encode the material to be printed.

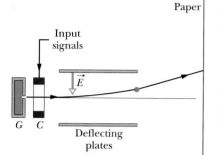

Fig. 23-14 The essential features of an ink-jet printer. Drops are shot out from generator G and receive a charge in charging unit C. An input signal from a computer controls the charge given to each drop and thus the effect of field $\vec{E}$ on the drop and the position on the paper at which the drop lands. About 100 tiny drops are needed to form a single character.

Volcanic Lightning

When the Sakurajima volcano erupts, as seen in this chapter's opening photograph, it spews ash into the air. That ash results when liquid water within the volcano, suddenly converted to steam by the flow of hot lava, shatters rock, which is then burnt. The liquid-to-steam conversion and the explosion of rock cause positive and negative charges to separate. Then, as the steam and ash are spewed into the air, they form a cloud that contains pockets of positive charge and pockets of negative charge.

As these pockets grow, the electric fields between adjacent pockets and between pockets and the volcano crater increase in magnitude. Whenever the magnitude of about 3×10^6 N/C is reached, the air undergoes *electric breakdown* and begins to conduct current. These momentary conducting paths appear in the air where the electric field has ionized air molecules, freeing some of their electrons. These electrons, propelled by the field, collide with air molecules in their way, which causes those molecules to emit light. We can see these brief paths, commonly called *sparks*, because of the light they emit. (A small-scale example of *sparking* can be seen around the charged metal cap in Fig. 23-15.)

The sparks above the volcano snake their way either down from a charge pocket to the crater wall or vice versa. You can tell the direction of a spark by how any dead-end branches on it are forked. If the branches fork downward, then the spark snaked its way downward. (See the bright spark extending from the right side of the photograph to the crater wall.) If the branches fork upward, then the spark snaked its way upward. (See the lower part of the central bright spark on the crater wall.) Sometimes a downward-snaking spark and an upward-snaking spark meet each other. Can you find an example in the photograph?

Fig. 23-15 The metal cap is so charged that the electric field it produces in the surrounding space causes the air there to undergo electric breakdown. The visible sparks reveal where momentary conducting paths are set up in the air, along which the electric field has removed electrons from their molecules and then accelerated them into collisions with the molecules.

Sample Problem 23-4

Figure 23-16 shows the deflecting plates of an ink-jet printer, with superimposed coordinate axes. An ink drop with a mass m of 1.3×10^{-10} kg and a negative charge of magnitude $Q = 1.5 \times 10^{-13}$ C enters the region between the plates, initially moving along the x axis with speed $v_x = 18$ m/s. The length L of the plates is 1.6 cm. The plates are charged and thus produce an electric field at all points between them. Assume that field $\vec{E}$ is downward directed, uniform, and has a magnitude of 1.4×10^6 N/C. What is the vertical deflection of the drop at the far edge of the plates? (The gravitational force on the drop is small relative to the electrostatic force acting on the drop and can be neglected.)

Fig. 23-16 Sample Problem 23-4. An ink drop of mass m and charge magnitude Q is deflected in the electric field of an ink-jet printer.

SOLUTION: The drop is negatively charged and the electric field is directed *downward*. The Key Idea here is that, from Eq. 23-28, a constant electrostatic force of magnitude QE acts *upward* on the charged drop. Thus, as the drop travels parallel to the x axis at constant speed v_x, it accelerates upward with some constant acceleration a_y. Applying Newton's second law ($F = ma$) for components along the y axis, we find that

$$a_y = \frac{F}{m} = \frac{QE}{m}. \qquad (23\text{-}30)$$

Let t represent the time required for the drop to pass through the region between the plates. During t the vertical and horizontal displacements of the drop are

$$y = \tfrac{1}{2}a_y t^2 \quad \text{and} \quad L = v_x t, \qquad (23\text{-}31)$$

respectively. Eliminating t between these two equations and substituting Eq. 23-30 for a_y, we find

$$y = \frac{QEL^2}{2mv_x^2}$$

$$= \frac{(1.5 \times 10^{-13}\ \text{C})(1.4 \times 10^6\ \text{N/C})(1.6 \times 10^{-2}\ \text{m})^2}{(2)(1.3 \times 10^{-10}\ \text{kg})(18\ \text{m/s})^2}$$

$$= 6.4 \times 10^{-4}\ \text{m}$$

$$= 0.64\ \text{mm}. \qquad \text{(Answer)}$$

Fig. 23-17 A molecule of H₂O, showing the three nuclei (represented by dots) and the regions in which the electrons can be located. The electric dipole moment $\vec{p}$ points from the (negative) oxygen side to the (positive) hydrogen side of the molecule.

23-9 A Dipole in an Electric Field

We have defined the electric dipole moment $\vec{p}$ of an electric dipole to be a vector that points from the negative to the positive end of the dipole. As you will see, the behavior of a dipole in a uniform external electric field $\vec{E}$ can be described completely in terms of the two vectors $\vec{E}$ and $\vec{p}$, with no need of any details about the dipole's structure.

A molecule of water (H₂O) is an electric dipole; Fig. 23-17 shows why. There the black dots represent the oxygen nucleus (having eight protons) and the two hydrogen nuclei (having one proton each). The colored enclosed areas represent the regions in which electrons can be located around the nuclei.

In a water molecule, the two hydrogen atoms and the oxygen atom do not lie on a straight line but form an angle of about 105°, as shown in Fig. 23-17. As a result, the molecule has a definite "oxygen side" and "hydrogen side." Moreover, the 10 electrons of the molecule tend to remain closer to the oxygen nucleus than to the hydrogen nuclei. This makes the oxygen side of the molecule slightly more negative than the hydrogen side and creates an electric dipole moment $\vec{p}$ that points along the symmetry axis of the molecule as shown. If the water molecule is placed in an external electric field, it behaves as would be expected of the more abstract electric dipole of Fig. 23-8.

To examine this behavior, we now consider such an abstract dipole in a uniform external electric field $\vec{E}$, as shown in Fig. 23-18a. We assume that the dipole is a rigid structure that consists of two centers of opposite charge, each of magnitude q, separated by a distance d. The dipole moment $\vec{p}$ makes an angle θ with field $\vec{E}$.

Electrostatic forces act on the charged ends of the dipole. Because the electric field is uniform, those forces act in opposite directions (as shown in Fig. 23-18) and with the same magnitude $F = qE$. Thus, *because the field is uniform,* the net force on the dipole from the field is zero and the center of mass of the dipole does not move. However, the forces on the charged ends do produce a net torque $\vec{\tau}$ on the dipole about its center of mass. The center of mass lies on the line connecting the charged ends, at some distance x from one end and thus a distance $d - x$ from the other end. From Eq. 11-31 ($\tau = rF \sin \phi$), we can write the magnitude of the net torque $\vec{\tau}$ as

$$\tau = Fx \sin \theta + F(d - x) \sin \theta = Fd \sin \theta. \tag{23-32}$$

We can also write the magnitude of $\vec{\tau}$ in terms of the magnitudes of the electric field E and the dipole moment $p = qd$. To do so, we substitute qE for F and p/q for d in Eq. 23-32, finding that the magnitude of $\vec{\tau}$ is

$$\tau = pE \sin \theta. \tag{23-33}$$

We can generalize this equation to vector form as

$$\vec{\tau} = \vec{p} \times \vec{E} \quad \text{(torque on a dipole)}. \tag{23-34}$$

(a)

(b)

Fig. 23-18 (a) An electric dipole in a uniform electric field $\vec{E}$. Two centers of equal but opposite charge are separated by distance d. The line between them represents their rigid connection. (b) Field $\vec{E}$ causes a torque τ on the dipole. The direction of $\vec{\tau}$ is into the page, as represented by the symbol $\otimes$.

Vectors $\vec{p}$ and $\vec{E}$ are shown in Fig. 23-18b. The torque acting on a dipole tends to rotate $\vec{p}$ (hence the dipole) into the direction of field $\vec{E}$, thereby reducing θ. In Fig. 23-18, such rotation is clockwise. As we discussed in Chapter 11, we can represent a torque that gives rise to a clockwise rotation by including a minus sign with the magnitude of the torque. With that notation, the torque of Fig. 23-18 is

$$\tau = -pE \sin \theta. \tag{23-35}$$

Potential Energy of an Electric Dipole

Potential energy can be associated with the orientation of an electric dipole in an electric field. The dipole has its least potential energy when it is in its equilibrium orientation, which is when its moment $\vec{p}$ is lined up with the field $\vec{E}$ (then $\vec{\tau} = \vec{p} \times \vec{E} = 0$). It has greater potential energy in all other orientations. Thus the dipole is like a pendulum, which has *its* least gravitational potential energy in *its* equilibrium orientation—at its lowest point. To rotate the dipole or the pendulum to any other orientation requires work by some external agent.

In any situation involving potential energy, we are free to define the zero-potential-energy configuration in a perfectly arbitrary way, because only differences in potential energy have physical meaning. It turns out that the expression for the potential energy of an electric dipole in an external electric field is simplest if we choose the potential energy to be zero when the angle θ in Fig. 23-18 is 90°. We then can find the potential energy U of the dipole at any other value of θ with Eq. 8-1 ($\Delta U = -W$) by calculating the work W done by the field on the dipole when the dipole is rotated to that value of θ from 90°. With the aid of Eq. 11-45 ($W = \int \tau \, d\theta$) and Eq. 23-35, we find that the potential energy U at any angle θ is

$$U = -W = -\int_{90°}^{\theta} \tau \, d\theta = \int_{90°}^{\theta} pE \sin \theta \, d\theta. \tag{23-36}$$

Evaluating the integral leads to

$$U = -pE \cos \theta. \tag{23-37}$$

We can generalize this equation to vector form as

$$U = -\vec{p} \cdot \vec{E} \qquad \text{(potential energy of a dipole).} \tag{23-38}$$

Equations 23-37 and 23-38 show us that the potential energy of the dipole is least ($U = -pE$) when $\theta = 0$, which is when $\vec{p}$ and $\vec{E}$ are in the same direction; the potential energy is greatest ($U = pE$) when $\theta = 180°$, which is when $\vec{p}$ and $\vec{E}$ are in opposite directions.

When a dipole rotates from an initial orientation θ_i to another orientation θ_f, the work W done on the dipole by the electric field is

$$W = -\Delta U = -(U_f - U_i), \tag{23-39}$$

where U_f and U_i are calculated with Eq. 23-38. If the change in orientation is caused by an applied torque (commonly said to be due to an external agent), then the work W_a done on the dipole by the applied torque is the negative of the work done on the dipole by the field; that is,

$$W_a = -W = (U_f - U_i). \tag{23-40}$$

✔**CHECKPOINT 5:**   The figure shows four orientations of an electric dipole in an external electric field. Rank the orientations according to (a) the magnitude of the torque on the dipole and (b) the potential energy of the dipole, greatest first.

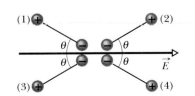

Sample Problem 23-5

A neutral water molecule (H_2O) in its vapor state has an electric dipole moment of magnitude 6.2×10^{-30} C·m.

(a) How far apart are the molecule's centers of positive and negative charge?

SOLUTION: The Key Idea here is that a molecule's dipole moment depends on the magnitude q of the molecule's positive or negative charge and the charge separation d. There are 10 electrons and 10 protons in a neutral water molecule, so the magnitude of its dipole moment is

$$p = qd = (10e)(d),$$

in which d is the separation we are seeking and e is the elementary charge. Thus,

$$d = \frac{p}{10e} = \frac{6.2 \times 10^{-30} \text{ C·m}}{(10)(1.60 \times 10^{-19} \text{ C})}$$
$$= 3.9 \times 10^{-12} \text{ m} = 3.9 \text{ pm.} \qquad \text{(Answer)}$$

This distance is not only small, but it is actually smaller than the radius of a hydrogen atom.

(b) If the molecule is placed in an electric field of 1.5×10^4 N/C,

what maximum torque can the field exert on it? (Such a field can easily be set up in the laboratory.)

SOLUTION: The Key Idea here is that the torque on a dipole is maximum when the angle θ between $\vec{p}$ and $\vec{E}$ is 90°. Substituting this value in Eq. 23-33 yields

$$\tau = pE \sin \theta$$
$$= (6.2 \times 10^{-30} \text{ C·m})(1.5 \times 10^4 \text{ N/C})(\sin 90°)$$
$$= 9.3 \times 10^{-26} \text{ N·m.} \qquad \text{(Answer)}$$

(c) How much work must an *external agent* do to turn this molecule end for end in this field, starting from its fully aligned position, for which $\theta = 0$?

SOLUTION: The Key Idea here is that the work done by an external agent (by means of a torque applied to the molecule) is equal to the change in the molecule's potential energy due to the change in orientation. From Eq. 23-40, we find

$$W_a = U(180°) - U(0)$$
$$= (-pE \cos 180°) - (-pE \cos 0)$$
$$= 2pE = (2)(6.2 \times 10^{-30} \text{ C·m})(1.5 \times 10^4 \text{ N/C})$$
$$= 1.9 \times 10^{-25} \text{ J.} \qquad \text{(Answer)}$$

REVIEW & SUMMARY

Electric Field   One way to explain the electrostatic force between two charges is to assume that each charge sets up an electric field in the space around it. The electrostatic force acting on any one charge is then due to the electric field set up at its location by the other charge.

Definition of Electric Field   The *electric field* $\vec{E}$ at any point is defined in terms of the electrostatic force $\vec{F}$ that would be exerted on a positive test charge q_0 placed there:

$$\vec{E} = \frac{\vec{F}}{q_0}. \qquad (23\text{-}1)$$

Electric Field Lines   *Electric field lines* provide a means for visualizing the direction and magnitude of electric fields. The electric field vector at any point is tangent to a field line through that point. The density of field lines in any region is proportional to the magnitude of the electric field in that region. Field lines originate on positive charges and terminate on negative charges.

Field Due to a Point Charge   The magnitude of the electric field $\vec{E}$ set up by a point charge q at a distance r from the charge is

$$E = \frac{1}{4\pi\varepsilon_0} \frac{|q|}{r^2}. \qquad (23\text{-}3)$$

The direction of $\vec{E}$ is away from the point charge if the charge is positive and toward the point charge if the charge is negative.

Field Due to an Electric Dipole   An *electric dipole* consists of two particles with charges of equal magnitude q but opposite sign, separated by a small distance d. Their **dipole moment** $\vec{p}$ has magnitude qd and points from the negative charge to the positive charge. The magnitude of the electric field set up by the dipole at a distant point on the dipole axis (which runs through both charges) is

$$E = \frac{1}{2\pi\varepsilon_0} \frac{p}{z^3}, \qquad (23\text{-}9)$$

where z is the distance between the point and the dipole center.

Field Due to a Continuous Charge Distribution   The electric field due to a *continuous charge distribution* is found by treating charge elements as point charges and then summing, via integration, the electric field vectors produced by all the charge elements.

Force on a Point Charge in an Electric Field   When a point charge q is placed in an electric field $\vec{E}$ set up by other charges, the electrostatic force $\vec{F}$ that acts on the point charge is

$$\vec{F} = q\vec{E}. \qquad (23\text{-}28)$$

Force $\vec{F}$ has the same direction as $\vec{E}$ if q is positive and the opposite direction if q is negative.

Dipole in an Electric Field   When an electric dipole of dipole moment $\vec{p}$ is placed in an electric field $\vec{E}$, the field exerts a torque $\vec{\tau}$ on the dipole:

$$\vec{\tau} = \vec{p} \times \vec{E}. \qquad (23\text{-}34)$$

The dipole has a potential energy U associated with its orientation in the field:

$$U = -\vec{p} \cdot \vec{E}. \qquad (23\text{-}38)$$

This potential energy is defined to be zero when $\vec{p}$ is perpendicular to $\vec{E}$; it is least ($U = -pE$) when $\vec{p}$ is aligned with $\vec{E}$, and most ($U = pE$) when $\vec{p}$ is directed opposite $\vec{E}$.

QUESTIONS

1. Figure 23-19 shows three electric field lines. What is the direction of the electrostatic force on a positive test charge placed at (a) point A and (b) point B? (c) At which point, A or B, will the acceleration of the test charge be greater if the charge is released?

Fig. 23-19 Question 1.

2. Figure 23-20a shows two charged particles on an axis. (a) Where on the axis (other than at an infinite distance) is there a point at which their net electric field is zero: between the charges, to their left, or to their right? (b) Is there a point of zero electric field off the axis (other than at an infinite distance)?

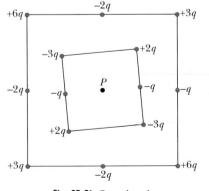

Fig. 23-20 Questions 2 and 3.

3. Figure 23-20b shows two protons and an electron that are evenly spaced on an axis. Where on the axis (other than at an infinite distance) is there a point at which their net electric field is zero: to the left of the particles, to their right, between the two protons, or between the electron and the nearer proton?

4. Figure 23-21 shows two square arrays of charged particles. The squares, which are centered on point P, are misaligned. The particles are separated by either d or $d/2$ along the perimeters of the squares. What are the magnitude and direction of the net electric field at P?

5. In Fig. 23-22, two particles of charge $-q$ are arranged symmetrically about the y axis; each produces an electric field at point P on that axis. (a) Are the magnitudes of the fields at P equal? (b) Is each electric field directed toward or away from the charge producing it? (c) Is the magnitude of the net electric field at P equal to the sum of the magnitudes E of the two field vectors (is it equal to $2E$)? (d) Do the x components of those two field vectors add or cancel? (e) Do their y components add or cancel? (f) Is the direction of the net field at P that of the canceling components or the adding components? (g) What is the direction of the net field?

Fig. 23-22 Question 5.

6. Three circular nonconducting rods of the same radius of curvature have uniform charges. Rod A has charge $+2Q$ and subtends an arc of $30°$, rod B has charge $+6Q$ and subtends $90°$, and rod C has charge $+4Q$ and subtends $60°$. Rank the rods according to their linear charge density, greatest first.

7. In Fig. 23-23a, a circular plastic rod with uniform charge $+Q$ produces an electric field of magnitude E at the center of curvature (at the origin). In Figs. 23-23b, c, and d, more circular rods with

Fig. 23-21 Question 4.

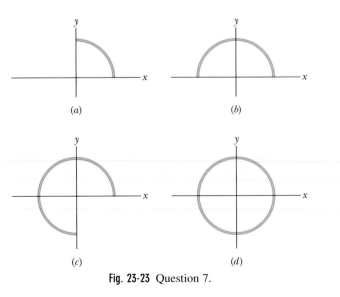

Fig. 23-23 Question 7.

identical uniform charges $+Q$ are added until the circle is complete. A fifth arrangement (which would be labeled e) is like that in d except that the rod in the fourth quadrant has charge $-Q$. Rank the five arrangements according to the magnitude of the electric field at the center of curvature, greatest first.

8. In Fig. 23-24, an electron e travels through a small hole in plate A and then toward plate B. A uniform electric field in the region between the plates then slows the electron without deflecting it. (a) What is the direction of the field? (b) Four other particles similarly travel through small holes in either plate A or plate B and then into the region between the plates. Three have charges $+q_1$, $+q_2$, and $-q_3$. The fourth (labeled n) is a neutron, which is electrically neutral. Does the speed of each of those four other particles increase, decrease, or remain the same in the region between the plates?

Fig. 23-24 Question 8.

9. Figure 23-25 shows the path of negatively charged particle 1 through a rectangular region of uniform electric field; the particle is deflected toward the top of the page. (a) Is the field directed leftward, rightward, toward the top of the page, or toward the bottom? (b) Three other charged particles are shown approaching the region of electric field. Which are deflected toward the top of the page and which toward the bottom?

Fig. 23-25 Question 9.

10. (a) In Checkpoint 5, if the dipole rotates from orientation 1 to orientation 2, is the work done on the dipole by the field positive, negative, or zero? (b) If, instead, the dipole rotates from orientation 1 to orientation 4, is the work done by the field more than, less than, or the same as in (a)?

11. The potential energies associated with four orientations of an electric dipole in an electric field are (1) $-5U_0$, (2) $-7U_0$, (3) $3U_0$, and (4) $5U_0$, where U_0 is positive. Rank the orientations according to (a) the angle between the electric dipole moment $\vec{p}$ and the electric field $\vec{E}$, and (b) the magnitude of the torque on the electric dipole, greatest first.

12. If you walk across some types of carpet on a dry day and then reach for a metal doorknob or (for more fun) the back of someone's neck, you might produce a spark. Why does the spark occur? (You can increase the brightness and noise of the spark if you reach with a pointed finger or, even better, a metal key with the pointed end forward.)

EXERCISES & PROBLEMS

ssm Solution is in the Student Solutions Manual.
www Solution is available on the World Wide Web at:
http://www.wiley.com/college/hrw
ilw Solution is available on the Interactive LearningWare.

SEC. 23-3 Electric Field Lines

1E. In Fig. 23-26 the electric field lines on the left have twice the separation of those on the right.
(a) If the magnitude of the field at A is 40 N/C, what force acts on a proton at A? (b) What is the magnitude of the field at B?

Fig. 23-26 Exercise 1.

2E. Sketch qualitatively the electric field lines both between and outside two concentric conducting spherical shells when a uniform positive charge q_1 is on the inner shell and a uniform negative charge $-q_2$ is on the outer. Consider the cases $q_1 > q_2$, $q_1 = q_2$, and $q_1 < q_2$.

3E. Sketch qualitatively the electric field lines for a thin, circular, uniformly charged disk of radius R. (*Hint:* Consider as limiting cases points very close to the disk, where the electric field is directed perpendicular to the surface, and points very far from it, where the electric field is like that of a point charge.) **ssm**

SEC. 23-4 The Electric Field Due to a Point Charge

4E. What is the magnitude of a point charge that would create an electric field of 1.00 N/C at points 1.00 m away?

5E. What is the magnitude of a point charge whose electric field 50 cm away has the magnitude 2.0 N/C? **ssm**

6E. Two particles with equal charge magnitudes 2.0×10^{-7} C but opposite signs are held 15 cm apart. What are the magnitude and direction of $\vec{E}$ at the point midway between the charges?

7E. An atom of plutonium-239 has a nuclear radius of 6.64 fm and the atomic number $Z = 94$. Assuming that the positive charge is distributed uniformly within the nucleus, what are the magnitude and direction of the electric field at the surface of the nucleus due to the positive charge? **ssm**

8P. In Fig. 23-27, two fixed point charges $q_1 = +1.0 \times 10^{-6}$ C and $q_2 = +3.0 \times 10^{-6}$ C are separated by a distance $d = 10$ cm. Plot their net electric field $E(x)$ as a function of x for both positive and negative values of x, taking E to be positive when the vector $\vec{E}$ points to the right and negative when $\vec{E}$ points to the left.

Fig. 23-27 Problems 8 and 10.

9P. Two point charges $q_1 = 2.1 \times 10^{-8}$ C and $q_2 = -4.0q_1$ are fixed in place 50 cm apart. Find the point along the straight line passing through the two charges at which the electric field is zero. ssm www

10P. (a) In Fig. 23-27, two fixed point charges $q_1 = -5q$ and $q_2 = +2q$ are separated by distance d. Locate the point (or points) at which the net electric field due to the two charges is zero. (b) Sketch the net electric field lines qualitatively.

11P. In Fig. 23-28, what is the magnitude of the electric field at point P due to the four point charges shown?

12P. Calculate the direction and magnitude of the electric field at point P in Fig. 23-29, due to the three point charges.

13P. What are the magnitude and direction of the electric field at the center of the square of Fig. 23-30 if $q = 1.0 \times 10^{-8}$ C and $a = 5.0$ cm? ssm ilw

Fig. 23-28 Problem 11.

Fig. 23-29 Problem 12.

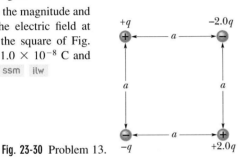

Fig. 23-30 Problem 13.

SEC. 23-5   The Electric Field Due to an Electric Dipole

14E. In Fig. 23-8, let both charges be positive. Assuming $z \gg d$, show that E at point P in that figure is then given by

$$E = \frac{1}{4\pi\varepsilon_0} \frac{2q}{z^2}.$$

15E. Calculate the electric dipole moment of an electron and a proton 4.30 nm apart. ssm

16P. Find the magnitude and direction of the electric field at point P due to the electric dipole in Fig. 23-31. P is located at a distance $r \gg d$ along the perpendicular bisector of the line joining the charges. Express your answer in terms of the magnitude and direction of the electric dipole moment $\vec{p}$.

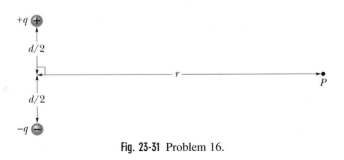

Fig. 23-31 Problem 16.

17P*. *Electric quadrupole.* Figure 23-32 shows an electric quadrupole. It consists of two dipoles with dipole moments that are equal in magnitude but opposite in direction. Show that the value of E on the axis of the quadrupole for a point P a distance z from its center (assume $z \gg d$) is given by

$$E = \frac{3Q}{4\pi\varepsilon_0 z^4},$$

in which Q ($= 2qd^2$) is known as the *quadrupole moment* of the charge distribution. ssm

Fig. 23-32 Problem 17.

SEC. 23-6   The Electric Field Due to a Line of Charge

18E. Figure 23-33 shows two parallel nonconducting rings arranged with their central axes along a common line. Ring 1 has uniform charge q_1 and radius R; ring 2 has uniform charge q_2 and the same radius R. The rings are separated by a distance $3R$. The net electric field at point P on the common line, at distance R from ring 1, is zero. What is the ratio q_1/q_2?

19P. An electron is constrained to the central axis of the ring of charge of radius R discussed in Section 23-6. Show that the electrostatic force on the electron can cause it to oscillate through the center of the ring with an angular frequency

Fig. 23-33 Exercise 18.

$$\omega = \sqrt{\frac{eq}{4\pi\varepsilon_0 mR^3}},$$

where q is the ring's charge and m is the electron's mass. ssm

20P. In Fig. 23-34a, two curved plastic rods, one of charge $+q$ and the other of charge $-q$, form a circle of radius R in an xy plane. The x axis passes through their connecting points, and the charge is distributed uniformly on both rods. What are the magnitude and direction of the electric field $\vec{E}$ produced at P, the center of the circle?

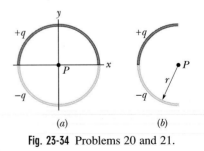

(a)   *(b)*

Fig. 23-34 Problems 20 and 21.

21P. A thin glass rod is bent into a semicircle of radius r. A charge $+q$ is uniformly distributed along the upper half, and a charge $-q$ is uniformly distributed along the lower half, as shown in Fig. 23-34b. Find the magnitude and direction of the electric field $\vec{E}$ at P, the center of the semicircle. ilw

22P. At what distance along the central axis of a ring of radius R and uniform charge is the magnitude of the electric field due to the ring's charge maximum?

23P. In Fig. 23-35, a nonconducting rod of length L has charge $-q$ uniformly distributed along its length. (a) What is the linear charge density of the rod? (b) What is the electric field at point P, a distance a from the end of the rod? (c) If P were very far from the rod compared to L, the rod would look like a point charge. Show that your answer to (b) reduces to the electric field of a point charge for $a \gg L$. ssm ilw www

Fig. 23-35 Problem 23.

24P. A thin nonconducting rod of finite length L has a charge q spread uniformly along it. Show that

$$E = \frac{q}{2\pi\varepsilon_0 y} \frac{1}{(L^2 + 4y^2)^{1/2}}$$

gives the magnitude E of the electric field at point P on the perpendicular bisector of the rod (Fig. 23-36).

Fig. 23-36 Problem 24.

25P*. In Fig. 23-37, a "semi-infinite" nonconducting rod (that is, infinite in one direction only) has uniform linear charge density λ. Show that the electric field at point P makes an angle of 45° with the rod and that this result is independent of the distance R. (*Hint:* Separately find the parallel and perpendicular (to the rod) components of the electric field at P, and then compare those components.) ssm

Fig. 23-37 Problem 25.

SEC. 23-7 The Electric Field Due to a Charged Disk

26E. A disk of radius 2.5 cm has a surface charge density of 5.3 μC/m^2 on its upper face. What is the magnitude of the electric field produced by the disk at a point on its central axis at distance $z = 12$ cm from the disk?

27P. At what distance along the central axis of a uniformly charged plastic disk of radius R is the magnitude of the electric field equal to one-half the magnitude of the field at the center of the surface of the disk? ssm

SEC. 23-8 A Point Charge in an Electric Field

28E. An electron is accelerated eastward at 1.80×10^9 m/s^2 by an electric field. Determine the magnitude and direction of the electric field.

29E. An electron is released from rest in a uniform electric field of magnitude 2.00×10^4 N/C. Calculate the acceleration of the electron. (Ignore gravitation.) ssm

30E. An alpha particle (the nucleus of a helium atom) has a mass of 6.64×10^{-27} kg and a charge of $+2e$. What are the magnitude and direction of the electric field that will balance the gravitational force on it?

31E. Calculate the magnitude of the force, due to an electric dipole of dipole moment 3.6×10^{-29} C·m, on an electron 25 nm from the center of the dipole, along the dipole axis. Assume that this distance is large relative to the dipole's charge separation. ilw

32E. Humid air breaks down (its molecules become ionized) in an electric field of 3.0×10^6 N/C. In that field, what is the magnitude of the electrostatic force on (a) an electron and (b) an ion with a single electron missing?

33E. A charged cloud system produces an electric field in the air near Earth's surface. A particle of charge -2.0×10^{-9} C is acted on by a downward electrostatic force of 3.0×10^{-6} N when placed in this field. (a) What is the magnitude of the electric field? (b) What are the magnitude and direction of the electrostatic force exerted on a proton placed in this field? (c) What is the gravitational force on the proton? (d) What is the ratio of the magnitude of the electrostatic force to the magnitude of the gravitational force in this case? ssm

34E. An electric field $\vec{E}$ with an average magnitude of about 150 N/C points downward in the atmosphere near Earth's surface. We wish to "float" a sulfur sphere weighing 4.4 N in this field by charging the sphere. (a) What charge (both sign and magnitude) must be used? (b) Why is the experiment impractical?

35E. Beams of high-speed protons can be produced in "guns" using electric fields to accelerate the protons. (a) What acceleration would a proton experience if the gun's electric field were 2.00×10^4 N/C? (b) What speed would the proton attain if the field accelerated the proton through a distance of 1.00 cm? ssm

36E. An electron with a speed of 5.00×10^8 cm/s enters an electric field of magnitude 1.00×10^3 N/C, traveling along the field lines in the direction that retards its motion. (a) How far will the electron travel in the field before stopping momentarily and (b) how much time will have elapsed? (c) If the region with the electric field is only 8.00 mm long (too short for the electron to stop within it), what fraction of the electron's initial kinetic energy will be lost in that region?

37E. In Millikan's experiment, an oil drop of radius 1.64 μm and density 0.851 g/cm^3 is suspended in chamber C (Fig. 23-13) when a downward-directed electric field of 1.92×10^5 N/C is applied. Find the charge on the drop, in terms of e. ssm

38P. In one of his experiments, Millikan observed that the following measured charges, among others, appeared at different times on a single drop:

6.563×10^{-19} C	13.13×10^{-19} C	19.71×10^{-19} C
8.204×10^{-19} C	16.48×10^{-19} C	22.89×10^{-19} C
11.50×10^{-19} C	18.08×10^{-19} C	26.13×10^{-19} C

What value for the elementary charge e can be deduced from these data?

39P. A uniform electric field exists in a region between two oppositely charged plates. An electron is released from rest at the surface of the negatively charged plate and strikes the surface of the opposite plate, 2.0 cm away, in a time 1.5×10^{-8} s. (a) What is the speed of the electron as it strikes the second plate? (b) What is the magnitude of the electric field $\vec{E}$? ilw

40P. At some instant the velocity components of an electron moving between two charged parallel plates are $v_x = 1.5 \times 10^5$ m/s and $v_y = 3.0 \times 10^3$ m/s. Suppose that the electric field between the plates is given by $\vec{E} = (120$ N/C$)\hat{j}$. (a) What is the acceleration of the electron? (b) What will be the velocity of the electron after its x coordinate has changed by 2.0 cm?

41P. Two large parallel copper plates are 5.0 cm apart and have a uniform electric field between them as depicted in Fig. 23-38. An electron is released from the negative plate at the same time that a proton is released from the positive plate. Neglect the force of the particles on each other and find their distance from the positive plate when they pass each other. (Does it surprise you that you need not know the electric field to solve this problem?) ssm www

Fig. 23-38 Problem 41.

42P. A 10.0 g block with a charge of $+8.00 \times 10^{-5}$ C is placed in electric field $\vec{E} = (3.00 \times 10^3)\hat{i} - 600\hat{j}$, where $\vec{E}$ is in newtons per coulomb. (a) What are the magnitude and direction of the force on the block? (b) If the block is released from rest at the origin at $t = 0$, what will be its coordinates at $t = 3.00$ s?

43P. In Fig. 23-39, a uniform, upward-directed electric field $\vec{E}$ of magnitude 2.00×10^3 N/C has been set up between two horizontal plates by charging the lower plate positively and the upper plate negatively. The plates have length $L = 10.0$ cm and separation $d = 2.00$ cm. An electron is then shot between the plates from the left edge of the lower plate. The initial velocity $\vec{v}_0$ of the electron makes an angle $\theta = 45.0°$ with the lower plate and has a magnitude of 6.00×10^6 m/s. (a) Will the electron strike one of the plates? (b) If so, which plate and how far horizontally from the left edge will the electron strike? ssm

Fig. 23-39 Problem 43.

SEC. 23-9   A Dipole in an Electric Field

44E. An electric dipole, consisting of charges of magnitude 1.50 nC separated by 6.20 μm, is in an electric field of strength

1100 N/C. (a) What is the magnitude of the electric dipole moment? (b) What is the difference between the potential energies corresponding to dipole orientations parallel to and antiparallel to the field?

45E. An electric dipole consists of charges $+2e$ and $-2e$ separated by 0.78 nm. It is in an electric field of strength 3.4×10^6 N/C. Calculate the magnitude of the torque on the dipole when the dipole moment is (a) parallel to, (b) perpendicular to, and (c) antiparallel to the electric field.

46P. Find the work required to turn an electric dipole end for end in a uniform electric field $\vec{E}$, in terms of the magnitude p of the dipole moment, the magnitude E of the field, and the initial angle θ_0 between $\vec{p}$ and $\vec{E}$.

47P. Find the frequency of oscillation of an electric dipole, of dipole moment $\vec{p}$ and rotational inertia I, for small amplitudes of oscillation about its equilibrium position in a uniform electric field of magnitude E. ssm

Additional Problem

48. The reproduction of flowers depends on insects carrying pollen grains from one flower to another. One way in which honeybees can do this is by collecting the grains electrically, because the bees are usually positively charged. When a bee hovers near a flower's anther (Fig. 23-40), which is electrically insulated, the pollen grains (which are moderately conducting) jump to the bee, where they cling during the flight to the next flower. As the bee nears that flower's stigma, which is electrically connected to ground through the flower's interior, the pollen grains jump from the bee to the stigma, fertilizing the flower.

(a) Assuming that a bee with a typical charge of 45 pC is a spherical conductor, find the magnitude of the bee's electric field at the location of a pollen grain 2.0 cm from the bee's center. (b) Is that field uniform or nonuniform? (c) Give a plausible explanation of why the pollen grains jump to the bee, cling to the bee during the flight, and then jump away from the bee to the grounded stigma. (*Hint:* Consider Fig. 22-5.) When a pollen grain reaches the bee, does it make electrical contact with it, so that the charge on the grain changes?

Fig. 23-40 Problem 48.

24 Gauss' Law

Lightning strikes Tucson in a brilliant display, each strike delivering about 10^{20} electrons from the cloud base to the ground.

How wide is a lightning strike? Since a strike can be seen from kilometers away, is it as wide as, say, a car?

The answer is in this chapter.

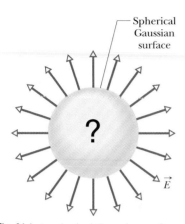

Fig. 24-1 A spherical Gaussian surface. If the electric field vectors are of uniform magnitude and point radially outward at all surface points, you can conclude that a net positive distribution of charge must lie within the surface and have spherical symmetry.

24-1 A New Look at Coulomb's Law

If you want to find the center of mass of a potato, you can do so by experiment or by laborious calculation, involving the numerical evaluation of a triple integral. However, if the potato happens to be a uniform ellipsoid, you know from its symmetry exactly where the center of mass is without calculation. Such are the advantages of symmetry. Symmetrical situations arise in all areas of physics; when possible, it makes sense to cast the laws of physics in forms that take full advantage of this fact.

Coulomb's law is the governing law in electrostatics, but it is not cast in a form that particularly simplifies the work in situations involving symmetry. In this chapter we introduce a new formulation of Coulomb's law, derived by German mathematician and physicist Carl Friedrich Gauss (1777–1855). This law, called **Gauss' law,** *can* be used to take advantage of special symmetry situations. For electrostatics problems, it is the full equivalent of Coulomb's law; which of them we choose to use depends only on the problem at hand.

Central to Gauss' law is a hypothetical closed surface called a **Gaussian surface.** The Gaussian surface can be of any shape you wish to make it, but the most useful surface is one that mimics the symmetry of the problem at hand. Thus, the Gaussian surface will often be a sphere, a cylinder, or some other symmetrical form. It must always be a *closed* surface, so that a clear distinction can be made between points that are inside the surface, on the surface, and outside the surface.

Imagine that you have established a Gaussian surface around a distribution of charges. Then Gauss' law comes into play:

> ▶ Gauss' law relates the electric fields at points on a (closed) Gaussian surface and the net charge enclosed by that surface.

Figure 24-1 shows a simple situation in which the Gaussian surface is a sphere. Suppose you know that there is an electric field at every point on the surface and that all the fields have the same magnitude and point radially outward. Without knowing anything about Gauss' law, you can guess that some net positive charge must be enclosed by the Gaussian surface. If you *do* know Gauss' law, you can calculate just how much net positive charge is enclosed. To make the calculation, you need know only "how much" electric field is intercepted by the surface—this "how much" involves the *flux* of the electric field through the surface.

24-2 Flux

Suppose that, as in Fig. 24-2a, you aim a wide airstream of uniform velocity $\vec{v}$ at a small square loop of area A. Let Φ represent the *volume flow rate* (volume per unit time) at which air flows through the loop. This rate depends on the angle between $\vec{v}$ and the plane of the loop. If $\vec{v}$ is perpendicular to the plane, the rate Φ is equal to vA.

If $\vec{v}$ is parallel to the plane of the loop, no air moves through the loop, so Φ is zero. For an intermediate angle θ, the rate Φ depends on the component of $\vec{v}$ that is normal to the plane (Fig. 24-2b). Since that component is $v \cos \theta$, the rate of volume flow through the loop is

$$\Phi = (v \cos \theta)A. \tag{24-1}$$

This rate of flow through an area is an example of a **flux**—a *volume flux* in this situation. Before we discuss a flux that is involved in electrostatics, we need to rewrite Eq. 24-1 in terms of vectors.

Fig. 24-2 (*a*) A uniform airstream of velocity $\vec{v}$ is perpendicular to the plane of a square loop of area *A*. (*b*) The component of $\vec{v}$ perpendicular to the plane of the loop is *v* cos *θ*, where *θ* is the angle between $\vec{v}$ and a normal to the plane. (*c*) The area vector $\vec{A}$ is perpendicular to the plane of the loop and makes an angle *θ* with $\vec{v}$. (*d*) The velocity field intercepted by the area of the loop.

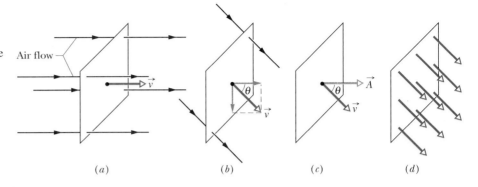

(*a*)   (*b*)   (*c*)   (*d*)

To do this, we first define an *area vector* $\vec{A}$ as being a vector whose magnitude is equal to an area (here the area of the loop) and whose direction is normal to the plane of the area (Fig. 24-2*c*). We then rewrite Eq. 24-1 as the scalar (or dot) product of the velocity vector $\vec{v}$ of the airstream and the area vector $\vec{A}$ of the loop:

$$\Phi = vA \cos \theta = \vec{v} \cdot \vec{A}, \tag{24-2}$$

where *θ* is the angle between $\vec{v}$ and $\vec{A}$.

The word "flux" comes from the Latin word meaning "to flow." That meaning makes sense if we talk about the flow of air volume through the loop. However, Eq. 24-2 can be regarded in a more abstract way. To see it, note that we can assign a velocity vector to each point in the airstream passing through the loop (Fig. 24-2*d*). The composite of all those vectors is a *velocity field*, so we can interpret Eq. 24-2 as giving the *flux of the velocity field through the loop*. With this interpretation, flux no longer means the actual flow of something through an area—rather it means the product of an area and the field across that area.

24-3 Flux of an Electric Field

To define the flux of an electric field, consider Fig. 24-3, which shows an arbitrary (asymmetric) Gaussian surface immersed in a nonuniform electric field. Let us divide the surface into small squares of area Δ*A*, each square being small enough to permit us to neglect any curvature and to consider the individual square to be flat. We represent each such element of area with an area vector $\Delta\vec{A}$, whose magnitude is the area Δ*A*. Each vector $\Delta\vec{A}$ is perpendicular to the Gaussian surface and directed away from the interior of the surface.

Because the squares have been taken to be arbitrarily small, the electric field $\vec{E}$ may be taken as constant over any given square. The vectors $\Delta\vec{A}$ and $\vec{E}$ for each square then make some angle *θ* with each other. Figure 24-3 shows an enlarged view of three squares (1, 2, and 3) on the Gaussian surface, and the angle *θ* for each.

A provisional definition for the flux of the electric field for the Gaussian surface of Fig. 24-3 is

$$\Phi = \sum \vec{E} \cdot \Delta\vec{A}. \tag{24-3}$$

This equation instructs us to visit each square on the Gaussian surface, to evaluate the scalar product $\vec{E} \cdot \Delta\vec{A}$ for the two vectors $\vec{E}$ and $\Delta\vec{A}$ that we find there, and to sum the results algebraically (that is, with signs included) for all the squares that make up the surface. The sign or a zero resulting from each scalar product determines whether the flux through its square is positive, negative, or zero. Squares like 1, in which $\vec{E}$ points inward, make a negative contribution to the sum of Eq. 24-3. Squares like 2, in which $\vec{E}$ lies in the surface, make zero contribution. Squares like 3, in which $\vec{E}$ points outward, make a positive contribution.

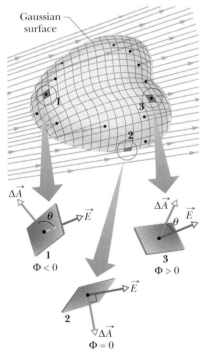

Fig. 24-3 A Gaussian surface of arbitrary shape immersed in an electric field. The surface is divided into small squares of area Δ*A*. The electric field vectors $\vec{E}$ and the area vectors $\Delta\vec{A}$ for three representative squares, marked 1, 2, and 3, are shown.

The exact definition of the flux of the electric field through a closed surface is found by allowing the area of the squares shown in Fig. 24-3 to become smaller and smaller, approaching a differential limit dA. The area vectors then approach a differential limit $d\vec{A}$. The sum of Eq. 24-3 then becomes an integral and we have, for the definition of electric flux,

$$\Phi = \oint \vec{E} \cdot d\vec{A} \qquad \text{(electric flux through a Gaussian surface).} \qquad (24\text{-}4)$$

The circle on the integral sign indicates that the integration is to be taken over the entire (closed) surface. The flux of the electric field is a scalar, and its SI unit is the newton–square-meter per coulomb ($\text{N} \cdot \text{m}^2/\text{C}$).

We can interpret Eq. 24-4 in the following way: First recall that we can use the density of electric field lines passing through an area as a proportional measure of an electric field $\vec{E}$ there. Specifically, the magnitude E is proportional to the number of electric field lines per unit area. Thus, the scalar product $\vec{E} \cdot d\vec{A}$ in Eq. 24-4 is proportional to the number of electric field lines passing through area $d\vec{A}$. Then, because the integration in Eq. 24-4 is carried out over a Gaussian surface, which is closed, we see that

> The electric flux Φ through a Gaussian surface is proportional to the net number of electric field lines passing through that surface.

Sample Problem 24-1

Figure 24-4 shows a Gaussian surface in the form of a cylinder of radius R immersed in a uniform electric field $\vec{E}$, with the cylinder axis parallel to the field. What is the flux Φ of the electric field through this closed surface?

SOLUTION: The Key Idea here is that we can find the flux Φ through the surface by integrating the scalar product $\vec{E} \cdot d\vec{A}$ over the Gaussian surface. We can do this by writing the flux as the sum of three terms: integrals over the left cylinder cap a, the cylindrical surface b, and the right cap c. Thus, from Eq. 24-4,

$$\Phi = \oint \vec{E} \cdot d\vec{A}$$
$$= \int_a \vec{E} \cdot d\vec{A} + \int_b \vec{E} \cdot d\vec{A} + \int_c \vec{E} \cdot d\vec{A}. \qquad (24\text{-}5)$$

For all points on the left cap, the angle θ between $\vec{E}$ and $d\vec{A}$ is 180° and the magnitude E of the field is constant. Thus,

$$\int_a \vec{E} \cdot d\vec{A} = \int E(\cos 180°) \, dA = -E \int dA = -EA,$$

where $\int dA$ gives the cap's area, $A \ (= \pi R^2)$. Similarly, for the right cap, where $\theta = 0$ for all points,

$$\int_c \vec{E} \cdot d\vec{A} = \int E(\cos 0) \, dA = EA.$$

Finally, for the cylindrical surface, where the angle θ is 90° at all points,

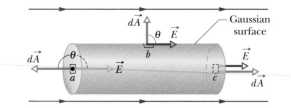

Fig. 24-4 Sample Problem 24-1. A cylindrical Gaussian surface, closed by end caps, is immersed in a uniform electric field. The cylinder axis is parallel to the field direction.

$$\int_b \vec{E} \cdot d\vec{A} = \int E(\cos 90°) \, dA = 0.$$

Substituting these results into Eq. 24-5 leads us to

$$\Phi = -EA + 0 + EA = 0. \qquad \text{(Answer)}$$

This result is perhaps not surprising because the field lines that represent the electric field all pass entirely through the Gaussian surface, entering through the left end cap, leaving through the right end cap, and giving a net flux of zero.

✔**CHECKPOINT 1:**   The figure here shows a Gaussian cube of face area A immersed in a uniform electric field $\vec{E}$ that has the positive direction of the z axis. In terms of E and A, what is the flux through (a) the front face (which is in the xy plane), (b) the rear face, (c) the top face, and (d) the whole cube?

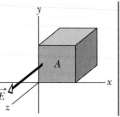

Sample Problem 24-2

A *nonuniform* electric field given by $\vec{E} = 3.0x\hat{i} + 4.0\hat{j}$ pierces the Gaussian cube shown in Fig. 24-5. (E is in newtons per coulomb and x is in meters.) What is the electric flux through the right face, the left face, and the top face?

SOLUTION: The Key Idea here is that we can find the flux Φ through the surface by integrating the scalar product $\vec{E} \cdot d\vec{A}$ over each face.

    Right face: An area vector $\vec{A}$ is always perpendicular to its surface and always points away from the interior of a Gaussian surface. Thus, the vector $d\vec{A}$ for the right face of the cube must point in the positive x direction. In unit vector notation, then,

$$d\vec{A} = dA\hat{i}.$$

From Eq. 24-4, the flux Φ_r through the right face is then

$$\Phi_r = \int \vec{E} \cdot d\vec{A} = \int (3.0x\hat{i} + 4.0\hat{j}) \cdot (dA\hat{i})$$

$$= \int [(3.0x)(dA)\hat{i} \cdot \hat{i} + (4.0)(dA)\hat{j} \cdot \hat{i})]$$

$$= \int (3.0x\, dA + 0) = 3.0 \int x\, dA.$$

We are about to integrate over the right face, but we note that x has the same value everywhere on that face—namely, $x = 3.0$ m. This means we can substitute that constant value for x. Then

$$\Phi_r = 3.0 \int (3.0)\, dA = 9.0 \int dA.$$

Now the integral merely gives us the area $A = 4.0$ m² of the right face, so

$$\Phi_r = (9.0 \text{ N/C})(4.0 \text{ m}^2) = 36 \text{ N} \cdot \text{m}^2/\text{C}. \quad \text{(Answer)}$$

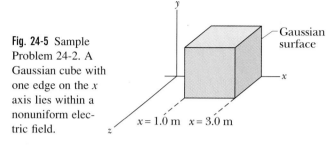

Fig. 24-5 Sample Problem 24-2. A Gaussian cube with one edge on the x axis lies within a nonuniform electric field.

    Left face: The procedure for finding the flux through the left face is the same as that for the right face. However, two factors change. (1) The differential area vector $d\vec{A}$ points in the negative x direction and thus $d\vec{A} = -dA\hat{i}$. (2) The term x again appears in our integration, and it is again constant over the face being considered. However, on the left face, $x = 1.0$ m. With these two changes, we find that the flux Φ_l through the left face is

$$\Phi_l = -12 \text{N} \cdot \text{m}^2/\text{C}. \quad \text{(Answer)}$$

    Top face: The differential area vector $d\vec{A}$ points in the positive y direction and thus $d\vec{A} = dA\hat{j}$. The flux Φ_t through the top face is then

$$\Phi_t = \int (3.0x\hat{i} + 4.0\hat{j}) \cdot (dA\hat{j})$$

$$= \int [(3.0x)(dA)\hat{i} \cdot \hat{j} + (4.0)(dA)\hat{j} \cdot \hat{j})]$$

$$= \int (0 + 4.0\, dA) = 4.0 \int dA$$

$$= 16 \text{ N} \cdot \text{m}^2/\text{C}. \quad \text{(Answer)}$$

24-4 Gauss' Law

Gauss' law relates the net flux Φ of an electric field through a closed surface (a Gaussian surface) to the *net* charge q_{enc} that is *enclosed* by that surface. It tells us that

$$\varepsilon_0 \Phi = q_{enc} \quad \text{(Gauss' law).} \quad (24\text{-}6)$$

By substituting Eq. 24-4, the definition of flux, we can also write Gauss' law as

$$\varepsilon_0 \oint \vec{E} \cdot d\vec{A} = q_{enc} \quad \text{(Gauss' law).} \quad (24\text{-}7)$$

Equations 24-6 and 24-7 hold only when the net charge is located in a vacuum or (what is the same for most practical purposes) in air. In Section 26-8, we modify Gauss' law to include situations in which a material such as mica, oil, or glass is present.

    In Eqs. 24-6 and 24-7, the net charge q_{enc} is the algebraic sum of all the *enclosed* positive and negative charges, and it can be positive, negative, or zero. We include the sign, rather than just use the magnitude of the enclosed charge, because the sign

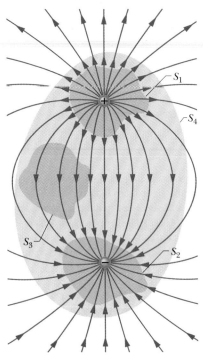

Fig. 24-6 Two point charges, equal in magnitude but opposite in sign, and the field lines that represent their net electric field. Four Gaussian surfaces are shown in cross section. Surface S_1 encloses the positive charge. Surface S_2 encloses the negative charge. Surface S_3 encloses no charge. Surface S_4 encloses both charges, and thus no net charge.

tells us something about the net flux through the Gaussian surface: If q_{enc} is positive, the net flux is *outward*; if q_{enc} is negative, the net flux is *inward*.

Charge outside the surface, no matter how large or how close it may be, is not included in the term q_{enc} in Gauss' law. The exact form or location of the charges inside the Gaussian surface is also of no concern; the only things that matter, on the right side of Eq. 24-7, are the magnitude and sign of the net enclosed charge. The quantity $\vec{E}$ on the left side of Eq. 24-7, however, is the electric field resulting from *all* charges, both those inside and those outside the Gaussian surface. This may seem to be inconsistent, but keep in mind what we saw in Sample Problem 24-1: The electric field due to a charge outside the Gaussian surface contributes zero net flux *through* the surface, because as many field lines due to that charge enter the surface as leave it.

Let us apply these ideas to Fig. 24-6, which shows two point charges, equal in magnitude but opposite in sign, and the field lines describing the electric fields that they set up in the surrounding space. Four Gaussian surfaces are also shown, in cross section. Let us consider each in turn.

Surface S_1.   The electric field is outward for all points on this surface. Thus, the flux of the electric field through this surface is positive, and so is the net charge within the surface, as Gauss' law requires. (That is, in Eq. 24-6, if Φ is positive, q_{enc} must be also.)

Surface S_2.   The electric field is inward for all points on this surface. Thus, the flux of the electric field is negative and so is the enclosed charge, as Gauss' law requires.

Surface S_3.   This surface encloses no charge, and thus $q_{enc} = 0$. Gauss' law (Eq. 24-6) requires that the net flux of the electric field through this surface be zero. That is reasonable because all the field lines pass entirely through the surface, entering it at the top and leaving at the bottom.

Surface S_4.   This surface encloses no *net* charge, because the enclosed positive and negative charges have equal magnitudes. Gauss' law requires that the net flux of the electric field through this surface be zero. That is reasonable because there are as many field lines leaving surface S_4 as entering it.

What would happen if we were to bring an enormous charge Q up close to surface S_4 in Fig. 24-6? The pattern of the field lines would certainly change, but the net flux for each of the four Gaussian surfaces would not change. We can understand this because the field lines associated with the added Q would pass entirely through each of the four Gaussian surfaces, making no contribution to the net flux through any of them. The value of Q would not enter Gauss' law in any way, because Q lies outside all four of the Gaussian surfaces that we are considering.

✔**CHECKPOINT 2:**   The figure shows three situations in which a Gaussian cube sits in an electric field. The arrows and the values indicate the directions of the field lines and the magnitudes (in $N \cdot m^2/C$) of the flux through the six sides of each cube. (The lighter arrows are for the hidden faces.) In which situations does the cube enclose (a) a positive net charge, (b) a negative net charge, and (c) zero net charge?

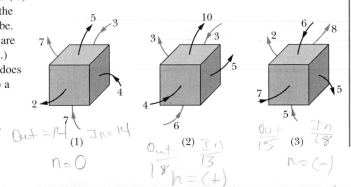

Sample Problem 24-3

Figure 24-7 shows five charged lumps of plastic and an electrically neutral coin. The cross section of a Gaussian surface S is indicated. What is the net electric flux through the surface if $q_1 = q_4 = +3.1$ nC, $q_2 = q_5 = -5.9$ nC, and $q_3 = -3.1$ nC?

Fig. 24-7 Sample Problem 24-3. Five plastic objects, each with an electric charge, and a coin, which has no net charge. A Gaussian surface, shown in cross section, encloses three of the plastic objects and the coin.

SOLUTION: The Key Idea here is that the *net* flux Φ through the surface depends on the *net* charge q_{enc} enclosed by surface S. This means that the coin and charges q_4 and q_5 do not contribute to Φ. The coin does not contribute because it is neutral and thus contains equal amounts of positive and negative charge. Charges q_4 and q_5 do not contribute because they are outside surface S. Thus, q_{enc} is $q_1 + q_2 + q_3$ and Eq. 24-6 gives us

$$\Phi = \frac{q_{enc}}{\varepsilon_0} = \frac{q_1 + q_2 + q_3}{\varepsilon_0}$$

$$= \frac{+3.1 \times 10^{-9}\,\text{C} - 5.9 \times 10^{-9}\,\text{C} - 3.1 \times 10^{-9}\,\text{C}}{8.85 \times 10^{-12}\,\text{C}^2/\text{N} \cdot \text{m}^2}$$

$$= -670\,\text{N} \cdot \text{m}^2/\text{C}. \qquad \text{(Answer)}$$

The minus sign shows that the net flux through the surface is inward and thus that the net charge within the surface is negative.

24-5 Gauss' Law and Coulomb's Law

If Gauss' law and Coulomb's law are equivalent, we should be able to derive each from the other. Here we derive Coulomb's law from Gauss' law and some symmetry considerations.

Figure 24-8 shows a positive point charge q, around which we have drawn a concentric spherical Gaussian surface of radius r. Let us divide this surface into differential areas dA. By definition, the area vector $d\vec{A}$ at any point is perpendicular to the surface and directed outward from the interior. From the symmetry of the situation, we know that at any point the electric field $\vec{E}$ is also perpendicular to the surface and directed outward from the interior. Thus, since the angle θ between $\vec{E}$ and $d\vec{A}$ is zero, we can rewrite Eq. 24-7 for Gauss' law as

$$\varepsilon_0 \oint \vec{E} \cdot d\vec{A} = \varepsilon_0 \oint E\,dA = q_{enc}. \qquad (24\text{-}8)$$

Here $q_{enc} = q$. Although E varies radially with the distance from q, it has the same value everywhere on the spherical surface. Since the integral in Eq. 24-8 is taken over that surface, E is a constant in the integration and can be brought out in front of the integral sign. That gives us

$$\varepsilon_0 E \oint dA = q. \qquad (24\text{-}9)$$

The integral is now merely the sum of all the differential areas dA on the sphere and thus is just the surface area, $4\pi r^2$. Substituting this, we have

$$\varepsilon_0 E(4\pi r^2) = q$$

or

$$E = \frac{1}{4\pi\varepsilon_0} \frac{q}{r^2}. \qquad (24\text{-}10)$$

Fig. 24-8 A spherical Gaussian surface centered on a point charge q.

Gaussian surface

This is exactly the electric field due to a point charge (Eq. 23-3), which we found using Coulomb's law. Thus, Gauss' law is equivalent to Coulomb's law.

✔CHECKPOINT 3:   There is a certain net flux Φ_i through a Gaussian sphere of radius r enclosing an isolated charged particle. Suppose the enclosing Gaussian surface is changed to (a) a larger Gaussian sphere, (b) a Gaussian cube with edge length equal to r, and (c) a Gaussian cube with edge length equal to $2r$. In each case, is the net flux through the new Gaussian surface greater than, less than, or equal to Φ_i?

PROBLEM-SOLVING TACTICS

Tactic 1: *Choosing a Gaussian Surface*

The derivation of Eq. 24-10 using Gauss' law is a warm-up for derivations of electric fields produced by other charge configurations, so let us go back over the steps involved. We started with a given positive point charge q; we know that electric field lines extend radially outward from q in a spherically symmetric pattern.

To find the magnitude E of the electric field at a distance r by Gauss' law (Eq. 24-7), we had to place a hypothetical closed Gaussian surface around q, through a point that is a distance r from q. Then we had to sum via integration the values of $\vec{E} \cdot d\vec{A}$ over the full Gaussian surface. To make this integration as simple as possible, we chose a spherical Gaussian surface (to mimic the spherical symmetry of the electric field). That choice produced three simplifying features. (1) The dot product $\vec{E} \cdot d\vec{A}$ became simple, because

at all points on the Gaussian surface the angle between $\vec{E}$ and $d\vec{A}$ is zero, and so at all points we have $\vec{E} \cdot d\vec{A} = E\, dA$. (2) The electric field magnitude E is the same at all points on the spherical Gaussian surface, so E was a constant in the integration and could be brought out in front of the integral sign. (3) The result was a very simple integration—a summation of the differential areas of the sphere, which we could immediately write as $4\pi r^2$.

Note that Gauss' law holds regardless of the shape of the Gaussian surface we choose to place around charge q_{enc}. However, if we had chosen, say, a cubical Gaussian surface, our three simplifying features would have disappeared and the integration of $\vec{E} \cdot d\vec{A}$ over the cubical surface would have been very difficult. The moral here is to choose the Gaussian surface that most simplifies the integration in Gauss' law.

24-6 A Charged Isolated Conductor

Gauss' law permits us to prove an important theorem about isolated conductors:

> ▶ If an excess charge is placed on an isolated conductor, that amount of charge will move entirely to the surface of the conductor. None of the excess charge will be found within the body of the conductor.

This might seem reasonable, considering that charges with the same sign repel each other. You might imagine that, by moving to the surface, the added charges are getting as far away from each other as they can. We turn to Gauss' law for verification of this speculation.

Figure 24-9a shows, in cross section, an isolated lump of copper hanging from an insulating thread and having an excess charge q. We place a Gaussian surface just inside the actual surface of the conductor.

The electric field inside this conductor must be zero. If this were not so, the field would exert forces on the conduction (free) electrons, which are always present in a conductor, and thus current would always exist within a conductor. (That is, charge would flow from place to place within the conductor.) Of course, there are no such perpetual currents in an isolated conductor, and so the internal electric field is zero.

(An internal electric field *does* appear as a conductor is being charged. However, the added charge quickly distributes itself in such a way that the net internal electric field—the vector sum of the electric fields due to all the charges, both inside and outside—is zero. The movement of charge then ceases, because the net force on each charge is zero; the charges are then in *electrostatic equilibrium*.)

If $\vec{E}$ is zero everywhere inside our copper conductor, it must be zero for all points on the Gaussian surface because that surface, though close to the surface of the conductor, is definitely inside the conductor. This means that the flux through the Gaussian surface must be zero. Gauss' law then tells us that the net charge in-

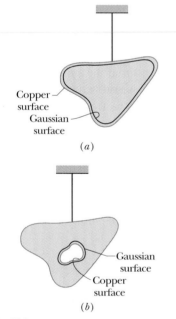

Fig. 24-9 (a) A lump of copper with a charge q hangs from an insulating thread. A Gaussian surface is placed within the metal, just inside the actual surface. (b) The lump of copper now has a cavity within it. A Gaussian surface lies within the metal, close to the cavity surface.

side the Gaussian surface must also be zero. Then because the excess charge is not inside the Gaussian surface, it must be outside that surface, which means it must lie on the actual surface of the conductor.

An Isolated Conductor with a Cavity

Figure 24-9b shows the same hanging conductor, but now with a cavity that is totally within the conductor. It is perhaps reasonable to suppose that when we scoop out the electrically neutral material to form the cavity, we do not change the distribution of charge or the pattern of the electric field that exists in Fig. 24-9a. Again, we must turn to Gauss' law for a quantitative proof.

We draw a Gaussian surface surrounding the cavity, close to its surface but inside the conducting body. Because $\vec{E} = 0$ inside the conductor, there can be no flux through this new Gaussian surface. Therefore, from Gauss' law, that surface can enclose no net charge. We conclude that there is no net charge on the cavity walls; all the excess charge remains on the outer surface of the conductor, as in Fig. 24-9a.

The Conductor Removed

Suppose that, by some magic, the excess charges could be "frozen" into position on the conductor's surface, perhaps by embedding them in a thin plastic coating, and suppose that then the conductor could be removed completely. This is equivalent to enlarging the cavity of Fig. 24-9b until it consumes the entire conductor, leaving only the charges. The electric field would not change at all; it would remain zero inside the thin shell of charge and would remain unchanged for all external points. This shows us that the electric field is set up by the charges and not by the conductor. The conductor simply provides an initial pathway for the charges to take up their positions.

The External Electric Field

You have seen that the excess charge on an isolated conductor moves entirely to the conductor's surface. However, unless the conductor is spherical, the charge does not distribute itself uniformly. Put another way, the surface charge density σ (charge per unit area) varies over the surface of any nonspherical conductor. Generally, this variation makes the determination of the electric field set up by the surface charges very difficult.

However, the electric field just outside the surface of a conductor is easy to determine using Gauss' law. To do this, we consider a section of the surface that is small enough to permit us to neglect any curvature and thus to take the section to be flat. We then imagine a tiny cylindrical Gaussian surface to be embedded in the section as in Fig. 24-10: One end cap is fully inside the conductor, the other is fully outside, and the cylinder is perpendicular to the conductor's surface.

The electric field $\vec{E}$ at and just outside the conductor's surface must also be perpendicular to that surface. If it were not, then it would have a component along the conductor's surface that would exert forces on the surface charges, causing them to move. However, such motion would violate our implicit assumption that we are dealing with electrostatic equilibrium. Therefore, $\vec{E}$ is perpendicular to the conductor's surface.

We now sum the flux through the Gaussian surface. There is no flux through the internal end cap, because the electric field within the conductor is zero. There is no flux through the curved surface of the cylinder, because internally (in the conductor) there is no electric field and externally the electric field is parallel to the

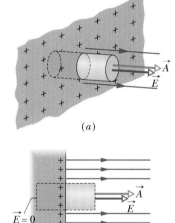

(a)

(b)

Fig. 24-10 Perspective view (a) and side view (b) of a tiny portion of a large, isolated conductor with excess positive charge on its surface. A (closed) cylindrical Gaussian surface, embedded perpendicularly in the conductor, encloses some of the charge. Electric field lines pierce the external end cap of the cylinder, but not the internal end cap. The external end cap has area A and area vector $\vec{A}$.

curved portion of the Gaussian surface. The only flux through the Gaussian surface is that through the external end cap, where $\vec{E}$ is perpendicular to the plane of the cap. We assume that the cap area A is small enough that the field magnitude E is constant over the cap. Then the flux through the cap is EA, and that is the net flux Φ through the Gaussian surface.

The charge q_{enc} enclosed by the Gaussian surface lies on the conductor's surface in an area A. If σ is the charge per unit area, then q_{enc} is equal to σA. When we substitute σA for q_{enc} and EA for Φ, Gauss' law (Eq. 24-6) becomes

$$\varepsilon_0 EA = \sigma A,$$

from which we find

$$E = \frac{\sigma}{\varepsilon_0} \quad \text{(conducting surface).} \qquad (24\text{-}11)$$

Thus, the magnitude of the electric field at a location just outside a conductor is proportional to the surface charge density at that location on the conductor. If the charge on the conductor is positive, the electric field is directed away from the conductor as in Fig. 24-10. It is directed toward the conductor if the charge is negative.

The field lines in Fig. 24-10 must terminate on negative charges somewhere in the environment. If we bring those charges near the conductor, the charge density at any given location on the conductor's surface changes, and so does the magnitude of the electric field. However, the relation between σ and E is still given by Eq. 24-11.

Sample Problem 24-4

Figure 24-11a shows a cross section of a spherical metal shell of inner radius R. A point charge of -5.0 μC is located at a distance $R/2$ from the center of the shell. If the shell is electrically neutral, what are the (induced) charges on its inner and outer surfaces? Are those charges uniformly distributed? What is the field pattern inside and outside the shell?

SOLUTION: Figure 24-11b shows a cross section of a spherical Gaussian surface within the metal, just outside the inner wall of the shell. One Key Idea here is that the electric field must be zero inside the metal (and thus on the Gaussian surface inside the metal). This means that the electric flux through the Gaussian surface must also be zero. Gauss' law then tells us that the *net* charge enclosed by the Gaussian surface must be zero. With a point charge of -5.0 μC within the shell, a charge of $+5.0$ μC must lie on the inner wall of the shell.

If the point charge were centered, this positive charge would be uniformly distributed along the inner wall. However, since the point charge is off-center, the distribution of positive charge is skewed, as suggested by Fig. 24-11b, because the positive charge tends to collect on the section of the inner wall nearest the (negative) point charge.

A second Key Idea is that because the shell is electrically neutral, its inner wall can have a charge of $+5.0$ μC only if electrons, with a total charge of -5.0 μC, leave the inner wall and move to the outer wall. There they spread out uniformly, as is also suggested

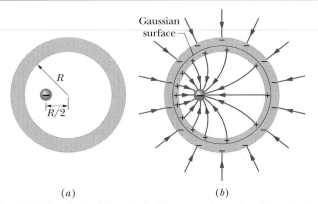

Fig. 24-11 Sample Problem 24-4. (a) A negative point charge is located within a spherical metal shell that is electrically neutral. (b) As a result, positive charge is nonuniformly distributed on the inner wall of the shell, and an equal amount of negative charge is uniformly distributed on the outer wall.

by Fig. 24-11b. This distribution of negative charge is uniform because the shell is spherical and because the skewed distribution of positive charge on the inner wall cannot produce an electric field in the shell to affect the distribution of charge on the outer wall.

The field lines inside and outside the shell are shown approximately in Fig. 24-11b. All the field lines intersect the shell and the point charge perpendicularly. Inside the shell the pattern of field lines is skewed owing to the skew of the positive charge distribution. Outside the shell the pattern is the same as if the point charge were centered and the shell were missing. In fact, this would be true no matter where inside the shell the point charge happened to be located.

✔CHECKPOINT 4:   A ball of charge $-50e$ lies at the center of a hollow spherical metal shell that has a net charge of $-100e$. What is the charge on (a) the shell's inner surface and (b) its outer surface?

24-7 Applying Gauss' Law: Cylindrical Symmetry

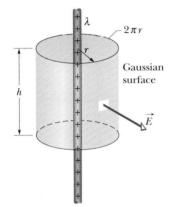

Fig. 24-12 A Gaussian surface in the form of a closed cylinder surrounds a section of a very long, uniformly charged, cylindrical plastic rod.

Figure 24-12 shows a section of an infinitely long cylindrical plastic rod with a uniform positive linear charge density λ. Let us find an expression for the magnitude of the electric field $\vec{E}$ at a distance r from the axis of the rod.

Our Gaussian surface should match the symmetry of the problem, which is cylindrical. We choose a circular cylinder of radius r and length h, coaxial with the rod. The Gaussian surface must be closed, so we include two end caps as part of the surface.

Imagine now that, while you are not watching, someone rotates the plastic rod around its longitudinal axis or turns it end for end. When you look again at the rod, you will not be able to detect any change. We conclude from this symmetry that the only uniquely specified direction in this problem is along a radial line. Thus, at every point on the cylindrical part of the Gaussian surface, $\vec{E}$ must have the same magnitude E and (for a positively charged rod) must be directed radially outward.

Since $2\pi r$ is the cylinder's circumference and h is its height, the area A of the cylindrical surface is $2\pi rh$. The flux of $\vec{E}$ through this cylindrical surface is then

$$\Phi = EA \cos \theta = E(2\pi rh) \cos 0 = E(2\pi rh).$$

There is no flux through the end caps because $\vec{E}$, being radially directed, is parallel to the end caps at every point.

The charge enclosed by the surface is λh, so Gauss' law,

$$\varepsilon_0 \Phi = q_{\text{enc}},$$

reduces to

$$\varepsilon_0 E(2\pi rh) = \lambda h,$$

yielding

$$E = \frac{\lambda}{2\pi\varepsilon_0 r} \qquad \text{(line of charge).} \qquad (24\text{-}12)$$

This is the electric field due to an infinitely long, straight line of charge, at a point that is a radial distance r from the line. The direction of $\vec{E}$ is radially outward from the line of charge if the charge is positive, and radially inward if it is negative. Equation 24-12 also approximates the field of a *finite* line of charge, at points that are not too near the ends (compared with the distance from the line).

Sample Problem 24-5

The visible portion of a lightning strike is preceded by an invisible stage in which a column of electrons extends downward from a cloud to the ground. These electrons come from the cloud and from air molecules that are ionized within the column. The linear charge density λ along the column is typically -1×10^{-3} C/m. Once the column reaches the ground, electrons within it are rapidly dumped to the ground. During the dumping, collisions between the moving electrons and the air within the column result in a brilliant flash of light. If air molecules break down (ionize) in an electric field exceeding 3×10^6 N/C, what is the radius of the column?

SOLUTION: One Key Idea here is that, although the column is not straight or infinitely long, we can approximate it as being a line of charge as in Fig. 24-12. (Since it contains a net negative charge,

Fig. 24-13 Lightning strikes a 20-m-high sycamore. Because the tree was wet, most of the charge traveled through the water on it and the tree was unharmed.

its electric field $\vec{E}$ points radially inward.) Then, according to Eq. 24-12, the field's magnitude E decreases with distance from the axis of the column of charge.

A second Key Idea is that the surface of the column of charge must be at the radius r where the magnitude of $\vec{E}$ is 3×10^6 N/C, because air molecules within that radius ionize while those farther out do not. Solving Eq. 24-12 for r and inserting the known data, we find the radius of the column to be

$$r = \frac{\lambda}{2\pi\varepsilon_0 E}$$

$$= \frac{1 \times 10^{-3} \text{ C/m}}{(2\pi)(8.85 \times 10^{-12} \text{ C}^2/\text{N} \cdot \text{m}^2)(3 \times 10^6 \text{ N/C})}$$

$$= 6 \text{ m.} \tag{Answer}$$

(The radius of the luminous portion of a lightning strike is smaller, perhaps only 0.5 m. You can get an idea of the width from Fig. 24-13.) Although the radius of the column may be only 6 m, do not assume that you are safe if you are at a somewhat greater distance from the strike point, because the electrons dumped by the strike travel along the ground. Such *ground currents* are lethal. Figure 24-14 shows evidence of ground currents.

Fig. 24-14 Ground currents from a lightning strike have burned grass off this golf green, exposing the soil.

24-8 Applying Gauss' Law: Planar Symmetry

Nonconducting Sheet

Figure 24-15 shows a portion of a thin, infinite, nonconducting sheet with a uniform (positive) surface charge density σ. A sheet of thin plastic wrap, uniformly charged on one side, can serve as a simple model. Let us find the electric field $\vec{E}$ a distance r in front of the sheet.

A useful Gaussian surface is a closed cylinder with end caps of area A, arranged to pierce the sheet perpendicularly as shown. From symmetry, $\vec{E}$ must be perpendicular to the sheet and hence to the end caps. Furthermore, since the charge is positive, $\vec{E}$ is directed *away* from the sheet, and thus the electric field lines pierce the two Gaussian end caps in an outward direction. Because the field lines do not pierce the curved surface, there is no flux through this portion of the Gaussian

Fig. 24-15 Perspective view (a) and side view (b) of a portion of a very large, thin plastic sheet, uniformly charged on one side to surface charge density σ. A closed cylindrical Gaussian surface passes through the sheet and is perpendicular to it.

surface. Thus $\vec{E} \cdot d\vec{A}$ is simply $E\,dA$; then Gauss' law,

$$\varepsilon_0 \oint \vec{E} \cdot d\vec{A} = q_{enc},$$

becomes

$$\varepsilon_0(EA + EA) = \sigma A,$$

where σA is the charge enclosed by the Gaussian surface. This gives

$$E = \frac{\sigma}{2\varepsilon_0} \qquad \text{(sheet of charge).} \qquad (24\text{-}13)$$

Since we are considering an infinite sheet with uniform charge density, this result holds for any point at a finite distance from the sheet. Equation 24-13 agrees with Eq. 23-27, which we found by integration of the electric field components that are produced by individual charges. (Look back to that time-consuming and challenging integration, and note how much more easily we obtain the result with Gauss' law. That is one reason for devoting a whole chapter to that law: for certain symmetric arrangements of charge, it is very much easier to use than integration of field components.)

Two Conducting Plates

Figure 24-16a shows a cross section of a thin, infinite conducting plate with excess positive charge. From Section 24-6 we know that this excess charge lies on the surface of the plate. Since the plate is thin and very large, we can assume that essentially all the excess charge is on the two large faces of the plate.

If there is no external electric field to force the positive charge into some particular distribution, it will spread out on the two faces with a uniform surface charge density of magnitude σ_1. From Eq. 24-11 we know that just outside the plate this charge sets up an electric field of magnitude $E = \sigma_1/\varepsilon_0$. Because the excess charge is positive, the field is directed away from the plate.

Figure 24-16b shows an identical plate with excess negative charge having the same magnitude of surface charge density σ_1. The only difference is that now the electric field is directed toward the plate.

Suppose we arrange for the plates of Figs. 24-16a and b to be close to each other and parallel (Fig. 24-16c). Since the plates are conductors, when we bring them into this arrangement, the excess charge on one plate attracts the excess charge on the other plate, and all the excess charge moves onto the inner faces of the plates as in Fig. 24-16c. With twice as much charge now on each inner face, the new surface charge density (call it σ) on each inner face is twice σ_1. Thus, the electric field at any point between the plates has the magnitude

$$E = \frac{2\sigma_1}{\varepsilon_0} = \frac{\sigma}{\varepsilon_0}. \qquad (24\text{-}14)$$

Fig. 24-16 (a) A thin, very large conducting plate with excess positive charge. (b) An identical plate with excess negative charge. (c) The two plates arranged so they are parallel and close.

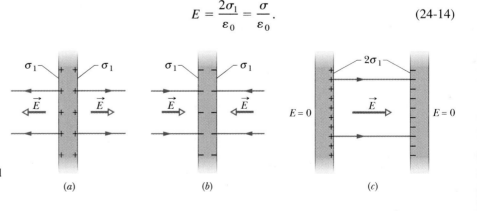

(a)   (b)   (c)

This field is directed away from the positively charged plate and toward the negatively charged plate. Since no excess charge is left on the outer faces, the electric field to the left and right of the plates is zero.

Because the charges on the plates moved when we brought the plates close to each other, Fig. 24-16c is *not* the superposition of Figs. 24-16a and b; that is, the charge distribution of the two-plate system is not merely the sum of the charge distributions of the individual plates.

You may wonder why we discuss such seemingly unrealistic situations as the field set up by an infinite line of charge, an infinite sheet of charge, or a pair of infinite plates of charge. One reason is that analyzing such situations with Gauss' law is easy. More important is that analyses for "infinite" situations yield good approximations to many real-world problems. Thus, Eq. 24-13 holds well for a finite nonconducting sheet as long as we are dealing with points close to the sheet and not too near its edges. Equation 24-14 holds well for a pair of finite conducting plates as long as we consider points that are not too close to their edges.

The trouble with the edges of a sheet or a plate, and the reason we take care not to deal with them, is that near an edge we can no longer use planar symmetry to find expressions for the fields. In fact, the field lines there are curved (said to be an *edge effect* or *fringing*), and the fields can be very difficult to express algebraically.

Sample Problem 24-6

Figure 24-17a shows portions of two large, parallel, nonconducting sheets, each with a fixed uniform charge on one side. The magnitudes of the surface charge densities are $\sigma_{(+)} = 6.8 \ \mu C/m^2$ for the positively charged sheet and $\sigma_{(-)} = 4.3 \ \mu C/m^2$ for the negatively charged sheet.

Find the electric field $\vec{E}$ (a) to the left of the sheets, (b) between the sheets, and (c) to the right of the sheets.

SOLUTION: The Key Idea here is that with the charges fixed in place, we can find the electric field of the sheets in Fig. 24-17a by (1) finding the field of each sheet as if that sheet were isolated and (2) algebraically adding the fields of the isolated sheets via the superposition principle. (We can add the fields algebraically because they are parallel to each other.) From Eq. 24-13, the magnitude $E_{(+)}$ of the electric field due to the positive sheet at any point is

$$E_{(+)} = \frac{\sigma_{(+)}}{2\varepsilon_0} = \frac{6.8 \times 10^{-6} \ C/m^2}{(2)(8.85 \times 10^{-12} \ C^2/N \cdot m^2)}$$

$$= 3.84 \times 10^5 \ N/C.$$

Similarly, the magnitude $E_{(-)}$ of the electric field at any point due to the negative sheet is

$$E_{(-)} = \frac{\sigma_{(-)}}{2\varepsilon_0} = \frac{4.3 \times 10^{-6} \ C/m^2}{(2)(8.85 \times 10^{-12} \ C^2/N \cdot m^2)}$$

$$= 2.43 \times 10^5 \ N/C.$$

Figure 24-17b shows the fields set up by the sheets to the left of the sheets (L), between them (B), and to their right (R).

The resultant fields in these three regions follow from the superposition principle. To the left, the field magnitude is

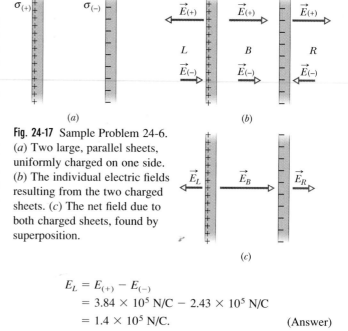

Fig. 24-17 Sample Problem 24-6. (a) Two large, parallel sheets, uniformly charged on one side. (b) The individual electric fields resulting from the two charged sheets. (c) The net field due to both charged sheets, found by superposition.

$$E_L = E_{(+)} - E_{(-)}$$

$$= 3.84 \times 10^5 \ N/C - 2.43 \times 10^5 \ N/C$$

$$= 1.4 \times 10^5 \ N/C. \qquad \text{(Answer)}$$

Because $E_{(+)}$ is larger than $E_{(-)}$, the net electric field $\vec{E}_L$ in this region is directed to the left, as Fig. 24-17c shows. To the right of the sheets, the electric field $\vec{E}_R$ has the same magnitude but is directed to the right, as Fig. 24-17c shows.

Between the sheets, the two fields add and we have

$$E_B = E_{(+)} + E_{(-)}$$

$$= 3.84 \times 10^5 \ N/C + 2.43 \times 10^5 \ N/C$$

$$= 6.3 \times 10^5 \ N/C. \qquad \text{(Answer)}$$

The electric field $\vec{E}_B$ is directed to the right.

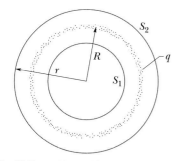

Fig. 24-18 A thin, uniformly charged, spherical shell with total charge q, in cross section. Two Gaussian surfaces S_1 and S_2 are also shown in cross section. Surface S_2 encloses the shell, and S_1 encloses only the empty interior of the shell.

24-9 Applying Gauss' Law: Spherical Symmetry

Here we use Gauss' law to prove the two shell theorems presented without proof in Section 22-4:

▶ A shell of uniform charge attracts or repels a charged particle that is outside the shell as if all the shell's charge were concentrated at the center of the shell.

▶ A shell of uniform charge exerts no electrostatic force on a charged particle that is located inside the shell.

Figure 24-18 shows a charged spherical shell of total charge q and radius R and two concentric spherical Gaussian surfaces, S_1 and S_2. If we followed the procedure of Section 24-5 as we applied Gauss' law to surface S_2, for which $r \geq R$, we would find that

$$E = \frac{1}{4\pi\varepsilon_0}\frac{q}{r^2} \qquad \text{(spherical shell, field at } r \geq R\text{).} \qquad (24\text{-}15)$$

This is the same field that would be set up by a point charge q at the center of the shell of charge. Thus, a shell of charge q would produce the same force on a charged particle placed outside the shell as would a point charge q located at the center of the shell. This proves the first shell theorem.

Applying Gauss' law to surface S_1, for which $r < R$, leads directly to

$$E = 0 \qquad \text{(spherical shell, field at } r < R\text{),} \qquad (24\text{-}16)$$

because this Gaussian surface encloses no charge. Thus, if a charged particle were enclosed by the shell, the shell would exert no net electrostatic force on it. This proves the second shell theorem.

Any spherically symmetric charge distribution, such as that of Fig. 24-19, can be constructed with a nest of concentric spherical shells. For purposes of applying the two shell theorems, the volume charge density ρ should have a single value for each shell but need not be the same from shell to shell. Thus, for the charge distribution as a whole, ρ can vary, but only with r, the radial distance from the center. We can then examine the effect of the charge distribution "shell by shell."

In Fig. 24-19a the entire charge lies within a Gaussian surface with $r > R$. The charge produces an electric field on the Gaussian surface as if the charge were a point charge located at the center, and Eq. 24-15 holds.

Figure 24-19b shows a Gaussian surface with $r < R$. To find the electric field at points on this Gaussian surface, we consider two sets of charged shells—one set inside the Gaussian surface and one set outside. Equation 24-16 says that the charge lying *outside* the Gaussian surface does not set up a net electric field on the Gaussian surface. Equation 24-15 says that the charge *enclosed* by the surface sets up an electric field as if that enclosed charge were concentrated at the center. Letting q' represent that enclosed charge, we can then rewrite Eq. 24-15 as

$$E = \frac{1}{4\pi\varepsilon_0}\frac{q'}{r^2} \qquad \text{(spherical distribution, field at } r \leq R\text{).} \qquad (24\text{-}17)$$

If the full charge q enclosed within radius R is uniform, then q' enclosed within

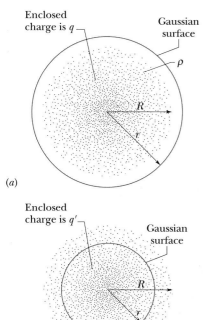

Fig. 24-19 The dots represent a spherically symmetric distribution of charge of radius R, whose volume charge density ρ is a function only of distance from the center. The charged object is not a conductor, and the charge is assumed to be fixed in position. A concentric spherical Gaussian surface with $r > R$ is shown in (a). A similar Gaussian surface with $r < R$ is shown in (b).

radius r in Fig. 24-19b is proportional to q:

$$\frac{\text{charge enclosed by } r}{\text{volume enclosed by } r} = \frac{\text{full charge}}{\text{full volume}}$$

or

$$\frac{q'}{\frac{4}{3}\pi r^3} = \frac{q}{\frac{4}{3}\pi R^3}. \qquad (24\text{-}18)$$

This gives us

$$q' = q\frac{r^3}{R^3}. \qquad (24\text{-}19)$$

Substituting this into Eq. 24-17 yields

$$E = \left(\frac{q}{4\pi\varepsilon_0 R^3}\right)r \qquad \text{(uniform charge, field at } r \le R). \qquad (24\text{-}20)$$

✔**CHECKPOINT 5:**   The figure shows two large, parallel, nonconducting sheets with identical (positive) uniform surface charge densities, and a sphere with a uniform (positive) volume charge density. Rank the four numbered points according to the magnitude of the net electric field there, greatest first.

REVIEW & SUMMARY

Gauss' Law   *Gauss' law* and Coulomb's law, although expressed in different forms, are equivalent ways of describing the relation between charge and electric field in static situations. Gauss' law is

$$\varepsilon_0\Phi = q_{\text{enc}} \qquad \text{(Gauss' law)}, \qquad (24\text{-}6)$$

in which q_{enc} is the net charge inside an imaginary closed surface (a **Gaussian surface**) and Φ is the net **flux** of the electric field through the surface:

$$\Phi = \oint \vec{E} \cdot d\vec{A} \qquad \begin{array}{l}\text{(electric flux through a} \\ \text{Gaussian surface).}\end{array} \qquad (24\text{-}4)$$

Coulomb's law can readily be derived from Gauss' law.

Applications of Gauss' Law   Using Gauss' law and, in some cases, symmetry arguments, we can derive several important results in electrostatic situations. Among these are:

1. An excess charge on a *conductor* is located entirely on the outer surface of the conductor.

2. The external electric field near the *surface of a charged conductor* is perpendicular to the surface and has magnitude

$$E = \frac{\sigma}{\varepsilon_0} \qquad \text{(conducting surface)}. \qquad (24\text{-}11)$$

Within the conductor, $E = 0$.

3. The electric field at any point due to an infinite *line of charge* with uniform linear charge density λ is perpendicular to the line

of charge and has magnitude

$$E = \frac{\lambda}{2\pi\varepsilon_0 r} \qquad \text{(line of charge)}, \qquad (24\text{-}12)$$

where r is the perpendicular distance from the line of charge to the point.

4. The electric field due to an *infinite nonconducting sheet* with uniform surface charge density σ is perpendicular to the plane of the sheet and has magnitude

$$E = \frac{\sigma}{2\varepsilon_0} \qquad \text{(sheet of charge)}. \qquad (24\text{-}13)$$

5. The electric field *outside a spherical shell of charge* with radius R and total charge q is directed radially and has magnitude

$$E = \frac{1}{4\pi\varepsilon_0}\frac{q}{r^2} \qquad \text{(spherical shell, for } r \ge R). \qquad (24\text{-}15)$$

Here r is the distance from the center of the shell to the point at which E is measured. (The charge behaves, for external points, as if it were all located at the center of the sphere.) The field *inside* a uniform spherical shell of charge is exactly zero:

$$E = 0 \qquad \text{(spherical shell, for } r < R). \qquad (24\text{-}16)$$

6. The electric field *inside a uniform sphere of charge* is directed radially and has magnitude

$$E = \left(\frac{q}{4\pi\varepsilon_0 R^3}\right)r. \qquad (24\text{-}20)$$

QUESTIONS

1. A surface has the area vector $\vec{A} = (2\hat{i} + 3\hat{j})$ m^2. What is the flux of an electric field through it if the field is (a) $\vec{E} = 4\hat{i}$ N/C and (b) $\vec{E} = 4\hat{k}$ N/C?

2. What is $\int dA$ for (a) a square of edge length a, (b) a circle of radius r, and (c) the curved surface of a cylinder of length h and radius r?

3. In Fig. 24-20, a full Gaussian surface encloses two of the four positively charged particles. (a) Which of the particles contribute to the electric field at point P on the surface? (b) Which net flux of electric field through the surface is greater (if either): that due to q_1 and q_2 or that due to all four charges?

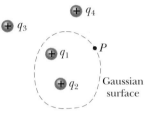

Fig. 24-20 Question 3.

4. Figure 24-21 shows, in cross section, a central metal ball, two spherical metal shells, and three spherical Gaussian surfaces of radii R, $2R$, and $3R$, all with the same center. The uniform charges on the three objects are: ball, Q; smaller shell, $3Q$; larger shell, $5Q$. Rank the Gaussian surfaces according to the magnitude of the electric field at any point on the surface, greatest first.

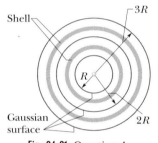

Fig. 24-21 Question 4.

5. Figure 24-22 shows three Gaussian surfaces, each half-submerged in a large, thick metal plate with a uniform surface charge density. Gaussian surface S_1 is the tallest and has the smallest square end caps; surface S_3 is shortest and has the largest square end caps; and S_2 has intermediate values. Rank the surfaces according to (a) the charge they enclose, (b) the magnitude of the electric field at points on their top end cap, (c) the net electric flux through that top end cap, and (d) the net electric flux through their bottom end cap, greatest first.

Fig. 24-22 Question 5.

6. Figure 24-23 shows, in cross section, three cylinders, each of uniform charge Q. Concentric with each cylinder is a cylindrical Gaussian surface, all three with the same radius. Rank the Gaussian surfaces according to the electric field at any point on the surface, greatest first.

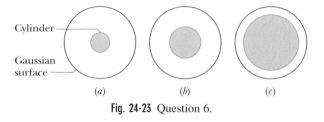

Fig. 24-23 Question 6.

7. Three infinite nonconducting sheets, with uniform surface charge densities σ, 2σ, and 3σ, are arranged to be parallel like the two sheets in Fig. 24-17a. What is their order, from left to right, if the electric field $\vec{E}$ produced by the arrangement has magnitude $E = 0$ in one region and $E = 2\sigma/\varepsilon_0$ in another region?

8. A small charged ball lies within the hollow of a metallic spherical shell of radius R. Here, for three situations, are the net charges on the ball and shell, respectively: (1) $+4q$, 0; (2) $-6q$, $+10q$; (3) $+16q$, $-12q$. Rank the situations according to the charge on (a) the inner surface of the shell and (b) the outer surface, most positive first.

9. Rank the situations of Question 8 according to the magnitude of the electric field (a) halfway through the shell and (b) at a point $2R$ from the center of the shell, greatest first.

10. Figure 24-24 shows four spheres, each with charge Q uniformly distributed through its volume. (a) Rank the spheres according to their volume charge density, greatest first. The figure also shows a point P for each sphere, all at the same distance from the center of the sphere. (b) Rank the spheres according to the magnitude of the electric field they produce at point P, greatest first.

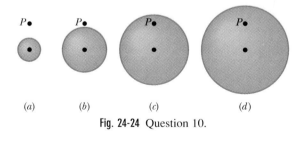

Fig. 24-24 Question 10.

EXERCISES & PROBLEMS

SEC. 24-2 Flux

1E. Water in an irrigation ditch of width $w = 3.22$ m and depth $d = 1.04$ m flows with a speed of 0.207 m/s. The *mass flux* of the flowing water through an imaginary surface is the product of the water's density (1000 kg/m^3) and its volume flux through that surface. Find the mass flux through the following imaginary surfaces: (a) a surface of area wd, entirely in the water, perpendicular to the flow; (b) a surface with area $3wd/2$, of which wd is in the water, perpendicular to the flow; (c) a surface of area $wd/2$, entirely in the water, perpendicular to the flow; (d) a surface of area wd, half in the water and half out, perpendicular to the flow; (e) a surface of area wd, entirely in the water, with its normal 34° from the direction of flow.

SEC. 24-3   Flux of an Electric Field

2E. The square surface shown in Fig. 24-25 measures 3.2 mm on each side. It is immersed in a uniform electric field with magnitude $E = 1800$ N/C. The field lines make an angle of 35° with a normal to the surface, as shown. Take that normal to be directed "outward," as though the surface were one face of a box. Calculate the electric flux through the surface. ssm

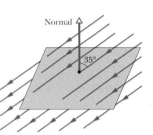

Fig. 24-25 Exercise 2.

3E. The cube in Fig. 24-26 has edge lengths of 1.40 m and is oriented as shown in a region of uniform electric field. Find the electric flux through the right face if the electric field, in newtons per coulomb, is given by (a) $6.00\hat{i}$, (b) $-2.00\hat{j}$, and (c) $-3.00\hat{i} + 4.00\hat{k}$. (d) What is the total flux through the cube for each of these fields?

Fig. 24-26 Exercise 3 and Problems 7 and 10.

SEC. 24-4   Gauss' Law

4E. You have four point charges, $2q$, q, $-q$, and $-2q$. If possible, describe how you would place a closed surface that encloses at least the charge $2q$ (and perhaps other charges) and through which the net electric flux is (a) 0, (b) $+3q/\varepsilon_0$, and (c) $-2q/\varepsilon_0$.

5E. A point charge of 1.8 μC is at the center of a cubical Gaussian surface 55 cm on edge. What is the net electric flux through the surface? ssm

6E. In Fig. 24-27, a butterfly net is in a uniform electric field of magnitude E. The rim, a circle of radius a, is aligned perpendicular to the field. Find the electric flux through the netting.

Fig. 24-27 Exercise 6.

7P. Find the net flux through the cube given in Exercise 3 and Fig. 24-26 if the electric field is given by (a) $\vec{E} = 3.00y\hat{j}$ and (b) $\vec{E} = -4.00\hat{i} + (6.00 + 3.00y)\hat{j}$. E is in newtons per coulomb, and y is in meters. (c) In each case, how much charge is enclosed by the cube? ilw

8P. When a shower is turned on in a closed bathroom, the splashing of the water on the bare tub can fill the room's air with negatively charged ions and produce an electric field in the air as great as 1000 N/C. Consider a bathroom with dimensions of 2.5 m × 3.0 m × 2.0 m. Along the ceiling, floor, and four walls, approximate the electric field in the air as being directed perpendicular to the surface and as having a uniform magnitude of 600 N/C. Also, treat those surfaces as forming a closed Gaussian surface around the room's air. What are (a) the volume charge density ρ and (b) the number of excess elementary charges e per cubic meter in the room's air?

9P. It is found experimentally that the electric field in a certain region of Earth's atmosphere is directed vertically down. At an altitude of 300 m the field has magnitude 60.0 N/C; at an altitude of 200 m, the magnitude is 100 N/C. Find the net amount of charge contained in a cube 100 m on edge, with horizontal faces at altitudes of 200 and 300 m. Neglect the curvature of Earth. ssm

10P. At each point on the surface of the cube shown in Fig. 24-26, the electric field is in the positive direction of y. The length of each edge of the cube is 3.0 m. On the top surface of the cube $\vec{E} = -34\hat{j}$ N/C, and on the bottom face of the cube $\vec{E} = +20\hat{j}$ N/C. Determine the net charge contained within the cube.

11P. A point charge q is placed at one corner of a cube of edge a. What is the flux through each of the cube faces? (*Hint:* Use Gauss' law and symmetry arguments.) ssm

SEC. 24-6   A Charged Isolated Conductor

12E. The electric field just above the surface of the charged drum of a photocopying machine has a magnitude E of 2.3×10^5 N/C. What is the surface charge density on the drum, assuming that the drum is a conductor?

13E. A uniformly charged conducting sphere of 1.2 m diameter has a surface charge density of 8.1 μC/m^2. (a) Find the net charge on the sphere. (b) What is the total electric flux leaving the surface of the sphere? ssm

14E. Space vehicles traveling through Earth's radiation belts can intercept a significant number of electrons. The resulting charge buildup can damage electronic components and disrupt operations. Suppose a spherical metallic satellite 1.3 m in diameter accumulates 2.4 μC of charge in one orbital revolution. (a) Find the resulting surface charge density. (b) Calculate the magnitude of the electric field just outside the surface of the satellite, due to the surface charge.

15P. An isolated conductor of arbitrary shape has a net charge of $+10 \times 10^{-6}$ C. Inside the conductor is a cavity within which is a point charge $q = +3.0 \times 10^{-6}$ C. What is the charge (a) on the cavity wall and (b) on the outer surface of the conductor? ssm www

SEC. 24-7   Applying Gauss' Law: Cylindrical Symmetry

16E. (a) The drum of the photocopying machine in Exercise 12 has a length of 42 cm and a diameter of 12 cm. What is the total charge on the drum? (b) The manufacturer wishes to produce a desktop version of the machine. This requires reducing the size of the drum to a length of 28 cm and a diameter of 8.0 cm. The electric field at the drum surface must remain unchanged. What must be the charge on this new drum?

17E. An infinite line of charge produces a field of 4.5×10^4 N/C at a distance of 2.0 m. Calculate the linear charge density. ssm

18P. Figure 24-28 shows a section of a long, thin-walled metal tube of radius R, with a charge per unit length λ on its surface. Derive expressions for E in terms of the distance r from the tube axis, considering both (a) $r > R$ and (b) $r < R$. Plot your results for the range $r = 0$ to $r = 5.0$ cm, assuming that $\lambda = 2.0 \times 10^{-8}$ C/m and $R = 3.0$ cm. (*Hint:* Use cylindrical Gaussian surfaces, coaxial with the metal tube.)

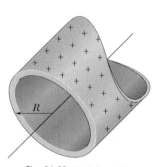
Fig. 24-28 Problem 18.

19P. A very long conducting cylindrical rod of length L with a total charge $+q$ is surrounded by a conducting cylindrical shell (also of length L) with total charge $-2q$, as shown in Fig. 24-29. Use Gauss' law to find (a) the electric field at points outside the conducting shell, (b) the distribution of charge on the shell, and (c) the electric field in the region between the shell and rod. ssm

Fig. 24-29 Problem 19.

20P. A long, straight wire has fixed negative charge with a linear charge density of magnitude 3.6 nC/m. The wire is to be enclosed by a thin, nonconducting cylinder of outside radius 1.5 cm, coaxial with the wire. The cylinder is to have positive charge on its outside surface with a surface charge density σ such that the net external electric field is zero. Calculate the required σ.

21P. Two long, charged, concentric cylinders have radii of 3.0 and 6.0 cm. The charge per unit length is 5.0×10^{-6} C/m on the inner cylinder and -7.0×10^{-6} C/m on the outer cylinder. Find the electric field at (a) $r = 4.0$ cm and (b) $r = 8.0$ cm, where r is the radial distance from the common central axis. ilw

22P. A long, nonconducting, solid cylinder of radius 4.0 cm has a nonuniform volume charge density ρ that is a function of the radial distance r from the axis of the cylinder, as given by $\rho = Ar^2$, with $A = 2.5$ μC/m^5. What is the magnitude of the electric field at a radial distance of (a) 3.0 cm and (b) 5.0 cm from the axis of the cylinder?

23P. Figure 24-30 shows a Geiger counter, a device used to detect ionizing radiation (radiation that causes ionization of atoms). The counter consists of a thin, positively charged central wire surrounded by a concentric, circular, conducting cylinder with an equal negative charge. Thus, a strong radial electric field is set up inside the cylinder. The cylinder contains a low-pressure inert gas. When a particle of radiation enters the device through the cylinder wall, it ionizes a few of the gas atoms. The resulting free electrons (label e) are drawn to the positive wire. However, the electric field is so intense that, between collisions with other gas atoms, the free electrons gain energy sufficient to ionize these atoms also. More free electrons are thereby created, and the process is repeated until the electrons reach the wire. The resulting "avalanche" of electrons is collected by the wire, generating a signal that is used to record the passage of the original particle of radiation. Suppose that the radius of the central wire is 25 μm, the radius of the cylinder 1.4 cm, and the length of the tube 16 cm. If the electric field at the cylinder's inner wall is 2.9×10^4 N/C, what is the total positive charge on the central wire? ssm

Fig. 24-30 Problem 23.

24P. A charge of uniform linear density 2.0 nC/m is distributed along a long, thin, nonconducting rod. The rod is coaxial with a long, hollow, conducting cylinder (inner radius = 5.0 cm, outer radius = 10 cm). The net charge on the conductor is zero. (a) What is the magnitude of the electric field 15 cm from the axis of the cylinder? What is the surface charge density on (b) the inner surface and (c) the outer surface of the conductor?

25P. Charge is distributed uniformly throughout the volume of an infinitely long cylinder of radius R. (a) Show that, at a distance r from the cylinder axis (for $r < R$),

$$E = \frac{\rho r}{2\varepsilon_0},$$

where ρ is the volume charge density. (b) Write an expression for E when $r > R$. ssm www

SEC. 24-8   Applying Gauss' Law: Planar Symmetry

26E. Figure 24-31 shows cross-sections through two large, parallel, nonconducting sheets with identical distributions of positive charge with surface charge density σ. What is $\vec{E}$ at points (a) above the sheets, (b) between them, and (c) below them?

Fig. 24-31 Exercise 26.

27E. A square metal plate of edge length 8.0 cm and negligible thickness has a total charge of 6.0×10^{-6} C. (a) Estimate the magnitude E of the electric field just off the center of the plate (at, say, a distance of 0.50 mm) by assuming that the charge is spread uniformly over the two faces of the plate. (b) Estimate E at a distance of 30 m (large relative to the plate size) by assuming that the plate is a point charge. ssm

28E. A large, flat, nonconducting surface has a uniform charge density σ. A small circular hole of radius R has been cut in the middle of the surface, as shown in Fig. 24-32. Ignore fringing of the field lines around all edges, and calculate the electric field at point P, a distance z from the center of the hole along its axis. (*Hint:* See Eq. 23-26 and use superposition.)

Fig. 24-32 Exercise 28.

29P. In Fig. 24-33, a small, nonconducting ball of mass $m = 1.0$ mg and charge $q = 2.0 \times 10^{-8}$ C (distributed uniformly through its volume) hangs from an insulating thread that makes an angle $\theta = 30°$ with a vertical, uniformly charged nonconducting sheet (shown in cross section). Considering the gravitational force on the ball and assuming that the sheet extends far vertically and into and out of the page, calculate the surface charge density σ of the sheet. ssm

Fig. 24-33 Problem 29.

30P. Two large, thin metal plates are parallel and close to each other, as in Fig. 24-16c. On their inner faces, the plates have excess surface charge densities of opposite signs and with a magnitude 7.0×10^{-22} C/m^2, with the negatively charged plate on the left. What are the magnitude and direction of the electric field $\vec{E}$ (a) to the left of the plates, (b) to the right of the plates, and (c) between the plates?

31P. An electron is shot directly toward the center of a large metal plate that has excess negative charge with surface charge density 2.0×10^{-6} C/m^2. If the initial kinetic energy of the electron is 100 eV and if the electron is to stop (owing to electrostatic repulsion from the plate) just as it reaches the plate, how far from the plate must it be shot? ssm

32P. Two large metal plates of area 1.0 m^2 face each other. They are 5.0 cm apart and have equal but opposite charges on their inner surfaces. If the magnitude E of the electric field between the plates is 55 N/C, what is the magnitude of the charge on each plate? Neglect edge effects.

33P*. A planar slab of thickness d has a uniform volume charge density ρ. Find the magnitude of the electric field at all points in space both (a) inside and (b) outside the slab, in terms of x, the distance measured from the central plane of the slab. ssm

SEC. 24-9   Applying Gauss' Law: Spherical Symmetry

34E. A point charge causes an electric flux of -750 N $\cdot$ m^2/C to pass through a spherical Gaussian surface of 10.0 cm radius centered on the charge. (a) If the radius of the Gaussian surface were doubled, how much flux would pass through the surface? (b) What is the value of the point charge?

35E. A conducting sphere of radius 10 cm has an unknown charge. If the electric field 15 cm from the center of the sphere has the magnitude 3.0×10^3 N/C and is directed radially inward, what is the net charge on the sphere? ssm

36E. Two charged concentric spheres have radii of 10.0 cm and 15.0 cm. The charge on the inner sphere is 4.00×10^{-8} C, and that on the outer sphere is 2.00×10^{-8} C. Find the electric field (a) at $r = 12.0$ cm and (b) at $r = 20.0$ cm.

37E. In a 1911 paper, Ernest Rutherford said: "In order to form some idea of the forces required to deflect an α particle through a large angle, consider an atom [as] containing a point positive charge Ze at its centre and surrounded by a distribution of negative electricity $-Ze$ uniformly distributed within a sphere of radius R. The electric field E . . . at a distance r from the center for a point *inside* the atom [is]

$$E = \frac{Ze}{4\pi\varepsilon_0}\left(\frac{1}{r^2} - \frac{r}{R^3}\right)."$$

Verify this equation. ssm

38E. Equation 24-11 ($E = \sigma/\varepsilon_0$) gives the electric field at points near a charged conducting surface. Apply this equation to a conducting sphere of radius r and charge q, and show that the electric field outside the sphere is the same as the field of a point charge located at the center of the sphere.

39P. A proton with speed $v = 3.00 \times 10^5$ m/s orbits just outside a charged sphere of radius $r = 1.00$ cm. What is the charge on the sphere? ssm www

40P. A point charge $+q$ is placed at the center of an electrically neutral, spherical conducting shell with inner radius a and outer radius b. What charge appears on (a) the inner surface of the shell and (b) the outer surface? What is the net electric field at a distance r from the center of the shell if (c) $r < a$, (d) $b > r > a$, and (e) $r > b$? Sketch field lines for those three regions. For $r > b$, what is the net electric field due to (f) the central point charge plus the inner surface charge and (g) the outer surface charge? A point charge $-q$ is now placed outside the shell. Does this point charge change the charge distribution on (h) the outer surface and (i) the inner surface? Sketch the field lines now. (j) Is there an electrostatic force on the second point charge? (k) Is there a net electrostatic force on the first point charge? (l) Does this situation violate Newton's third law?

41P. A solid nonconducting sphere of radius R has a nonuniform charge distribution of volume charge density $\rho = \rho_s r/R$, where ρ_s is a constant and r is the distance from the center of the sphere. Show (a) that the total charge on the sphere is $Q = \pi\rho_s R^3$ and (b) that

$$E = \frac{1}{4\pi\varepsilon_0}\frac{Q}{R^4}r^2$$

gives the magnitude of the electric field inside the sphere. ilw

42P. A hydrogen atom can be considered as having a central point-like proton of positive charge $+e$ and an electron of negative charge $-e$ that is distributed about the proton according to the volume charge density $\rho = A \exp(-2r/a_0)$. Here A is a constant, $a_0 = 0.53 \times 10^{-10}$ m is the *Bohr radius*, and r is the distance from the center of the atom. (a) Using the fact that hydrogen is electrically neutral, find A. (b) Then find the electric field produced by the atom at the Bohr radius.

43P. In Fig. 24-34 a sphere, of radius a and charge $+q$ uniformly distributed throughout its volume, is concentric with a spherical conducting shell of inner radius b and outer radius c. This shell has a net charge of $-q$. Find expressions for the electric field, as a function of the radius r, (a) within the sphere ($r < a$), (b) between the sphere and the shell ($a < r < b$), (c) inside the shell ($b < r < c$), and (d) outside the shell ($r > c$). (e) What are the charges on the inner and outer surfaces of the shell? ssm

Fig. 24-34 Problem 43.

44P. Figure 24-35a shows a spherical shell of charge with uniform volume charge density ρ. Plot E due to the shell for distances r from the center of the shell ranging from zero to 30 cm. Assume that $\rho = 1.0 \times 10^{-6}$ C/m³, $a = 10$ cm, and $b = 20$ cm.

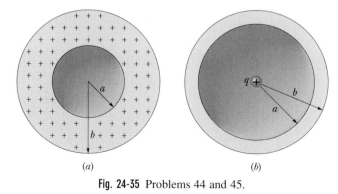

Fig. 24-35 Problems 44 and 45.

45P. In Fig. 24-35b, a nonconducting spherical shell, of inner radius a and outer radius b, has a positive volume charge density $\rho = A/r$ (within its thickness), where A is a constant and r is the distance from the center of the shell. In addition, a positive point charge q is located at that center. What value should A have if the electric field in the shell ($a \le r \le b$) is to be uniform? (*Hint:* The constant A depends on a but not on b.) ssm

46P*. A nonconducting sphere has a uniform volume charge density ρ. Let $\vec{r}$ be the vector from the center of the sphere to a general point P within the sphere. (a) Show that the electric field at P is given by $\vec{E} = \rho\vec{r}/3\varepsilon_0$. (Note that the result is independent of the radius of the sphere.) (b) A spherical cavity is hollowed out of the sphere, as shown in Fig. 24-36. Using superposition concepts, show that the electric field at all points within the cavity is uniform and equal to $\vec{E} = \rho\vec{a}/3\varepsilon_0$, where $\vec{a}$ is the position vector from the center of the sphere to the center of the cavity. (Note that this result is independent of the radius of the sphere and the radius of the cavity.)

Fig. 24-36 Problem 46.

47P*. A spherically symmetrical but nonuniform volume distribution of charge produces an electric field of magnitude $E = Kr^4$, directed radially outward from the center of the sphere. Here r is the radial distance from that center, and K is a constant. What is the volume density ρ of the charge distribution?

Additional Problem

48. *The chocolate crumb mystery.* Explosions ignited by electrostatic discharges (sparks) constitute a serious danger in facilities handling grain or powder. Such an explosion occurred in chocolate crumb powder at a biscuit factory in the 1970s. At the factory, workers usually emptied newly delivered sacks of the powder into a loading bin, from which it was blown through grounded PVC pipes to a silo for storage. Somewhere along this route, two conditions for an explosion were met: (1) The magnitude of an electric field became 3.0×10^6 N/C or greater, so that electric breakdown and thus sparking could occur. (2) The energy of a spark was 150 mJ or greater so that it could ignite the powder explosively. Let us check for the first condition in the powder flow through the PVC pipes.

Suppose a stream of *negatively* charged chocolate crumb powder is blown through a cylindrical PVC pipe with radius $R = 5.0$ cm. Assume that the powder and its charge are spread uniformly through the pipe with a volume charge density ρ. (a) Using Gauss' law, find an expression for the magnitude of the electric field $\vec{E}$ in the pipe as a function of the radial distance r from the center of the pipe. (b) Does the magnitude increase or decrease with r? (c) Is the electric field $\vec{E}$ directed radially inward or outward? (d) Assuming a volume charge density ρ of magnitude 1.1×10^{-3} C/m³ (which was typical at the biscuit factory), find the maximum magnitude of the electric field and determine where that maximum field occurs. (e) Could sparking occur, and if so, where? (The story continues with Problem 57 in Chapter 25.)

25 Electric Potential

While enjoying the Sequoia National Park from a lookout platform, this woman found her hair rising from her head. Amused, her brother took her photograph. Five minutes after they left, lightning struck the platform, killing one person and injuring seven.

What had caused the woman's hair to rise?

The answer is in this chapter.

25-1 Electric Potential Energy

Newton's law for the gravitational force and Coulomb's law for the electrostatic force are mathematically identical. Thus, the general features we have discussed for the gravitational force should apply to the electrostatic force.

In particular, we can infer, correctly, that the electrostatic force is a *conservative force*. Thus, when an electrostatic force acts between two or more charged particles within a system of particles, we can assign an **electric potential energy** U to the system. Moreover, if the system changes its configuration from an initial state i to a different final state f, the electrostatic force does work W on the particles. From Eq. 8-1, we then know that the resulting change ΔU in the potential energy of the system is

$$\Delta U = U_f - U_i = -W. \tag{25-1}$$

As with other conservative forces, the work done by the electrostatic force is *path independent*. Suppose a charged particle within the system moves from point i to point f while an electrostatic force between it and the rest of the system acts on it. Provided the rest of the system does not change, the work W done by the force is the same for *all* paths between points i and f.

For convenience, we usually take the *reference configuration* of a system of charged particles to be that in which the particles are all infinitely separated from each other. Also, we usually set the corresponding *reference potential energy* to be zero. Suppose that several charged particles come together from initially infinite separations (state i) to form a system of nearby particles (state f). Let the initial potential energy U_i be zero, and let W_∞ represent the work done by the electrostatic forces between the particles during the move in from infinity. Then from Eq. 25-1, the final potential energy U of the system is

$$U = -W_\infty. \tag{25-2}$$

As is true of other kinds of potential energy, electric potential energy is considered to be a type of mechanical energy. Recall from Chapter 8 that if only conservative forces act within a (closed) system, the mechanical energy of the system is conserved. We shall use this fact extensively in the rest of this chapter.

PROBLEM-SOLVING TACTICS

Tactic 1: *Electric Potential Energy; Work Done by a Field*
An electric potential energy is associated with a system of particles as a whole. However, you will see statements (starting with Sample Problem 25-1) that associate it with only one particle within a system. For example, you might read, "An electron in an electric field has a potential energy of 10^{-7} J." Such statements are often acceptable, but you should always keep in mind that the potential energy is actually associated with a system—here the electron plus the charged particles that set up the electric field. Also keep in mind that it makes sense to assign a particular potential energy value, such as 10^{-7} J here, to a particle or even a system *only* if the reference potential energy value is known.

When the potential energy is associated with only one particle within a system, you often will read that work is done on the particle *by the electric field*. What is meant is that work is done by the force on the particle due to the charges that set up the field.

Sample Problem 25-1

Electrons are continually being knocked out of air molecules in the atmosphere by cosmic-ray particles coming in from space. Once released, each electron experiences an electrostatic force $\vec{F}$ due to the electric field $\vec{E}$ that is produced in the atmosphere by charged particles already on Earth. Near Earth's surface the electric field has the magnitude $E = 150$ N/C and is directed downward. What

is the change ΔU in the electric potential energy of a released electron when the electrostatic force causes it to move vertically upward through a distance $d = 520$ m (Fig. 25-1)?

SOLUTION: We need three **Key Ideas** here. One is that the change ΔU in the electric potential energy of the electron is related to the work

Fig. 25-1 Sample Problem 25-1. An electron in the atmosphere is moved upward through displacement $\vec{d}$ by an electrostatic force $\vec{F}$ due to an electric field $\vec{E}$.

W done on the electron by the electric field. Equation 25-1 ($\Delta U = -W$) gives the relation. A second Key Idea is that the work done by a constant force $\vec{F}$ on a particle undergoing a displacement $\vec{d}$ is

$$W = \vec{F} \cdot \vec{d}. \qquad (25\text{-}3)$$

Finally, the third Key Idea is that the electrostatic force and the electric field are related by $\vec{F} = q\vec{E}$, where here q is the charge of an electron (= -1.6×10^{-19} C). Substituting for $\vec{F}$ in Eq. 25-3 and taking the dot product yield

$$W = q\vec{E} \cdot \vec{d} = qEd \cos \theta, \qquad (25\text{-}4)$$

where θ is the angle between the directions of $\vec{E}$ and $\vec{d}$. The field

$\vec{E}$ is directed downward and the displacement $\vec{d}$ is directed upward, so $\theta = 180°$. Substituting this and other data into Eq. 25-4, we find

$$W = (-1.6 \times 10^{-19} \text{ C})(150 \text{ N/C})(520 \text{ m}) \cos 180°$$
$$= 1.2 \times 10^{-14} \text{ J}.$$

Equation 25-1 then yields

$$\Delta U = -W = -1.2 \times 10^{-14} \text{ J}. \qquad \text{(Answer)}$$

This result tells us that during the 520 m ascent, the electric potential energy of the electron *decreases* by 1.2×10^{-14} J.

✔CHECKPOINT 1: In the figure, a proton moves from point i to point f in a uniform electric field directed as shown. (a) Does the electric field do positive or negative work on the proton? (b) Does the electric potential energy of the proton increase or decrease?

25-2 Electric Potential

As you can infer from Sample Problem 25-1, the potential energy of a charged particle in an electric field depends on the magnitude of the charge. However, the potential energy *per unit charge* has a unique value at any point in an electric field.

For an example of this, suppose we place a test particle of positive charge 1.60×10^{-19} C at a point in an electric field where the particle has an electric potential energy of 2.40×10^{-17} J. Then the potential energy per unit charge is

$$\frac{2.40 \times 10^{-17} \text{ J}}{1.60 \times 10^{-19} \text{ C}} = 150 \text{ J/C}.$$

Next, suppose we replace that test particle with one having twice as much positive charge, 3.20×10^{-19} C. We would find that the second particle has an electric potential energy of 4.80×10^{-17} J, twice that of the first particle. However, the potential energy per unit charge would be the same, still 150 J/C.

Thus, the potential energy per unit charge, which can be symbolized as U/q, is independent of the charge q of the particle we happen to use and is *characteristic only of the electric field* we are investigating. The potential energy per unit charge at a point in an electric field is called the **electric potential** V (or simply the **potential**) at that point. Thus,

$$V = \frac{U}{q}. \qquad (25\text{-}5)$$

Note that electric potential is a scalar, not a vector.

The *electric potential difference* ΔV between any two points i and f in an electric field is equal to the difference in potential energy per unit charge between the two points:

$$\Delta V = V_f - V_i = \frac{U_f}{q} - \frac{U_i}{q} = \frac{\Delta U}{q}. \qquad (25\text{-}6)$$

Using Eq. 25-1 to substitute $-W$ for ΔU in Eq. 25-6, we can define the potential difference between points i and f as

$$\Delta V = V_f - V_i = -\frac{W}{q} \quad \text{(potential difference defined).} \quad (25\text{-}7)$$

The potential difference between two points is thus the negative of the work done by the electrostatic force to move a unit charge from one point to the other. A potential difference can be positive, negative, or zero, depending on the signs and magnitudes of q and W.

If we set $U_i = 0$ at infinity as our reference potential energy, then by Eq. 25-5, the electric potential must also be zero there. Then from Eq. 25-7, we can define the electric potential V at any point in an electric field to be

$$V = -\frac{W_\infty}{q} \quad \text{(potential defined),} \quad (25\text{-}8)$$

where W_∞ is the work done by the electric field on a charged particle as that particle moves in from infinity to point f. A potential V can be positive, negative, or zero, depending on the signs and magnitudes of q and W_∞.

The SI unit for potential that follows from Eq. 25-8 is the joule per coulomb. This combination occurs so often that a special unit, the *volt* (abbreviated V) is used to represent it. Thus,

$$1 \text{ volt} = 1 \text{ joule per coulomb.} \quad (25\text{-}9)$$

This new unit allows us to adopt a more conventional unit for the electric field $\vec{E}$, which we have measured up to now in newtons per coulomb. With two unit conversions, we obtain

$$1 \text{ N/C} = \left(1\frac{\text{N}}{\text{C}}\right)\left(\frac{1 \text{ V} \cdot \text{C}}{1 \text{ J}}\right)\left(\frac{1 \text{ J}}{1 \text{ N} \cdot \text{m}}\right)$$
$$= 1 \text{ V/m.} \quad (25\text{-}10)$$

The conversion factor in the second set of parentheses comes from Eq. 25-9; that in the third set of parentheses is derived from the definition of the joule. From now on, we shall express values of the electric field in volts per meter rather than in newtons per coulomb.

Finally, we can now define an energy unit that is a convenient one for energy measurements in the atomic and subatomic domain: One *electron-volt* (eV) is the energy equal to the work required to move a single elementary charge e, such as that of the electron or the proton, through a potential difference of exactly one volt. Equation 25-7 tells us that the magnitude of this work is $q \, \Delta V$, so

$$1 \text{ eV} = e(1 \text{ V})$$
$$= (1.60 \times 10^{-19} \text{ C})(1 \text{ J/C}) = 1.60 \times 10^{-19} \text{ J.}$$

PROBLEM-SOLVING TACTICS

Tactic 2: *Electric Potential and Electric Potential Energy*
Electric potential V and electric potential energy U are quite different quantities and should not be confused.

▶ *Electric potential* is a property of an electric field, regardless of whether a charged object has been placed in that field; it is measured in joules per coulomb, or volts.

▶ *Electric potential energy* is an energy of a charged object in an external electric field (or more precisely, an energy of the system consisting of the object and the external electric field); it is measured in joules.

Work Done by an Applied Force

Suppose we move a particle of charge q from point i to point f in an electric field by applying a force to it. During the move, our applied force does work W_{app} on the charge while the electric field does work W on it. By the work–kinetic energy theorem of Eq. 7-10, the change ΔK in the kinetic energy of the particle is

$$\Delta K = K_f - K_i = W_{app} + W. \qquad (25\text{-}11)$$

Now suppose the particle is stationary before and after the move. Then K_f and K_i are both zero, and Eq. 25-11 reduces to

$$W_{app} = -W. \qquad (25\text{-}12)$$

In words, the work W_{app} done by our applied force during the move is equal to the negative of the work W done by the electric field—provided there is no change in kinetic energy.

By using Eq. 25-12 to substitute W_{app} into Eq. 25-1, we can relate the work done by our applied force to the change in the potential energy of the particle during the move. We find

$$\Delta U = U_f - U_i = W_{app}. \qquad (25\text{-}13)$$

By similarly using Eq. 25-12 to substitute W_{app} into Eq. 25-7, we can relate our work W_{app} to the electric potential difference ΔV between the initial and final locations of the particle. We find

$$W_{app} = q \, \Delta V. \qquad (25\text{-}14)$$

W_{app} can be positive, negative, or zero depending on the signs and magnitudes of q and ΔV. It is the work we must do to move a particle of charge q through a potential difference ΔV with no change in the particle's kinetic energy.

✔**CHECKPOINT 2:**   In the figure of Checkpoint 1, we move a proton from point i to point f in a uniform electric field directed as shown. (a) Does our force do positive or negative work? (b) Does the proton move to a point of higher or lower potential?

25-3 Equipotential Surfaces

Adjacent points that have the same electric potential form an **equipotential surface,** which can be either an imaginary surface or a real, physical surface. No net work W is done on a charged particle by an electric field when the particle moves between two points i and f on the same equipotential surface. This follows from Eq. 25-7, which tells us that W must be zero if $V_f = V_i$. Because of the path independence of work (and thus of potential energy and potential), $W = 0$ for *any* path connecting points i and f, regardless of whether that path lies entirely on the equipotential surface.

Figure 25-2 shows a *family* of equipotential surfaces, associated with the electric field due to some distribution of charges. The work done by the electric field on a charged particle as the particle moves from one end to the other of paths I and II is zero because each of these paths begins and ends on the same equipotential surface. The work done as the charged particle moves from one end to the other of paths III and IV is not zero but has the same value for both these paths because the initial and final potentials are identical for the two paths; that is, paths III and IV connect the same pair of equipotential surfaces.

From symmetry, the equipotential surfaces produced by a point charge or a spherically symmetrical charge distribution are a family of concentric spheres. For a uniform electric field, the surfaces are a family of planes perpendicular to the field

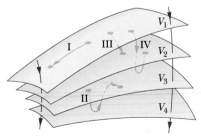

Fig. 25-2 Portions of four equipotential surfaces at electric potentials $V_1 = 100$ V, $V_2 = 80$ V, $V_3 = 60$ V, and $V_4 = 40$ V. Four paths along which a test charge may move are shown. Two electric field lines are also indicated.

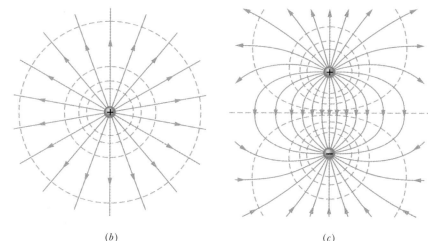

Equipotential surface
Field line

(a)   (b)   (c)

Fig. 25-3 Electric field lines (purple) and cross sections of equipotential surfaces (gold) for (a) a uniform field, (b) the field of a point charge, and (c) the field of an electric dipole.

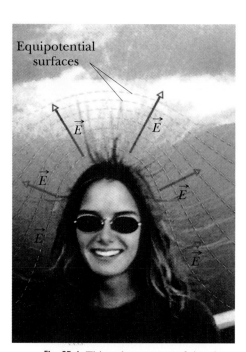

Equipotential surfaces

$\vec{E}$

Fig. 25-4 This enhancement of the chapter's opening photograph shows the result of overhead clouds creating a strong electric field $\vec{E}$ near a woman's head. Many of the hair strands extended along the field, which was perpendicular to the equipotential surfaces and greatest where those surfaces were closest, near the top of her head.

lines. In fact, equipotential surfaces are always perpendicular to electric field lines and thus to $\vec{E}$, which is always tangent to these lines. If $\vec{E}$ were *not* perpendicular to an equipotential surface, it would have a component lying along that surface. This component would then do work on a charged particle as it moved along the surface. However, by Eq. 25-7 work cannot be done if the surface is truly an equipotential surface; the only possible conclusion is that $\vec{E}$ must be everywhere perpendicular to the surface. Figure 25-3 shows electric field lines and cross sections of the equipotential surfaces for a uniform electric field and for the field associated with a point charge and with an electric dipole.

We now return to the woman in the opening photograph for this chapter. Because she was standing on a platform that was connected to the mountainside, she was at about the same potential as the mountainside. Overhead, a highly charged cloud system had moved in and created a strong electric field around her and the mountainside, with $\vec{E}$ pointing outward from her and the mountain. Electrostatic forces due to this field drove some of the conduction electrons in the woman downward through her body, leaving strands of her hair positively charged. The magnitude of $\vec{E}$ was apparently large, but less than the value of about 3×10^6 V/m that would have caused electrical breakdown of the air molecules. (That value was exceeded shortly later when lightning struck the platform.)

The equipotential surfaces surrounding the woman on the mountainside platform can be inferred from her hair; the strands are extended along the direction of $\vec{E}$ and thus are perpendicular to the equipotential surfaces, so the surfaces must be as drawn in Fig. 25-4. The field magnitude E was apparently greatest (the equipotential surfaces were apparently most closely spaced) just above her head, because the hair there extended farther than the hair around the sides.

The lesson here is simple. If an electric field causes the hairs on your head to stand up, you would do better to run for shelter than to pose for a snapshot.

25-4 Calculating the Potential from the Field

We can calculate the potential difference between any two points i and f in an electric field if we know the electric field vector $\vec{E}$ all along any path connecting those points. To make the calculation, we find the work done on a positive test charge by the field as the charge moves from i to f, and then use Eq. 25-7.

Fig. 25-5 A test charge q_0 moves from point i to point f along the path shown in a nonuniform electric field. During a displacement $d\vec{s}$, an electrostatic force $q_0\vec{E}$ acts on the test charge. This force points in the direction of the field line at the location of the test charge.

Consider an arbitrary electric field, represented by the field lines in Fig. 25-5, and a positive test charge q_0 that moves along the path shown from point i to point f. At any point on the path, an electrostatic force $q_0\vec{E}$ acts on the charge as it moves through a differential displacement $d\vec{s}$. From Chapter 7, we know that the differential work dW done on a particle by a force $\vec{F}$ during a displacement $d\vec{s}$ is

$$dW = \vec{F} \cdot d\vec{s}. \qquad (25\text{-}15)$$

For the situation of Fig. 25-5, $\vec{F} = q_0\vec{E}$ and Eq. 25-15 becomes

$$dW = q_0\vec{E} \cdot d\vec{s}. \qquad (25\text{-}16)$$

To find the total work W done on the particle by the field as the particle moves from point i to point f, we sum—via integration—the differential works done on the charge as it moves through all the differential displacements $d\vec{s}$ along the path:

$$W = q_0 \int_i^f \vec{E} \cdot d\vec{s}. \qquad (25\text{-}17)$$

If we substitute the total work W from Eq. 25-17 into Eq. 25-7, we find

$$V_f - V_i = -\int_i^f \vec{E} \cdot d\vec{s}. \qquad (25\text{-}18)$$

Thus, the potential difference $V_f - V_i$ between any two points i and f in an electric field is equal to the negative of the *line integral* (meaning the integral along a particular path) of $\vec{E} \cdot d\vec{s}$ from i to f. However, because the electrostatic force is conservative, all paths (whether easy or difficult to use) yield the same result.

If the electric field is known throughout a certain region, Eq. 25-18 allows us to calculate the difference in potential between any two points in the field. If we choose the potential V_i at point i to be zero, then Eq. 25-18 becomes

$$V = -\int_i^f \vec{E} \cdot d\vec{s}, \qquad (25\text{-}19)$$

in which we have dropped the subscript f on V_f. Equation 25-19 gives us the potential V at any point f in the electric field *relative to the zero potential* at point i. If we let point i be at infinity, then Eq. 25-19 gives us the potential V at any point f relative to the zero potential at infinity.

✔**CHECKPOINT 3:**   The figure shows a family of parallel equipotential surfaces (in cross section) and five paths along which we shall move an electron from one surface to another. (a) What is the direction of the electric field associated with the surfaces? (b) For each path, is the work we do positive, negative, or zero? (c) Rank the paths according to the work we do, greatest first.

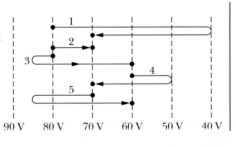

Sample Problem 25-2

(a) Figure 25-6a shows two points i and f in a uniform electric field $\vec{E}$. The points lie on the same electric field line (not shown) and are separated by a distance d. Find the potential difference $V_f - V_i$ by moving a positive test charge q_0 from i to f along the path shown, which is parallel to the field direction.

SOLUTION: The Key Idea here is that we can find the potential difference between any two points in an electric field by integrating $\vec{E} \cdot d\vec{s}$ along a path connecting those two points according to Eq. 25-18. We do this by mentally moving a test charge q_0 along that path, from initial point i to final point f. As we move such a test charge along the path in Fig. 25-6a, its differential displacement $d\vec{s}$ always has the same direction as $\vec{E}$. Thus, the angle θ between $\vec{E}$ and $d\vec{s}$ is zero and the dot product in Eq. 25-18 is

$$\vec{E} \cdot d\vec{s} = E \, ds \cos \theta = E \, ds. \qquad (25\text{-}20)$$

Equations 25-18 and 25-20 then give us

$$V_f - V_i = - \int_i^f \vec{E} \cdot d\vec{s} = - \int_i^f E \, ds. \qquad (25\text{-}21)$$

Since the field is uniform, E is constant over the path and can be moved outside the integral, giving us

$$V_f - V_i = -E \int_i^f ds = -Ed, \qquad \text{(Answer)}$$

in which the integral is simply the length d of the path. The minus sign in the result shows that the potential at point f in Fig. 25-6a is lower than the potential at point i. This is a general result: The potential always decreases along a path that extends in the direction of the electric field lines.

(b) Now find the potential difference $V_f - V_i$ by moving the positive test charge q_0 from i to f along the path icf shown in Fig. 25-6b.

SOLUTION: The Key Idea of (a) applies here too, except now we move the test charge along a path that consists of two lines: ic and cf. At all points along line ic, the displacement $d\vec{s}$ of the test charge is perpendicular to $\vec{E}$. Thus, the angle θ between $\vec{E}$ and $d\vec{s}$ is 90°, and the dot product $\vec{E} \cdot d\vec{s}$ is 0. Equation 25-18 then tells us that points i and c are at the same potential: $V_c - V_i = 0$.

For line cf we have $\theta = 45°$ and, from Eq. 25-18,

$$V_f - V_i = - \int_c^f \vec{E} \cdot d\vec{s} = - \int_c^f E(\cos 45°) \, ds$$

$$= -E(\cos 45°) \int_c^f ds.$$

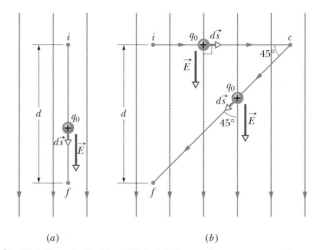

Fig. 25-6 Sample Problem 25-2. (a) A test charge q_0 moves in a straight line from point i to point f, along the direction of a uniform electric field. (b) Charge q_0 moves along path icf in the same electric field.

The integral in this equation is just the length of line cf; from Fig. 25-6b, that length is $d/\sin 45°$. Thus,

$$V_f - V_i = -E(\cos 45°) \frac{d}{\sin 45°} = -Ed. \qquad \text{(Answer)}$$

This is the same result we obtained in (a), as it must be; the potential difference between two points does not depend on the path connecting them. Moral: When you want to find the potential difference between two points by moving a test charge between them, you can save time and work by choosing a path that simplifies the use of Eq. 25-18.

25-5 Potential Due to a Point Charge

We will now use Eq. 25-18 to derive an expression for the electric potential V in the space around a charged particle, relative to the zero potential at infinity. Consider a point P at a distance R from a fixed particle of positive charge q (Fig. 25-7). To use Eq. 25-18, we imagine that we move a positive test charge q_0 from point P to infinity. Because the path we take does not matter, let us choose the simplest one— a line that extends radially from the fixed particle through P to infinity.

To use Eq. 25-18, we must evaluate the dot product

$$\vec{E} \cdot d\vec{s} = E \cos \theta \, ds. \qquad (25\text{-}22)$$

The electric field $\vec{E}$ in Fig. 25-7 is directed radially outward from the fixed particle. Thus, the differential displacement $d\vec{s}$ of the test particle along its path has the same direction as $\vec{E}$. That means that in Eq. 25-22, angle $\theta = 0$ and $\cos \theta = 1$. Because

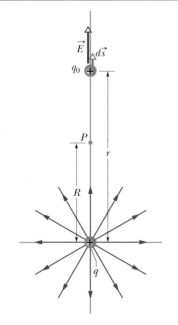

Fig. 25-7 The positive point charge q produces an electric field $\vec{E}$ and an electric potential V at point P. We find the potential by moving a test charge q_0 from P to infinity. The test charge is shown at distance r from the point charge, during differential displacement $d\vec{s}$.

$V(r)$

x

y

Fig. 25-8 A computer-generated plot of the electric potential $V(r)$ due to a positive point charge located at the origin of an xy plane. The potentials at points in that plane are plotted vertically. (Curved lines have been added to help you visualize the plot.) The infinite value of V predicted by Eq. 25-26 for $r = 0$ is not plotted.

the path is radial, let us write ds as dr. Then, substituting the limits R and ∞, we can write Eq. 25-18 as

$$V_f - V_i = -\int_R^\infty E \, dr. \tag{25-23}$$

Next, we set $V_f = 0$ (at ∞) and $V_i = V$ (at R). Then, for the magnitude of the electric field at the site of the test charge, we substitute from Eq. 23-3:

$$E = \frac{1}{4\pi\varepsilon_0} \frac{q}{r^2}. \tag{25-24}$$

With these changes, Eq. 25-23 then gives us

$$0 - V = -\frac{q}{4\pi\varepsilon_0} \int_R^\infty \frac{1}{r^2} \, dr = \frac{q}{4\pi\varepsilon_0} \left[\frac{1}{r}\right]_R^\infty$$

$$= -\frac{1}{4\pi\varepsilon_0} \frac{q}{R}. \tag{25-25}$$

Solving for V and switching R to r, we then have

$$V = \frac{1}{4\pi\varepsilon_0} \frac{q}{r} \tag{25-26}$$

as the electric potential V due to a particle of charge q, at any radial distance r from the particle.

Although we have derived Eq. 25-26 for a positively charged particle, the derivation holds also for a negatively charged particle, in which case, q is a negative quantity. Note that the sign of V is the same as the sign of q:

▶ A positively charged particle produces a positive electric potential. A negatively charged particle produces a negative electric potential.

Figure 25-8 shows a computer-generated plot of Eq. 25-26 for a positively charged particle; the magnitude of V is plotted vertically. Note that the magnitude increases as $r \to 0$. In fact, according to Eq. 25-26, V is infinite at $r = 0$, although Fig. 25-8 shows a finite, smoothed-off value there.

Equation 25-26 also gives the electric potential *outside or on the external surface of* a spherically symmetric charge distribution. We can prove this by using one of the shell theorems of Sections 22-4 and 24-9 to replace the actual spherical charge distribution with an equal charge concentrated at its center. Then the derivation leading to Eq. 25-26 follows, provided we do not consider a point within the actual distribution.

PROBLEM-SOLVING TACTICS

Tactic 3: *Finding a Potential Difference*
To find the potential difference ΔV between any two points in the field of an isolated point charge, we can evaluate Eq. 25-26 at each point and then subtract the results. The value of ΔV will be the same for any choice of reference potential energy because that choice is eliminated by the subtraction.

25-6 Potential Due to a Group of Point Charges

We can find the net potential at a point due to a group of point charges with the help of the superposition principle. We calculate the potential resulting from each charge at the given point separately, using Eq. 25-26 with the sign of the charge included. Then we sum the potentials. For n charges, the net potential is

$$V = \sum_{i=1}^{n} V_i = \frac{1}{4\pi\varepsilon_0} \sum_{i=1}^{n} \frac{q_i}{r_i} \qquad (n \text{ point charges}). \qquad (25\text{-}27)$$

Here q_i is the value of the ith charge, and r_i is the radial distance of the given point from the ith charge. The sum in Eq. 25-27 is an *algebraic sum*, not a vector sum like the sum that would be used to calculate the electric field resulting from a group of point charges. Herein lies an important computational advantage of potential over electric field: It is a lot easier to sum several scalar quantities than to sum several vector quantities whose directions and components must be considered.

✔**CHECKPOINT 4:** The figure here shows three arrangements of two protons. Rank the arrangements according to the net electric potential produced at point P by the protons, greatest first.

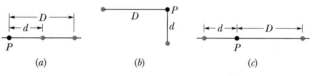

(a)   (b)   (c)

Sample Problem 25-3

What is the electric potential at point P, located at the center of the square of point charges shown in Fig. 25-9a? The distance d is 1.3 m, and the charges are

$$q_1 = +12 \text{ nC}, \qquad q_3 = +31 \text{ nC},$$

$$q_2 = -24 \text{ nC}, \qquad q_4 = +17 \text{ nC}.$$

SOLUTION: The Key Idea here is that the electric potential V at P is the algebraic sum of the electric potentials contributed by the four point charges. (Because electric potential is a scalar, the orientations of the point charges do not matter.) Thus, from Eq. 25-27, we have

$$V = \sum_{i=1}^{4} V_i = \frac{1}{4\pi\varepsilon_0} \left(\frac{q_1}{r} + \frac{q_2}{r} + \frac{q_3}{r} + \frac{q_4}{r} \right).$$

The distance r is $d/\sqrt{2}$, which is 0.919 m, and the sum of the charges is

$$q_1 + q_2 + q_3 + q_4 = (12 - 24 + 31 + 17) \times 10^{-9} \text{ C}$$
$$= 36 \times 10^{-9} \text{ C}.$$

Thus, $$V = \frac{(8.99 \times 10^9 \text{ N} \cdot \text{m}^2/\text{C}^2)(36 \times 10^{-9} \text{ C})}{0.919 \text{ m}}$$

$$\approx 350 \text{ V}. \qquad \text{(Answer)}$$

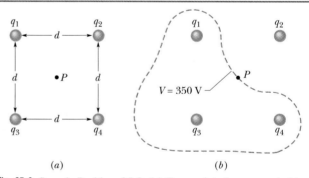

Fig. 25-9 Sample Problem 25-3. (*a*) Four point charges are held fixed at the corners of a square. (*b*) The closed curve is a cross section, in the plane of the figure, of the equipotential surface that contains point P. (The curve is only roughly drawn.)

Close to any of the three positive charges in Fig. 25-9a, the potential has very large positive values. Close to the single negative charge, the potential has very large negative values. Therefore, there must be points within the square that have the same intermediate potential as that at point P. The curve in Fig. 25-9b shows the intersection of the plane of the figure with the equipotential surface that contains point P. Any point along that curve has the same potential as point P.

Sample Problem 25-4

(a) In Fig. 25-10a, 12 electrons (of charge $-e$) are equally spaced and fixed around a circle of radius R. Relative to $V = 0$ at infinity, what are the electric potential and electric field at the center C of the circle due to these electrons?

SOLUTION: The Key Idea here is that the electric potential V at C is the algebraic sum of the electric potentials contributed by all the electrons. (Because electric potential is a scalar, the orientations of the electrons do not matter.) Because the electrons all have the same negative charge $-e$ and are all the same distance R from C, Eq. 25-27 gives us

$$V = -12 \frac{1}{4\pi\varepsilon_0} \frac{e}{R}. \qquad \text{(Answer)} \quad (25\text{-}28)$$

For the electric field at C, the Key Idea is that electric field is a vector quantity and thus the orientation of the electrons is important. Because of the symmetry of the arrangement in Fig. 25-10a, the electric field vector at C due to any given electron is canceled by the field vector due to the electron that is diametrically opposite it. Thus, at C,

$$\vec{E} = 0. \qquad \text{(Answer)}$$

(b) If the electrons are moved along the circle until they are non-

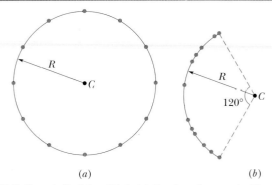

(a)   (b)

Fig. 25-10 Sample Problem 25-4. (a) Twelve electrons uniformly spaced around a circle. (b) Those electrons are now nonuniformly spaced along an arc of the original circle.

uniformly spaced over a 120° arc (Fig. 25-10b), what then is the potential at C? How does the electric field at C change (if at all)?

SOLUTION: The potential is still given by Eq. 25-28, because the distance between C and each electron is unchanged and orientation is irrelevant. The electric field is no longer zero, because the arrangement is no longer symmetric. There is now a net field that is directed toward the charge distribution.

25-7 Potential Due to an Electric Dipole

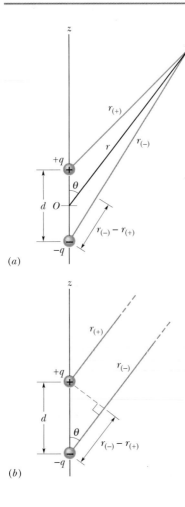

Now let us apply Eq. 25-27 to an electric dipole to find the potential at an arbitrary point P in Fig. 25-11a. At P, the positive point charge (at distance $r_{(+)}$) sets up potential $V_{(+)}$ and the negative point charge (at distance $r_{(-)}$) sets up potential $V_{(-)}$. Then the net potential at P is given by Eq. 25-27 as

$$V = \sum_{i=1}^{2} V_i = V_{(+)} + V_{(-)} = \frac{1}{4\pi\varepsilon_0} \left(\frac{q}{r_{(+)}} + \frac{-q}{r_{(-)}} \right)$$

$$= \frac{q}{4\pi\varepsilon_0} \frac{r_{(-)} - r_{(+)}}{r_{(-)}r_{(+)}}. \qquad (25\text{-}29)$$

Naturally occurring dipoles—such as those possessed by many molecules—are quite small, so we are usually interested only in points that are relatively far from the dipole, such that $r \gg d$, where d is the distance between the charges. Under those conditions, the approximations that follow from Fig. 25-11b are

$$r_{(-)} - r_{(+)} \approx d \cos\theta \quad \text{and} \quad r_{(-)}r_{(+)} \approx r^2.$$

If we substitute these quantities into Eq. 25-29, we can approximate V to be

$$V = \frac{q}{4\pi\varepsilon_0} \frac{d \cos\theta}{r^2},$$

where θ is measured from the dipole axis as shown in Fig. 25-11a. We can now

Fig. 25-11 (a) Point P is a distance r from the midpoint O of a dipole. The line OP makes an angle θ with the dipole axis. (b) If P is far from the dipole, the lines of lengths $r_{(+)}$ and $r_{(-)}$ are approximately parallel to the line of length r, and the dashed black line is approximately perpendicular to the line of length $r_{(-)}$.

write V as

$$V = \frac{1}{4\pi\varepsilon_0} \frac{p \cos\theta}{r^2} \qquad \text{(electric dipole)}, \qquad (25\text{-}30)$$

in which $p \; (= qd)$ is the magnitude of the electric dipole moment $\vec{p}$ defined in Section 23-5. The vector $\vec{p}$ is directed along the dipole axis, from the negative to the positive charge. (Thus, θ is measured from the direction of $\vec{p}$.)

✔**CHECKPOINT 5:** Suppose that three points are set at equal (large) distances r from the center of the dipole in Fig. 25-11: Point a is on the dipole axis above the positive charge, point b is on the axis below the negative charge, and point c is on a perpendicular bisector through the line connecting the two charges. Rank the points according to the electric potential of the dipole there, greatest (most positive) first.

(a)

(b)

Fig. 25-12 (a) An atom, showing the positively charged nucleus (green) and the negatively charged electrons (gold shading). The centers of positive and negative charge coincide. (b) If the atom is placed in an external electric field $\vec{E}$, the electron orbits are distorted so that the centers of positive and negative charge no longer coincide. An induced dipole moment $\vec{p}$ appears. The distortion is greatly exaggerated here.

Induced Dipole Moment

Many molecules such as water have *permanent* electric dipole moments. In other molecules (called *nonpolar molecules*) and in every isolated atom, the centers of the positive and negative charges coincide (Fig. 25-12a) and thus no dipole moment is set up. However, if we place an atom or a nonpolar molecule in an external electric field, the field distorts the electron orbits and separates the centers of positive and negative charge (Fig. 25-12b). Because the electrons are negatively charged, they tend to be shifted in a direction opposite the field. This shift sets up a dipole moment $\vec{p}$ that points in the direction of the field. This dipole moment is said to be *induced* by the field, and the atom or molecule is then said to be *polarized* by the field (it has a positive side and a negative side). When the field is removed, the induced dipole moment and the polarization disappear.

25-8 Potential Due to a Continuous Charge Distribution

When a charge distribution q is continuous (as on a uniformly charged thin rod or disk), we cannot use the summation of Eq. 25-27 to find the potential V at a point P. Instead, we must choose a differential element of charge dq, determine the potential dV at P due to dq, and then integrate over the entire charge distribution.

Let us again take the zero of potential to be at infinity. If we treat the element of charge dq as a point charge, then we can use Eq. 25-26 to express the potential dV at point P due to dq:

$$dV = \frac{1}{4\pi\varepsilon_0} \frac{dq}{r} \qquad \text{(positive or negative dq)}. \qquad (25\text{-}31)$$

Here r is the distance between P and dq. To find the total potential V at P, we integrate to sum the potentials due to all the charge elements:

$$V = \int dV = \frac{1}{4\pi\varepsilon_0} \int \frac{dq}{r}. \qquad (25\text{-}32)$$

The integral must be taken over the entire charge distribution. Note that because the electric potential is a scalar, there are *no vector components* to consider in Eq. 25-32.

We now examine two continuous charge distributions, a line of charge and a charged disk.

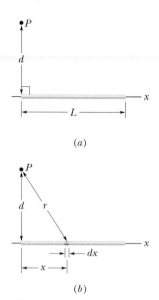

(a)

(b)

Fig. 25-13 (a) A thin, uniformly charged rod produces an electric potential V at point P. (b) An element of charge produces a differential potential dV at P.

Line of Charge

In Fig. 25-13a, a thin nonconducting rod of length L has a positive charge of uniform linear density λ. Let us determine the electric potential V due to the rod at point P, a perpendicular distance d from the left end of the rod.

We consider a differential element dx of the rod as shown in Fig. 25-13b. This (or any other) element of the rod has a differential charge of

$$dq = \lambda \, dx. \tag{25-33}$$

This element produces an electric potential dV at point P, which is a distance $r = (x^2 + d^2)^{1/2}$ from the element. Treating the element as a point charge, we can use Eq. 25-31 to write the potential dV as

$$dV = \frac{1}{4\pi\varepsilon_0} \frac{dq}{r} = \frac{1}{4\pi\varepsilon_0} \frac{\lambda \, dx}{(x^2 + d^2)^{1/2}}. \tag{25-34}$$

Since the charge on the rod is positive and we have taken $V = 0$ at infinity, we know from Section 25-5 that dV in Eq. 25-34 must be positive.

We now find the total potential V produced by the rod at point P by integrating Eq. 25-34 along the length of the rod, from $x = 0$ to $x = L$, using integral 17 in Appendix E. We find

$$
\begin{aligned}
V = \int dV &= \int_0^L \frac{1}{4\pi\varepsilon_0} \frac{\lambda}{(x^2 + d^2)^{1/2}} \, dx \\
&= \frac{\lambda}{4\pi\varepsilon_0} \int_0^L \frac{dx}{(x^2 + d^2)^{1/2}} \\
&= \frac{\lambda}{4\pi\varepsilon_0} \left[\ln\left(x + (x^2 + d^2)^{1/2} \right) \right]_0^L \\
&= \frac{\lambda}{4\pi\varepsilon_0} \left[\ln\left(L + (L^2 + d^2)^{1/2} \right) - \ln d \right].
\end{aligned}
$$

We can simplify this result by using the general relation $\ln A - \ln B = \ln(A/B)$. We then find

$$V = \frac{\lambda}{4\pi\varepsilon_0} \ln\left[\frac{L + (L^2 + d^2)^{1/2}}{d} \right]. \tag{25-35}$$

Because V is the sum of positive values of dV, it should be positive—but does Eq. 25-35 give a positive V? Since the argument of the logarithm is greater than one, the logarithm is a positive number and V is indeed positive.

Charged Disk

In Section 23-7, we calculated the magnitude of the electric field at points on the central axis of a plastic disk of radius R that has a uniform charge density σ on one surface. Here we derive an expression for $V(z)$, the electric potential at any point on the central axis.

In Fig. 25-14, consider a differential element consisting of a flat ring of radius R' and radial width dR'. Its charge has magnitude

$$dq = \sigma(2\pi R')(dR'),$$

in which $(2\pi R')(dR')$ is the upper surface area of the ring. All parts of this charged element are the same distance r from point P on the disk's axis. With the aid of Fig. 25-14, we can use Eq. 25-31 to write the contribution of this ring to the electric

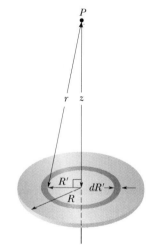

Fig. 25-14 A plastic disk of radius R, charged on its top surface to a uniform surface charge density σ. We wish to find the potential V at point P on the central axis of the disk.

potential at P as

$$dV = \frac{1}{4\pi\varepsilon_0} \frac{dq}{r} = \frac{1}{4\pi\varepsilon_0} \frac{\sigma(2\pi R')(dR')}{\sqrt{z^2 + R'^2}}. \tag{25-36}$$

We find the net potential at P by adding (via integration) the contributions of all the strips from $R' = 0$ to $R' = R$:

$$V = \int dV = \frac{\sigma}{2\varepsilon_0} \int_0^R \frac{R'\, dR'}{\sqrt{z^2 + R'^2}} = \frac{\sigma}{2\varepsilon_0}(\sqrt{z^2 + R^2} - z). \tag{25-37}$$

Note that the variable in the second integral of Eq. 25-37 is R' and not z, which remains constant while the integration over the surface of the disk is carried out. (Note also that, in evaluating the integral, we have assumed that $z \geq 0$.)

PROBLEM-SOLVING TACTICS

Tactic 4: *Signs of Trouble with Electric Potential*
When you calculate the potential V at some point P due to a line of charge or any other continuous charge configuration, the signs can cause you trouble. Here is a generic guide to sort out the signs.

If the charge is negative, should the symbols dq and λ represent negative quantities, or should you explicitly show the signs, using $-dq$ and $-\lambda$? You can do either as long as you remember what your notation means, so that when you get to the final step, you can correctly interpret the sign of V.

Another approach, which can be used when the entire charge distribution is of a single sign, is to let the symbols dq and λ represent magnitudes only. The result of the calculation will give

you the magnitude of V at P. Then add a sign to V based on the sign of the charge. (If the zero potential is at infinity, positive charge gives a positive potential and negative charge gives a negative potential.)

If you happen to reverse the limits on the integral used to calculate a potential, you will obtain a negative value for V. The magnitude will be correct, but discard the minus sign. Then determine the proper sign for V from the sign of the charge. As an example, we would have obtained a minus sign in Eq. 25-35 if we had reversed the limits in the integral above that equation. We would then have discarded that minus sign and noted that the potential is positive because the charge producing it is positive.

25-9 Calculating the Field from the Potential

In Section 25-4, you saw how to find the potential at a point f if you know the electric field along a path from a reference point to point f. In this section, we propose to go the other way—that is, to find the electric field when we know the potential. As Fig. 25-3 shows, solving this problem graphically is easy: If we know the potential V at all points near an assembly of charges, we can draw in a family of equipotential surfaces. The electric field lines, sketched perpendicular to those surfaces, reveal the variation of $\vec{E}$. What we are seeking here is the mathematical equivalent of this graphical procedure.

Figure 25-15 shows cross sections of a family of closely spaced equipotential surfaces, the potential difference between each pair of adjacent surfaces being dV. As the figure suggests, the field $\vec{E}$ at any point P is perpendicular to the equipotential surface through P.

Suppose that a positive test charge q_0 moves through a displacement $\vec{ds}$ from one equipotential surface to the adjacent surface. From Eq. 25-7, we see that the work the electric field does on the test charge during the move is $-q_0\, dV$. From Eq. 25-16 and Fig. 25-15, we see that the work done by the electric field may also be written as the scalar product $(q_0\vec{E}) \cdot \vec{ds}$, or $q_0 E(\cos\theta)\, ds$. Equating these two expressions for the work yields

$$-q_0\, dV = q_0 E(\cos\theta)\, ds, \tag{25-38}$$

or

$$E\cos\theta = -\frac{dV}{ds}. \tag{25-39}$$

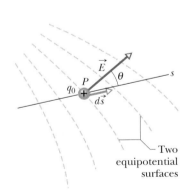

Fig. 25-15 A test charge q_0 moves a distance $\vec{ds}$ from one equipotential surface to another. (The separation between the surfaces has been exaggerated for clarity.) The displacement $\vec{ds}$ makes an angle θ with the direction of the electric field $\vec{E}$.

Since $E \cos \theta$ is the component of $\vec{E}$ in the direction of $d\vec{s}$, Eq. 25-39 becomes

$$E_s = -\frac{\partial V}{\partial s}. \qquad (25\text{-}40)$$

We have added a subscript to E and switched to the partial derivative symbols to emphasize that Eq. 25-40 involves only the variation of V along a specified axis (here called the s axis) and only the component of $\vec{E}$ along that axis. In words, Eq. 25-40 (which is essentially the inverse of Eq. 25-18) states:

▶ The component of $\vec{E}$ in any direction is the negative of the rate of change of the electric potential with distance in that direction.

If we take the s axis to be, in turn, the x, y, and z axes, we find that the x, y, and z components of $\vec{E}$ at any point are

$$E_x = -\frac{\partial V}{\partial x}; \qquad E_y = -\frac{\partial V}{\partial y}; \qquad E_z = -\frac{\partial V}{\partial z}. \qquad (25\text{-}41)$$

Thus, if we know V for all points in the region around a charge distribution—that is, if we know the function $V(x, y, z)$—we can find the components of $\vec{E}$, and thus $\vec{E}$ itself, at any point by taking partial derivatives.

For the simple situation in which the electric field $\vec{E}$ is uniform, Eq. 25-40 becomes

$$E = -\frac{\Delta V}{\Delta s}, \qquad (25\text{-}42)$$

where s is perpendicular to the equipotential surfaces. The component of the electric field is zero in any direction parallel to the equipotential surfaces.

✔CHECKPOINT 6:   The figure shows three pairs of parallel plates with the same separation, and the electric potential of each plate. The electric field between the plates is uniform and perpendicular to the plates. (a) Rank the pairs according to the magnitude of the electric field between the plates, greatest first. (b) For which pair is the electric field pointing rightward? (c) If an electron is released midway between the third pair of plates, does it remain there, move rightward at constant speed, move leftward at constant speed, accelerate rightward, or accelerate leftward?

−50 V +150 V   −20 V +200 V   −200 V −400 V
    (1)   (2)   (3)

Sample Problem 25-5

The electric potential at any point on the central axis of a uniformly charged disk is given by Eq. 25-37,

$$V = \frac{\sigma}{2\varepsilon_0}(\sqrt{z^2 + R^2} - z).$$

Starting with this expression, derive an expression for the electric field at any point on the axis of the disk.

SOLUTION: We want the electric field $\vec{E}$ as a function of distance z along the axis of the disk. For any value of z, the direction of $\vec{E}$ must be along that axis because the disk has circular symmetry about that axis. Thus, we want the component E_z of $\vec{E}$ in the direction of z. Then the Key Idea is that this component is the negative of the rate of change of the electric potential with distance z. Thus, from the last of Eqs. 25-41, we can write

$$E_z = -\frac{\partial V}{\partial z} = -\frac{\sigma}{2\varepsilon_0}\frac{d}{dz}(\sqrt{z^2 + R^2} - z)$$

$$= \frac{\sigma}{2\varepsilon_0}\left(1 - \frac{z}{\sqrt{z^2 + R^2}}\right). \qquad \text{(Answer)}$$

This is the same expression that we derived in Section 23-7 by integration, using Coulomb's law.

25-10 Electric Potential Energy of a System of Point Charges

In Section 25-1, we discussed the electric potential energy of a charged particle as an electrostatic force does work on it. In that section, we assumed that the charges that produced the force were fixed in place, so that neither the force nor the corresponding electric field could be influenced by the presence of the test charge. In this section we can take a broader view, to find the electric potential energy of a *system* of charges due to the electric field produced *by* those same charges.

For a simple example, suppose you push together two bodies that have charges of the same electrical sign. The work that you must do is stored as electric potential energy in the two-body system (provided the kinetic energy of the bodies does not change). If you later release the charges, you can recover this stored energy, in whole or in part, as kinetic energy of the charged bodies as they rush away from each other.

We define the electric potential energy *of a system of point charges*, held in fixed positions by forces not specified, as follows:

> The electric potential energy of a system of fixed point charges is equal to the work that must be done by an external agent to assemble the system, bringing each charge in from an infinite distance.

We assume that the charges are stationary both in their initial infinitely distant positions and in their final assembled configuration.

Figure 25-16 shows two point charges q_1 and q_2, separated by a distance r. To find the electric potential energy of this two-charge system, we must mentally build the system, starting with both charges infinitely far away and at rest. When we bring q_1 in from infinity and put it in place we do no work, because no electrostatic force acts on q_1. However, when we next bring q_2 in from infinity and put it in place, we must do work, because q_1 exerts an electrostatic force on q_2 during the move.

We can calculate that work with Eq. 25-8 by dropping the minus sign (so that the equation gives the work *we* do rather than the field's work) and substituting q_2 for the general charge q. Our work is then equal to q_2V, where V is the potential that has been set up by q_1 at the point where we put q_2. From Eq. 25-26, that potential is

$$V = \frac{1}{4\pi\varepsilon_0} \frac{q_1}{r}.$$

Thus, from our definition, the electric potential energy of the pair of point charges of Fig. 25-16 is

$$U = W = q_2V = \frac{1}{4\pi\varepsilon_0} \frac{q_1 q_2}{r}. \tag{25-43}$$

If the charges have the same sign, we have to do positive work to push them together against their mutual repulsion. Hence, as Eq. 25-43 shows, the potential energy of the system is then positive. If the charges have opposite signs, we have to do negative work against their mutual attraction to bring them together if they are to be stationary. The potential energy of the system is then negative. Sample Problem 25-6 shows how to extend this process to more than two charges.

Fig. 25-16 Two charges held a fixed distance r apart.

Sample Problem 25-6

Figure 25-17 shows three point charges held in fixed positions by forces that are not shown. What is the electric potential energy U of this system of charges? Assume that $d = 12$ cm and that

$$q_1 = +q, \quad q_2 = -4q, \quad \text{and} \quad q_3 = +2q,$$

in which $q = 150$ nC.

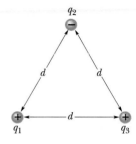

Fig. 25-17 Sample Problem 25-6. Three charges are fixed at the vertices of an equilateral triangle. What is the electric potential energy of the system?

SOLUTION: The Key Idea here is that the potential energy U of the system is equal to the work we must do to assemble the system, bringing in each charge from an infinite distance. Therefore, let's mentally build the system of Fig. 25-17, starting with one of the point charges, say q_1, in place and the others at infinity. Then we bring another one, say q_2, in from infinity and put it in place. From Eq. 25-43 with d substituted for r, the potential energy U_{12} associated with the pair of point charges q_1 and q_2 is

$$U_{12} = \frac{1}{4\pi\varepsilon_0} \frac{q_1 q_2}{d}.$$

We then bring the last point charge q_3 in from infinity and put it in place. The work that we must do in this last step is equal to the sum of the work we must do to bring q_3 near q_1 and the work we must do to bring it near q_2. From Eq. 25-43, with d substituted for r, that sum is

$$W_{13} + W_{23} = U_{13} + U_{23} = \frac{1}{4\pi\varepsilon_0} \frac{q_1 q_3}{d} + \frac{1}{4\pi\varepsilon_0} \frac{q_2 q_3}{d}.$$

The total potential energy U of the three-charge system is the sum of the potential energies associated with the three pairs of charges.

This sum (which is actually independent of the order in which the charges are brought together) is

$$
\begin{aligned}
U &= U_{12} + U_{13} + U_{23} \\
&= \frac{1}{4\pi\varepsilon_0} \left(\frac{(+q)(-4q)}{d} + \frac{(+q)(+2q)}{d} + \frac{(-4q)(+2q)}{d} \right) \\
&= -\frac{10q^2}{4\pi\varepsilon_0 d} \\
&= -\frac{(8.99 \times 10^9 \ \text{N} \cdot \text{m}^2/\text{C}^2)(10)(150 \times 10^{-9} \ \text{C})^2}{0.12 \ \text{m}} \\
&= -1.7 \times 10^{-2} \ \text{J} = -17 \ \text{mJ}. \qquad \text{(Answer)}
\end{aligned}
$$

The negative potential energy means that negative work would have to be done to assemble this structure, starting with the three charges infinitely separated and at rest. Put another way, an external agent would have to do 17 mJ of work to disassemble the structure completely, ending with the three charges infinitely far apart.

25-11 Potential of a Charged Isolated Conductor

In Section 24-6, we concluded that $\vec{E} = 0$ for all points inside an isolated conductor. We then used Gauss' law to prove that an excess charge placed on an isolated conductor lies entirely on its surface. (This is true even if the conductor has an empty internal cavity.) Here we use the first of these facts to prove an extension of the second:

> An excess charge placed on an isolated conductor will distribute itself on the surface of that conductor so that all points of the conductor—whether on the surface or inside—come to the same potential. This is true even if the conductor has an internal cavity and even if that cavity contains a net charge.

Our proof follows directly from Eq. 25-18, which is

$$V_f - V_i = -\int_i^f \vec{E} \cdot d\vec{s}.$$

Since $\vec{E} = 0$ for all points within a conductor, it follows directly that $V_f = V_i$ for all possible pairs of points i and f in the conductor.

Figure 25-18a is a plot of potential against radial distance r from the center for an isolated spherical conducting shell of 1.0 m radius, having a charge of 1.0 μC.

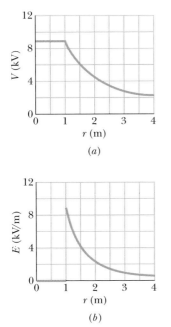

Fig. 25-18 (a) A plot of V(r) both inside and outside a charged spherical shell of radius 1.0 m. (b) A plot of E(r) for the same shell.

For points outside the shell, we can calculate $V(r)$ from Eq. 25-26 because the charge q behaves for such external points as if it were concentrated at the center of the shell. That equation holds right up to the surface of the shell. Now let us push a small test charge through the shell—assuming a small hole exists—to its center. No extra work is needed to do this because no net electric force acts on the test charge once it is inside the shell. Thus, the potential at all points inside the shell has the same value as that on the surface, as Fig. 25-18a shows.

Figure 25-18b shows the variation of electric field with radial distance for the same shell. Note that $E = 0$ everywhere inside the shell. The curves of Fig. 25-18b can be derived from the curve of Fig. 25-18a by differentiating with respect to r, using Eq. 25-40 (recall that the derivative of any constant is zero). The curve of Fig. 25-18a can be derived from the curves of Fig. 25-18b by integrating with respect to r, using Eq. 25-19.

On nonspherical conductors, a surface charge does not distribute itself uniformly over the surface of the conductor. At sharp points or edges, the surface charge density—and thus the external electric field, which is proportional to it—may reach very high values. The air around such sharp points may become ionized, producing the corona discharge that golfers and mountaineers see on the tips of bushes, golf clubs, and rock hammers when thunderstorms threaten. Such corona discharges, like hair that stands on end, are often the precursors of lightning strikes. In such circumstances, it is wise to enclose yourself in a cavity inside a conducting shell, where the electric field is guaranteed to be zero. A car (unless it is a convertible or made with a plastic body) is almost ideal (Fig. 25-19).

If an isolated conductor is placed in an *external electric field,* as in Fig. 25-20, all points of the conductor still come to a single potential regardless of whether the conductor has an excess charge. The free conduction electrons distribute themselves on the surface in such a way that the electric field they produce at interior points cancels the external electric field that would otherwise be there. Furthermore, the electron distribution causes the net electric field at all points on the surface to be perpendicular to the surface. If the conductor in Fig. 25-20 could be somehow removed, leaving the surface charges frozen in place, the pattern of the electric field would remain absolutely unchanged, for both exterior and interior points.

Fig. 25-19 A large spark jumps to a car's body and then exits by moving across the insulating left front tire (note the flash there), leaving the person inside unharmed.

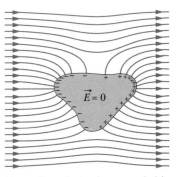

Fig. 25-20 An uncharged conductor is suspended in an external electric field. The free electrons in the conductor distribute themselves on the surface as shown, so as to reduce the net electric field inside the conductor to zero and make the net field at the surface perpendicular to the surface.

REVIEW & SUMMARY

Electric Potential Energy The change ΔU in the electric potential energy U of a point charge as the charge moves from an initial point i to a final point f in an electric field is

$$\Delta U = U_f - U_i = -W, \qquad (25\text{-}1)$$

where W is the work done by the electrostatic force (due to the electric field) on the point charge during the move from i to f. If the potential energy is defined to be zero at infinity, the **electric potential energy** U of the point charge at a particular point is

$$U = -W_\infty. \qquad (25\text{-}2)$$

Here W_∞ is the work done by the electrostatic force on the point charge as the charge moves from infinity to the particular point.

Electric Potential Difference and Electric Potential We define the **potential difference** ΔV between two points i and f in an electric field as

$$\Delta V = V_f - V_i = -\frac{W}{q}, \qquad (25\text{-}7)$$

where q is the charge of a particle on which work is done by the field. The **potential** at a point is

$$V = -\frac{W_\infty}{q}. \qquad (25\text{-}8)$$

The SI unit of potential is the *volt:* 1 volt = 1 joule per coulomb.

Potential and potential difference can also be written in terms of the electric potential energy U of a particle of charge q in an electric field:

$$V = \frac{U}{q}, \qquad (25\text{-}5)$$

$$\Delta V = V_f - V_i = \frac{U_f}{q} - \frac{U_i}{q} = \frac{\Delta U}{q}. \qquad (25\text{-}6)$$

Equipotential Surfaces The points on an **equipotential surface** all have the same electric potential. The work done on a test charge in moving it from one such surface to another is independent of the locations of the initial and final points on these surfaces and of the path that joins the points. The electric field $\vec{E}$ is always directed perpendicularly to corresponding equipotential surfaces.

Finding V from $\vec{E}$ The electric potential difference between two points i and f is

$$V_f - V_i = -\int_i^f \vec{E} \cdot d\vec{s}, \qquad (25\text{-}18)$$

where the integral is taken over any path connecting the points. If we choose $V_i = 0$ we have, for the potential at a particular point,

$$V = -\int_i^f \vec{E} \cdot d\vec{s}. \qquad (25\text{-}19)$$

Potential Due to Point Charges The electric potential due to a single point charge at a distance r from that point charge is

$$V = \frac{1}{4\pi\varepsilon_0} \frac{q}{r}. \qquad (25\text{-}26)$$

V has the same sign as q. The potential due to a collection of point charges is

$$V = \sum_{i=1}^n V_i = \frac{1}{4\pi\varepsilon_0} \sum_{i=1}^n \frac{q_i}{r_i}. \qquad (25\text{-}27)$$

Potential Due to an Electric Dipole At a distance r from an electric dipole with dipole moment magnitude $p = qd$, the electric potential of the dipole is

$$V = \frac{1}{4\pi\varepsilon_0} \frac{p \cos\theta}{r^2} \qquad (25\text{-}30)$$

for $r \gg d$; the angle θ is defined in Fig. 25-11.

Potential Due to a Continuous Charge Distribution For a continuous distribution of charge, Eq. 25-27 becomes

$$V = \frac{1}{4\pi\varepsilon_0} \int \frac{dq}{r}, \qquad (25\text{-}32)$$

in which the integral is taken over the entire distribution.

Calculating $\vec{E}$ from V The component of $\vec{E}$ in any direction is the negative of the rate of change of the potential with distance in that direction:

$$E_s = -\frac{\partial V}{\partial s}. \qquad (25\text{-}40)$$

The x, y, and z components of $\vec{E}$ may be found from

$$E_x = -\frac{\partial V}{\partial x}; \quad E_y = -\frac{\partial V}{\partial y}; \quad E_z = -\frac{\partial V}{\partial z}. \qquad (25\text{-}41)$$

When $\vec{E}$ is uniform, Eq. 25-40 reduces to

$$E = -\frac{\Delta V}{\Delta s}, \qquad (25\text{-}42)$$

where s is perpendicular to the equipotential surfaces. The electric field is zero in the direction parallel to an equipotential surface.

Electric Potential Energy of a System of Point Charges The electric potential energy of a system of point charges is equal to the work needed to assemble the system with the charges initially at rest and infinitely distant from each other. For two charges at separation r,

$$U = W = \frac{1}{4\pi\varepsilon_0} \frac{q_1 q_2}{r}. \qquad (25\text{-}43)$$

Potential of a Charged Conductor An excess charge placed on a conductor will, in the equilibrium state, be located entirely on the outer surface of the conductor. The charge will distribute itself so that the entire conductor, including interior points, is at a uniform potential.

QUESTIONS

1. Figure 25-21 shows three paths along which we can move positively charged sphere *A* closer to positively charged sphere *B*, which is fixed in place. (a) Would sphere *A* be moved to a higher or lower electric potential? Is the work done (b) by our force and (c) by the electric field (due to the second sphere) positive, negative, or zero? (d) Rank the paths according to the work our force does, greatest first.

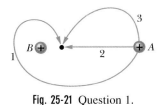

Fig. 25-21 Question 1.

2. Figure 25-22 shows four pairs of charged particles. Let $V = 0$ at infinity. For which pairs is there another point of zero net electric potential *on the axis shown,* (a) between the particles and (b) to their right? (c) Where such a zero potential point exists, is the net electric field $\vec{E}$ due to the particles equal to zero? (d) For each pair, are there points off the axis (other than at infinity, of course) where $V = 0$?

Fig. 25-22 Questions 2 and 8.

3. Figure 25-23 shows a square array of charged particles, with distance *d* between adjacent particles. What is the electric potential at point *P* at the center of the square if the electric potential is zero at infinity?

4. Figure 25-24 shows four arrangements of charged particles, all the same distance from the origin. Rank the situations according to the net electric potential at the origin, most positive first. Take the potential to be zero at infinity.

Fig. 25-23 Question 3.

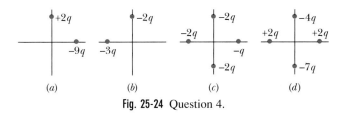

(a)   *(b)*   *(c)*   *(d)*

Fig. 25-24 Question 4.

5. (a) In Fig. 25-25*a*, what is the potential at point *P* due to charge *Q* at distance *R* from *P*? Set $V = 0$ at infinity. (b) In Fig. 25-25*b*, the same charge *Q* has been spread uniformly over a circular arc of radius *R* and central angle 40°. What is the potential at point *P*,

the center of curvature of the arc? (c) In Fig. 25-25*c*, the same charge *Q* has been spread uniformly over a circle of radius *R*. What is the potential at point *P*, the center of the circle? (d) Rank the three situations according to the magnitude of the electric field that is set up at *P*, greatest first.

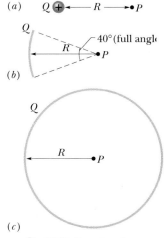

Fig. 25-25 Question 5.

6. Figure 25-26 shows three sets of cross sections of equipotential surfaces; all three cover the same size region of space. (a) Rank the arrangements according to the magnitude of the electric field present in the region, greatest first. (b) In which is the electric field directed down the page?

Fig. 25-26 Question 6.

7. Figure 25-27 gives the electric potential *V* as a function of *x*. (a) Rank the five regions according to the magnitude of the *x* component of the electric field within them, greatest first. What is the direction of the field along the *x* axis in (b) region 2 and (c) region 4?

8. Figure 25-22 shows four pairs of charged particles with identical separations. (a) Rank the pairs according to their electric potential energy, greatest (most positive) first. (b) For each pair, if the separation between the particles is increased, does the potential energy of the pair increase or decrease?

Fig. 25-27 Question 7.

9. Figure 25-28 shows a system of three charged particles. If you move the particle of charge $+q$ from point *A* to point *D*, are the following positive, negative, or zero: (a) the change in the electric potential energy of the three-particle system, (b) the work done by the net electrostatic force on the particle you moved, and (c) the work done by your force? (d) What are the answers to (a) through (c) if, instead, the move is from point *B* to point *C*?

10. In the situation of Question 9, is the work done by your force positive, negative, or zero if the move is (a) from *A* to *B*, (b) from *A* to *C*, and (c) from *B* to *D*? (d) Rank those moves according to the magnitude of the work done by your force, greatest first.

Fig. 25-28 Questions 9 and 10.

EXERCISES & PROBLEMS

ssm Solution is in the Student Solutions Manual.
www Solution is available on the World Wide Web at:
http://www.wiley.com/college/hrw
ilw Solution is available on the Interactive LearningWare.

SEC. 25-2 Electric Potential

1E. A particular 12 V car battery can send a total charge of 84 A · h (ampere-hours) through a circuit, from one terminal to the other. (a) How many coulombs of charge does this represent? (*Hint:* See Eq. 22-3.) (b) If this entire charge undergoes a potential difference of 12 V, how much energy is involved? ssm

2E. The electric potential difference between the ground and a cloud in a particular thunderstorm is 1.2×10^9 V. What is the magnitude of the change in the electric potential energy (in multiples of the electron-volt) of an electron that moves between the ground and the cloud?

3P. In a given lightning flash, the potential difference between a cloud and the ground is 1.0×10^9 V and the quantity of charge transferred is 30 C. (a) What is the decrease in energy of that transferred charge? (b) If all that energy could be used to accelerate a 1000 kg automobile from rest, what would be the automobile's final speed? (c) If the energy could be used to melt ice, how much ice would it melt at 0°C? The heat of fusion of ice is 3.33×10^5 J/kg. ssm

SEC. 25-4 Calculating the Potential from the Field

4E. When an electron moves from A to B along an electric field line in Fig. 25-29, the electric field does 3.94×10^{-19} J of work on it. What are the electric potential differences (a) $V_B - V_A$, (b) $V_C - V_A$, and (c) $V_C - V_B$?

5E. An infinite nonconducting sheet has a surface charge density $\sigma = 0.10 \ \mu C/m^2$ on one side. How far apart are equipotential surfaces whose potentials differ by 50 V? ssm

Fig. 25-29 Exercise 4.

6E. Two large, parallel, conducting plates are 12 cm apart and have charges of equal magnitude and opposite sign on their facing surfaces. An electrostatic force of 3.9×10^{-15} N acts on an electron placed anywhere between the two plates. (Neglect fringing.) (a) Find the electric field at the position of the electron. (b) What is the potential difference between the plates?

7P. A Geiger counter has a metal cylinder 2.00 cm in diameter along whose axis is stretched a wire 1.30×10^{-4} cm in diameter. If the potential difference between the wire and the cylinder is 850 V, what is the electric field at the surface of (a) the wire and (b) the cylinder? (*Hint:* Use the result of Problem 23 of Chapter 24.) ssm

8P. The electric field inside a nonconducting sphere of radius R, with charge spread uniformly throughout its volume, is radially directed and has magnitude

$$E(r) = \frac{qr}{4\pi\varepsilon_0 R^3}.$$

Here q (positive or negative) is the total charge within the sphere, and r is the distance from the sphere's center. (a) Taking $V = 0$ at the center of the sphere, find the electric potential $V(r)$ inside the sphere. (b) What is the difference in electric potential between a point on the surface and the sphere's center? (c) If q is positive, which of those two points is at the higher potential?

9P*. A charge q is distributed uniformly throughout a spherical volume of radius R. (a) Setting $V = 0$ at infinity, show that the potential at a distance r from the center, where $r < R$, is given by

$$V = \frac{q(3R^2 - r^2)}{8\pi\varepsilon_0 R^3}.$$

(*Hint:* See Section 24-9.) (b) Why does this result differ from that in (a) of Problem 8? (c) What is the potential difference between a point on the surface and the sphere's center? (d) Why doesn't this result differ from that of (b) of Problem 8? ssm

10P. Figure 25-30 shows, edge-on, an infinite nonconducting sheet with positive surface charge density σ on one side. (a) Use Eq. 25-18 and Eq. 24-13 to show that the electric potential of an infinite sheet of charge can be written $V = V_0 - (\sigma/2\varepsilon_0)z$, where V_0 is the electric potential at the surface of the sheet and z is the perpendicular distance from the sheet. (b) How much work is done by the electric field of the sheet as a small positive test charge q_0 is moved from an initial position on the sheet to a final position located a distance z from the sheet?

Fig. 25-30 Problem 10.

11P*. A thick spherical shell of charge Q and uniform volume charge density ρ is bounded by radii r_1 and r_2, where $r_2 > r_1$. With $V = 0$ at infinity, find the electric potential V as a function of the distance r from the center of the distribution, considering the regions (a) $r > r_2$, (b) $r_2 > r > r_1$, and (c) $r < r_1$. (d) Do these solutions agree at $r = r_2$ and $r = r_1$? (*Hint:* See Section 24-9.) ssm

SEC. 25-6 Potential Due to a Group of Point Charges

12E. As a space shuttle moves through the dilute ionized gas of Earth's ionosphere, its potential is typically changed by -1.0 V during one revolution. By assuming that the shuttle is a sphere of radius 10 m, estimate the amount of charge it collects.

13E. Consider a point charge $q = 1.0 \ \mu C$, point A at distance $d_1 = 2.0$ m from q, and point B at distance $d_2 = 1.0$ m. (a) If these points are diametrically opposite each other, as in Fig. 25-31a, what is the electric potential difference $V_A - V_B$? (b) What is that electric potential difference if points A and B are located as in Fig. 25-31b?

Fig. 25-31 Exercise 13.

14E. Figure 25-32 shows two charged particles on an axis. Sketch the electric field lines and the equipotential surfaces in the plane of the page for (a) $q_1 = +q$ and $q_2 = +2q$ and (b) $q_1 = +q$ and $q_2 = -3q$.

15E. In Fig. 25-32, set $V = 0$ at infinity and let the particles have charges $q_1 = +q$ and $q_2 = -3q$. Then locate (in terms of the separation distance d) any point on the x axis (other than at infinity) at which the net potential due to the two particles is zero.

Fig. 25-32 Exercises 14, 15, and 16.

16E. Two particles, of charges q_1 and q_2, are separated by distance d in Fig. 25-32. The net electric field of the particles is zero at $x = d/4$. With $V = 0$ at infinity, locate (in terms of d) any point on the x axis (other than at infinity) at which the electric potential due to the two particles is zero.

17P. A spherical drop of water carrying a charge of 30 pC has a potential of 500 V at its surface (with $V = 0$ at infinity). (a) What is the radius of the drop? (b) If two such drops of the same charge and radius combine to form a single spherical drop, what is the potential at the surface of the new drop? ssm ilw www

18P. What are (a) the charge and (b) the charge density on the surface of a conducting sphere of radius 0.15 m whose potential is 200 V (with $V = 0$ at infinity)?

19P. An electric field of approximately 100 V/m is often observed near the surface of Earth. If this were the field over the entire surface, what would be the electric potential of a point on the surface? (Set $V = 0$ at infinity.) ssm

20P. In Fig. 25-33, point P is at the center of the rectangle. With $V = 0$ at infinity, what is the net electric potential at P due to the six charged particles?

Fig. 25-33 Problem 20.

21P. In Fig. 25-34, what is the net potential at point P due to the four point charges, if $V = 0$ at infinity? ssm

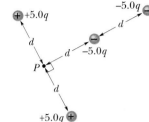

Fig. 25-34 Problem 21.

SEC. 25-7 Potential Due to an Electric Dipole

22E. The ammonia molecule NH_3 has a permanent electric dipole moment equal to 1.47 D, where 1 D = 1 debye unit = 3.34×10^{-30} C·m. Calculate the electric potential due to an ammonia molecule at a point 52.0 nm away along the axis of the dipole. (Set $V = 0$ at infinity.) ilw

23P. Figure 25-35 shows three charged particles located on a horizontal axis. For points (such as P) on the axis with $r \gg d$, show that the electric potential $V(r)$ is given by

$$V = \frac{1}{4\pi\varepsilon_0} \frac{q}{r}\left(1 + \frac{2d}{r}\right).$$

(Hint: The charge configuration can be viewed as the sum of an isolated charge and a dipole.) ssm www

Fig. 25-35 Problem 23.

SEC. 25-8 Potential Due to a Continuous Charge Distribution

24E. (a) Figure 25-36a shows a positively charged plastic rod of length L and uniform linear charge density λ. Setting $V = 0$ at infinity and considering Fig. 25-13 and Eq. 25-35, find the electric potential at point P without written calculation. (b) Figure 25-36b shows an identical rod, except that it is split in half and the right half is negatively charged; the left and right halves have the same magnitude λ of uniform linear charge density. With V still zero at infinity, what is the electric potential at point P in Fig. 25-36b?

Fig. 25-36 Exercise 24.

25E. A plastic rod has been formed into a circle of radius R. It has a positive charge +Q uniformly distributed along one-quarter of its circumference and a negative charge of $-6Q$ uniformly distributed along the rest of the circumference (Fig. 25-37). With $V = 0$ at infinity, what is the electric potential (a) at the center C of the circle and (b) at point P, which is on the central axis of the circle at distance z from the center? ssm

Fig. 25-37 Exercise 25.

26E. In Fig. 25-38, a plastic rod having a uniformly distributed charge $-Q$ has been bent into a circular arc of radius R and central angle $120°$. With $V = 0$ at infinity, what is the electric potential at P, the center of curvature of the rod?

27E. A plastic disk is charged on one side with a uniform surface charge density σ, and then three quadrants of the disk are removed. The remaining quadrant is shown in Fig. 25-39. With $V = 0$ at infinity, what is the potential due to the remaining quadrant at point P, which is on the central axis of the original disk at a distance z from the original center? ssm

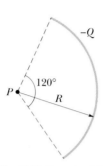

Fig. 25-38 Exercise 26.

28P. Figure 25-40 shows a plastic rod of length L and uniform positive charge Q lying on an x axis. With $V = 0$ at infinity, find the electric potential at point P_1 on the axis, at distance d from one end of the rod.

29P. The plastic rod shown in Fig. 25-40 has length L and a nonuniform linear charge density $\lambda = cx$, where c is a positive constant. With $V = 0$ at infinity, find the electric potential at point P_1 on the axis, at distance d from one end.

Fig. 25-39 Exercise 27.

Fig. 25-40 Problems 28, 29, 34, and 35.

SEC. 25-9 Calculating the Field from the Potential

30E. Two large parallel metal plates are 1.5 cm apart and have equal but opposite charges on their facing surfaces. Take the potential of the negative plate to be zero. If the potential halfway between the plates is then $+5.0$ V, what is the electric field in the region between the plates?

31E. The electric potential at points in an xy plane is given by $V = (2.0 \text{ V/m}^2)x^2 - (3.0 \text{ V/m}^2)y^2$. What are the magnitude and direction of the electric field at the point $(3.0 \text{ m}, 2.0 \text{ m})$?

32E. The electric potential V in the space between two flat parallel plates is given by $V = 1500x^2$, where V is in volts if x, the distance

from one of the plates, is in meters. Calculate the magnitude and direction of the electric field at $x = 1.3$ cm.

33P. (a) Using Eq. 25-32, show that the electric potential at a point on the central axis of a thin ring of charge of radius R and a distance z from the ring is

$$V = \frac{1}{4\pi\varepsilon_0} \frac{q}{\sqrt{z^2 + R^2}}.$$

(b) From this result, derive an expression for E at points on the ring's axis; compare your result with the calculation of E in Section 23-6. ssm www

34P. The plastic rod of length L in Fig. 25-40 has the nonuniform linear charge density $\lambda = cx$, where c is a positive constant. (a) With $V = 0$ at infinity, find the electric potential at point P_2 on the y axis, a distance y from one end of the rod. (b) From that result, find the electric field component E_y at P_2. (c) Why cannot the field component E_x at P_2 be found using the result of (a)?

35P. (a) Use the result of Problem 28 to find the electric field component E_x at point P_1 in Fig. 25-40. (*Hint:* First substitute the variable x for the distance d in the result.) (b) Use symmetry to determine the electric field component E_y at P_1. ssm

SEC. 25-10 Electric Potential Energy of a System of Point Charges

36E. (a) What is the electric potential energy of two electrons separated by 2.00 nm? (b) If the separation increases, does the potential energy increase or decrease?

37E. Derive an expression for the work required to set up the four-charge configuration of Fig. 25-41, assuming the charges are initially infinitely far apart. ilw

38E. What is the electric potential energy of the charge configuration of Fig. 25-9a? Use the numerical values provided in Sample Problem 25-3.

39P. In the rectangle of Fig. 25-42, the sides have lengths 5.0 cm and 15 cm, $q_1 = -5.0$ μC, and $q_2 = +2.0$ μC. With $V = 0$ at infinity, what are the electric potentials (a) at corner A and (b) at

Fig. 25-41 Exercise 37.

Fig. 25-42 Problem 39.

corner B? (c) How much work is required to move a third charge $q_3 = +3.0$ μC from B to A along a diagonal of the rectangle? (d) Does this work increase or decrease the electric energy of the three-charge system? Is more, less, or the same work required if q_3 is moved along paths that are (e) inside the rectangle but not on a diagonal and (f) outside the rectangle? ssm

40P. In Fig. 25-43, how much work is required to bring the charge of $+5q$ in from infinity along the dashed line and place it

Fig. 25-43 Problem 40.

as shown near the two fixed charges $+4q$ and $-2q$? Take distance $d = 1.40$ cm and charge $q = 1.6 \times 10^{-19}$ C.

41P. A particle of positive charge Q is fixed at point P. A second particle of mass m and negative charge $-q$ moves at constant speed in a circle of radius r_1, centered at P. Derive an expression for the work W that must be done by an external agent on the second particle to increase the radius of the circle of motion to r_2. ssm www

42P. Calculate (a) the electric potential established by the nucleus of a hydrogen atom at the average distance ($r = 5.29 \times 10^{-11}$ m) of the atom's electron (take $V = 0$ at infinite distance), (b) the electric potential energy of the atom when the electron is at this radius, and (c) the kinetic energy of the electron, assuming it to be moving in a circular orbit of this radius centered on the nucleus. (d) How much energy is required to ionize the hydrogen atom (that is, to remove the electron from the nucleus so that the separation is effectively infinite)? Express all energies in electron-volts.

43P. A particle of charge q is fixed at point P, and a second particle of mass m and the same charge q is initially held a distance r_1 from P. The second particle is then released. Determine its speed when it is a distance r_2 from P. Let $q = 3.1$ μC, $m = 20$ mg, $r_1 = 0.90$ mm, and $r_2 = 2.5$ mm. ilw

44P. A charge of -9.0 nC is uniformly distributed around a thin plastic ring of radius 1.5 m that lies in the yz plane with its center at the origin. A point charge of -6.0 pC is located on the x axis at $x = 3.0$ m. Calculate the work done on the point charge by an external force to move the point charge to the origin.

45P. Two tiny metal spheres A and B of mass $m_A = 5.00$ g and $m_B = 10.0$ g have equal positive charges $q = 5.00$ μC. The spheres are connected by a massless nonconducting string of length $d = 1.00$ m, which is much greater than the radii of the spheres. (a) What is the electric potential energy of the system? (b) Suppose you cut the string. At that instant, what is the acceleration of each sphere? (c) A long time after you cut the string, what is the speed of each sphere? ssm

46P. A thin, spherical, conducting shell of radius R is mounted on an isolating support and charged to a potential of $-V$. An electron is then fired from point P at distance r from the center of the shell ($r \gg R$) with initial speed v_0 and directly toward the shell's center. What value of v_0 is needed for the electron to just reach the shell before reversing direction?

47P. Two electrons are fixed 2.0 cm apart. Another electron is shot from infinity and stops midway between the two. What is its initial speed? ssm

48P. Two charged, parallel, flat conducting surfaces are spaced $d = 1.00$ cm apart and produce a potential difference $\Delta V = 625$ V between them. An electron is projected from one surface directly toward the second. What is the initial speed of the electron if it stops just at the second surface?

49P. An electron is projected with an initial speed of 3.2×10^5 m/s directly toward a proton that is fixed in place. If the electron is initially a great distance from the proton, at what distance from the

proton is the speed of the electron instantaneously equal to twice the initial value?

SEC. 25-11 Potential of a Charged Isolated Conductor

50E. An empty hollow metal sphere has a potential of $+400$ V with respect to ground (defined to be at $V = 0$) and has a charge of 5.0×10^{-9} C. Find the electric potential at the center of the sphere.

51E. What is the excess charge on a conducting sphere of radius $r = 0.15$ m if the potential of the sphere is 1500 V and $V = 0$ at infinity? ssm

52E. Consider two widely separated conducting spheres, 1 and 2, the second having twice the diameter of the first. The smaller sphere initially has a positive charge q, and the larger one is initially uncharged. You now connect the spheres with a long thin wire. (a) How are the final potentials V_1 and V_2 of the spheres related? (b) What are the final charges q_1 and q_2 on the spheres, in terms of q? (c) What is the ratio of the final surface charge density of sphere 1 to that of sphere 2?

53P. Two metal spheres, each of radius 3.0 cm, have a center-to-center separation of 2.0 m. One has a charge of $+1.0 \times 10^{-8}$ C; the other has a charge of -3.0×10^{-8} C. Assume that the separation is large enough relative to the size of the spheres to permit us to consider the charge on each to be uniformly distributed (the spheres do not affect each other). With $V = 0$ at infinity, calculate (a) the potential at the point halfway between their centers and (b) the potential of each sphere. ssm

54P. A charged metal sphere of radius 15 cm has a net charge of 3.0×10^{-8} C. (a) What is the electric field at the sphere's surface? (b) If $V = 0$ at infinity, what is the electric potential at the sphere's surface? (c) At what distance from the sphere's surface has the electric potential decreased by 500 V?

55P. (a) If Earth had a net surface charge density of 1.0 electron per square meter (a very artificial assumption), what would its potential be? (Set $V = 0$ at infinity.) (b) What would be the electric field due to Earth just outside its surface?

56P. Two thin, isolated, concentric conducting spheres of radii R_1 and R_2 (with $R_1 < R_2$) have charges q_1 and q_2. With $V = 0$ at infinity, derive expressions for $E(r)$ and $V(r)$, where r is distance from the center of the spheres. Plot $E(r)$ and $V(r)$ from $r = 0$ to $r = 4.0$ m for $R_1 = 0.50$ m, $R_2 = 1.0$ m, $q_1 = +2.0$ μC, and $q_2 = +1.0$ μC.

Additional Problem

57. *The chocolate crumb mystery.* This story begins with Problem 48 in Chapter 24. (a) From the answer to part (a) of that problem, find an expression for the electric potential as a function of the radial distance r from the center of the pipe. (The electric potential is zero on the grounded pipe wall.) (b) For the typical volume charge density $\rho = -1.1 \times 10^{-3}$ C/m^3, what is the difference in the electric potential between the pipe's center and its inside wall? (The story continues with Problem 48 in Chapter 26.)

26 Capacitance

During ventricular fibrillation, a common type of heart attack, the chambers of the heart fail to pump blood because their muscle fibers randomly contract and relax. To save a victim of ventricular fibrillation, the heart muscle must be shocked to reestablish its normal rhythm. For that, 20 A of current must be sent through the chest cavity to transfer 200 J of electrical energy in about 2.0 ms. This requires about 100 kW of electric power. Such a requirement may easily be met in a hospital, but not by, say, the electrical system of an ambulance arriving to help the victim.

What, then, can provide the power needed for defibrillation at remote locations?

The answer is in this chapter.

Fig. 26-1 An assortment of capacitors.

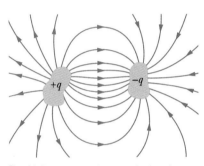

Fig. 26-2 Two conductors, isolated electrically from each other and from their surroundings, form a *capacitor*. When the capacitor is charged, the conductors, or *plates* as they are called, have equal but opposite charges of magnitude q.

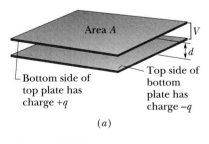

Area A

Bottom side of top plate has charge $+q$

Top side of bottom plate has charge $-q$

(a)

Electric field lines

(b)

26-1 The Uses of Capacitors

You can store energy as potential energy by pulling a bowstring, stretching a spring, compressing a gas, or lifting a book. You can also store energy as potential energy in an electric field, and a **capacitor** is a device you can use to do exactly that.

There is a capacitor in a portable battery-operated photoflash unit, for example. It accumulates charge relatively slowly during the readying process between flashes, building up an electric field as it does so. It holds this field and the associated energy until the energy is rapidly released to initiate the flash.

Capacitors have many uses in our electronic and microelectronic age beyond serving as storehouses for potential energy. As one example, they are vital elements in the circuits with which we tune radio and television transmitters and receivers. As another example, microscopic capacitors form the memory banks of computers. These tiny devices are not as important for their stored energy as for the ON–OFF information that the presence or absence of their electric fields provides.

26-2 Capacitance

Figure 26-1 shows some of the many sizes and shapes of capacitors. Figure 26-2 shows the basic elements of *any* capacitor—two isolated conductors of any shape. No matter what their geometry, flat or not, we call these conductors *plates*.

Figure 26-3a shows a less general but more conventional arrangement, called a *parallel-plate capacitor*, consisting of two parallel conducting plates of area A separated by a distance d. The symbol that we use to represent a capacitor (⊣⊢) is based on the structure of a parallel-plate capacitor but is used for capacitors of all geometries. We assume for the time being that no material medium (such as glass or plastic) is present in the region between the plates. In Section 26-6, we shall remove this restriction.

When a capacitor is *charged*, its plates have equal but opposite charges of $+q$ and $-q$. However, we refer to the *charge of a capacitor* as being q, the absolute value of these charges on the plates. (Note that q is not the net charge on the capacitor, which is zero.)

Because the plates are conductors, they are equipotential surfaces; all points on a plate are at the same electric potential. Moreover, there is a potential difference between the two plates. For historical reasons, we represent the absolute value of this potential difference with V rather than with ΔV as we would with previous notation.

The charge q and the potential difference V for a capacitor are proportional to each other; that is,

$$q = CV. \qquad (26\text{-}1)$$

The proportionality constant C is called the **capacitance** of the capacitor. Its value depends only on the geometry of the plates and *not* on their charge or potential difference. The capacitance is a measure of how much charge must be put on the

Fig. 26-3 (a) A parallel-plate capacitor, made up of two plates of area A separated by a distance d. The plates have equal and opposite charges of magnitude q on their facing surfaces. (b) As the field lines show, the electric field due to the charged plates is uniform in the central region between the plates. The field is not uniform at the edges of the plates, as indicated by the 'fringing' of the field lines there.

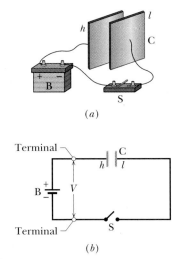

(a)

Terminal

C

h l

B   V

Terminal   S

(b)

Fig. 26-4 (a) Battery B, switch S, and plates h and l of capacitor C, connected in a circuit. (b) A schematic diagram with the *circuit elements* represented by their symbols.

plates to produce a certain potential difference between them: The *greater the capacitance, the more charge is required.*

The SI unit of capacitance that follows from Eq 26-1 is the coulomb per volt. This unit occurs so often that it is given a special name, the *farad* (F):

$$1 \text{ farad} = 1 \text{ F} = 1 \text{ coulomb per volt} = 1 \text{ C/V.} \qquad (26\text{-}2)$$

As you will see, the farad is a very large unit. Submultiples of the farad, such as the microfarad (1 μF = 10^{-6} F) and the picofarad (1 pF = 10^{-12} F), are more convenient units in practice.

Charging a Capacitor

One way to charge a capacitor is to place it in an electric circuit with a battery. An *electric circuit* is a path through which charge can flow. A *battery* is a device that maintains a certain potential difference between its *terminals* (points at which charge can enter or leave the battery) by means of internal electrochemical reactions in which electric forces can move internal charge.

In Fig. 26-4a, a battery B, a switch S, an uncharged capacitor C, and interconnecting wires form a circuit. The same circuit is shown in the *schematic diagram* of Fig. 26-4b, in which the symbols for a battery, a switch, and a capacitor represent those devices. The battery maintains potential difference V between its terminals. The terminal of higher potential is labeled $+$ and is often called the *positive* terminal; the terminal of lower potential is labeled $-$ and is often called the *negative* terminal.

The circuit shown in Figs. 26-4a and b is said to be *incomplete* because switch S is *open*; that is, it does not electrically connect the wires attached to it. When the switch is *closed,* electrically connecting those wires, the circuit is complete and charge can then flow through the switch and the wires. As we discussed in Chapter 22, the charge that can flow through a conductor, such as a wire, is that of electrons. When the circuit of Fig. 26-4 is completed, electrons are driven through the wires by an electric field that the battery sets up in the wires. The field drives electrons from capacitor plate h to the positive terminal of the battery; thus, plate h, losing electrons, becomes positively charged. The field drives just as many electrons from the negative terminal of the battery to capacitor plate l; thus, plate l, gaining electrons, becomes negatively charged *just as much* as plate h, losing electrons, becomes positively charged.

Initially, when the plates are uncharged, the potential difference between them is zero. As the plates become oppositely charged, that potential difference increases until it equals the potential difference V between the terminals of the battery. Then plate h and the positive terminal of the battery are at the same potential, and there is no longer an electric field in the wire between them. Similarly, plate l and the negative terminal reach the same potential and there is then no electric field in the wire between them. Thus, with the field zero, there is no further drive of electrons. The capacitor is then said to be *fully charged,* with a potential difference V and charge q that are related by Eq. 26-1.

In this book we assume that during the charging of a capacitor and afterward, charge cannot pass from one plate to the other across the gap separating them. Also, we assume that a capacitor can retain (or *store*) charge indefinitely, until it is put into a circuit where it can be *discharged.*

✔**CHECKPOINT 1:**   Does the capacitance C of a capacitor increase, decrease, or remain the same (a) when the charge q on it is doubled and (b) when the potential difference V across it is tripled?

Tactic 1: *The Symbol V and Potential Difference*

In previous chapters, the symbol V represents an electric potential at a point or along an equipotential surface. However, in matters concerning electrical devices, V often represents a *potential difference* between two points or two equipotential surfaces. Equation 26-1 is an example of this second use of the symbol. In Section 26-3, you will see a mixture of the two meanings of V. There and in later chapters, you need to be alert as to the intent of this symbol.

You will also be seeing, in this book and elsewhere, a variety of phrases regarding potential difference. A potential difference or a "potential" or a "voltage" may be *applied* to a device, or it may be *across* a device. A capacitor can be charged to a potential difference, as in "a capacitor is charged to 12 V." Also, a battery can be characterized by the potential difference across it, as in "a 12 V battery." Always keep in mind what is meant by such phrases: There is a potential difference between two points, such as two points in a circuit or at the terminals of a device such as a battery.

26-3 Calculating the Capacitance

Our task here is to calculate the capacitance of a capacitor once we know its geometry. Because we will consider a number of different geometries, it seems wise to develop a general plan to simplify the work. In brief our plan is as follows: (1) Assume a charge q on the plates; (2) calculate the electric field $\vec{E}$ between the plates in terms of this charge, using Gauss' law; (3) knowing $\vec{E}$, calculate the potential difference V between the plates from Eq. 25-18; (4) calculate C from Eq. 26-1.

Before we start, we can simplify the calculation of both the electric field and the potential difference by making certain assumptions. We discuss each in turn.

Calculating the Electric Field

To relate the electric field $\vec{E}$ between the plates of a capacitor to the charge q on either plate, we shall use Gauss' law:

$$\varepsilon_0 \oint \vec{E} \cdot d\vec{A} = q. \tag{26-3}$$

Here q is the charge enclosed by a Gaussian surface, and $\oint \vec{E} \cdot d\vec{A}$ is the net electric flux through that surface. In all cases that we shall consider, the Gaussian surface will be such that whenever electric flux passes through it, $\vec{E}$ will have a uniform magnitude E and the vectors $\vec{E}$ and $d\vec{A}$ will be parallel. Equation 26-3 then will reduce to

$$q = \varepsilon_0 E A \qquad \text{(special case of Eq. 26-3),} \tag{26-4}$$

in which A is the area of that part of the Gaussian surface through which flux passes. For convenience, we shall always draw the Gaussian surface in such a way that it completely encloses the charge on the positive plate; see Fig. 26-5 for an example.

Calculating the Potential Difference

In the notation of Chapter 25 (Eq. 25-18), the potential difference between the plates of a capacitor is related to the field $\vec{E}$ by

$$V_f - V_i = - \int_i^f \vec{E} \cdot d\vec{s}, \tag{26-5}$$

in which the integral is to be evaluated along any path that starts on one plate and ends on the other. We shall always choose a path that follows an electric field line, from the negative plate to the positive plate. For this path, the vectors $\vec{E}$ and $d\vec{s}$ will have opposite directions, so the dot product $\vec{E} \cdot d\vec{s}$ will be equal to $-E\,ds$. Thus,

Fig. 26-5 A charged parallel-plate capacitor. A Gaussian surface encloses the charge on the positive plate. The integration of Eq. 26-6 is taken along a path extending directly from the negative plate to the positive plate.

the right side of Eq. 26-5 will then be positive. Letting V represent the difference $V_f - V_i$, we can then recast Eq. 26-5 as

$$V = \int_-^+ E \, ds \qquad \text{(special case of Eq. 26-5)}, \qquad (26\text{-}6)$$

in which the − and + remind us that our path of integration starts on the negative plate and ends on the positive plate.

We are now ready to apply Eqs. 26-4 and 26-6 to some particular cases.

A Parallel-Plate Capacitor

We assume, as Fig. 26-5 suggests, that the plates of our parallel-plate capacitor are so large and so close together that we can neglect the fringing of the electric field at the edges of the plates, taking $\vec{E}$ to be constant throughout the region between the plates.

We draw a Gaussian surface that encloses just the charge q on the positive plate, as in Fig. 26-5. From Eq. 26-4 we can then write

$$q = \varepsilon_0 E A, \qquad (26\text{-}7)$$

where A is the area of the plate.

Equation 26-6 yields

$$V = \int_-^+ E \, ds = E \int_0^d ds = Ed. \qquad (26\text{-}8)$$

In Eq. 26-8, E can be placed outside the integral because it is a constant; the second integral then is simply the plate separation d.

If we now substitute q from Eq. 26-7 and V from Eq. 26-8 into the relation $q = CV$ (Eq. 26-1), we find

$$C = \frac{\varepsilon_0 A}{d} \qquad \text{(parallel-plate capacitor)}. \qquad (26\text{-}9)$$

Thus, the capacitance does indeed depend only on geometrical factors—namely, the plate area A and the plate separation d. Note that C increases as we increase the plate area A or decrease the separation d.

As an aside, we point out that Eq. 26-9 suggests one of our reasons for writing the electrostatic constant in Coulomb's law in the form $1/4\pi\varepsilon_0$. If we had not done so, Eq. 26-9—which is used more often in engineering practice than Coulomb's law—would have been less simple in form. We note further that Eq. 26-9 permits us to express the permittivity constant ε_0 in a unit more appropriate for use in problems involving capacitors; namely,

$$\varepsilon_0 = 8.85 \times 10^{-12} \text{ F/m} = 8.85 \text{ pF/m}. \qquad (26\text{-}10)$$

We have previously expressed this constant as

$$\varepsilon_0 = 8.85 \times 10^{-12} \text{ C}^2/\text{N} \cdot \text{m}^2. \qquad (26\text{-}11)$$

Total charge $+q$ Total charge $-q$

Path of integration

Gaussian surface

Fig. 26-6 A cross section of a long cylindrical capacitor, showing a cylindrical Gaussian surface of radius r (that encloses the positive plate) and the radial path of integration along which Eq. 26-6 is to be applied. This figure also serves to illustrate a spherical capacitor in a cross section through its center.

A Cylindrical Capacitor

Figure 26-6 shows, in cross section, a cylindrical capacitor of length L formed by two coaxial cylinders of radii a and b. We assume that $L \gg b$ so that we can neglect the fringing of the electric field that occurs at the ends of the cylinders. Each plate contains a charge of magnitude q.

As a Gaussian surface, we choose a cylinder of length L and radius r, closed by end caps and placed as is shown in Fig. 26-6. Equation 26-4 then yields

$$q = \varepsilon_0 EA = \varepsilon_0 E(2\pi rL),$$

in which $2\pi rL$ is the area of the curved part of the Gaussian surface. There is no flux through the end caps. Solving for E yields

$$E = \frac{q}{2\pi\varepsilon_0 Lr}. \qquad (26\text{-}12)$$

Substitution of this result into Eq. 26-6 yields

$$V = \int_-^+ E\, ds = -\frac{q}{2\pi\varepsilon_0 L} \int_b^a \frac{dr}{r} = \frac{q}{2\pi\varepsilon_0 L} \ln\!\left(\frac{b}{a}\right), \qquad (26\text{-}13)$$

where we have used the fact that here $ds = -dr$ (we integrated radially inward). From the relation $C = q/V$, we then have

$$C = 2\pi\varepsilon_0 \frac{L}{\ln(b/a)} \qquad \text{(cylindrical capacitor)}. \qquad (26\text{-}14)$$

We see that the capacitance of a cylindrical capacitor, like that of a parallel-plate capacitor, depends only on geometrical factors, in this case L, b, and a.

A Spherical Capacitor

Figure 26-6 can also serve as a central cross section of a capacitor that consists of two concentric spherical shells, of radii a and b. As a Gaussian surface we draw a sphere of radius r concentric with the two shells; then Eq. 26-4 yields

$$q = \varepsilon_0 EA = \varepsilon_0 E(4\pi r^2),$$

in which $4\pi r^2$ is the area of the spherical Gaussian surface. We solve this equation for E, obtaining

$$E = \frac{1}{4\pi\varepsilon_0} \frac{q}{r^2}, \qquad (26\text{-}15)$$

which we recognize as the expression for the electric field due to a uniform spherical charge distribution (Eq. 24-15).

If we substitute this expression into Eq. 26-6, we find

$$V = \int_-^+ E\, ds = -\frac{q}{4\pi\varepsilon_0} \int_b^a \frac{dr}{r^2} = \frac{q}{4\pi\varepsilon_0} \left(\frac{1}{a} - \frac{1}{b}\right) = \frac{q}{4\pi\varepsilon_0} \frac{b-a}{ab}, \qquad (26\text{-}16)$$

where again we have substituted $-dr$ for ds. If we now substitute Eq. 26-16 into Eq. 26-1 and solve for C, we find

$$C = 4\pi\varepsilon_0 \frac{ab}{b-a} \qquad \text{(spherical capacitor)}. \qquad (26\text{-}17)$$

An Isolated Sphere

We can assign a capacitance to a *single* isolated spherical conductor of radius R by assuming that the "missing plate" is a conducting sphere of infinite radius. After all, the field lines that leave the surface of a positively charged isolated conductor must end somewhere; the walls of the room in which the conductor is housed can serve effectively as our sphere of infinite radius.

To find the capacitance of the isolated conductor, we first rewrite Eq. 26-17 as

$$C = 4\pi\varepsilon_0 \frac{a}{1 - a/b}.$$

If we then let $b \rightarrow \infty$ and substitute R for a, we find

$$C = 4\pi\varepsilon_0 R \qquad \text{(isolated sphere).} \qquad (26\text{-}18)$$

Note that this formula and the others we have derived for capacitance (Eqs. 26-9, 26-14, and 26-17) involve the constant ε_0 multiplied by a quantity that has the dimensions of a length.

✔ **CHECKPOINT 2:**   For capacitors charged by the same battery, does the charge stored by the capacitor increase, decrease, or remain the same in each of the following situations? (a) The plate separation of a parallel-plate capacitor is increased. (b) The radius of the inner cylinder of a cylindrical capacitor is increased. (c) The radius of the outer spherical shell of a spherical capacitor is increased.

Sample Problem 26-1

A storage capacitor on a random access memory (RAM) chip has a capacitance of 55 fF. If the capacitor is charged to 5.3 V, how many excess electrons are on its negative plate?

SOLUTION: One Key Idea here is that we can find the number n of excess electrons on the negative plate if we know q, the total *amount* of excess charge on that plate. Then $n = q/e$, where e is the magnitude of the charge on each electron. A second Key Idea is that q is related to the potential difference V to which the capacitor

is charged, according to Eq. 26-1 ($q = CV$). Combining these two ideas then gives us

$$n = \frac{q}{e} = \frac{CV}{e} = \frac{(55 \times 10^{-15}\ \text{F})(5.3\ \text{V})}{1.60 \times 10^{-19}\ \text{C}}$$
$$= 1.8 \times 10^6\ \text{electrons.} \qquad \text{(Answer)}$$

For electrons, this is a very small number. A speck of household dust, so tiny that it essentially never settles, contains about 10^{17} electrons (and the same number of protons).

26-4 Capacitors in Parallel and in Series

When there is a combination of capacitors in a circuit, we can sometimes replace that combination with an **equivalent capacitor**—that is, a single capacitor that has the same capacitance as the actual combination of capacitors. With such a replacement, we can simplify the circuit, affording easier solutions for unknown quantities of the circuit. Here we discuss two basic combinations of capacitors that allow such a replacement.

Capacitors in Parallel

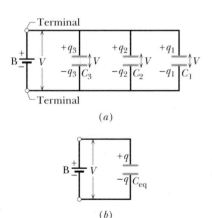

Fig. 26-7 (a) Three capacitors connected in parallel to battery B. The battery maintains potential difference V across its terminals and thus across *each* capacitor. (b) The equivalent capacitor, with capacitance C_{eq}, replaces the parallel combination.

Figure 26-7a shows an electric circuit in which three capacitors are connected *in parallel* to battery B. This description has little to do with how the capacitor plates are drawn. Rather, "in parallel" means that the capacitors are directly wired together at one plate and directly wired together at the other plate, and that a potential difference V is applied across the two groups of wired-together plates. Thus, each capacitor has the same potential difference V, which produces charge on the capacitor. (In Fig. 26-7a, the applied potential V is maintained by the battery.) In general,

▶ When a potential difference V is applied across several capacitors connected in parallel, that potential difference V is applied across each capacitor. The total charge q stored on the capacitors is the sum of the charges stored on all the capacitors.

When we analyze a circuit of capacitors in parallel, we can simplify it with this mental replacement:

> Capacitors connected in parallel can be replaced with an equivalent capacitor that has the same *total* charge q and the same potential difference V as the actual capacitors.

(You might remember this result with the nonsense word "par-V," which is close to "party.") Figure 26-7b shows the equivalent capacitor (with equivalent capacitance C_{eq}) that has replaced the three capacitors (with actual capacitances C_1, C_2, and C_3) of Fig. 26-7a.

To derive an expression for C_{eq} in Fig. 26-7b, we first use Eq. 26-1 to find the charge on each actual capacitor:

$$q_1 = C_1 V, \quad q_2 = C_2 V, \quad \text{and} \quad q_3 = C_3 V.$$

The total charge on the parallel combination of Fig. 26-7a is then

$$q = q_1 + q_2 + q_3 = (C_1 + C_2 + C_3)V.$$

The equivalent capacitance, with the same total charge q and applied potential difference V as the combination, is then

$$C_{eq} = \frac{q}{V} = C_1 + C_2 + C_3,$$

a result that we can easily extend to any number n of capacitors, as

$$C_{eq} = \sum_{j=1}^{n} C_j \qquad (n \text{ capacitors in parallel}). \tag{26-19}$$

Thus, to find the equivalent capacitance of a parallel combination, we simply add the individual capacitances.

Capacitors in Series

Figure 26-8a shows three capacitors connected *in series* to battery B. This description has little to do with how the capacitors are drawn. Rather, "in series" means that the capacitors are wired serially, one after the other, and that a potential difference V is applied across the two ends of the series. (In Fig. 26-8a, this potential difference V is maintained by battery B.) The potential differences that then exist across the capacitors in the series produce identical charges q on them.

> When a potential difference V is applied across several capacitors connected in series, the capacitors have identical charges q. The sum of the potential differences across all the capacitors is equal to the applied potential difference V.

We can explain how the capacitors end up with identical charges by following a *chain reaction* of events, in which the charging of each capacitor causes the charging of the next capacitor. We start with capacitor 3 and work upward to capacitor 1. When the battery is first connected to the series of capacitors, it produces charge $-q$ on the bottom plate of capacitor 3. That charge then repels negative charge from the top plate of capacitor 3 (leaving it with charge $+q$). The repelled negative charge moves to the bottom plate of capacitor 2 (giving it charge $-q$). That charge on the bottom plate of capacitor 2 then repels negative charge from the top plate of capacitor 2 (leaving it with charge $+q$) to the bottom plate of capacitor 1 (giving it charge $-q$). Finally the charge on the bottom plate of capacitor 1 helps move negative

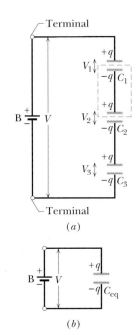

Fig. 26-8 (a) Three capacitors connected in series to battery B. The battery maintains potential difference V between the top and bottom plates of the series combination. (b) The equivalent capacitor, with capacitance C_{eq}, replaces the series combination.

charge from the top plate of capacitor 1 to the battery, leaving that top plate with charge $+q$.

Here are two important points about capacitors in series:

1. When charge is shifted from one capacitor to another in a series of capacitors, it can move along only one route, such as from capacitor 3 to capacitor 2 in Fig. 26-8a. If there are additional routes, the capacitors are not in series. An example is given in Sample Problem 26-2.

2. The battery directly produces charges on only the two plates to which it is connected (the bottom plate of capacitor 3 and the top plate of capacitor 1 in Fig. 26-8a). Charges that are produced on the other plates are due merely to the shifting of charge already there. For example, in Fig. 26-8, the part of the circuit enclosed by dashed lines is electrically isolated from the rest of the circuit. Thus, the net charge of that part cannot be changed by the battery—its charge can only be redistributed.

When we analyze a circuit of capacitors in series, we can simplify it with this mental replacement:

> ► Capacitors that are connected in series can be replaced with an equivalent capacitor that has the same charge q and the same *total* potential difference V as the actual series capacitors.

(You might remember this with the nonsense word "seri-q.") Figure 26-8b shows the equivalent capacitor (with equivalent capacitance C_{eq}) that has replaced the three actual capacitors (with actual capacitances C_1, C_2, and C_3) of Fig. 26-8a.

To derive an expression for C_{eq} in Fig. 26-8b, we first use Eq. 26-1 to find the potential difference of each actual capacitor:

$$V_1 = \frac{q}{C_1}, \quad V_2 = \frac{q}{C_2}, \quad \text{and} \quad V_3 = \frac{q}{C_3}.$$

The total potential difference V due to the battery is the sum of these three potential differences. Thus,

$$V = V_1 + V_2 + V_3 = q\left(\frac{1}{C_1} + \frac{1}{C_2} + \frac{1}{C_3}\right).$$

The equivalent capacitance is then

$$C_{eq} = \frac{q}{V} = \frac{1}{1/C_1 + 1/C_2 + 1/C_3},$$

or

$$\frac{1}{C_{eq}} = \frac{1}{C_1} + \frac{1}{C_2} + \frac{1}{C_3}.$$

We can easily extend this to any number n of capacitors as

$$\frac{1}{C_{eq}} = \sum_{j=1}^{n} \frac{1}{C_j} \qquad (n \text{ capacitors in series}). \tag{26-20}$$

Using Eq. 26-20 you can show that the equivalent of a series of capacitances is always *less* than the least capacitance in the series.

✔**CHECKPOINT 3:**   A battery of potential V stores charge q on a combination of two identical capacitors. What are the potential difference across and the charge on either capacitor if the capacitors are (a) in parallel and (b) in series?

Sample Problem 26-2

(a) Find the equivalent capacitance for the combination of capacitances shown in Fig. 26-9a, across which potential difference V is applied. Assume

$$C_1 = 12.0 \ \mu F, \quad C_2 = 5.30 \ \mu F, \quad \text{and} \quad C_3 = 4.50 \ \mu F.$$

SOLUTION: The Key Idea here is that any capacitors connected in series can be replaced with their equivalent capacitor, and any capacitors connected in parallel can be replaced with their equivalent capacitor. Therefore, we should first check whether any of the capacitors in Fig. 26-9a are in parallel or series.

Capacitors 1 and 3 are connected one after the other, but are they in series? No. The potential V that is applied to the capacitors produces charge on the bottom plate of capacitor 3. That charge causes charge to shift from the top plate of capacitor 3. However, note that the shifting charge can move to the bottom plates of both capacitor 1 and capacitor 2. Because there is more than one route for the shifting charge, capacitor 3 is *not* in series with capacitor 1 (or capacitor 2).

Are capacitor 1 and capacitor 2 in parallel? Yes. Their top plates are directly wired together and their bottom plates are directly wired together, and electric potential is applied between the top-plate pair and the bottom-plate pair. Thus, capacitor 1 and capacitor 2 are in parallel, and Eq. 26-19 tells us that their equivalent capacitance C_{12} is

$$C_{12} = C_1 + C_2 = 12.0 \ \mu F + 5.30 \ \mu F = 17.3 \ \mu F.$$

In Fig. 26-9b, we have replaced capacitors 1 and 2 with their equivalent capacitor, call it capacitor 12 (say "one two"). (The connections at points A and B are exactly the same in Figs. 26-9a and b.)

Is capacitor 12 in series with capacitor 3? Again applying the test for series capacitances, we see that the charge that shifts from the top plate of capacitor 3 must entirely go to the bottom plate of capacitor 12. Thus, capacitor 12 and capacitor 3 are in series, and we can replace them with their equivalent C_{123}, as shown in Fig. 26-9c. From Eq. 26-20, we have

$$\frac{1}{C_{123}} = \frac{1}{C_{12}} + \frac{1}{C_3} = \frac{1}{17.3 \ \mu F} + \frac{1}{4.50 \ \mu F} = 0.280 \ \mu F^{-1},$$

from which

$$C_{123} = \frac{1}{0.280 \ \mu F^{-1}} = 3.57 \ \mu F. \qquad \text{(Answer)}$$

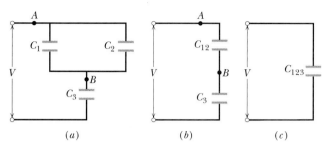

Fig. 26-9 Sample Problem 26-2. (*a*) Three capacitors. (*b*) C_1 and C_2, a parallel combination, are replaced by C_{12}. (*c*) C_{12} and C_3, a series combination, are replaced by the equivalent capacitance C_{123}.

(b) The potential difference that is applied to the input terminals in Fig. 26-9a is $V = 12.5$ V. What is the charge on C_1?

SOLUTION: One Key Idea here is that, to get the charge q_1 on capacitor 1, we now have to work backward to that capacitor, starting with the equivalent capacitor 123. Since the given potential difference V (= 12.5 V) is applied across the actual combination of three capacitors in Fig. 26-9a, it is also applied across capacitor 123 in Fig. 26-9c. Thus, Eq. 26-1 ($q = CV$) gives us

$$q_{123} = C_{123}V = (3.57 \ \mu F)(12.5 \ V) = 44.6 \ \mu C.$$

A second Key Idea is that the series capacitors 12 and 3 in Fig. 26-9b have the same charge as their equivalent capacitor 123 (recall "seri-q"). Thus, capacitor 12 has charge $q_{12} = q_{123} = 44.6 \ \mu C$. From Eq. 26-1, the potential difference across capacitor 12 must be

$$V_{12} = \frac{q_{12}}{C_{12}} = \frac{44.6 \ \mu C}{17.3 \ \mu F} = 2.58 \ V.$$

A third Key Idea is that the parallel capacitors 1 and 2 both have the same potential difference as their equivalent capacitor 12 (recall "par-V"). Thus, capacitor 1 has the potential difference $V_1 = V_{12} = 2.58$ V. Thus, from Eq. 26-1, the charge on capacitor 1 must be

$$q_1 = C_1V_1 = (12.0 \ \mu F)(2.58 \ V)$$
$$= 31.0 \ \mu C. \qquad \text{(Answer)}$$

Sample Problem 26-3

Capacitor 1, with $C_1 = 3.55 \ \mu F$, is charged to a potential difference $V_0 = 6.30$ V, using a 6.30 V battery. The battery is then removed and the capacitor is connected as in Fig. 26-10 to an uncharged capacitor 2, with $C_2 = 8.95 \ \mu F$. When switch S is closed, charge flows between the capacitors until they have the same potential difference V. Find V.

SOLUTION: The situation here differs from the previous example because an applied electric potential is *not* maintained across a combination of capacitors by a battery or some other source. Here, just

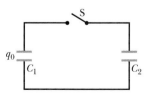

Fig. 26-10 Sample Problem 26-3. A potential difference V_0 is applied to capacitor 1 and the charging battery is removed. Switch S is then closed so that the charge on capacitor 1 is shared with capacitor 2.

after switch S is closed, the only applied electric potential is that of capacitor 1 on capacitor 2, and that potential is decreasing. Thus, although the capacitors in Fig. 26-10 are connected end to end, in this situation they are not *in series*; and although they are drawn parallel, in this situation they are not *in parallel*.

To find the final electric potential (when the system comes to equilibrium and charge stops flowing), we use this Key Idea: After the switch is closed, the original charge q_0 on capacitor 1 is redistributed (shared) between capacitor 1 and capacitor 2. When equilibrium is reached, we can relate the original charge q_0 with the final charges q_1 and q_2 by writing

$$q_0 = q_1 + q_2.$$

Applying the relation $q = CV$ to each term of this equation yields

$$C_1 V_0 = C_1 V + C_2 V,$$

from which

$$V = V_0 \frac{C_1}{C_1 + C_2} = \frac{(6.30 \text{ V})(3.55 \text{ }\mu\text{F})}{3.55 \text{ }\mu\text{F} + 8.95 \text{ }\mu\text{F}}$$

$$= 1.79 \text{ V.} \qquad \text{(Answer)}$$

When the capacitors reach this value of electric potential difference, the charge flow stops.

✔**CHECKPOINT 4:**   In this sample problem, suppose capacitor 2 is replaced by a series combination of capacitors 3 and 4. (a) After the switch is closed and charge has stopped flowing, what is the relation between the initial charge q_0, the charge q_1 then on capacitor 1, and the charge q_{34} then on the equivalent capacitor 34? (b) If $C_3 > C_4$, is the charge q_3 on capacitor 3 more than, less than, or equal to the charge q_4 on capacitor 4?

PROBLEM-SOLVING TACTICS

Tactic 2: *Multiple-Capacitor Circuits*

Let us review the procedure used in the solution of Sample Problem 26-2, in which several capacitors are connected to a battery. To find a single equivalent capacitance, we simplify the given arrangement of capacitances by replacing them, in steps, with equivalent capacitances, using Eq. 26-19 when we find capacitances in parallel and Eq. 26-20 when we find capacitances in series. Then, to find the charge stored by that single equivalent capacitance, we use Eq. 26-1 and the potential difference V applied by the battery.

That result tells us the net charge stored on the actual arrangement of capacitors. However, to find the charge on, or the potential difference across, any particular capacitor in the actual arrangement, we need to reverse our steps of simplification. With each reversed step, we use these two rules: When capacitances are in parallel, they have the same potential difference as their equivalent capacitance, and we use Eq. 26-1 to find the charge on each capacitance; when they are in series, they have the same charge as

their equivalent capacitance, and we use Eq. 26-1 to find the potential difference across each capacitance.

Tactic 3: *Batteries and Capacitors*

A battery maintains a certain potential difference across its terminals. Thus, when capacitor 1 of Sample Problem 26-3 is connected to the 6.30 V battery, charge flows between the capacitor and the battery until the capacitor has the same potential difference across it as the battery.

A capacitor differs from a battery in that a capacitor lacks the internal electrochemical reactions needed to release charged particles (electrons) from internal atoms and molecules. Thus, when the charged capacitor 1 of Sample Problem 26-3 is disconnected from the battery and then connected to the uncharged capacitor 2 with switch S closed, the potential difference across capacitor 1 is not maintained. The quantity that *is* maintained is the charge q_0 of the two-capacitor system; that is, charge obeys a conservation law, *not* electric potential.

26-5   Energy Stored in an Electric Field

Work must be done by an external agent to charge a capacitor. Starting with an uncharged capacitor, for example, imagine that—using "magic tweezers"—you remove electrons from one plate and transfer them one at a time to the other plate. The electric field that builds up in the space between the plates has a direction that tends to oppose further transfer. Thus, as charge accumulates on the capacitor plates, you have to do increasingly larger amounts of work to transfer additional electrons. In practice, this work is done not by "magic tweezers" but by a battery, at the expense of its store of chemical energy.

We visualize the work required to charge a capacitor as being stored in the form of **electric potential energy** U in the electric field between the plates. You can recover this energy at will, by discharging the capacitor in a circuit, just as you can recover the potential energy stored in a stretched bow by releasing the bowstring to transfer the energy to the kinetic energy of an arrow.

Suppose that, at a given instant, a charge q' has been transferred from one plate of a capacitor to the other. The potential difference V' between the plates at that instant will be q'/C. If an extra increment of charge dq' is then transferred, the increment of work required will be, from Eq. 25-7,

$$dW = V' \, dq' = \frac{q'}{C} \, dq'.$$

The work required to bring the total capacitor charge up to a final value q is

$$W = \int dW = \frac{1}{C} \int_0^q q' \, dq' = \frac{q^2}{2C}.$$

This work is stored as potential energy U in the capacitor, so that

$$U = \frac{q^2}{2C} \qquad \text{(potential energy).} \qquad (26\text{-}21)$$

From Eq. 26-1, we can also write this as

$$U = \tfrac{1}{2}CV^2 \qquad \text{(potential energy).} \qquad (26\text{-}22)$$

Equations 26-21 and 26-22 hold no matter what the geometry of the capacitor is.

To gain some physical insight into energy storage, consider two parallel-plate capacitors that are identical except that capacitor 1 has twice the plate separation of capacitor 2. Then capacitor 1 has twice the volume between its plates and also, from Eq. 26-9, half the capacitance of capacitor 2. Equation 26-4 tells us that if both capacitors have the same charge q, the electric fields between their plates are identical. And Eq. 26-21 tells us that capacitor 1 has twice the stored potential energy of capacitor 2. Thus, of two otherwise identical capacitors with the same charge and same electric field, the one with twice the volume between its plates has twice the stored potential energy. Arguments like this tend to verify our earlier assumption:

▶ The potential energy of a charged capacitor may be viewed as being stored in the electric field between its plates.

The Medical Defibrillator

The ability of a capacitor to store potential energy is the basis of *defibrillator* devices, which are used by emergency medical teams to stop the fibrillation of heart attack victims. In the portable version, a battery charges a capacitor to a high potential difference, storing a large amount of energy in less than a minute. The battery maintains only a modest potential difference; an electronic circuit repeatedly uses that potential difference to greatly increase the potential difference of the capacitor. The power, or rate of energy transfer, during this process is also modest.

Conducting leads ("paddles") are placed on the victim's chest. When a control switch is closed, the capacitor sends a portion of its stored energy from paddle to paddle through the victim. As an example, when a 70 μF capacitor in a defibrillator is charged to 5000 V, Eq. 26-22 gives the energy stored in the capacitor as

$$U = \tfrac{1}{2}CV^2 = \tfrac{1}{2}(70 \times 10^{-6} \text{ F})(5000 \text{ V})^2 = 875 \text{ J}.$$

About 200 J of this energy is sent through the victim during a pulse of about 2.0 ms. The power of the pulse is

$$P = \frac{U}{t} = \frac{200 \text{ J}}{2.0 \times 10^{-3} \text{ s}} = 100 \text{ kW},$$

Fig. 26-11 To photograph a bullet blowing apart a banana, Harold Edgerton, the inventor of the stroboscope, used a capacitor to dump electrical energy into one of his stroboscopic lamps, which then brightly illuminated the banana for only 0.3 μs.

which is much greater than the power of the battery itself. This same technique of slowly charging a capacitor with a battery and then discharging the capacitor at a much higher power is commonly used in flash photography and stroboscopic photography (Fig. 26-11).

Energy Density

In a parallel-plate capacitor, neglecting fringing, the electric field has the same value at all points between the plates. Thus, the **energy density** u—that is, the potential energy per unit volume between the plates—should also be uniform. We can find u by dividing the total potential energy by the volume Ad of the space between the plates. Using Eq. 26-22, we obtain

$$u = \frac{U}{Ad} = \frac{CV^2}{2Ad}.$$

With Eq. 26-9 ($C = \varepsilon_0 A/d$), this result becomes

$$u = \tfrac{1}{2}\varepsilon_0 \left(\frac{V}{d}\right)^2.$$

However, from Eq. 25-42, V/d equals the electric field magnitude E, so

$$u = \tfrac{1}{2}\varepsilon_0 E^2 \qquad \text{(energy density).} \tag{26-23}$$

Although we derived this result for the special case of a parallel-plate capacitor, it holds generally, whatever may be the source of the electric field. If an electric field $\vec{E}$ exists at any point in space, we can think of that point as a site of electric potential energy whose amount per unit volume is given by Eq. 26-23.

Sample Problem 26-4

An isolated conducting sphere whose radius R is 6.85 cm has a charge $q = 1.25$ nC.

(a) How much potential energy is stored in the electric field of this charged conductor?

SOLUTION: The Key Idea here is that the energy U stored in a capacitor depends on the charge q on the capacitor and the capacitance C of the capacitor, according to Eq. 26-21. Substituting from Eq. 26-18 for C, Eq. 26-21 gives us

$$U = \frac{q^2}{2C} = \frac{q^2}{8\pi\varepsilon_0 R}$$

$$= \frac{(1.25 \times 10^{-9}\text{ C})^2}{(8\pi)(8.85 \times 10^{-12}\text{ F/m})(0.0685\text{ m})}$$

$$= 1.03 \times 10^{-7}\text{ J} = 103\text{ nJ.} \qquad \text{(Answer)}$$

(b) What is the energy density at the surface of the sphere?

SOLUTION: The Key Idea here is that the density u of the energy stored in an electric field depends on the magnitude E of the field, according to Eq. 26-23 ($u = \tfrac{1}{2}\varepsilon_0 E^2$), so we must first find E at the surface of the sphere. This is given by Eq. 24-15:

$$E = \frac{1}{4\pi\varepsilon_0}\frac{q}{R^2}.$$

The energy density is then

$$u = \tfrac{1}{2}\varepsilon_0 E^2 = \frac{q^2}{32\pi^2\varepsilon_0 R^4}$$

$$= \frac{(1.25 \times 10^{-9}\text{ C})^2}{(32\pi^2)(8.85 \times 10^{-12}\text{ C}^2/\text{N}\cdot\text{m}^2)(0.0685\text{ m})^4}$$

$$= 2.54 \times 10^{-5}\text{ J/m}^3 = 25.4\ \mu\text{J/m}^3. \qquad \text{(Answer)}$$

26-6 Capacitor with a Dielectric

If you fill the space between the plates of a capacitor with a *dielectric*, which is an insulating material such as mineral oil or plastic, what happens to the capacitance? Michael Faraday—to whom the whole concept of capacitance is largely due and for whom the SI unit of capacitance is named—first looked into this matter in 1837. Using simple equipment much like that shown in Fig. 26-12, he found that the

Fig. 26-12 The simple electrostatic apparatus used by Faraday. An assembled apparatus (second from left) forms a spherical capacitor consisting of a central brass ball and a concentric brass shell. Faraday placed dielectric materials in the space between the ball and the shell.

TABLE 26-1   **Some Properties of Dielectrics**[a]

Material	Dielectric Constant κ	Dielectric Strength (kV/mm)
Air (1 atm)	1.00054	3
Polystyrene	2.6	24
Paper	3.5	16
Transformer oil	4.5	
Pyrex	4.7	14
Ruby mica	5.4	
Porcelain	6.5	
Silicon	12	
Germanium	16	
Ethanol	25	
Water (20°C)	80.4	
Water (25°C)	78.5	
Titania ceramic	130	
Strontium titanate	310	8

For a vacuum, κ = unity.

[a]Measured at room temperature, except for the water.

V = a constant

(a)

q = a constant

(b)

capacitance *increased* by a numerical factor κ, which he called the **dielectric constant** of the insulating material. Table 26-1 shows some dielectric materials and their dielectric constants. The dielectric constant of a vacuum is unity by definition. Because air is mostly empty space, its measured dielectric constant is only slightly greater than unity.

Another effect of the introduction of a dielectric is to limit the potential difference that can be applied between the plates to a certain value V_{max}, called the *breakdown potential*. If this value is substantially exceeded, the dielectric material will break down and form a conducting path between the plates. Every dielectric material has a characteristic *dielectric strength,* which is the maximum value of the electric field that it can tolerate without breakdown. A few such values are listed in Table 26-1.

As we discussed in connection with Eq. 26-18, the capacitance of any capacitor can be written in the form

$$C = \varepsilon_0 \mathscr{L}, \qquad (26\text{-}24)$$

in which $\mathscr{L}$ has the dimensions of a length. For example, $\mathscr{L} = A/d$ for a parallel-plate capacitor. Faraday's discovery was that, with a dielectric *completely* filling the space between the plates, Eq. 26-24 becomes

$$C = \kappa \varepsilon_0 \mathscr{L} = \kappa C_{air}, \qquad (26\text{-}25)$$

where C_{air} is the value of the capacitance with only air between the plates.

Figure 26-13 provides some insight into Faraday's experiments. In Fig. 26-13a the battery ensures that the potential difference V between the plates will remain

Fig. 26-13 (a) If the potential difference between the plates of a capacitor is maintained, as by battery B, the effect of a dielectric is to increase the charge on the plates. (b) If the charge on the capacitor plates is maintained, as in this case, the effect of a dielectric is to reduce the potential difference between the plates. The scale shown is that of a *potentiometer*, a device used to measure potential difference (here, between the plates). A capacitor cannot discharge through a potentiometer.

constant. When a dielectric slab is inserted between the plates, the charge q on the plates increases by a factor of κ; the additional charge is delivered to the capacitor plates by the battery. In Fig. 26-13b there is no battery and therefore the charge q must remain constant when the dielectric slab is inserted; then the potential difference V between the plates decreases by a factor of κ. Both these observations are consistent (through the relation $q = CV$) with the increase in capacitance caused by the dielectric.

Comparison of Eqs. 26-24 and 26-25 suggests that the effect of a dielectric can be summed up in more general terms:

▶ In a region completely filled by a dielectric material of dielectric constant κ, all electrostatic equations containing the permittivity constant ε_0 are to be modified by replacing ε_0 with $\kappa\varepsilon_0$.

Thus, a point charge inside a dielectric produces an electric field that, by Coulomb's law, has the magnitude

$$E = \frac{1}{4\pi\kappa\varepsilon_0}\frac{q}{r^2}. \tag{26-26}$$

Also, the expression for the electric field just outside an isolated conductor immersed in a dielectric (see Eq. 24-11) becomes

$$E = \frac{\sigma}{\kappa\varepsilon_0}. \tag{26-27}$$

Both these equations show that *for a fixed distribution of charges, the effect of a dielectric is to weaken the electric field* that would otherwise be present.

Sample Problem 26-5

A parallel-plate capacitor whose capacitance C is 13.5 pF is charged by a battery to a potential difference $V = 12.5$ V between its plates. The charging battery is now disconnected and a porcelain slab ($\kappa = 6.50$) is slipped between the plates. What is the potential energy of the capacitor–slab device, both before and after the slab is put into place?

SOLUTION: The Key Idea here is that we can relate the potential energy U of the capacitor to the capacitance C and either the potential V (with Eq. 26-22) or the charge q (with Eq. 26-21):

$$U_i = \tfrac{1}{2}CV^2 = \frac{q^2}{2C}.$$

Because we are given the initial potential V (= 12.5 V), we use Eq. 26-22 to find the initial stored energy:

$$U_i = \tfrac{1}{2}CV^2 = \tfrac{1}{2}(13.5 \times 10^{-12}\text{ F})(12.5\text{ V})^2$$
$$= 1.055 \times 10^{-9}\text{ J} = 1055\text{ pJ} \approx 1100\text{ pJ.} \quad \text{(Answer)}$$

To find the final potential energy U_f (after the slab is introduced), we need another Key Idea: Because the battery has been disconnected, the charge on the capacitor cannot change when the dielectric is inserted. However, the potential *does* change. Thus, we must now use Eq. 26-21 (based on q) to write the final potential energy U_f, but now that the slab is within the capacitor, the capac-

itance is κC. We then have

$$U_f = \frac{q^2}{2\kappa C} = \frac{U_i}{\kappa} = \frac{1055\text{ pJ}}{6.50} = 162\text{ pJ} \approx 160\text{ pJ.} \quad \text{(Answer)}$$

When the slab is introduced, the potential energy decreases by a factor of κ.

The "missing" energy, in principle, would be apparent to the person who introduced the slab. The capacitor would exert a tiny tug on the slab and would do work on it, in amount

$$W = U_i - U_f = (1055 - 162)\text{ pJ} = 893\text{ pJ.}$$

If the slab were allowed to slide between the plates with no restraint and if there were no friction, the slab would oscillate back and forth between the plates with a (constant) mechanical energy of 893 pJ, and this system energy would transfer back and forth between kinetic energy of the moving slab and potential energy stored in the electric field.

✔CHECKPOINT 5:   If the battery in this sample problem remains connected, do the following increase, decrease, or remain the same when the slab is introduced: (a) the potential difference between the capacitor plates, (b) the capacitance, (c) the charge on the capacitor, (d) the potential energy of the device, (e) the electric field between the plates? (*Hint:* For (e), note that the charge is not fixed.)

26-7 Dielectrics: An Atomic View

What happens, in atomic and molecular terms, when we put a dielectric in an electric field? There are two possibilities, depending on the nature of the molecules:

1. *Polar dielectrics.* The molecules of some dielectrics, like water, have permanent electric dipole moments. In such materials (called *polar dielectrics*), the electric dipoles tend to line up with an external electric field as in Fig. 26-14. Because the molecules are continuously jostling each other as a result of their random thermal motion, this alignment is not complete, but it becomes more complete as the magnitude of the applied field is increased (or as the temperature, and thus the jostling, is decreased). The alignment of the electric dipoles produces an electric field that is directed opposite the applied field and smaller in magnitude.

2. *Nonpolar dielectrics.* Regardless of whether they have permanent electric dipole moments, molecules acquire dipole moments by induction when placed in an external electric field. In Section 25-7 (see Fig. 25-12), we saw that this occurs because the external field tends to "stretch" the molecules, slightly separating the centers of negative and positive charge.

Figure 26-15a shows a nonpolar dielectric slab with no external electric field applied. In Fig. 26-15b, an electric field $\vec{E}_0$ is applied via a capacitor, whose plates are charged as shown. The result is a slight separation of the centers of the positive and negative charge distributions within the slab, producing positive charge on one face of the slab (due to the positive ends of dipoles there) and negative charge on the opposite face (due to the negative ends of dipoles there). The slab as a whole remains electrically neutral and—within the slab—there is no excess charge in any volume element.

Figure 26-15c shows that the induced surface charges on the faces produce an electric field $\vec{E}'$ in the direction opposite that of the applied electric field $\vec{E}_0$. The resultant field $\vec{E}$ inside the dielectric (the vector sum of fields $\vec{E}_0$ and $\vec{E}'$) has the direction of $\vec{E}_0$ but is smaller in magnitude.

Both the field $\vec{E}'$ produced by the surface charges in Fig. 26-15c and the electric field produced by the permanent electric dipoles in Fig. 26-14 act in the same way— they oppose the applied field $\vec{E}$. Thus, the effect of both polar and nonpolar dielectrics is to weaken any applied field within them, as between the plates of a capacitor.

We can now see why the dielectric porcelain slab in Sample Problem 26-5 is pulled into the capacitor: As it enters the space between the plates, the surface charge that appears on each slab face has the sign opposite that of the charge on the nearby capacitor plate. Thus, slab and plates attract each other.

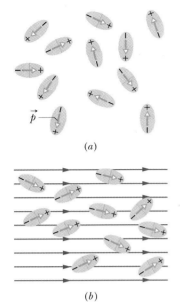

Fig. 26-14 (a) Molecules with a permanent electric dipole moment, showing their random orientation in the absence of an external electric field. (b) An electric field is applied, producing partial alignment of the dipoles. Thermal agitation prevents complete alignment.

Fig. 26-15 (a) A nonpolar dielectric slab. The circles represent the electrically neutral atoms within the slab. (b) An electric field is applied via charged capacitor plates; the field slightly stretches the atoms, separating the centers of positive and negative charge. (c) The separation produces surface charges on the slab faces. These charges set up a field $\vec{E}'$, which opposes the applied field $\vec{E}_0$. The resultant field $\vec{E}$ inside the dielectric (the vector sum of $\vec{E}_0$ and $\vec{E}'$) has the same direction as $\vec{E}_0$ but less magnitude.

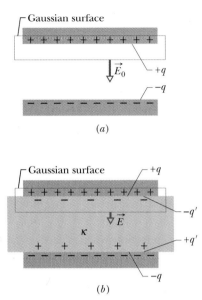

Fig. 26-16 A parallel-plate capacitor (*a*) without and (*b*) with a dielectric slab inserted. The charge *q* on the plates is assumed to be the same in both cases.

26-8 Dielectrics and Gauss' Law

In our discussion of Gauss' law in Chapter 24, we assumed that the charges existed in a vacuum. Here we shall see how to modify and generalize that law if dielectric materials, such as those listed in Table 26-1, are present. Figure 26-16 shows a parallel-plate capacitor of plate area *A*, both with and without a dielectric. We assume that the charge *q* on the plates is the same in both situations. Note that the field between the plates induces charges on the faces of the dielectric by one of the methods of Section 26-7.

For the situation of Fig. 26-16*a*, without a dielectric, we can find the electric field $\vec{E}_0$ between the plates as we did in Fig. 26-5: We enclose the charge $+q$ on the top plate with a Gaussian surface and then apply Gauss' law. Letting E_0 represent the magnitude of the field, we find

$$\varepsilon_0 \oint \vec{E} \cdot d\vec{A} = \varepsilon_0 E_0 A = q, \tag{26-28}$$

or

$$E_0 = \frac{q}{\varepsilon_0 A}. \tag{26-29}$$

In Fig. 26-16*b*, with the dielectric in place, we can find the electric field between the plates (and within the dielectric) by using the same Gaussian surface. However, now the surface encloses two types of charge: It still encloses charge $+q$ on the top plate, but it now also encloses the induced charge $-q'$ on the top face of the dielectric. The charge on the conducting plate is said to be *free charge* because it can move if we change the electric potential of the plate; the induced charge on the surface of the dielectric is not free charge because it cannot move from that surface.

The net charge enclosed by the Gaussian surface in Fig. 26-16*b* is $q - q'$, so Gauss' law now gives

$$\varepsilon_0 \oint \vec{E} \cdot d\vec{A} = \varepsilon_0 E A = q - q', \tag{26-30}$$

or

$$E = \frac{q - q'}{\varepsilon_0 A}. \tag{26-31}$$

The effect of the dielectric is to weaken the original field E_0 by a factor of κ, so we may write

$$E = \frac{E_0}{\kappa} = \frac{q}{\kappa \varepsilon_0 A}. \tag{26-32}$$

Comparison of Eqs. 26-31 and 26-32 shows that

$$q - q' = \frac{q}{\kappa}. \tag{26-33}$$

Equation 26-33 shows correctly that the magnitude q' of the induced surface charge is less than that of the free charge q and is zero if no dielectric is present (then, $\kappa = 1$ in Eq. 26-33).

By substituting for $q - q'$ from Eq. 26-33 in Eq. 26-30, we can write Gauss' law in the form

$$\varepsilon_0 \oint \kappa \vec{E} \cdot d\vec{A} = q \qquad \text{(Gauss' law with dielectric)}. \tag{26-34}$$

This important equation, although derived for a parallel-plate capacitor, is true generally and is the most general form in which Gauss' law can be written. Note the following:

1. The flux integral now involves $\kappa\vec{E}$, not just $\vec{E}$. (The vector $\varepsilon_0\kappa\vec{E}$ is sometimes called the *electric displacement* $\vec{D}$, so that Eq. 26-34 can be written in the form $\oint \vec{D} \cdot d\vec{A} = q$.)

2. The charge q enclosed by the Gaussian surface is now taken to be the *free charge only*. The induced surface charge is deliberately ignored on the right side of Eq. 26-34, having been taken fully into account by introducing the dielectric constant κ on the left side.

3. Equation 26-34 differs from Eq. 24-7, our original statement of Gauss' law, only in that ε_0 in the latter equation has been replaced by $\kappa\varepsilon_0$. We keep κ inside the integral of Eq. 26-34 to allow for cases in which κ is not constant over the entire Gaussian surface.

Sample Problem 26-6

Figure 26-17 shows a parallel-plate capacitor of plate area A and plate separation d. A potential difference V_0 is applied between the plates. The battery is then disconnected, and a dielectric slab of thickness b and dielectric constant κ is placed between the plates as shown. Assume

$$A = 115 \text{ cm}^2, \qquad d = 1.24 \text{ cm}, \qquad V_0 = 85.5 \text{ V},$$

$$b = 0.780 \text{ cm}, \qquad \kappa = 2.61.$$

(a) What is the capacitance C_0 before the dielectric slab is inserted?

SOLUTION: From Eq. 26-9 we have

$$C_0 = \frac{\varepsilon_0 A}{d} = \frac{(8.85 \times 10^{-12} \text{ F/m})(115 \times 10^{-4} \text{ m}^2)}{1.24 \times 10^{-2} \text{ m}}$$

$$= 8.21 \times 10^{-12} \text{ F} = 8.21 \text{ pF}. \qquad \text{(Answer)}$$

(b) What free charge appears on the plates?

SOLUTION: From Eq. 26-1,

$$q = C_0 V_0 = (8.21 \times 10^{-12} \text{ F})(85.5 \text{ V})$$

$$= 7.02 \times 10^{-10} \text{ C} = 702 \text{ pC}. \qquad \text{(Answer)}$$

Because the charging battery was disconnected before the slab was introduced, the free charge remains unchanged as the slab is put into place.

(c) What is the electric field E_0 in the gaps between the plates and the dielectric slab?

SOLUTION: A Key Idea here is to apply Gauss' law, in the form of Eq. 26-34, to Gaussian surface I in Fig. 26-17—that surface passes through the gap, and so it encloses *only* the free charge on the upper capacitor plate. Because the area vector $d\vec{A}$ and the field vector $\vec{E}_0$ are both directed downward, the dot product in Eq. 26-34 becomes

$$\vec{E}_0 \cdot d\vec{A} = E_0 \, dA \cos 0° = E_0 \, dA.$$

Equation 26-34 then becomes

$$\varepsilon_0 \kappa E_0 \oint dA = q.$$

The integration now simply gives the surface area A of the plate.

Fig. 26-17 Sample Problem 26-6. A parallel-plate capacitor containing a dielectric slab that only partially fills the space between the plates.

Thus, we obtain

$$\varepsilon_0 \kappa E_0 A = q,$$

or

$$E_0 = \frac{q}{\varepsilon_0 \kappa A}.$$

One more Key Idea is needed before we evaluate E_0; that is, we must put $\kappa = 1$ here because Gaussian surface I does not pass through the dielectric. Thus, we have

$$E_0 = \frac{q}{\varepsilon_0 \kappa A} = \frac{7.02 \times 10^{-10} \text{ C}}{(8.85 \times 10^{-12} \text{ F/m})(1)(115 \times 10^{-4} \text{ m}^2)}$$

$$= 6900 \text{ V/m} = 6.90 \text{ kV/m}. \qquad \text{(Answer)}$$

Note that the value of E_0 does not change when the slab is introduced because the amount of charge enclosed by Gaussian surface I in Fig. 26-17 does not change.

(d) What is the electric field E_1 in the dielectric slab?

SOLUTION: The Key Idea here is to apply Eq. 26-34 to Gaussian surface II in Fig. 26-17. That surface encloses free charge $-q$ and induced charge $+q'$, but we ignore the latter when we use Eq. 26-34. We find

$$\varepsilon_0 \oint \kappa \vec{E}_1 \cdot d\vec{A} = -\varepsilon_0 \kappa E_1 A = -q. \qquad (26\text{-}35)$$

(The first minus sign in this equation comes from the dot product $\vec{E}_1 \cdot d\vec{A}$, because now the field vector $\vec{E}_1$ is directed downward and the area vector $d\vec{A}$ is directed upward.) Equation 26-35 gives us

$$E_1 = \frac{q}{\varepsilon_0 \kappa A} = \frac{E_0}{\kappa} = \frac{6.90 \text{ kV/m}}{2.61} = 2.64 \text{ kV/m}. \qquad \text{(Answer)}$$

(e) What is the potential difference V between the plates after the slab has been introduced?

SOLUTION: The **Key Idea** here is to find V by integrating along a straight-line path extending directly from the bottom plate to the top plate. Within the dielectric, the path length is b and the electric field is E_1. Within the two gaps above and below the dielectric, the total path length is $d - b$ and the electric field is E_0. Equation 26-6 then yields

$$V = \int_-^+ E\,ds = E_0(d - b) + E_1b$$

$$= (6900\ \text{V/m})(0.0124\ \text{m} - 0.00780\ \text{m})$$
$$\quad + (2640\ \text{V/m})(0.00780\ \text{m})$$
$$= 52.3\ \text{V}. \qquad \text{(Answer)}$$

This is less than the original potential difference of 85.5 V.

(f) What is the capacitance with the slab in place?

SOLUTION: The **Key Idea** now is that the capacitance C is related to the free charge q and the potential difference V via Eq. 26-1, just as when a dielectric is not in place. Taking q from (b) and V from (e), we have

$$C = \frac{q}{V} = \frac{7.02 \times 10^{-10}\ \text{C}}{52.3\ \text{V}}$$

$$= 1.34 \times 10^{-11}\ \text{F} = 13.4\ \text{pF}. \qquad \text{(Answer)}$$

This is greater than the original capacitance of 8.21 pF.

✔**CHECKPOINT 6:** In this sample problem, if the thickness b of the slab increases, do the following increase, decrease, or remain the same: (a) the electric field E_1, (b) the potential difference between the plates, and (c) the capacitance of the capacitor?

REVIEW & SUMMARY

Capacitor; Capacitance A **capacitor** consists of two isolated conductors (the *plates*) with equal and opposite charges $+q$ and q. Its **capacitance** C is defined from

$$q = CV, \qquad (26\text{-}1)$$

where V is the potential difference between the plates. The SI unit of capacitance is the farad (1 farad = 1 F = 1 coulomb per volt).

Determining Capacitance We generally determine the capacitance of a particular capacitor configuration by (1) assuming a charge q to have been placed on the plates, (2) finding the electric field $\vec{E}$ due to this charge, (3) evaluating the potential difference V, and (4) calculating C from Eq. 26-1. Some specific results are the following:

A *parallel-plate capacitor* with flat parallel plates of area A and spacing d has capacitance

$$C = \frac{\varepsilon_0 A}{d}. \qquad (26\text{-}9)$$

A *cylindrical capacitor* (two long coaxial cylinders) of length L and radii a and b has capacitance

$$C = 2\pi\varepsilon_0 \frac{L}{\ln(b/a)}. \qquad (26\text{-}14)$$

A *spherical capacitor* with concentric spherical plates of radii a and b has capacitance

$$C = 4\pi\varepsilon_0 \frac{ab}{b - a}. \qquad (26\text{-}17)$$

If we let $b \to \infty$ and $a = R$ in Eq. 26-17, we obtain the capacitance of an *isolated sphere* of radius R:

$$C = 4\pi\varepsilon_0 R. \qquad (26\text{-}18)$$

Capacitors in Parallel and in Series The **equivalent capacitances** C_{eq} of combinations of individual capacitors connected in **parallel** and in **series** can be found from

$$C_{eq} = \sum_{j=1}^n C_j \qquad (n \text{ capacitors in parallel}) \qquad (26\text{-}19)$$

and

$$\frac{1}{C_{eq}} = \sum_{j=1}^n \frac{1}{C_j} \qquad (n \text{ capacitors in series}). \qquad (26\text{-}20)$$

Equivalent capacitances can be used to calculate the capacitances of more complicated series-parallel combinations.

Potential Energy and Energy Density The **electric potential energy** U of a charged capacitor,

$$U = \frac{q^2}{2C} = \tfrac{1}{2}CV^2, \qquad (26\text{-}21, 26\text{-}22)$$

is equal to the work required to charge it. This energy can be associated with the capacitor's electric field $\vec{E}$. By extension we can associate stored energy with an electric field. In vacuum, the **energy density** u, or potential energy per unit volume, within an electric field of magnitude E is given by

$$u = \tfrac{1}{2}\varepsilon_0 E^2. \qquad (26\text{-}23)$$

Capacitance with a Dielectric If the space between the plates of a capacitor is completely filled with a dielectric material, the capacitance C is increased by a factor κ, called the **dielectric constant,** which is characteristic of the material. In a region that is completely filled by a dielectric, all electrostatic equations containing ε_0 must be modified by replacing ε_0 with $\kappa\varepsilon_0$.

The effects of adding a dielectric can be understood physically in terms of the action of an electric field on the permanent or induced electric dipoles in the dielectric slab. The result is the formation of induced charges on the surfaces of the dielectric, which results in a weakening of the field within the dielectric for the same free charge on the plates.

Gauss' Law with a Dielectric   When a dielectric is present, Gauss' law may be generalized to

$$\varepsilon_0 \oint \kappa \vec{E} \cdot d\vec{A} = q. \tag{26-34}$$

Here q is the free charge; any induced surface charge is accounted for by including the dielectric constant κ inside the integral.

QUESTIONS

1. Figure 26-18 shows plots of charge versus potential difference for three parallel-plate capacitors, which have the plate areas and separations given in the table. Which of the plots goes with which of the capacitors?

Capacitor	Area	Separation
1	A	d
2	$2A$	d
3	A	$2d$

Fig. 26-18 Question 1.

2. Figure 26-19 shows an open switch, a battery of potential difference V, a current-measuring meter A, and three identical uncharged capacitors of capacitance C. When the switch is closed and the circuit reaches equilibrium, what are (a) the potential difference across each capacitor and (b) the charge on the left-hand plate of each capacitor? (c) During the charging process, what is the net charge that passes through the meter?

Fig. 26-19 Question 2.

3. For each circuit in Fig. 26-20, are the capacitors connected in series, in parallel, or in neither mode?

Fig. 26-20 Question 3.

4. (a) In Fig. 26-21a, are capacitors C_1 and C_3 in series? (b) In the same figure, are capacitors C_1 and C_2 in parallel? (c) Rank the equivalent capacitances of the four circuits shown in Fig. 26-21, greatest first.

Fig. 26-21 Question 4.

5. What is the equivalent capacitance of three capacitors, each of capacitance C, if they are connected to a battery (a) in series with one another and (b) in parallel? (c) In which arrangement is there more charge on the equivalent capacitance?

6. You are to connect capacitances C_1 and C_2, with $C_1 > C_2$, to a battery, first individually, then in series, and then in parallel. Rank those arrangements according to the amount of charge stored, greatest first.

7. Initially, a single capacitance C_1 is wired to a battery. Then capacitance C_2 is added in parallel. Are (a) the potential difference across C_1 and (b) the charge q_1 on C_1 now more than, less than, or the same as previously? (c) Is the equivalent capacitance C_{12} of C_1 and C_2 more than, less than, or equal to C_1? (d) Is the total charge stored on C_1 and C_2 together more than, less than, or equal to the charge stored previously on C_1?

8. Repeat Question 7 for C_2 added in series, not in parallel.

9. Figure 26-22 shows three circuits, each consisting of a switch

Fig. 26-22 Question 9.

and two capacitors, initially charged as indicated. After the switches have been closed, in which circuit (if any) will the charge on the left-hand capacitor (a) increase, (b) decrease, and (c) remain the same?

10. Two isolated metal spheres A and B have radii R and $2R$, respectively, and the same charge q. (a) Is the capacitance of A more than, less than, or equal to that of B? (b) Is the energy density just outside the surface of A more than, less than, or equal to that of B? (c) Is the energy density at distance $3R$ from the center of A more than, less than, or equal to that at the same distance from the center of B? (d) Is the total energy of the electric field due to A more than, less than, or equal to that of B?

11. When a dielectric slab is inserted between the plates of one of the two identical capacitors in Fig. 26-23, do the following properties of that capacitor increase, decrease, or remain the same: (a) capacitance, (b) charge, (c) potential difference, and (d) potential energy? (e) How about the same properties of the other capacitor?

Fig. 26-23 Question 11.

EXERCISES & PROBLEMS

ssm Solution is in the Student Solutions Manual.
www Solution is available on the World Wide Web at:
http://www.wiley.com/college/hrw
ilw Solution is available on the Interactive LearningWare.

SEC. 26-2 Capacitance

1E. An electrometer is a device used to measure static charge—an unknown charge is placed on the plates of the meter's capacitor, and the potential difference is measured. What minimum charge can be measured by an electrometer with a capacitance of 50 pF and a voltage sensitivity of 0.15 V?

2E. The two metal objects in Fig. 26-24 have net charges of +70 pC and −70 pC, which result in a 20 V potential difference between them. (a) What is the capacitance of the system? (b) If the charges are changed to +200 pC and −200 pC, what does the capacitance become? (c) What does the potential difference become?

Fig. 26-24 Exercise 2.

3E. The capacitor in Fig. 26-25 has a capacitance of 25 μF and is initially uncharged. The battery provides a potential difference of 120 V. After switch S is closed, how much charge will pass through it? ssm

Fig. 26-25 Exercise 3.

SEC. 26-3 Calculating the Capacitance

4E. If we solve Eq. 26-9 for ε_0, we see that its SI unit is the farad per meter. Show that this unit is equivalent to that obtained earlier for ε_0—namely, the coulomb squared per newton-meter squared ($C^2/N \cdot m^2$).

5E. A parallel-plate capacitor has circular plates of 8.2 cm radius and 1.3 mm separation. (a) Calculate the capacitance. (b) What charge will appear on the plates if a potential difference of 120 V is applied? ssm

6E. You have two flat metal plates, each of area 1.00 m^2, with which to construct a parallel-plate capacitor. If the capacitance of the device is to be 1.00 F, what must be the separation between the plates? Could this capacitor actually be constructed?

7E. A spherical drop of mercury of radius R has a capacitance given by $C = 4\pi\varepsilon_0 R$. If two such drops combine to form a single larger drop, what is its capacitance? ssm

8E. The plates of a spherical capacitor have radii 38.0 mm and 40.0 mm. (a) Calculate the capacitance. (b) What must be the plate area of a parallel-plate capacitor with the same plate separation and capacitance?

9P. Suppose that the two spherical shells of a spherical capacitor have approximately equal radii. Under these conditions the device approximates a parallel-plate capacitor with $b - a = d$. Show that Eq. 26-17 does indeed reduce to Eq. 26-9 in this case. ssm

SEC. 26-4 Capacitors in Parallel and in Series

10E. In Fig. 26-26, find the equivalent capacitance of the combination. Assume that $C_1 = 10.0$ μF, $C_2 = 5.00$ μF, and $C_3 = 4.00$ μF.

11E. How many 1.00 μF capacitors must be connected in parallel to store a charge of 1.00 C with a potential of 110 V across the capacitors? ssm

12E. Each of the uncharged capacitors in Fig. 26-27 has a capacitance of 25.0 μF. A potential difference of 4200 V is established when the switch is closed. How many coulombs of charge then pass through meter A?

13E. In Fig. 26-28 find the equivalent capacitance of the combination. Assume that $C_1 = 10.0$ μF, $C_2 = 5.00$ μF, and $C_3 = 4.00$ μF. ilw

14P. In Fig. 26-28 suppose that capacitor 3 breaks down electrically, becoming equivalent to a conducting path. What *changes* in (a) the charge and (b) the potential difference occur for capacitor 1? Assume that $V = 100$ V.

15P. Figure 26-29 shows two capacitors in series; the center section of length b is movable vertically. Show that the equivalent capacitance of this series combination is independent of the position of the center section and is given by $C = \varepsilon_0 A/(a - b)$, where A is the plate area. ssm www

16P. In Fig. 26-30, the battery has a potential difference of 10 V and the five capacitors each have a capacitance of 10 μF. What is the charge on (a) capacitor 1 and (b) capacitor 2?

17P. A 100 pF capacitor is charged to a potential difference of 50 V, and the charging battery is disconnected. The capacitor is then connected in parallel with a second (initially uncharged) capacitor. If the potential difference across the first capacitor drops to 35 V, what is the capacitance of this second capacitor? ssm ilw

Fig. 26-26 Exercise 10 and Problem 30.

Fig. 26-27 Exercise 12.

Fig. 26-28 Exercise 13, Problems 14 and 28.

Fig. 26-29 Problem 15.

Fig. 26-30 Problem 16.

18P. In Fig. 26-31, the battery has a potential difference of 20 V. Find (a) the equivalent capacitance of all the capacitors and (b) the charge stored on that equivalent capacitance. Find the potential across and charge on (c) capacitor 1, (d) capacitor 2, and (e) capacitor 3.

Fig. 26-31 Problem 18.

19P. In Fig. 26-32, the capacitances are $C_1 = 1.0 \ \mu F$ and $C_2 = 3.0 \ \mu F$ and both capacitors are charged to a potential difference of $V = 100$ V but with opposite polarity as shown. Switches S_1 and S_2 are now closed. (a) What is now the potential difference between points a and b? What are now the charges on capacitors (b) 1 and (c) 2? ssm www

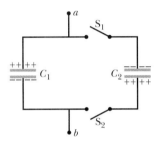

Fig. 26-32 Problem 19.

20P. In Fig. 26-33, battery B supplies 12 V. Find the charge on each capacitor (a) first when only switch S_1 is closed and (b) later when switch S_2 is also closed. Take $C_1 = 1.0 \ \mu F$, $C_2 = 2.0 \ \mu F$, $C_3 = 3.0 \ \mu F$, and $C_4 = 4.0 \ \mu F$.

Fig. 26-33 Problem 20.

21P. When switch S is thrown to the left in Fig. 26-34, the plates of capacitor 1 acquire a potential difference V_0. Capacitors 2 and 3 are initially uncharged. The switch is now thrown to the right. What are the final charges q_1, q_2, and q_3 on the capacitors? ssm

Fig. 26-34 Problem 21.

SEC. 26-5 Energy Stored in an Electric Field

22E. How much energy is stored in one cubic meter of air due to the "fair weather" electric field of magnitude 150 V/m?

23E. What capacitance is required to store an energy of 10 kW · h at a potential difference of 1000 V? ssm

24E. A parallel-plate air-filled capacitor having area 40 cm² and plate spacing 1.0 mm is charged to a potential difference of 600 V. Find (a) the capacitance, (b) the magnitude of the charge on each plate, (c) the stored energy, (d) the electric field between the plates, and (e) the energy density between the plates.

25E. Two capacitors, of 2.0 and 4.0 μF capacitance, are connected in parallel across a 300 V potential difference. Calculate the total energy stored in the capacitors. ssm

26P. A parallel-connected bank of 5.00 μF capacitors is used to store electric energy. What does it cost to charge the 2000 capacitors of the bank to 50 000 V, assuming 3.0¢/kW · h?

27P. One capacitor is charged until its stored energy is 4.0 J. A second uncharged capacitor is then connected to it in parallel. (a) If the charge distributes equally, what is now the total energy stored in the electric fields? (b) Where did the excess energy go? ilw

28P. In Fig. 26-28 find (a) the charge, (b) the potential difference, and (c) the stored energy for each capacitor. Assume the numerical values of Exercise 13, with $V = 100$ V.

29P. A parallel-plate capacitor has plates of area A and separation d and is charged to a potential difference V. The charging battery is then disconnected, and the plates are pulled apart until their separation is $2d$. Derive expressions in terms of A, d, and V for (a) the new potential difference; (b) the initial and final stored energies, U_i and U_f; and (c) the work required to separate the plates. ssm ilw

30P. In Fig. 26-26, find (a) the charge, (b) the potential difference, and (c) the stored energy for each capacitor. Assume the numerical values of Exercise 10, with $V = 100$ V.

31P. A cylindrical capacitor has radii a and b as in Fig. 26-6. Show that half the stored electric potential energy lies within a cylinder whose radius is $r = \sqrt{ab}$. ssm www

32P. A charged isolated metal sphere of diameter 10 cm has a potential of 8000 V relative to $V = 0$ at infinity. Calculate the energy density in the electric field near the surface of the sphere.

33P. (a) Show that the plates of a parallel-plate capacitor attract each other with a force given by $F = q^2/2\varepsilon_0 A$. Do so by calculating the work needed to increase the plate separation from x to $x + dx$, with the charge q remaining constant. (b) Next show that the force per unit area (the *electrostatic stress*) acting on either capacitor plate is given by $\frac{1}{2}\varepsilon_0 E^2$. (Actually, this is the force per unit area on *any* conductor of *any* shape with an electric field $\vec{E}$ at its surface.) ssm

SEC. 26-6 Capacitor with a Dielectric

34E. An air-filled parallel-plate capacitor has a capacitance of 1.3 pF. The separation of the plates is doubled and wax is inserted between them. The new capacitance is 2.6 pF. Find the dielectric constant of the wax.

35E. Given a 7.4 pF air-filled capacitor, you are asked to convert it to a capacitor that can store up to 7.4 μJ with a maximum potential difference of 652 V. What dielectric in Table 26-1 should you use to fill the gap in the air capacitor if you do not allow for a margin of error? ssm

36E. A parallel-plate air-filled capacitor has a capacitance of 50 pF. (a) If each of its plates has an area of 0.35 m^2, what is the separation? (b) If the region between the plates is now filled with material having $\kappa = 5.6$, what is the capacitance?

37E. A coaxial cable used in a transmission line has an inner radius of 0.10 mm and an outer radius of 0.60 mm. Calculate the capacitance per meter for the cable. Assume that the space between the conductors is filled with polystyrene. ssm

38P. You are asked to construct a capacitor having a capacitance near 1 nF and a breakdown potential in excess of 10 000 V. You think of using the sides of a tall Pyrex drinking glass as a dielectric, lining the inside and outside curved surfaces with aluminum foil to act as the plates. The glass is 15 cm tall with an inner radius of 3.6 cm and an outer radius of 3.8 cm. What are the (a) capacitance and (b) breakdown potential of this capacitor?

39P. A certain substance has a dielectric constant of 2.8 and a dielectric strength of 18 MV/m. If it is used as the dielectric material in a parallel-plate capacitor, what minimum area should the plates of the capacitor have to obtain a capacitance of 7.0×10^{-2} μF and to ensure that the capacitor will be able to withstand a potential difference of 4.0 kV? ssm ilw

40P. A parallel-plate capacitor of plate area A is filled with two dielectrics as in Fig. 26-35a. Show that the capacitance is

$$C = \frac{\varepsilon_0 A}{d} \frac{\kappa_1 + \kappa_2}{2}.$$

Check this formula for limiting cases. (*Hint:* Can you justify this arrangement as being two capacitors in parallel?)

(a)   (b)

Fig. 26-35 Problems 40 and 41.

41P. A parallel-plate capacitor of plate area A is filled with two dielectrics as in Fig. 26-35b. Show that the capacitance is

$$C = \frac{2\varepsilon_0 A}{d} \frac{\kappa_1 \kappa_2}{\kappa_1 + \kappa_2}.$$

Check this formula for limiting cases. (*Hint:* Can you justify this arrangement as being two capacitors in series?) ssm

42P. What is the capacitance of the capacitor, of plate area A, shown in Fig. 26-36? (*Hint:* See Problems 40 and 41.)

Fig. 26-36 Problem 42.

SEC. 26-8 Dielectrics and Gauss' Law

43E. A parallel-plate capacitor has a capacitance of 100 pF, a plate area of 100 cm^2, and a mica dielectric ($\kappa = 5.4$) completely filling the space between the plates. At 50 V potential difference, calculate (a) the electric field magnitude E in the mica, (b) the magnitude of the free charge on the plates, and (c) the magnitude of the induced surface charge on the mica. ssm

44E. In Sample Problem 26-6, suppose that the battery remains connected while the dielectric slab is being introduced. Calculate (a) the capacitance, (b) the charge on the capacitor plates, (c) the electric field in the gap, and (d) the electric field in the slab, after the slab is in place.

45P. The space between two concentric conducting spherical shells of radii b and a (where $b > a$) is filled with a substance of dielectric constant κ. A potential difference V exists between the inner and outer shells. Determine (a) the capacitance of the device, (b) the free charge q on the inner shell, and (c) the charge q' induced along the surface of the inner shell. ssm www

46P. Two parallel plates of area 100 cm^2 are given charges of equal magnitudes 8.9×10^{-7} C but opposite signs. The electric field within the dielectric material filling the space between the plates is 1.4×10^6 V/m. (a) Calculate the dielectric constant of the material. (b) Determine the magnitude of the charge induced on each dielectric surface.

47P. A dielectric slab of thickness b is inserted between the plates of a parallel-plate capacitor of plate separation d. Show that the capacitance is then given by

$$C = \frac{\kappa \varepsilon_0 A}{\kappa d - b(\kappa - 1)}.$$

(*Hint:* You can derive the formula following the procedure outlined in Sample Problem 26-6.) Does this formula predict the correct numerical result of Sample Problem 26-6? Verify that the formula gives reasonable results for the special cases of $b = 0$, $\kappa = 1$, and $b = d$. ssm

Additional Problem

48. *The chocolate crumb mystery.* This story begins with Problem 48 in Chapter 24 and Problem 57 in Chapter 25. As part of the investigation of the biscuit factory explosion, the electric potentials of the workers were measured as they emptied sacks of chocolate crumb powder into the loading bin, stirring up a cloud of the powder around themselves. Each worker had an electric potential of about 7.0 kV relative to the ground, which was taken as zero potential. (a) Assuming that each worker was effectively a capacitor with a typical capacitance of 200 pF, find the energy stored in that effective capacitor. If a single spark between the worker and any conducting object connected to the ground neutralized the worker, that energy would be transferred to the spark. According to measurements, a spark that could ignite a cloud of chocolate crumb powder, and thus set off an explosion, had to have an energy of at least 150 mJ. (b) Could a spark from a worker have set off an explosion in the cloud of powder in the loading bin? (The story continues with Problem 44 in Chapter 27.)

27 Current and Resistance

The pride of Germany and a wonder of its time, the zeppelin *Hindenburg*—almost the length of three football fields—was the largest flying machine that had ever been built. Although it was kept aloft by 16 cells of highly flammable hydrogen gas, it made many trans-Atlantic trips without incident. In fact, German zeppelins, which all depended on hydrogen, had never suffered an accident due to the hydrogen. However, shortly after 7:21 p.m. on May 6, 1937, as the *Hindenburg* was ready to land at the U.S. Naval Air Station at Lakehurst, New Jersey, the ship burst into flames. Its crew had been waiting for a rainstorm to diminish, and handling ropes had just been let down to a navy ground crew, when ripples were sighted on the outer fabric of the ship about one-third of the way forward from the stern. Seconds later a flame erupted from that region, and a red glow illuminated the interior of the ship. Within 32 seconds the burning ship fell to the ground.

After so many successful flights of hydrogen-floated zeppelins, why did this zeppelin burst into flames?

The answer is in this chapter.

27-1 Moving Charges and Electric Currents

Chapters 22 through 26 deal largely with *electrostatics*—that is, with charges at rest. With this chapter we begin to focus on **electric currents**—that is, charges in motion.

Examples of electric currents abound, ranging from the large currents that constitute lightning strokes to the tiny nerve currents that regulate our muscular activity. The currents in household wiring, in lightbulbs, and in electrical appliances are familiar to all. A beam of electrons—a current—moves through an evacuated space in the picture tube of a common television set. Charged particles of *both* signs flow in the ionized gases of fluorescent lamps, in the batteries of radios, and in car batteries. Electric currents can also be found in the semiconductors in calculators and in the chips that control microwave ovens and electric dishwashers.

On a global scale, charged particles trapped in the Van Allen radiation belts surge back and forth above the atmosphere between Earth's north and south magnetic poles. On the scale of the solar system, enormous currents of protons, electrons, and ions fly radially outward from the Sun as the *solar wind*. On the galactic scale, cosmic rays, which are largely energetic protons, stream through our Milky Way galaxy, some reaching Earth.

Although an electric current is a stream of moving charges, not all moving charges constitute an electric current. If there is to be an electric current through a given surface, there must be a net flow of charge through that surface. Two examples clarify our meaning.

1. The free electrons (conduction electrons) in an isolated length of copper wire are in random motion at speeds of the order of 10^6 m/s. If you pass a hypothetical plane through such a wire, conduction electrons pass through it *in both directions* at the rate of many billions per second—but there is *no net transport* of charge and thus *no current* through the wire. However, if you connect the ends of the wire to a battery, you slightly bias the flow in one direction, with the result that there now is a net transport of charge and thus an electric current through the wire.

2. The flow of water through a garden hose represents the directed flow of positive charge (the protons in the water molecules) at a rate of perhaps several million coulombs per second. There is no net transport of charge, however, because there is a parallel flow of negative charge (the electrons in the water molecules) of exactly the same amount moving in exactly the same direction.

In this chapter we restrict ourselves largely to the study—within the framework of classical physics—of *steady* currents of *conduction electrons* moving through *metallic conductors* such as copper wires.

27-2 Electric Current

As Fig. 27-1a reminds us, an isolated conducting loop—regardless of whether it has an excess charge—is all at the same potential. No electric field can exist within it or along its surface. Although conduction electrons are available, no net electric force acts on them and thus there is no current.

If, as in Fig. 27-1b, we insert a battery in the loop, the conducting loop is no longer at a single potential. Electric fields act inside the material making up the loop, exerting forces on the conduction electrons, causing them to move, and thus establishing a current. After a very short time, the electron flow reaches a constant value and the current is in its *steady state* (it does not vary with time).

(a)

(b)

Fig. 27-1 (a) A loop of copper in electrostatic equilibrium. The entire loop is at a single potential, and the electric field is zero at all points inside the copper. (b) Adding a battery imposes an electric potential difference between the ends of the loop that are connected to the terminals of the battery. The battery thus produces an electric field within the loop, from terminal to terminal, and the field causes charges to move around the loop. This movement of charges is a current *i*.

Fig. 27-2 The current i through the conductor has the same value at planes aa', bb', and cc'.

Figure 27-2 shows a section of a conductor, part of a conducting loop in which current has been established. If charge dq passes through a hypothetical plane (such as aa') in time dt, then the current i through that plane is defined as

$$i = \frac{dq}{dt} \quad \text{(definition of current).} \tag{27-1}$$

We can find the charge that passes through the plane in a time interval extending from 0 to t by integration:

$$q = \int dq = \int_0^t i \, dt, \tag{27-2}$$

in which the current i may vary with time.

Under steady-state conditions, the current is the same for planes aa', bb', and cc' and indeed for all planes that pass completely through the conductor, no matter what their location or orientation. This follows from the fact that charge is conserved. Under the steady-state conditions assumed here, an electron must pass through plane aa' for every electron that passes through plane cc'. In the same way, if we have a steady flow of water through a garden hose, a drop of water must leave the nozzle for every drop that enters the hose at the other end. The amount of water in the hose is a conserved quantity.

The SI unit for current is the coulomb per second, also called the *ampere* (A):

1 ampere = 1 A = 1 coulomb per second = 1 C/s.

The ampere is an SI base unit; the coulomb is defined in terms of the ampere, as we discussed in Chapter 22. The formal definition of the ampere is discussed in Chapter 30.

Current, as defined by Eq. 27-1, is a scalar because both charge and time in that equation are scalars. Yet, as in Fig. 27-1b, we often represent a current with an arrow to indicate that charge is moving. Such arrows are not vectors, however, and they do not require vector addition. Figure 27-3a shows a conductor with current i_0 splitting at a junction into two branches. Because charge is conserved, the magnitudes of the currents in the branches must add to yield the magnitude of the current in the original conductor, so that

$$i_0 = i_1 + i_2. \tag{27-3}$$

As Fig. 27-3b suggests, bending or reorienting the wires in space does not change the validity of Eq. 27-3. Current arrows show only a direction (or sense) of flow along a conductor, not a direction in space.

The Directions of Currents

In Fig. 27-1b we drew the current arrows in the direction in which positively charged particles would be forced to move through the loop by the electric field. Such positive *charge carriers*, as they are often called, would move away from the positive battery terminal and toward the negative terminal. Actually, the charge carriers in the copper loop of Fig. 27-1b are electrons and thus are negatively charged. The electric field forces them to move in the direction opposite the current arrows, from the negative terminal to the positive terminal. For historical reasons, however, we use the following convention:

▶ A current arrow is drawn in the direction in which positive charge carriers would move, even if the actual charge carriers are negative and move in the opposite direction.

i_1

i_0

i_2

a

(a)

i_1

i_0

i_2

a

(b)

Fig. 27-3 The relation $i_0 = i_1 + i_2$ is true at junction a no matter what the orientation in space of the three wires. Currents are scalars, not vectors.

We can use this convention because in *most* situations, the assumed motion of positive charge carriers in one direction has the same effect as the actual motion of negative charge carriers in the opposite direction. (When the effect is not the same, we shall, of course, drop the convention and describe the actual motion.)

✔**CHECKPOINT 1:**   The figure here shows a portion of a circuit. What are the magnitude and direction of the current i in the lower right-hand wire?

Sample Problem 27-1

Water flows through a garden hose at a volume flow rate dV/dt of 450 cm³/s. What is the current of negative charge?

SOLUTION: The current i of negative charge is due to the electrons in the water molecules moving through the hose. The current is the rate at which that negative charge passes through any plane that cuts completely across the hose. Thus, the Key Idea here is that we can write the current in terms of the number of molecules that pass through such a plane per second as

$$i = \left(\begin{array}{c}\text{charge} \\ \text{per} \\ \text{electron}\end{array}\right)\left(\begin{array}{c}\text{electrons} \\ \text{per} \\ \text{molecule}\end{array}\right)\left(\begin{array}{c}\text{molecules} \\ \text{per} \\ \text{second}\end{array}\right)$$

or

$$i = (e)(10)\frac{dN}{dt}.$$

We substitute 10 electrons per molecule because a water (H_2O) molecule contains 8 electrons in the single oxygen atom and 1 electron in each of the two hydrogen atoms.

We can express the rate dN/dt in terms of the given volume flow rate dV/dt by first writing

$$\left(\begin{array}{c}\text{molecules} \\ \text{per} \\ \text{second}\end{array}\right) = \left(\begin{array}{c}\text{molecules} \\ \text{per} \\ \text{mole}\end{array}\right)\left(\begin{array}{c}\text{moles} \\ \text{per unit} \\ \text{mass}\end{array}\right)\left(\begin{array}{c}\text{mass} \\ \text{per unit} \\ \text{volume}\end{array}\right)\left(\begin{array}{c}\text{volume} \\ \text{per} \\ \text{second}\end{array}\right).$$

"Molecules per mole" is Avogadro's number N_A. "Moles per unit

mass" is the inverse of the mass per mole, which is the molar mass M of water. "Mass per unit volume" is the (mass) density ρ_{mass} of water. The volume per second is the volume flow rate dV/dt. Thus, we have

$$\frac{dN}{dt} = N_A\left(\frac{1}{M}\right)\rho_{\text{mass}}\left(\frac{dV}{dt}\right) = \frac{N_A\rho_{\text{mass}}}{M}\frac{dV}{dt}.$$

Substituting this into the equation for i, we find

$$i = 10eN_A M^{-1}\rho_{\text{mass}}\frac{dV}{dt}.$$

N_A is 6.02×10^{23} molecules/mol, or 6.02×10^{23} mol^{-1}, and ρ_{mass} is 1000 kg/m³. We can get the molar mass of water from the molar masses listed in Appendix F: We add the molar mass of oxygen (16 g/mol) to twice the molar mass of hydrogen (1 g/mol), obtaining 18 g/mol = 0.018 kg/mol. Then

$$\begin{aligned} i &= (10)(1.6 \times 10^{-19} \text{ C})(6.02 \times 10^{23} \text{ mol}^{-1}) \\ &\quad \times (0.018 \text{ kg/mol})^{-1}(1000 \text{ kg/m}^3)(450 \times 10^{-6} \text{ m}^3/\text{s}) \\ &= 2.41 \times 10^7 \text{ C/s} = 2.41 \times 10^7 \text{ A} \\ &= 24.1 \text{ MA.} \quad\quad\quad\quad\quad\quad\quad\quad\quad\text{(Answer)} \end{aligned}$$

This current of negative charge is exactly compensated by a current of positive charge associated with the nuclei of the three atoms that make up the water molecule. Thus, there is no net flow of charge through the hose.

27-3 Current Density

Sometimes we are interested in the current i in a particular conductor. At other times we take a localized view and study the flow of charge through a cross section of the conductor at a particular point. To describe this flow, we can use the **current density** $\vec{J}$, which has the same direction as the velocity of the moving charges if they are positive and the opposite direction if they are negative. For each element of the cross section, the magnitude J is equal to the current per unit area through that element. We can write the amount of current through the element as $\vec{J} \cdot d\vec{A}$, where $d\vec{A}$ is the

Fig. 27-4 Streamlines representing current density in the flow of charge through a constricted conductor.

area vector of the element, perpendicular to the element. The total current through the surface is then

$$i = \int \vec{J} \cdot d\vec{A}. \qquad (27\text{-}4)$$

If the current is uniform across the surface and parallel to $d\vec{A}$, then $\vec{J}$ is also uniform and parallel to $d\vec{A}$. Then Eq. 27-4 becomes

$$i = \int J \, dA = J \int dA = JA,$$

so

$$J = \frac{i}{A}, \qquad (27\text{-}5)$$

where A is the total area of the surface. From Eq. 27-4 or 27-5 we see that the SI unit for current density is the ampere per square meter (A/m^2).

In Chapter 23 we saw that we can represent an electric field with electric field lines. Figure 27-4 shows how current density can be represented with a similar set of lines, which we can call *streamlines*. The current, which is toward the right in Fig. 27-4, makes a transition from the wider conductor at the left to the narrower conductor at the right. Because charge is conserved during the transition, the amount of charge and thus the amount of current cannot change. However, the current density does change—it is greater in the narrower conductor. The spacing of the streamlines suggests this increase in current density; streamlines that are closer together imply greater current density.

Drift Speed

When a conductor does not have a current through it, its conduction electrons move randomly, with no net motion in any direction. When the conductor does have a current through it, these electrons actually still move randomly, but now they tend to *drift* with a **drift speed** v_d in the direction opposite that of the applied electric field that causes the current. The drift speed is tiny compared to the speeds in the random motion. For example, in the copper conductors of household wiring, electron drift speeds are perhaps 10^{-5} or 10^{-4} m/s, whereas the random-motion speeds are around 10^6 m/s.

We can use Fig. 27-5 to relate the drift speed v_d of the conduction electrons in a current through a wire to the magnitude J of the current density in the wire. For convenience, Fig. 27-5 shows the equivalent drift of *positive* charge carriers in the direction of the applied electric field $\vec{E}$. Let us assume that these charge carriers all move with the same drift speed v_d and that the current density J is uniform across the wire's cross-sectional area A. The number of charge carriers in a length L of the wire is nAL, where n is the number of carriers per unit volume. The total charge of the carriers in the length L, each with charge e, is then

$$q = (nAL)e.$$

Because the carriers all move along the wire with speed v_d, this total charge moves through any cross section of the wire in the time interval

$$t = \frac{L}{v_d}.$$

Equation 27-1 tells us that the current i is the time rate of transfer of charge across

Fig. 27-5 Positive charge carriers drift at speed v_d in the direction of the applied electric field $\vec{E}$. By convention, the direction of the current density $\vec{J}$ and the sense of the current arrow are drawn in that same direction.

a cross section, so here we have

$$i = \frac{q}{t} = \frac{nALe}{L/v_d} = nAev_d. \tag{27-6}$$

Solving for v_d and recalling Eq. 27-5 ($J - i/A$), we obtain

$$v_d = \frac{i}{nAe} = \frac{J}{ne}$$

or, extended to vector form,

$$\vec{J} = (ne)\vec{v}_d. \tag{27-7}$$

Here the product ne, whose SI unit is the coulomb per cubic meter (C/m^3), is the *carrier charge density*. For positive carriers, ne is positive and Eq. 27-7 predicts that $\vec{J}$ and $\vec{v}_d$ have the same direction. For negative carriers, ne is negative and $\vec{J}$ and $\vec{v}_d$ have opposite directions.

✔**CHECKPOINT 2:**   The figure here shows conduction electrons moving leftward through a wire. Are the following leftward or rightward: (a) the current i, (b) the current density $\vec{J}$, (c) the electric field $\vec{E}$ in the wire?

Sample Problem 27-2

(a) The current density in a cylindrical wire of radius $R = 2.0$ mm is uniform across a cross section of the wire and is $J = 2.0 \times 10^5$ A/m². What is the current through the outer portion of the wire between radial distances $R/2$ and R (Fig. 27-6a)?

SOLUTION: The **Key Idea** here is that, because the current density is uniform across the cross section, the current density J, the current i, and the cross-sectional area A are related by Eq. 27-5 ($J = i/A$). However, we want only the current through a reduced cross-sectional area A' of the wire (rather than the entire area), where

$$A' = \pi R^2 - \pi \left(\frac{R}{2}\right)^2 = \pi \left(\frac{3R^2}{4}\right)$$

$$= \frac{3\pi}{4}(0.002 \text{ m})^2 = 9.424 \times 10^{-6} \text{ m}^2.$$

We now rewrite Eq. 27-5 as

$$i = JA'$$

and then substitute the data to find

$$i = (2.0 \times 10^5 \text{ A/m}^2)(9.424 \times 10^{-6} \text{ m}^2)$$

$$= 1.9 \text{ A}. \qquad \text{(Answer)}$$

(b) Suppose, instead, that the current density through a cross section varies with radial distance r as $J = ar^2$, in which $a = 3.0 \times 10^{11}$ A/m⁴ and r is in meters. What now is the current through the same outer portion of the wire?

SOLUTION: The **Key Idea** here is that, because the current density is not uniform across a cross section of the wire, we must resort to Eq. 27-4 ($i = \int \vec{J} \cdot d\vec{A}$) and integrate the current density over the portion of the wire from $r = R/2$ to $r = R$. The current density vector $\vec{J}$ (along the wire's length) and the differential area vector

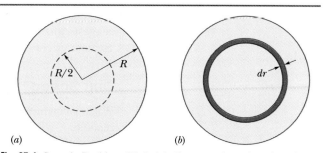

Fig. 27-6 Sample Problem 27-2. (*a*) Cross section of a wire of radius R. (*b*) A thin ring has width dr and circumference $2\pi r$, and thus a differential area $dA = 2\pi r \, dr$.

$d\vec{A}$ (perpendicular to a cross section of the wire) have the same direction. Thus,

$$\vec{J} \cdot d\vec{A} = J \, dA \cos 0 = J \, dA.$$

We need to replace the differential area dA with something we can actually integrate between the limits $r = R/2$ and $r = R$. The simplest replacement (because J is given as a function of r) is the area $2\pi r \, dr$ of a thin ring of circumference $2\pi r$ and width dr (Fig. 27-6*b*). We can then integrate with r as the variable of integration. Equation 27-4 then gives us

$$i = \int \vec{J} \cdot d\vec{A} = \int J \, dA$$

$$= \int_{R/2}^{R} ar^2 \, 2\pi r \, dr = 2\pi a \int_{R/2}^{R} r^3 \, dr$$

$$= 2\pi a \left[\frac{r^4}{4}\right]_{R/2}^{R} = \frac{\pi a}{2}\left[R^4 - \frac{R^4}{16}\right] = \frac{15}{32}\pi a R^4$$

$$= \frac{15}{32}\pi(3.0 \times 10^{11} \text{ A/m}^4)(0.002 \text{ m})^4 = 7.1 \text{ A}. \quad \text{(Answer)}$$

Sample Problem 27-3

What is the drift speed of the conduction electrons in a copper wire with radius $r = 900$ μm when it has a uniform current $i = 17$ mA? Assume that each copper atom contributes one conduction electron to the current and the current density is uniform across the wire's cross section.

SOLUTION: We need three Key Ideas here:

1. The drift speed v_d is related to the current density $\vec{J}$ and the number n of conduction electrons per unit volume according to Eq. 27-7, which we can write in magnitude form as $J = nev_d$.

2. Because the current density is uniform, its magnitude J is related to the given current i and wire size by Eq. 27-5 ($J = i/A$, where A is the cross-sectional area of the wire).

3. Because we assume one conduction electron per atom, the number n of conduction electrons per unit volume is the same as the number of atoms per unit volume.

Let us start with the third idea by writing

$$n = \begin{pmatrix} \text{atoms} \\ \text{per unit} \\ \text{volume} \end{pmatrix} = \begin{pmatrix} \text{atoms} \\ \text{per} \\ \text{mole} \end{pmatrix} \begin{pmatrix} \text{moles} \\ \text{per unit} \\ \text{mass} \end{pmatrix} \begin{pmatrix} \text{mass} \\ \text{per unit} \\ \text{volume} \end{pmatrix}.$$

The number of atoms per mole is just Avogadro's number N_A ($= 6.02 \times 10^{23}$ mol^{-1}). Moles per unit mass is the inverse of the mass per mole, which here is the molar mass M of copper. The mass per unit volume is the (mass) density ρ_{mass} of copper. Thus,

$$n = N_A \left(\frac{1}{M} \right) \rho_{\text{mass}} = \frac{N_A \rho_{\text{mass}}}{M}.$$

Taking copper's molar mass M and density ρ_{mass} from Appendix F, we then have (with some conversions of units)

$$n = \frac{(6.02 \times 10^{23} \text{ mol}^{-1})(8.96 \times 10^3 \text{ kg/m}^3)}{63.54 \times 10^{-3} \text{ kg/mol}}$$

$$= 8.49 \times 10^{28} \text{ electrons/m}^3$$

or

$$n = 8.49 \times 10^{28} \text{ m}^{-3}.$$

Next let us combine the first two key ideas by writing

$$\frac{i}{A} = nev_d.$$

Substituting for A with πr^2 ($= 2.54 \times 10^{-6}$ m^2), and solving for v_d, we then find

$$v_d = \frac{i}{ne(\pi r^2)}$$

$$= \frac{17 \times 10^{-3} \text{ A}}{(8.49 \times 10^{28} \text{ m}^{-3})(1.6 \times 10^{-19} \text{ C})(2.54 \times 10^{-6} \text{ m}^2)}$$

$$= 4.9 \times 10^{-7} \text{ m/s}, \qquad \text{(Answer)}$$

which is only 1.8 mm/h, slower than a sluggish snail.

You may well ask: "If the electrons drift so slowly, why do the room lights turn on so quickly when I throw the switch?" Confusion on this point results from not distinguishing between the drift speed of the electrons and the speed at which *changes* in the electric field configuration travel along wires. This latter speed is nearly that of light; electrons everywhere in the wire begin drifting almost at once, including into the lightbulbs. Similarly, when you open the valve on your garden hose, with the hose full of water, a pressure wave travels along the hose at the speed of sound in water. The speed at which the water itself moves through the hose—measured perhaps with a dye marker—is much slower.

27-4 Resistance and Resistivity

If we apply the same potential difference between the ends of geometrically similar rods of copper and of glass, very different currents result. The characteristic of the conductor that enters here is its electrical **resistance.** We determine the resistance between any two points of a conductor by applying a potential difference V between those points and measuring the current i that results. The resistance R is then

$$R = \frac{V}{i} \qquad \text{(definition of } R\text{)}. \qquad (27\text{-}8)$$

The SI unit for resistance that follows from Eq. 27-8 is the volt per ampere. This combination occurs so often that we give it a special name, the **ohm** (symbol Ω); that is,

$$1 \text{ ohm} = 1 \text{ } \Omega = 1 \text{ volt per ampere}$$

$$= 1 \text{ V/A}. \qquad (27\text{-}9)$$

A conductor whose function in a circuit is to provide a specified resistance is called a **resistor** (see Fig. 27-7). In a circuit diagram, we represent a resistor and a resistance

Fig. 27-7 An assortment of resistors. The circular bands are color-coding marks that identify the value of the resistance.

(a)

(b)

Fig. 27-8 Two ways of applying a potential difference to a conducting rod. The heavy gray connectors are assumed to have negligible resistance. When they are arranged as in (a), the measured resistance is larger than when they are arranged as in (b).

with the symbol ─WW─. If we write Eq. 27-8 as

$$i = \frac{V}{R},$$

we see that "resistance" is aptly named. For a given potential difference, the greater the resistance (to current), the smaller the current.

The resistance of a conductor depends on the manner in which the potential difference is applied to it. Figure 27-8, for example, shows a given potential difference applied in two different ways to the same conductor. As the current density streamlines suggest, the currents in the two cases—hence the measured resistances—will be different. Unless otherwise stated, we shall assume that any given potential difference is applied as in Fig. 27-8b.

As we have done several times in other connections, we often wish to take a general view and deal not with particular objects but with materials. Here we do so by focusing not on the potential difference V across a particular resistor but on the electric field $\vec{E}$ at a point in a resistive material. Instead of dealing with the current i through the resistor, we deal with the current density $\vec{J}$ at the point in question. Instead of the resistance R of an object, we deal with the **resistivity** ρ of the *material*:

$$\rho = \frac{E}{J} \qquad \text{(definition of } \rho \text{)}. \qquad (27\text{-}10)$$

(Compare this equation with Eq. 27-8.)

If we combine the SI units of E and J according to Eq. 27-10, we get, for the unit of ρ, the ohm-meter ($\Omega \cdot$ m):

$$\frac{\text{unit } (E)}{\text{unit } (J)} = \frac{\text{V/m}}{\text{A/m}^2} = \frac{\text{V}}{\text{A}}\text{m} = \Omega \cdot \text{m}.$$

(Do not confuse the *ohm-meter*, the unit of resistivity, with the *ohmmeter*, which is an instrument that measures resistance.) Table 27-1 lists the resistivities of some materials.

We can write Eq. 27-10 in vector form as

$$\vec{E} = \rho\vec{J}. \qquad (27\text{-}11)$$

Equations 27-10 and 27-11 hold only for *isotropic* materials—materials whose electrical properties are the same in all directions.

We often speak of the **conductivity** σ of a material. This is simply the reciprocal of its resistivity, so

$$\sigma = \frac{1}{\rho} \qquad \text{(definition of } \sigma \text{)}. \qquad (27\text{-}12)$$

The SI unit of conductivity is the reciprocal ohm-meter, $(\Omega \cdot \text{m})^{-1}$. The unit name mhos per meter is sometimes used (mho is ohm backwards). The definition of σ allows us to write Eq. 27-11 in the alternative form

$$\vec{J} = \sigma\vec{E}. \qquad (27\text{-}13)$$

Calculating Resistance from Resistivity

We have just made an important distinction:

> Resistance is a property of an object. Resistivity is a property of a material.

TABLE 27-1 **Resistivities of Some Materials at Room Temperature (20°C)**

Material	Resistivity, ρ ($\Omega \cdot$ m)	Temperature Coefficient of Resistivity, α (K^{-1})
Typical Metals		
Silver	1.62×10^{-8}	4.1×10^{-3}
Copper	1.69×10^{-8}	4.3×10^{-3}
Aluminum	2.75×10^{-8}	4.4×10^{-3}
Tungsten	5.25×10^{-8}	4.5×10^{-3}
Iron	9.68×10^{-8}	6.5×10^{-3}
Platinum	10.6×10^{-8}	3.9×10^{-3}
Manganin[a]	4.82×10^{-8}	0.002×10^{-3}
Typical Semiconductors		
Silicon, pure	2.5×10^{3}	-70×10^{-3}
Silicon, n-type[b]	8.7×10^{-4}	
Silicon, p-type[c]	2.8×10^{-3}	
Typical Insulators		
Glass	$10^{10} - 10^{14}$	
Fused quartz	$\sim 10^{16}$	

[a]An alloy specifically designed to have a small value of α.

[b]Pure silicon doped with phosphorus impurities to a charge carrier density of 10^{23} m^{-3}.

[c]Pure silicon doped with aluminum impurities to a charge carrier density of 10^{23} m^{-3}.

If we know the resistivity of a substance such as copper, we can calculate the resistance of a length of wire made of that substance. Let A be the cross-sectional area of the wire, let L be its length, and let a potential difference V exist between its ends (Fig. 27-9). If the streamlines representing the current density are uniform throughout the wire, the electric field and the current density will be constant for all points within the wire and, from Eqs. 25-42 and 27-5, will have the values

$$E = V/L \quad \text{and} \quad J = i/A. \tag{27-14}$$

We can then combine Eqs. 27-10 and 27-14 to write

$$\rho = \frac{E}{J} = \frac{V/L}{i/A}. \tag{27-15}$$

However, V/i is the resistance R, which allows us to recast Eq. 27-15 as

$$R = \rho \frac{L}{A}. \tag{27-16}$$

Equation 27-16 can be applied only to a homogeneous isotropic conductor of uniform cross section, with the potential difference applied as in Fig. 27-8b.

The macroscopic quantities V, i, and R are of greatest interest when we are making electrical measurements on specific conductors. They are the quantities that we read directly on meters. We turn to the microscopic quantities E, J, and ρ when we are interested in the fundamental electrical properties of materials.

Fig. 27-9 A potential difference V is applied between the ends of a wire of length L and cross section A, establishing a current i.

✔**CHECKPOINT 3:**   The figure here shows three cylindrical copper conductors along with their face areas and lengths. Rank them according to the current through them, greatest first, when the same potential difference V is placed across their lengths.

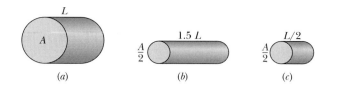

Variation with Temperature

The values of most physical properties vary with temperature, and resistivity is no exception. Figure 27-10, for example, shows the variation of this property for copper over a wide temperature range. The relation between temperature and resistivity for copper—and for metals in general—is fairly linear over a rather broad temperature range. For such linear relations we can write an empirical approximation that is good enough for most engineering purposes:

$$\rho - \rho_0 = \rho_0 \alpha (T - T_0). \tag{27-17}$$

Here T_0 is a selected reference temperature and ρ_0 is the resistivity at that temperature. Usually $T_0 = 293$ K (room temperature), for which $\rho_0 = 1.69 \times 10^{-8}\ \Omega \cdot$ m for copper.

Because temperature enters Eq. 27-17 only as a difference, it does not matter whether you use the Celsius or Kelvin scale in that equation because the sizes of degrees on these scales are identical. The quantity α in Eq. 27-17, called the *temperature coefficient of resistivity,* is chosen so that the equation gives good agreement with experiment for temperatures in the chosen range. Some values of α for metals are listed in Table 27-1.

The *Hindenburg*

When the zeppelin *Hindenburg* was preparing to land, the handling ropes were let down to the ground crew. Exposed to the rain, the ropes became wet (and thus were able to conduct a current). In this condition, the ropes "grounded" the metal framework of the zeppelin to which they were attached; that is, the wet ropes formed a conducting path between the framework and the ground, making the electric potential of the framework the same as that of the ground. This should have also grounded the outer fabric of the zeppelin. The *Hindenburg*, however, had been the first zeppelin to have its outer fabric painted with a sealant of large electrical resistivity. Thus, the fabric remained at the electric potential of the atmosphere at the zeppelin's altitude of about 43 m. Due to the rainstorm, that potential was large relative to the potential at ground level.

The handling of the ropes apparently ruptured one of the hydrogen cells and released hydrogen between that cell and the zeppelin's outer fabric, causing the reported rippling of the fabric. There was then a dangerous situation: The fabric was wet with conducting rainwater and was at a potential much different from that of the framework of the zeppelin. Apparently, charge flowed along the wet fabric and then sparked through the released hydrogen to reach the metal framework of the zeppelin, igniting the hydrogen in the process. The burning rapidly ignited the cells of hydrogen in the zeppelin and brought the ship down. If the sealant on the outer fabric of the *Hindenburg* had been of less resistivity (like that of earlier and later zeppelins), the *Hindenburg* disaster probably would not have occurred.

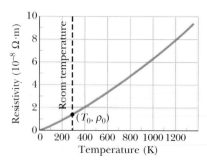

Fig. 27-10 The resistivity of copper as a function of temperature. The dot on the curve marks a convenient reference point at temperature $T_0 = 293$ K and resistivity $\rho_0 = 1.69 \times 10^{-8}\ \Omega \cdot$ m.

Sample Problem 27-4

A rectangular block of iron has dimensions 1.2 cm $\times$ 1.2 cm $\times$ 15 cm. A potential difference is to be applied to the block between parallel sides and in such a way that those sides are equipotential surfaces (as in Fig. 27-8b). What is the resistance of the block if the two parallel sides are (1) the square ends (with dimensions 1.2 cm $\times$ 1.2 cm) and (2) two rectangular sides (with dimensions 1.2 cm $\times$ 15 cm)?

SOLUTION: The Key Idea here is that the resistance R of an object depends on how the electric potential is applied to the object. In particular, it depends on the ratio L/A, according to Eq. 27-16 ($R = \rho L/A$), where A is the area of the surfaces to which the potential difference is applied and L is the distance between those surfaces. For arrangement 1, $L = 15$ cm $= 0.15$ m and

$$A = (1.2 \text{ cm})^2 = 1.44 \times 10^{-4} \text{ m}^2.$$

Substituting into Eq. 27-16 with the resistivity ρ from Table 27-1, we then find that for arrangement 1,

$$R = \frac{\rho L}{A} = \frac{(9.68 \times 10^{-8}\ \Omega \cdot \text{m})(0.15 \text{ m})}{1.44 \times 10^{-4} \text{ m}^2}$$
$$= 1.0 \times 10^{-4}\ \Omega = 100\ \mu\Omega. \qquad \text{(Answer)}$$

Similarly, for arrangement 2, with distance $L = 1.2$ cm and area $A = (1.2 \text{ cm})(15 \text{ cm})$, we obtain

$$R = \frac{\rho L}{A} = \frac{(9.68 \times 10^{-8}\ \Omega \cdot \text{m})(1.2 \times 10^{-2} \text{ m})}{1.80 \times 10^{-3} \text{ m}^2}$$
$$= 6.5 \times 10^{-7}\ \Omega = 0.65\ \mu\Omega. \qquad \text{(Answer)}$$

27-5 Ohm's Law

As we just discussed in Section 27-4, a resistor is a conductor with a specified resistance. It has that same resistance no matter what the magnitude and direction (*polarity*) of the applied potential difference. Other conducting devices, however, might have resistances that change with the applied potential difference.

Figure 27-11a shows how to distinguish such devices. A potential difference V is applied across the device being tested, and the resulting current i through the device is measured as V is varied in both magnitude and polarity. The polarity of V is arbitrarily taken to be positive when the left terminal of the device is at a higher potential than the right terminal. The direction of the resulting current (from left to right) is arbitrarily assigned a plus sign. The reverse polarity of V (with the right terminal at a higher potential) is then negative; the current it causes is assigned a minus sign.

Figure 27-11b is a plot of i versus V for one device. This plot is a straight line passing through the origin, so the ratio i/V (which is the slope of the straight line) is the same for all values of V. This means that the resistance $R = V/i$ of the device is independent of the magnitude and polarity of the applied potential difference V.

Figure 27-11c is a plot for another conducting device. Current can exist in this device only when the polarity of V is positive and the applied potential difference is more than about 1.5 V. When current does exist, the relation between i and V is not linear; it depends on the value of the applied potential difference V.

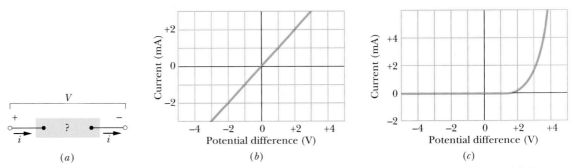

Fig. 27-11 (a) A potential difference V is applied to the terminals of a device, establishing a current i. (b) A plot of current i versus applied potential difference V when the device is a 1000 Ω resistor. (c) A plot when the device is a semiconducting pn junction diode.

We distinguish between the two types of device by saying that one obeys Ohm's law and the other does not.

▶**Ohm's law** is an assertion that the current through a device is *always* directly proportional to the potential difference applied to the device.

(This assertion is correct only in certain situations; still, for historical reasons, the term "law" is used.) The device of Fig. 27-11*b*—which turns out to be a 1000 Ω resistor—obeys Ohm's law. The device of Fig. 27-11*c*—which turns out to be a so-called *pn* junction diode—does not.

▶A conducting device obeys Ohm's law when the resistance of the device is independent of the magnitude and polarity of the applied potential difference.

Modern microelectronics—and therefore much of the character of our present technological civilization—depends almost totally on devices that do *not* obey Ohm's law. Your calculator, for example, is full of them.

It is often contended that $V = iR$ is a statement of Ohm's law. That is not true! This equation is the defining equation for resistance, and it applies to all conducting devices, whether they obey Ohm's law or not. If we measure the potential difference V across, and the current i through, any device, even a *pn* junction diode, we can find its resistance *at that value of V* as $R = V/i$. The essence of Ohm's law, however, is that a plot of i versus V is linear; that is, R is independent of V.

We can express Ohm's law in a more general way if we focus on conducting *materials* rather than on conducting *devices*. The relevant relation is then Eq. 27-11 ($\vec{E} = \rho \vec{J}$), which is the analog of $V = iR$.

▶A conducting material obeys Ohm's law when the resistivity of the material is independent of the magnitude and direction of the applied electric field.

All homogeneous materials, whether they are conductors like copper or semiconductors like pure silicon or silicon containing special impurities, obey Ohm's law within some range of values of the electric field. If the field is too strong, however, there are departures from Ohm's law in all cases.

✔**CHECKPOINT 4:**   The following table gives the current i (in amperes) through two devices for several values of potential difference V (in volts). From these data, determine which device does not obey Ohm's law.

Device 1		Device 2	
V	i	V	i
2.00	4.50	2.00	1.50
3.00	6.75	3.00	2.20
4.00	9.00	4.00	2.80

27-6   A Microscopic View of Ohm's Law

To find out *why* particular materials obey Ohm's law, we must look into the details of the conduction process at the atomic level. Here we consider only conduction in metals, such as copper. We base our analysis on the *free-electron model,* in which we assume that the conduction electrons in the metal are free to move throughout the volume of a sample, like the molecules of a gas in a closed container. We form. The power P associated with that transfer is the rate of transfer dU/dt, which is

$$P = iV \qquad \text{(rate of electric energy transfer).}} \qquad (27\text{-}21)$$

Moreover, this power P is also the rate that energy is transferred from the battery to the unspecified device. If that device is a motor connected to a mechanical load, the energy is transferred as work done on the load. If the device is a storage battery that is being charged, the energy is transferred to stored chemical energy in the storage battery. If the device is a resistor, the energy is transferred to internal thermal energy, tending to increase the resistor's temperature.

The unit of power that follows from Eq. 27-21 is the volt-ampere (V · A). We can write it as

$$1 \text{ V} \cdot \text{A} = \left(1 \frac{\text{J}}{\text{C}}\right)\left(1 \frac{\text{C}}{\text{s}}\right) = 1 \frac{\text{J}}{\text{s}} = 1 \text{ W.}$$

The course of an electron moving through a resistor at constant drift speed is much like that of a stone falling through water at constant terminal speed. The average kinetic energy of the electron remains constant, and its lost electric potential energy appears as thermal energy in the resistor and the surroundings. On a microscopic scale this energy transfer is due to collisions between the electron and the molecules of the resistor, which leads to an increase in the temperature of the resistor lattice. The mechanical energy thus transferred to thermal energy is *dissipated* (lost), because the transfer cannot be reversed.

For a resistor or some other device with resistance R, we can combine Eqs. 27-8 ($R = V/i$) and 27-21 to obtain, for the rate of electric energy dissipation due to a resistance, either

$$P = i^2 R \qquad \text{(resistive dissipation)} \qquad (27\text{-}22)$$

or

$$P = \frac{V^2}{R} \qquad \text{(resistive dissipation).} \qquad (27\text{-}23)$$

Caution: We must be careful to distinguish these two equations from Eq. 27-21: $P = iV$ applies to electric energy transfers of all kinds; $P = i^2R$ and $P = V^2/R$ apply only to the transfer of electric potential energy to thermal energy in a device with resistance.

The wire coils within a toaster have appreciable resistance. When there is a current through them, electric energy is transferred to thermal energy of the coils, increasing their temperature. The coils then emit infrared radiation and visible light that will toast (or burn) bread.

✔**CHECKPOINT 5:** A potential difference V is connected across a device with resistance R, causing current i through the device. Rank the following variations according to the change in the rate at which electric energy is converted to thermal energy due to the resistance, greatest change first: (a) V is doubled with R unchanged, (b) i is doubled with R unchanged, (c) R is doubled with V unchanged, (d) R is doubled with i unchanged.

Sample Problem 27-6

You are given a length of uniform heating wire made of a nickel–chromium–iron alloy called Nichrome; it has a resistance R of 72 Ω. At what rate is energy dissipated in each of the following situations? (1) A potential difference of 120 V is applied across the full length of the wire. (2) The wire is cut in half, and a potential difference of 120 V is applied across the length of each half.

SOLUTION: The Key Idea is that a current in a resistive material pro-

duces a transfer of mechanical energy to thermal energy; the rate of transfer (dissipation) is given by Eqs. 27-21 to 27-23. Because we know the potential V and resistance R, we use Eq. 27-23, which yields, for situation 1,

$$P = \frac{V^2}{R} = \frac{(120 \text{ V})^2}{72 \text{ Ω}} = 200 \text{ W.} \qquad \text{(Answer)}$$

In situation 2, the resistance of each half of the wire is (72 Ω)/2, or 36 Ω. Thus, the dissipation rate for each half is

$$P' = \frac{(120\ V)^2}{36\ \Omega} = 400\ W,$$

and that for the two halves is

$$P = 2P' = 800\ W. \qquad \text{(Answer)}$$

This is four times the dissipation rate of the full length of wire. Thus, you might conclude that you could buy a heating coil, cut it in half, and reconnect it to obtain four times the heat output. Why is this unwise? (What would happen to the amount of current in the coil?)

27-8 Semiconductors

Semiconducting devices are at the heart of the microelectronic revolution that ushered in the information age. Table 27-2 compares the properties of silicon—a typical semiconductor—and copper—a typical metallic conductor. We see that silicon has many fewer charge carriers, a much higher resistivity, and a temperature coefficient of resistivity that is both large and negative. Thus, although the resistivity of copper increases with temperature, that of pure silicon decreases.

Pure silicon has such a high resistivity that it is effectively an insulator and thus not of much direct use in microelectronic circuits. However, its resistivity can be greatly reduced in a controlled way by adding minute amounts of specific "impurity" atoms in a process called *doping*. Table 27-1 gives typical values of resistivity for silicon before and after doping with two different impurities.

We can roughly explain the differences in resistivity (and thus in conductivity) between semiconductors, insulators, and metallic conductors in terms of the energies of their electrons. (We need quantum physics to explain in more detail.) In a metallic conductor such as copper wire, most of the electrons are firmly locked into place within the molecules; much energy would be required to free them so they could move and participate in an electric current. However, there are also some electrons that, roughly speaking, are only loosely held in place and that require only little energy to become free. Thermal energy can supply that energy, as can an electric field applied across the conductor. The field would not only free these loosely held electrons but would also propel them along the wire; thus, the field would drive a current through the conductor.

In an insulator, significantly greater energy is required to free electrons so they can move through the material. Thermal energy cannot supply enough energy, and neither can any reasonable electric field applied to the insulator. Thus, no electrons are available to move through the insulator, and hence no current occurs even with an applied electric field.

A semiconductor is like an insulator *except* that the energy required to free some electrons is not quite so great. More important, doping can supply electrons or positive charge carriers that are very loosely held within the material and thus are easy to get moving. Moreover, by controlling the doping of a semiconductor, we can

TABLE 27-2   **Some Electrical Properties of Copper and Silicon**[a]

Property	Copper	Silicon
Type of material	Metal	Semiconductor
Charge carrier density, m^{-3}	9×10^{28}	1×10^{16}
Resistivity, Ω · m	2×10^{-8}	3×10^{3}
Temperature coefficient of resistivity, K^{-1}	$+4 \times 10^{-3}$	-70×10^{-3}

[a]Rounded to one significant figure for easy comparison.

Fig. 27-14 The resistance of mercury drops to zero at a temperature of about 4 K.

control the density of charge carriers that can participate in a current, and thereby can control some of its electrical properties. Most semiconducting devices, such as transistors and junction diodes, are fabricated by the selective doping of different regions of the silicon with impurity atoms of different kinds.

Let us now look again at Eq. 27-20 for the resistivity of a conductor:

$$\rho = \frac{m}{e^2 n \tau}, \qquad (27\text{-}24)$$

where n is the number of charge carriers per unit volume and τ is the mean time between collisions of the charge carriers. (We derived this equation for conductors, but it also applies to semiconductors.) Let us consider how the variables n and τ change as the temperature is increased.

In a conductor, n is large but very nearly constant with any change in temperature. The increase of resistivity with temperature for metals (Fig. 27-10) is due to an increase in the collision rate of the charge carriers, which shows up in Eq. 27-24 as a decrease in τ, the mean time between collisions.

In a semiconductor, n is small but increases very rapidly with temperature as the increased thermal agitation makes more charge carriers available. This causes a *decrease* of resistivity with increasing temperature, as indicated by the negative temperature coefficient of resistivity for silicon in Table 27-2. The same increase in collision rate that we noted for metals also occurs for semiconductors, but its effect is swamped by the rapid increase in the number of charge carriers.

27-9 Superconductors

In 1911, Dutch physicist Kamerlingh Onnes discovered that the resistivity of mercury absolutely disappears at temperatures below about 4 K (Fig. 27-14). This phenomenon of **superconductivity** is of vast potential importance in technology because it means that charge can flow through a superconducting conductor without producing thermal energy losses. Currents created in a superconducting ring, for example, have persisted for several years without diminution; the electrons making up the current require a force and a source of energy at start-up time, but not thereafter.

Prior to 1986, the technological development of superconductivity was throttled by the cost of producing the extremely low temperatures that were required to achieve the effect. In 1986, however, new ceramic materials were discovered that become superconducting at considerably higher (and thus cheaper to produce) temperatures. Practical application of superconducting devices at room temperature may eventually become feasible.

Superconductivity is a much different phenomenon from conductivity. In fact, the best of the normal conductors, such as silver and copper, cannot become superconducting at any temperature, and the new ceramic superconductors are actually good insulators when they are not at low enough temperatures to be in a superconducting state.

One explanation for superconductivity is that the electrons which make up the current move in coordinated pairs. One of the electrons in a pair may electrically distort the molecular structure of the superconducting material as it moves through, creating nearby a short-lived concentration of positive charge. The other electron in the pair may then be attracted toward this positive charge. According to the theory, such coordination between electrons would prevent them from colliding with the molecules of the material and thus would eliminate electrical resistance. The theory worked well to explain the pre-1986, lower temperature superconductors, but new theories appear to be needed for the newer, higher temperature superconductors.

A disk-shaped magnet is levitated above a superconducting material that has been cooled by liquid nitrogen. The goldfish is along for the ride.

REVIEW & SUMMARY

Current An **electric current** i in a conductor is defined by

$$i = \frac{dq}{dt}. \qquad (27\text{-}1)$$

Here dq is the amount of (positive) charge that passes in time dt through a hypothetical surface that cuts across the conductor. By convention, the direction of electric current is taken as the direction in which positive charge carriers would move. The SI unit of electric current is the **ampere** (A): 1 A = 1 C/s.

Current Density Current (a scalar) is related to **current density** $\vec{J}$ (a vector) by

$$i = \int \vec{J} \cdot d\vec{A}, \qquad (27\text{-}4)$$

where $d\vec{A}$ is a vector perpendicular to a surface element of area dA, and the integral is taken over any surface cutting across the conductor. $\vec{J}$ has the same direction as the velocity of the moving charges if they are positive and the opposite direction if they are negative.

Drift Speed of the Charge Carriers When an electric field $\vec{E}$ is established in a conductor, the charge carriers (assumed positive) acquire a **drift speed** v_d in the direction of $\vec{E}$; the velocity $\vec{v}_d$ is related to the current density by

$$\vec{J} = (ne)\vec{v}_d, \qquad (27\text{-}7)$$

where ne is the carrier charge density.

Resistance of a Conductor The **resistance** R of a conductor is defined as

$$R = \frac{V}{i} \qquad \text{(definition of } R\text{)}, \qquad (27\text{-}8)$$

where V is the potential difference across the conductor and i is the current. The SI unit of resistance is the **ohm** (Ω): 1 Ω = 1 V/A. Similar equations define the **resistivity** ρ and **conductivity** σ of a material:

$$\rho = \frac{1}{\sigma} = \frac{E}{J} \qquad \text{(definitions of } \rho \text{ and } \sigma\text{)}, \qquad (27\text{-}12, 27\text{-}10)$$

where E is the magnitude of the applied electric field. The SI unit of resistivity is the ohm-meter ($\Omega \cdot$ m). Equation 27-10 corresponds to the vector equation

$$\vec{E} = \rho \vec{J}. \qquad (27\text{-}11)$$

The resistance R of a conducting wire of length L and uniform cross section is

$$R = \rho \frac{L}{A}, \qquad (27\text{-}16)$$

where A is the cross-sectional area.

Change of ρ with Temperature The resistivity ρ for most materials changes with temperature. For many materials, including metals, the relation between ρ and temperature T is approximated by the equation

$$\rho - \rho_0 = \rho_0 \alpha (T - T_0). \qquad (27\text{-}17)$$

Here T_0 is a reference temperature, ρ_0 is the resistivity at T_0, and α is the temperature coefficient of resistivity for the material.

Ohm's Law A given device (conductor, resistor, or any other electrical device) obeys *Ohm's law* if its resistance R, defined by Eq. 27-8 as V/i, is independent of the applied potential difference V. A given *material* obeys Ohm's law if its resistivity, defined by Eq. 27-10, is independent of the magnitude and direction of the applied electric field $\vec{E}$.

Resistivity of a Metal By assuming that the conduction electrons in a metal are free to move like the molecules of a gas, it is possible to derive an expression for the resistivity of a metal:

$$\rho = \frac{m}{e^2 n \tau}. \qquad (27\text{-}20)$$

Here n is the number of free electrons per unit volume, and τ is the mean time between the collisions of an electron with the atoms of the metal. We can explain why metals obey Ohm's law by pointing out that τ is essentially independent of the magnitude E of any electric field applied to a metal.

Power The power P, or rate of energy transfer, in an electrical device across which a potential difference V is maintained is

$$P = iV \qquad \text{(rate of electric energy transfer)}. \qquad (27\text{-}21)$$

Resistive Dissipation If the device is a resistor, we can write Eq. 27-21 as

$$P = i^2 R = \frac{V^2}{R} \qquad \text{(resistive dissipation)}. \qquad (27\text{-}22, 27\text{-}23)$$

In a resistor, electric potential energy is converted to internal thermal energy via collisions between charge carriers and atoms.

Semiconductors *Semiconductors* are materials with few conduction electrons but can become conductors when they are *doped* with other atoms that contribute free electrons.

Superconductors *Superconductors* are materials that lose all electrical resistance at low temperatures. Recent research has discovered materials that are superconducting at surprisingly high temperatures.

QUESTIONS

1. Figure 27-15 shows plots of the current i through a certain cross section of a wire over four different time periods. Rank the periods according to the net charge that passes through the cross section during each, greatest first.

Fig. 27-15 Question 1.

2. Figure 27-16 shows four situations in which positive and negative charges move horizontally through a region and gives the rate at which each charge moves. Rank the situations according to the effective current through the regions, greatest first.

Fig. 27-16 Question 2.

3. Figure 27-17 shows cross sections through three wires of equal length and of the same material. The figure also gives the length of each side in millimeters. Rank the wires according to their resistances (measured end to end along each wire's length), greatest first.

Fig. 27-17 Question 3.

4. If you stretch a cylindrical wire and it remains cylindrical, does the resistance of the wire (measured end to end along its length) increase, decrease, or remain the same?

5. Figure 27-18 shows cross sections through three long conductors of the same length and material, with square cross sections of edge lengths as shown. Conductor B will fit snugly within conductor A, and conductor C will fit snugly within conductor B. Rank the fol-

Fig. 27-18 Question 5.

lowing according to their end-to-end resistances, greatest first: the individual conductors and the combinations of $A + B$, $B + C$, and $A + B + C$.

6. Figure 27-19 shows a rectangular solid conductor of edge lengths L, $2L$, and $3L$. A certain potential difference V is to be applied between pairs of opposite faces of the conductor as in Fig. 27-8b: left–right, top–bottom, and front–back. Rank those pairs according to (a) the magnitude of the electric field within the conductor, (b) the current density within the conductor, (c) the current through the conductor, and (d) the drift speed of the electrons through the conductor, greatest first.

Fig. 27-19 Question 6.

7. The following table gives the lengths of three copper rods, their diameters, and the potential differences between their ends. Rank the rods according to (a) the magnitude of the electric field within them, (b) the current density within them, and (c) the drift speed of electrons through them, greatest first.

Rod	Length	Diameter	Potential Difference
1	L	$3d$	V
2	$2L$	d	$2V$
3	$3L$	$2d$	$2V$

8. The following table gives the conductivity and the density of conduction electrons for materials A, B, C, and D. Rank the materials according to the average time between collisions of the conduction electrons in the materials, greatest first.

	A	B	C	D
Conductivity	σ	2σ	2σ	σ
Electrons/m^3	n	$2n$	n	$2n$

9. Three wires, of the same diameter, are connected in turn between two points maintained at a constant potential difference. Their resistivities and lengths are ρ and L (wire A), 1.2ρ and $1.2L$ (wire B), and 0.9ρ and L (wire C). Rank the wires according to the rate at which energy is transferred to thermal energy within them, greatest first.

10. Figure 27-20 gives the resistivities of four materials as a function of temperature. (a) Which materials are conductors, and which are semiconductors? In which materials does an increase in temperature result in (b) an increase in the number of conduction electrons per unit volume and (c) an increase in the collision rate of conduction electrons?

Fig. 27-20 Question 10.

EXERCISES & PROBLEMS

SEC. 27-2   Electric Current

1E. A current of 5.0 A exists in a 10 Ω resistor for 4.0 min. How many (a) coulombs and (b) electrons pass through any cross section of the resistor in this time? ssm

2P. A charged belt, 50 cm wide, travels at 30 m/s between a source of charge and a sphere. The belt carries charge into the sphere at a rate corresponding to 100 μA. Compute the surface charge density on the belt.

3P. An isolated conducting sphere has a 10 cm radius. One wire carries a current of 1.000 002 0 A into it. Another wire carries a current of 1.000 000 0 A out of it. How long would it take for the sphere to increase in potential by 1000 V? ssm

SEC. 27-3   Current Density

4E. A small but measurable current of 1.2×10^{-10} A exists in a copper wire whose diameter is 2.5 mm. Assuming the current is uniform, calculate (a) the current density and (b) the electron drift speed. (See Sample Problem 27-3.)

5E. A beam contains 2.0×10^8 doubly charged positive ions per cubic centimeter, all of which are moving north with a speed of 1.0×10^5 m/s. (a) What are the magnitude and direction of the current density $\vec{J}$? (b) Can you calculate the total current i in this ion beam? If not, what additional information is needed? ssm

6E. The (United States) National Electric Code, which sets maximum safe currents for insulated copper wires of various diameters, is given (in part) in the table. Plot the safe current density as a function of diameter. Which wire gauge has the maximum safe current density? ("Gauge" is a way of identifying wire diameters, and 1 mil = 10^{-3} in.)

Gauge	4	6	8	10	12	14	16	18
Diameter, mils	204	162	129	102	81	64	51	40
Safe current, A	70	50	35	25	20	15	6	3

7E. A fuse in an electric circuit is a wire that is designed to melt, and thereby open the circuit, if the current exceeds a predetermined value. Suppose that the material to be used in a fuse melts when the current density rises to 440 A/cm². What diameter of cylindrical wire should be used to make a fuse that will limit the current to 0.50 A? ssm

8P. Near Earth, the density of protons in the solar wind (a stream of particles from the Sun) is 8.70 cm⁻³, and their speed is 470 km/s. (a) Find the current density of these protons. (b) If Earth's magnetic field did not deflect them, the protons would strike the planet. What total current would Earth then receive?

9P. A steady beam of alpha particles ($q = +2e$) traveling with constant kinetic energy 20 MeV carries a current of 0.25 μA. (a) If the beam is directed perpendicular to a plane surface, how many alpha particles strike the surface in 3.0 s? (b) At any instant, how many alpha particles are there in a given 20 cm length of the beam? (c) Through what potential difference is it necessary to accelerate each alpha particle from rest to bring it to an energy of 20 MeV? ssm

10P. (a) The current density across a cylindrical conductor of radius R varies in magnitude according to the equation

$$J = J_0\left(1 - \frac{r}{R}\right),$$

where r is the distance from the central axis. Thus, the current density is a maximum J_0 at that axis ($r = 0$) and decreases linearly to zero at the surface ($r = R$). Calculate the current in terms of J_0 and the conductor's cross-sectional area $A = \pi R^2$. (b) Suppose that, instead, the current density is a maximum J_0 at the cylinder's surface and decreases linearly to zero at the axis: $J = J_0 r/R$. Calculate the current. Why is the result different from that in (a)?

11P. How long does it take electrons to get from a car battery to the starting motor? Assume the current is 300 A and the electrons travel through a copper wire with cross-sectional area 0.21 cm² and length 0.85 m. (See Sample Problem 27-3.) ilw

SEC. 27-4   Resistance and Resistivity

12E. A wire of Nichrome (a nickel–chromium–iron alloy commonly used in heating elements) is 1.0 m long and 1.0 mm² in cross-sectional area. It carries a current of 4.0 A when a 2.0 V potential difference is applied between its ends. Calculate the conductivity σ of Nichrome.

13E. A conducting wire has a 1.0 mm diameter, a 2.0 m length, and a 50 mΩ resistance. What is the resistivity of the material? ssm

14E. A steel trolley-car rail has a cross-sectional area of 56.0 cm². What is the resistance of 10.0 km of rail? The resistivity of the steel is 3.00×10^{-7} Ω·m.

15E. A human being can be electrocuted if a current as small as 50 mA passes near the heart. An electrician working with sweaty hands makes good contact with the two conductors he is holding, one in each hand. If his resistance is 2000 Ω, what might the fatal voltage be? ssm

16E. A wire 4.00 m long and 6.00 mm in diameter has a resistance of 15.0 mΩ. A potential difference of 23.0 V is applied between the ends. (a) What is the current in the wire? (b) What is the current density? (c) Calculate the resistivity of the wire material. Identify the material. (Use Table 27-1.)

17E. A coil is formed by winding 250 turns of insulated 16-gauge copper wire (diameter = 1.3 mm) in a single layer on a cylindrical form of radius 12 cm. What is the resistance of the coil? Neglect the thickness of the insulation. (Use Table 27-1.) ssm

18E. (a) At what temperature would the resistance of a copper conductor be double its resistance at 20.0°C? (Use 20.0°C as the reference point in Eq. 27-17; compare your answer with Fig. 27-10.)

(b) Does this same "doubling temperature" hold for all copper conductors, regardless of shape or size?

19E. A wire with a resistance of 6.0 Ω is drawn out through a die so that its new length is three times its original length. Find the resistance of the longer wire, assuming that the resistivity and density of the material are unchanged. ssm ilw

20E. A certain wire has a resistance R. What is the resistance of a second wire, made of the same material, that is half as long and has half the diameter?

21P. Two conductors are made of the same material and have the same length. Conductor A is a solid wire of diameter 1.0 mm. Conductor B is a hollow tube of outside diameter 2.0 mm and inside diameter 1.0 mm. What is the resistance ratio R_A/R_B, measured between their ends? ssm www

22P. An electrical cable consists of 125 strands of fine wire, each having 2.65 μΩ resistance. The same potential difference is applied between the ends of all the strands and results in a total current of 0.750 A. (a) What is the current in each strand? (b) What is the applied potential difference? (c) What is the resistance of the cable?

23P. When 115 V is applied across a wire that is 10 m long and has a 0.30 mm radius, the current density is 1.4×10^4 A/m². Find the resistivity of the wire. ssm

24P. A block in the shape of a rectangular solid has a cross-sectional area of 3.50 cm² across its width, a front-to-rear length of 15.8 cm, and a resistance of 935 Ω. The material of which the block is made has 5.33×10^{22} conduction electrons/m³. A potential difference of 35.8 V is maintained between its front and rear faces. (a) What is the current in the block? (b) If the current density is uniform, what is its value? (c) What is the drift velocity of the conduction electrons? (d) What is the magnitude of the electric field in the block?

25P. A common flashlight bulb is rated at 0.30 A and 2.9 V (the values of the current and voltage under operating conditions). If the resistance of the bulb filament at room temperature (20°C) is 1.1 Ω, what is the temperature of the filament when the bulb is on? The filament is made of tungsten. ilw

26P. Earth's lower atmosphere contains negative and positive ions that are produced by radioactive elements in the soil and cosmic rays from space. In a certain region, the atmospheric electric field strength is 120 V/m, directed vertically down. This field causes singly charged positive ions, at a density of 620/cm³, to drift downward and singly charged negative ions, at a density of 550/cm³, to drift upward (Fig. 27-21). The measured conductivity of the air in

that region is $2.70 \times 10^{-14}/\Omega \cdot$ m. Calculate (a) the ion drift speed, assumed to be the same for positive and negative ions, and (b) the current density.

27P. When a metal rod is heated, not only its resistance but also its length and its cross-sectional area change. The relation $R = \rho L/A$ suggests that all three factors should be taken into account in measuring ρ at various temperatures. (a) If the temperature changes by 1.0 C°, what percentage changes in R, L, and A occur for a copper conductor? (b) The coefficient of linear expansion for copper is 1.7×10^{-5}/K. What conclusion do you draw? ssm www

28P. If the gauge number of a wire is increased by 6, the diameter is halved; if a gauge number is increased by 1, the diameter decreases by the factor $2^{1/6}$ (see the table in Exercise 6). Knowing this, and knowing that 1000 ft of 10-gauge copper wire has a resistance of approximately 1.00 Ω, estimate the resistance of 25 ft of 22-gauge copper wire.

29P. A resistor has the shape of a truncated right-circular cone (Fig. 27-22). The end radii are a and b, and the altitude is L. If the taper is small, we may assume that the current density is uniform across any cross section. (a) Calculate the resistance of this object. (b) Show that your answer reduces to $\rho(L/A)$ for the special case of zero taper (that is, for $a = b$). ssm

Fig. 27-22 Problem 29.

SEC. 27-6 A Microscopic View of Ohm's Law

30P. Show that, according to the free-electron model of electrical conduction in metals and classical physics, the resistivity of metals should be proportional to $\sqrt{T}$, where T is the temperature in kelvins. (See Eq. 20-31.)

SEC. 27-7 Power in Electric Circuits

31E. A certain x-ray tube operates at a current of 7.0 mA and a potential difference of 80 kV. What is its power in watts? ssm

32E. A student kept his 9.0 V, 7.0 W radio turned on at full volume from 9:00 p.m. until 2:00 a.m. How much charge went through it?

33E. A 120 V potential difference is applied to a space heater whose resistance is 14 Ω when hot. (a) At what rate is electric energy transferred to heat? (b) At 5.0¢/kW · h, what does it cost to operate the device for 5.0 h? ssm

34E. Thermal energy is produced in a resistor at a rate of 100 W when the current is 3.00 A. What is the resistance?

35E. An unknown resistor is connected between the terminals of a 3.00 V battery. Energy is dissipated in the resistor at the rate of 0.540 W. The same resistor is then connected between the terminals of a 1.50 V battery. At what rate is energy now dissipated? ilw

36E. A 120 V potential difference is applied to a space heater that dissipates 500 W during operation. (a) What is its resistance during

Fig. 27-21 Problem 26.

operation? (b) At what rate do electrons flow through any cross section of the heater element?

37P. A 1250 W radiant heater is constructed to operate at 115 V. (a) What will be the current in the heater? (b) What is the resistance of the heating coil? (c) How much thermal energy is produced in 1.0 h by the heater? ssm ilw www

38P. A heating element is made by maintaining a potential difference of 75.0 V across the length of a Nichrome wire that has a 2.60×10^{-6} m^2 cross section. Nichrome has a resistivity of 5.00×10^{-7} $\Omega \cdot$ m. (a) If the element dissipates 5000 W, what is its length? (b) If a potential difference of 100 V is used to obtain the same dissipation rate, what should the length be?

39P. A Nichrome heater dissipates 500 W when the applied potential difference is 110 V and the wire temperature is 800°C. What would be the dissipation rate if the wire temperature were held at 200°C by immersing the wire in a bath of cooling oil? The applied potential difference remains the same, and α for Nichrome at 800°C is 4.0×10^{-4}/K. ssm

40P. A 100 W lightbulb is plugged into a standard 120 V outlet. (a) How much does it cost per month to leave the light turned on continuously? Assume electric energy costs 6¢/kW · h. (b) What is the resistance of the bulb? (c) What is the current in the bulb? (d) Is the resistance different when the bulb is turned off?

41P. A linear accelerator produces a pulsed beam of electrons. The pulse current is 0.50 A, and each pulse has a duration of 0.10 μs. (a) How many electrons are accelerated per pulse? (b) What is the average current for an accelerator operating at 500 pulses/s? (c) If the electrons are accelerated to an energy of 50 MeV, what are the average and peak powers of the accelerator? ssm

42P. A cylindrical resistor of radius 5.0 mm and length 2.0 cm is made of material that has a resistivity of 3.5×10^{-5} $\Omega \cdot$ m. What are (a) the current density and (b) the potential difference when the energy dissipation rate in the resistor is 1.0 W?

43P. A copper wire of cross-sectional area 2.0×10^{-6} m^2 and length 4.0 m has a current of 2.0 A uniformly distributed across that area. (a) What is the magnitude of the electric field along the wire? (b) How much electric energy is transferred to thermal energy in 30 min?

Additional Problems

44. *The chocolate crumb mystery.* This story begins with Problem 48 in Chapter 24 and continues through Chapters 25 and 26. The chocolate crumb powder moved to the silo through a pipe of radius R with uniform speed v and uniform charge density ρ. (a) Find an expression for the current i (the rate at which charge on the powder moved) through a perpendicular cross section of the pipe. (b) Evaluate i for the conditions at the factory: pipe radius $R = 5.0$ cm, speed $v = 2.0$ m/s, and charge density $\rho = 1.1 \times 10^{-3}$ C/m^3.

If the powder were to flow through a change V in electric potential, its energy could be transferred to a spark at the rate $P = iV$. (c) Could there be such a transfer within the pipe due to the radial potential difference discussed in Problem 57 of Chapter 25?

As the powder flowed from the pipe into the silo, the electric potential of the powder changed. The magnitude of that change

was at least equal to the radial potential difference within the pipe (as evaluated in Problem 57 of Chapter 25). (d) Assuming that value for the potential difference and using the current found in (b) above, find the rate at which energy could have been transferred from the powder to a spark as the powder exited the pipe. (e) If a spark did occur at the exit and lasted for 0.20 s (a reasonable expectation), how much energy would have been transferred to the spark?

Recall from Problem 48 in Chapter 24 that a minimum energy transfer of 150 mJ is needed to cause an explosion. (f) Where did the powder explosion most likely occur: in the powder cloud at the unloading bin (considered in Problem 48 of Chapter 26), within the pipe, or at the exit of the pipe into the silo?

45. *Heart attack or electrocution?* One morning a man walks barefooted away from a picnic onto moist ground near a tower that supports electric transmission lines. Suddenly he collapses. His relatives at the picnic table see him fall and, on reaching him a few seconds later, find that he is in ventricular fibrillation. The man dies before an emergency team can reach him with defibrillation equipment. Later the family files a lawsuit against the power company, claiming that the victim was electrocuted because of accidental current leakage from the tower. You are hired as part of the forensic team to investigate the death—was it due to heart attack or electrocution?

Investigation of the power company records reveals that there was indeed an *electrical fault* at the tower that morning—for about 1.0 s a current I leaked from a rod into the ground. Assume that the current spread uniformly (hemispherically) into the ground (Fig. 27-23). Let ρ be the resistivity of the ground and r be the distance from the rod. Find expressions for (a) the current density and (b) the electric field magnitude, both as functions of r. The lower end of the rod was spherical with a radius of b. (c) From your expression for the electric field magnitude, find an expression for the potential difference ΔV between the lower end of the rod and a point at distance r. Your investigation finds that $I = 100$ A, $\rho = 100$ $\Omega \cdot$ m, and $b = 1.0$ cm, and that the victim was located at $r = 10$ m. At the victim's location, what were (d) the current density, (e) the electric field magnitude, and (f) the potential difference ΔV? (This story continues with Problem 56 in Chapter 28.)

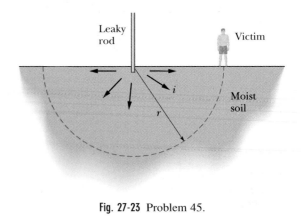

Fig. 27-23 Problem 45.

28 Circuits

The electric eel (*Electrophorus*) lurks in rivers of South America, killing the fish on which it preys with pulses of current. It does so by producing a potential difference of several hundred volts along its length; the result-

ing current in the surrounding water, from near the eel's head to the tail region, can be as much as one ampere. If you were to brush up against this eel while swimming, you might wonder (after recovering from the very painful stun):

How can the creature manage to produce a current that large without shocking itself?

The answer is in this chapter.

The world's largest battery, housed in Chino, California, has a power capability of 10 MW, which is put to use during peak power demands on the electric system served by Southern California Edison. Because the battery does work on charge carriers, it is an emf device.

28-1 "Pumping" Charges

If you want to make charge carriers flow through a resistor, you must establish a potential difference between the ends of the device. One way to do this is to connect each end of the resistor to one plate of a charged capacitor. The trouble with this scheme is that the flow of charge acts to discharge the capacitor, quickly bringing the plates to the same potential. When that happens, there is no longer an electric field in the resistor and thus the flow of charge stops.

To produce a steady flow of charge, you need a "charge pump," a device that—by doing work on the charge carriers—maintains a potential difference between a pair of terminals. We call such a device an **emf device,** and the device is said to provide an **emf** $\mathscr{E}$, which means that it does work on charge carriers. An emf device is sometimes called a *seat of emf.* The term *emf* comes from the outdated phrase *electromotive force,* which was adopted before scientists clearly understood the function of an emf device.

In Chapter 27, we discussed the motion of charge carriers through a circuit in terms of the electric field set up in the circuit—the field produces forces that move the charge carriers. In this chapter we take a different approach: We discuss the motion of the charge carriers in terms of the required energy—an emf device supplies the energy for the motion via the work it does.

A common emf device is the *battery,* used to power a wide variety of machines from wristwatches to submarines. The emf device that most influences our daily lives, however, is the *electric generator,* which, by means of electrical connections (wires) from a generating plant, creates a potential difference in our homes and workplaces. The emf devices known as *solar cells,* long familiar as the winglike panels on spacecraft, also dot the countryside for domestic applications. Less familiar emf devices are the *fuel cells* that power the space shuttles and the *thermopiles* that provide onboard electrical power for some spacecraft and for remote stations in Antarctica and elsewhere. An emf device does not have to be an instrument—living systems, ranging from electric eels and human beings to plants, have physiological emf devices.

Although the devices we have listed differ widely in their modes of operation, they all perform the same basic function—they do work on charge carriers and thus maintain a potential difference between their terminals.

28-2 Work, Energy, and Emf

Figure 28-1 shows an emf device (consider it to be a battery) that is part of a simple circuit containing a single resistance R (the symbol for resistance and a resistor is -⋀⋀-). The emf device keeps one of its terminals (called the positive terminal and often labeled +) at a higher electric potential than the other terminal (called the negative terminal and labeled −). We can represent the emf of the device with an arrow that points from the negative terminal toward the positive terminal as in Fig. 28-1. A small circle on the tail of the emf arrow distinguishes it from the arrows that indicate current direction.

When an emf device is not connected to a circuit, its internal chemistry does not cause any net flow of charge carriers within it. However, when it is connected to a circuit as in Fig. 28-1, its internal chemistry causes a net flow of positive charge carriers from the negative terminal to the positive terminal, in the direction of the emf arrow. This flow is part of the current that is set up around the circuit in that same direction (clockwise in Fig. 28-1).

Fig. 28-1 A simple electric circuit, in which a device of emf $\mathscr{E}$ does work on the charge carriers and maintains a steady current i in a resistor of resistance R.

Within the emf device, positive charge carriers move from a region of low electric potential and thus low electric potential energy (at the negative terminal) to a region of higher electric potential and higher electric potential energy (at the positive terminal). This motion is just the opposite of what the electric field between the terminals (which is directed from the positive terminal toward the negative terminal) would cause the charge carriers to do.

Thus, there must be some source of energy within the device, enabling it to do work on the charges by forcing them to move as they do. The energy source may be chemical, as in a battery or a fuel cell. It may involve mechanical forces, as in an electric generator. Temperature differences may supply the energy, as in a thermopile; or the Sun may supply it, as in a solar cell.

Let us now analyze the circuit of Fig. 28-1 from the point of view of work and energy transfers. In any time interval dt, a charge dq passes through any cross section of this circuit, such as aa'. This same amount of charge must enter the emf device at its low-potential end and leave at its high-potential end. The device must do an amount of work dW on the charge dq to force it to move in this way. We define the emf of the emf device in terms of this work:

$$\mathcal{E} = \frac{dW}{dq} \qquad \text{(definition of } \mathcal{E}\text{)}. \qquad (28\text{-}1)$$

In words, the emf of an emf device is the work per unit charge that the device does in moving charge from its low-potential terminal to its high-potential terminal. The SI unit for emf is the joule per coulomb; in Chapter 25 we defined that unit as the *volt*.

An **ideal emf device** is one that lacks any internal resistance to the internal movement of charge from terminal to terminal. The potential difference between the terminals of an ideal emf device is equal to the emf of the device. For example, an ideal battery with an emf of 12.0 V always has a potential difference of 12.0 V between its terminals.

A **real emf device,** such as any real battery, has internal resistance to the internal movement of charge. When a real emf device is not connected to a circuit, and thus does not have current through it, the potential difference between its terminals is equal to its emf. However, when that device has current through it, the potential difference between its terminals differs from its emf. We will discuss such real batteries in Section 28-4.

When an emf device is connected to a circuit, the device transfers energy to the charge carriers passing through it. This energy can then be transferred from the charge carriers to other devices in the circuit, for example, to light a bulb. Figure 28-2a shows a circuit containing two ideal rechargeable (*storage*) batteries A and B, a resistance R, and an electric motor M that can lift an object by using energy it obtains from charge carriers in the circuit. Note that the batteries are connected so that they tend to send charges around the circuit in opposite directions. The actual direction of the current in the circuit is determined by the battery with the larger emf, which happens to be battery B, so the chemical energy within battery B is decreasing as energy is transferred to the charge carriers passing through it. However, the chemical energy within battery A is increasing because the current in it is directed from the positive terminal to the negative terminal. Thus, battery B is charging battery A. Battery B is also providing energy to motor M and energy that is being dissipated by resistance R. Figure 28-2b shows all three energy transfers from battery B; each decreases that battery's chemical energy.

(a)

(b)

Fig. 28-2 (*a*) In the circuit, $\mathcal{E}_B > \mathcal{E}_A$; so battery B determines the direction of the current. (*b*) The energy transfers in the circuit, assuming that no dissipation occurs in the motor.

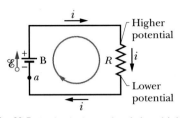

Fig. 28-3 A single-loop circuit in which a resistance R is connected across an ideal battery B with emf $\mathscr{E}$. The resulting current i is the same throughout the circuit.

28-3 Calculating the Current in a Single-Loop Circuit

We discuss here two equivalent ways to calculate the current in the simple *single-loop* circuit of Fig. 28-3; one method is based on energy conservation considerations, and the other on the concept of potential. The circuit consists of an ideal battery B with emf $\mathscr{E}$, a resistor of resistance R, and two connecting wires. (Unless otherwise indicated, we assume that wires in circuits have negligible resistance. Their function, then, is merely to provide pathways along which charge carriers can move.)

Energy Method

Equation 27-22 ($P = i^2R$) tells us that in a time interval dt an amount of energy given by $i^2R\,dt$ will appear in the resistor of Fig. 28-3 as thermal energy. (Since we assume the wires to have negligible resistance, no thermal energy will appear in them.) During the same interval, a charge $dq = i\,dt$ will have moved through battery B, and the work that the battery will have done on this charge, according to Eq. 28-1, is

$$dW = \mathscr{E}\,dq = \mathscr{E}i\,dt.$$

From the principle of conservation of energy, the work done by the (ideal) battery must equal the thermal energy that appears in the resistor:

$$\mathscr{E}i\,dt = i^2R\,dt.$$

This gives us

$$\mathscr{E} = iR.$$

The emf $\mathscr{E}$ is the energy per unit charge transferred to the moving charges by the battery. The quantity iR is the energy per unit charge transferred *from* the moving charges to thermal energy within the resistor. Therefore, this equation means that the energy per unit charge transferred to the moving charges is equal to the energy per unit charge transferred from them. Solving for i, we find

$$i = \frac{\mathscr{E}}{R}. \tag{28-2}$$

Potential Method

Suppose we start at any point in the circuit of Fig. 28-3 and mentally proceed around the circuit in either direction, adding algebraically the potential differences that we encounter. Then when we return to our starting point, we must also have returned to our starting potential. Before actually doing so, we shall formalize this idea in a statement that holds not only for single-loop circuits such as that of Fig. 28-3 but also for any complete loop in a *multiloop* circuit, as we shall discuss in Section 28-6:

> **LOOP RULE:** The algebraic sum of the changes in potential encountered in a complete traversal of any loop of a circuit must be zero.

This is often referred to as *Kirchhoff's loop rule* (or *Kirchhoff's voltage law*), after German physicist Gustav Robert Kirchhoff. This rule is equivalent to saying that each point on a mountain has only one elevation above sea level. If you start from any point and return to it after walking around the mountain, the algebraic sum of the changes in elevation that you encounter must be zero.

In Fig. 28-3, let us start at point a, whose potential is V_a, and mentally walk clockwise around the circuit until we are back at a, keeping track of potential changes as we move. Our starting point is at the low-potential terminal of the battery. Since the battery is ideal, the potential difference between its terminals is equal to $\mathscr{E}$. When we pass through the battery to the high-potential terminal, the change in potential is $+\mathscr{E}$.

As we walk along the top wire to the top end of the resistor, there is no potential change because the wire has negligible resistance; it is at the same potential as the high-potential terminal of the battery. So too is the top end of the resistor. When we pass through the resistor, however, the potential changes according to Eq. 27-8 (which we can rewrite as $V = iR$). Moreover, the potential must decrease because we are moving from the higher potential side of the resistor. Thus, the change in potential is $-iR$.

We return to point a by moving along the bottom wire. Since this wire also has negligible resistance, we again find no potential change. Back at point a, the potential is again V_a. Because we traversed a complete loop, our initial potential, as modified for potential changes along the way, must be equal to our final potential; that is,

$$V_a + \mathscr{E} - iR = V_a.$$

The value of V_a cancels from this equation, which becomes

$$\mathscr{E} - iR = 0.$$

Solving this equation for i gives us the same result, $i = \mathscr{E}/R$, as the energy method (Eq. 28-2).

If we apply the loop rule to a complete *counterclockwise* walk around the circuit, the rule gives us

$$-\mathscr{E} + iR = 0$$

and we again find that $i = \mathscr{E}/R$. Thus, you may mentally circle a loop in either direction to apply the loop rule.

To prepare for circuits more complex than that of Fig. 28-3, let us set down two rules for finding potential differences as we move around a loop:

▶ **RESISTANCE RULE:** For a move through a resistance in the direction of the current, the change in potential is $-iR$; in the opposite direction it is $+iR$.

▶ **EMF RULE:** For a move through an ideal emf device in the direction of the emf arrow, the change in potential is $+\mathscr{E}$; in the opposite direction it is $-\mathscr{E}$.

✔**CHECKPOINT 1:**   The figure shows the current i in a single-loop circuit with a battery B and a resistance R (and wires of negligible resistance). (a) Should the emf arrow at B be drawn pointing leftward or rightward? At points a, b, and c, rank (b) the magnitude of the current, (c) the electric potential, and (d) the electric potential energy of the charge carriers, greatest first.

28-4 Other Single-Loop Circuits

In this section we extend the simple circuit of Fig. 28-3 in two ways.

Internal Resistance

Figure 28-4a shows a real battery, with internal resistance r, wired to an external resistor of resistance R. The internal resistance of the battery is the electrical resistance of the conducting materials of the battery and thus is an unremovable feature of the battery. In Fig. 28-4a, however, the battery is drawn as if it could be separated into an ideal battery with emf $\mathscr{E}$ and a resistor of resistance r. The order in which the symbols for these separated parts are drawn does not matter.

If we apply the loop rule clockwise beginning at point a, the *changes* in potential give us

$$\mathscr{E} - ir - iR = 0. \tag{28-3}$$

Solving for the current, we find

$$i = \frac{\mathscr{E}}{R + r}. \tag{28-4}$$

Note that this equation reduces to Eq. 28-2 if the battery is ideal—that is, if $r = 0$.

Figure 28-4b shows graphically the changes in electric potential around the circuit. (To better link Fig. 28-4b with the *closed circuit* in Fig. 28-4a, imagine curling the graph into a cylinder with point a at the left overlapping point a at the right.) Note how traversing the circuit is like walking around a (potential) mountain and returning to your starting point—you also return to the starting elevation.

In this book, when a battery is not described as real or if no internal resistance is indicated, you can generally assume that it is ideal—but, of course, in the real world batteries are always real and have internal resistance.

Resistances in Series

Figure 28-5a shows three resistances connected **in series** to an ideal battery with emf $\mathscr{E}$. This description has little to do with how the resistances are drawn. Rather, "in series" means that the resistances are wired one after another and that a potential difference V is applied across the two ends of the series. In Fig. 28-5a, the resistances are connected one after another between a and b, and a potential difference is main-

(a)   (b)

Fig. 28-4 (a) A single-loop circuit containing a real battery having internal resistance r and emf $\mathscr{E}$. (b) The same circuit, now spread out in a line. The potentials encountered in traversing the circuit clockwise from a are also shown. The potential V_a is arbitrarily assigned a value of zero, and other potentials in the circuit are graphed relative to V_a.

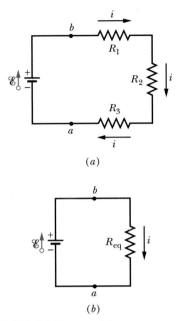

Fig. 28-5 (*a*) Three resistors are connected in series between points *a* and *b*. (*b*) An equivalent circuit, with the three resistors replaced with their equivalent resistance R_{eq}.

tained across *a* and *b* by the battery. The potential differences that then exist across the resistances in the series produce identical currents *i* in them. In general,

> When a potential difference *V* is applied across resistances connected in series, the resistances have identical currents *i*. The sum of the potential differences across the resistances is equal to the applied potential difference *V*.

Note that charge moving through the series resistances can move along only a single route. If there are additional routes, so that the currents in different resistances are different, the resistances are not connected in series.

> Resistances connected in series can be replaced with an equivalent resistance R_{eq} that has the same current *i* and the same *total* potential difference *V* as the actual resistances.

You might remember that R_{eq} and all the actual series resistances have the same current *i* with the nonsense word "ser-i." Figure 28-5*b* shows the equivalent resistance R_{eq} that can replace the three resistances of Fig. 28-5*a*.

To derive an expression for R_{eq} in Fig. 28-5*b*, we apply the loop rule to both circuits. For Fig. 28-5*a*, starting at terminal *a* and going clockwise around the circuit, we find

$$\mathscr{E} - iR_1 - iR_2 - iR_3 = 0,$$

or

$$i = \frac{\mathscr{E}}{R_1 + R_2 + R_3}. \tag{28-5}$$

For Fig. 28-5*b*, with the three resistances replaced with a single equivalent resistance R_{eq}, we find

$$\mathscr{E} - iR_{eq} = 0,$$

or

$$i = \frac{\mathscr{E}}{R_{eq}}. \tag{28-6}$$

Comparison of Eqs. 28-5 and 28-6 shows that

$$R_{eq} = R_1 + R_2 + R_3.$$

The extension to *n* resistances is straightforward and is

$$R_{eq} = \sum_{j=1}^{n} R_j \qquad (n \text{ resistances in series}). \tag{28-7}$$

Note that when resistances are in series, their equivalent resistance is greater than any of the individual resistances.

CHECKPOINT 2: In Fig. 28-5*a*, if $R_1 > R_2 > R_3$, rank the three resistances according to (a) the current through them and (b) the potential difference across them, greatest first.

28-5 Potential Differences

We often want to find the potential difference between two points in a circuit. In Fig. 28-4*a*, for example, what is the potential difference between points *b* and *a*? To find out, let us start at point *b* and traverse the circuit clockwise to point *a*, passing through resistor *R*. If V_a and V_b are the potentials at *a* and *b*, respectively, we have

$$V_b - iR = V_a$$

because (according to our resistance rule) we experience a decrease in potential in going through a resistance in the direction of the current. We rewrite this as

$$V_b - V_a = +iR, \tag{28-8}$$

which tells us that point b is at greater potential than point a. Combining Eq. 28-8 with Eq. 28-4, we have

$$V_b - V_a = \mathscr{E}\, \frac{R}{R + r}, \tag{28-9}$$

where again r is the internal resistance of the emf device.

> ▶ To find the potential difference between any two points in a circuit, start at one point and traverse the circuit to the other, following any path, and add algebraically the changes in potential that you encounter.

Let us again calculate $V_b - V_a$, starting again at point b but this time proceeding counterclockwise to a through the battery. We have

$$V_b + ir - \mathscr{E} = V_a,$$

or
$$V_b - V_a = \mathscr{E} - ir. \tag{28-10}$$

Combining this with Eq. 28-4 again leads to Eq. 28-9.

The quantity $V_b - V_a$ in Fig. 28-4 is the potential difference that the battery sets up across the battery terminals. As noted earlier, $V_b - V_a$ is equal to the emf $\mathscr{E}$ of the battery only if the battery has no internal resistance ($r = 0$ in Eq. 28-9) or if the circuit is open ($i = 0$ in Eq. 28-10).

Suppose that in Fig. 28-4, $\mathscr{E} = 12$ V, $R = 10\ \Omega$, and $r = 2.0\ \Omega$. Then Eq. 28-9 tells us that the potential across the battery's terminals is

$$V_b - V_a = (12\ \text{V})\, \frac{10\ \Omega}{10\ \Omega + 2.0\ \Omega} = 10\ \text{V}.$$

In "pumping" charge through itself, the battery does work per unit charge of $\mathscr{E} = 12$ J/C, or 12 V. However, because of the internal resistance of the battery, it produces a potential difference of only 10 J/C, or 10 V, across its terminals.

Power, Potential, and Emf

When a battery or some other type of emf device does work on the charge carriers to establish a current i, it transfers energy from its source of energy (such as the chemical source in a battery) to the charge carriers. Because a real emf device has an internal resistance r, it also transfers energy to internal thermal energy via resistive dissipation, discussed in Section 27-7. Let us relate these transfers.

The net rate P of energy transfer from the emf device to the charge carriers is given by Eq. 27-21:

$$P = iV, \tag{28-11}$$

where V is the potential across the terminals of the emf device. From Eq. 28-10, we can substitute $V = \mathscr{E} - ir$ into Eq. 28-11 to find

$$P = i(\mathscr{E} - ir) = i\mathscr{E} - i^2 r. \tag{28-12}$$

We see that the term $i^2 r$ in Eq. 28-12 is the rate P_r of energy transfer to thermal energy within the emf device:

$$P_r = i^2 r \qquad \text{(internal dissipation rate)}. \tag{28-13}$$

Then the term $i\mathscr{E}$ in Eq. 28-12 must be the rate P_{emf} at which the emf device transfers energy to *both* the charge carriers and to internal thermal energy. Thus,

$$P_{emf} = i\mathscr{E} \qquad \text{(power of emf device).}$$ (28-14)

If a battery is being *recharged*, with a "wrong way" current through it, the energy transfer is then *from* the charge carriers *to* the battery—both to the battery's chemical energy and to the energy dissipated in the internal resistance *r*. The rate of change of the chemical energy is given by Eq. 28-14, the rate of dissipation is given by Eq. 28-13, and the rate at which the carriers supply energy is given by Eq. 28-11.

Sample Problem 28-1

The emfs and resistances in the circuit of Fig. 28-6a have the following values:

$$\mathscr{E}_1 = 4.4 \text{ V}, \quad \mathscr{E}_2 = 2.1 \text{ V},$$

$$r_1 = 2.3 \ \Omega, \quad r_2 = 1.8 \ \Omega, \quad R = 5.5 \ \Omega.$$

(a) What is the current *i* in the circuit?

SOLUTION: The **Key Idea** here is that we can get an expression involving the current *i* in this single-loop circuit by applying the loop rule. Although knowing the direction of *i* is not necessary, we can easily determine it from the emfs of the two batteries. Because $\mathscr{E}_1$ is greater than $\mathscr{E}_2$, battery 1 controls the direction of *i*, so the direction is clockwise. Let us then apply the loop rule by going counterclockwise—against the current—and starting at point *a*. We find

$$-\mathscr{E}_1 + ir_1 + iR + ir_2 + \mathscr{E}_2 = 0.$$

Check that this equation also results if we apply the loop rule clockwise or start at some point other than *a*. Also, take the time to compare this equation term by term with Fig. 28-6b, which shows the potential changes graphically (with the potential at point *a* arbitrarily taken to be zero).

Solving the above loop equation for the current *i*, we obtain

$$i = \frac{\mathscr{E}_1 - \mathscr{E}_2}{R + r_1 + r_2} = \frac{4.4 \text{ V} - 2.1 \text{ V}}{5.5 \ \Omega + 2.3 \ \Omega + 1.8 \ \Omega}$$

$$= 0.2396 \text{ A} \approx 240 \text{ mA.} \qquad \text{(Answer)}$$

(b) What is the potential difference between the terminals of battery 1 in Fig. 28-6a?

SOLUTION: The **Key Idea** is to sum the potential differences between points *a* and *b*. Let us start at point *b* (effectively the negative terminal of battery 1) and travel clockwise through battery 1 to point *a* (effectively the positive terminal), keeping track of potential changes. We find that

$$V_b - ir_1 + \mathscr{E}_1 = V_a,$$

which gives us

$$V_a - V_b = -ir_1 + \mathscr{E}_1$$

$$= -(0.2396 \text{ A})(2.3 \ \Omega) + 4.4 \text{ V}$$

$$= +3.84 \text{ V} \approx 3.8 \text{ V}, \qquad \text{(Answer)}$$

which is less than the emf of the battery. You can verify this result

Fig. 28-6 Sample Problem 28-1. (*a*) A single-loop circuit containing two real batteries and a resistor. The batteries oppose each other; that is, they tend to send current in opposite directions through the resistor. (*b*) A graph of the potentials, counterclockwise from point *a*, with the potential at *a* arbitrarily taken to be zero. (To better link the circuit with the graph, mentally cut the circuit at *a* and then unfold the left side of the circuit toward the left and the right side of the circuit toward the right.)

by starting at point *b* in Fig. 28-6a and traversing the circuit counterclockwise to point *a*.

✔CHECKPOINT 3: A battery has an emf of 12 V and an internal resistance of 2 Ω. Is the terminal-to-terminal potential difference greater than, less than, or equal to 12 V if the current in the battery is (a) from the negative to the positive terminal, (b) from the positive to the negative terminal, and (c) zero?

Tactic 1: *Assuming the Direction of a Current*

In solving circuit problems, you do not need to know the direction of a current in advance. Instead, you can just assume its direction, although that may take some physics courage. To show this, assume the current in Fig. 28-6a is counterclockwise; that is, reverse the direction of the current arrows shown. Applying the loop rule counterclockwise from point *a* now yields

$$-\mathscr{E}_1 - ir_1 - iR - ir_2 + \mathscr{E}_2 = 0,$$

or

$$i = -\frac{\mathscr{E}_1 - \mathscr{E}_2}{R + r_1 + r_2}.$$

Substituting the numerical values of Sample Problem 28-1 yields $i = -240$ mA for the current. The minus sign is a signal that the current is opposite the direction we initially assumed.

28-6 Multiloop Circuits

Figure 28-7 shows a circuit containing more than one loop. For simplicity, we assume the batteries are ideal. There are two *junctions* in this circuit, at *b* and *d*, and there are three *branches* connecting these junctions. The branches are the left branch (*bad*), the right branch (*bcd*), and the central branch (*bd*). What are the currents in the three branches?

We arbitrarily label the currents, using a different subscript for each branch. Current i_1 has the same value everywhere in branch *bad*, i_2 has the same value everywhere in branch *bcd*, and i_3 is the current through branch *bd*. The directions of the currents are assumed arbitrarily.

Consider junction *d* for a moment: Charge comes into that junction via incoming currents i_1 and i_3, and it leaves via outgoing current i_2. Because there is no variation in the charge at the junction, the total incoming current must equal the total outgoing current:

$$i_1 + i_3 = i_2. \tag{28-15}$$

You can easily check that applying this condition to junction *b* leads to exactly the same equation. Equation 28-15 thus suggests a general principle:

> **JUNCTION RULE:** The sum of the currents entering any junction must be equal to the sum of the currents leaving that junction.

This rule is often called *Kirchhoff's junction rule* (or *Kirchhoff's current law*). It is simply a statement of the conservation of charge for a steady flow of charge—there is neither a build-up nor a depletion of charge at a junction. Thus, our basic tools for solving complex circuits are the *loop rule* (based on the conservation of energy) and the *junction rule* (based on the conservation of charge).

Equation 28-15 is a single equation involving three unknowns. To solve the circuit completely (that is, to find all three currents), we need two more equations involving those same unknowns. We obtain them by applying the loop rule twice. In the circuit of Fig. 28-7, we have three loops from which to choose: the left-hand loop (*badb*), the right-hand loop (*bcdb*), and the big loop (*badcb*). Which two loops we choose does not matter—let's choose the left-hand loop and the right-hand loop.

If we traverse the left-hand loop in a counterclockwise direction from point *b*, the loop rule gives us

$$\mathscr{E}_1 - i_1R_1 + i_3R_3 = 0. \tag{28-16}$$

If we traverse the right-hand loop in a counterclockwise direction from point *b*, the loop rule gives us

$$-i_3R_3 - i_2R_2 - \mathscr{E}_2 = 0. \tag{28-17}$$

We now have three equations (Eqs. 28-15, 28-16, and 28-17) in the three unknown currents, and they can be solved by a variety of techniques.

Fig. 28-7 A multiloop circuit consisting of three branches: left-hand branch *bad*, right-hand branch *bcd*, and central branch *bd*. The circuit also consists of three loops: left-hand loop *badb*, right-hand loop *bcdb*, and big loop *badcb*.

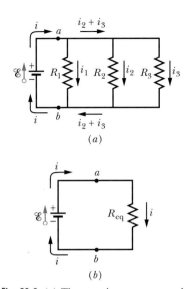

Fig. 28-8 (a) Three resistors connected in parallel across points a and b. (b) An equivalent circuit, with the three resistors replaced with their equivalent resistance R_{eq}.

If we had applied the loop rule to the big loop, we would have obtained (moving counterclockwise from b) the equation

$$\mathscr{E}_1 - i_1R_1 - i_2R_2 - \mathscr{E}_2 = 0.$$

This equation may look like fresh information, but in fact it is only the sum of Eqs. 28-16 and 28-17. (It would, however, yield the proper results when used with Eq. 28-15 and either 28-16 or 28-17.)

Resistances in Parallel

Figure 28-8a shows three resistances connected *in parallel* to an ideal battery of emf $\mathscr{E}$. The term "in parallel" means that the resistances are directly wired together on one side and directly wired together on the other side, and that a potential difference V is applied across the pair of connected sides. Thus, all three resistances have the same potential difference V across them, producing a current through each. In general,

> When a potential difference V is applied across resistances connected in parallel, the resistances all have that same potential difference V.

In Fig. 28-8a, the applied potential difference V is maintained by the battery. In Fig. 28-8b, the three parallel resistances have been replaced with an equivalent resistance R_{eq}.

> Resistances connected in parallel can be replaced with an equivalent resistance R_{eq} that has the same potential difference V and the same *total* current i as the actual resistances.

You might remember that R_{eq} and all the actual parallel resistances have the same potential difference V with the nonsense word "par-V."

To derive an expression for R_{eq} in Fig. 28-8b, we first write the current in each actual resistance in Fig. 28-8a as

$$i_1 = \frac{V}{R_1}, \quad i_2 = \frac{V}{R_2}, \quad \text{and} \quad i_3 = \frac{V}{R_3},$$

where V is the potential difference between a and b. If we apply the junction rule at point a in Fig. 28-8a and then substitute these values, we find

$$i = i_1 + i_2 + i_3 = V\left(\frac{1}{R_1} + \frac{1}{R_2} + \frac{1}{R_3}\right). \tag{28-18}$$

If we replaced the parallel combination with the equivalent resistance R_{eq} (Fig. 28-8b), we would have

$$i = \frac{V}{R_{eq}}. \tag{28-19}$$

Comparing Eqs. 28-18 and 28-19 leads to

$$\frac{1}{R_{eq}} = \frac{1}{R_1} + \frac{1}{R_2} + \frac{1}{R_3}. \tag{28-20}$$

Extending this result to the case of n resistances, we have

$$\frac{1}{R_{eq}} = \sum_{j=1}^{n} \frac{1}{R_j} \quad (n \text{ resistances in parallel}). \tag{28-21}$$

For the case of two resistances, the equivalent resistance is their product divided by

TABLE 28-1   **Series and Parallel Resistors and Capacitors**

Series	Parallel	Series	Parallel
Resistors		Capacitors	
$R_{eq} = \sum\limits_{j=1}^{n} R_j$ Eq. 28-7	$\dfrac{1}{R_{eq}} = \sum\limits_{j=1}^{n} \dfrac{1}{R_j}$ Eq. 28-21	$\dfrac{1}{C_{eq}} = \sum\limits_{j=1}^{n} \dfrac{1}{C_j}$ Eq 26-20	$C_{eq} = \sum\limits_{j=1}^{n} C_j$ Eq. 26-19
Same current through all resistors	Same potential difference across all resistors	Same charge on all capacitors	Same potential difference across all capacitors

their sum; that is,

$$R_{eq} = \frac{R_1 R_2}{R_1 + R_2}. \qquad (28\text{-}22)$$

If you accidentally took the equivalent resistance to be the sum divided by the product, you would notice at once that this result would be dimensionally incorrect.

Note that when two or more resistances are connected in parallel, the equivalent resistance is smaller than any of the combining resistances. Table 28-1 summarizes the equivalence relations for resistors and capacitors in series and in parallel.

✔**CHECKPOINT 4:** A battery, with potential V across it, is connected to a combination of two identical resistors and then has current i through it. What are the potential difference across and the current through either resistor if the resistors are (a) in series and (b) in parallel?

Sample Problem 28-2

Figure 28-9a shows a multiloop circuit containing one ideal battery and four resistances with the following values:

$$R_1 = 20 \ \Omega, \quad R_2 = 20 \ \Omega, \quad \mathscr{E} = 12 \ V,$$
$$R_3 = 30 \ \Omega, \quad R_4 = 8.0 \ \Omega.$$

(a) What is the current through the battery?

SOLUTION: First note that the current through the battery must also be the current through R_1. Thus, one Key Idea here is that we might find that current by applying the loop rule to a loop that includes R_1 because the current would be included in the potential difference across R_1. Either the left-hand loop or the big loop will do. Noting that the emf arrow of the battery points upward, so the current the battery supplies is clockwise, we might apply the loop rule to the left-hand loop, clockwise from point a. With i being the current through the battery, we would get

$$+\mathscr{E} - iR_1 - iR_2 - iR_4 = 0 \qquad \text{(incorrect)}.$$

However, this equation is incorrect because it assumes that R_1, R_2, and R_4 all have the same current i. Resistances R_1 and R_4 do have the same current, because the current passing through R_4 must pass through the battery and then through R_1 with no change in value. However, that current splits at junction point b—only part passes through R_2, the rest through R_3.

To distinguish the several currents in the circuit, we must label them individually as in Fig. 28-9b. Then, circling clockwise from a, we can write the loop rule for the left-hand loop as

$$+\mathscr{E} - i_1 R_1 - i_2 R_2 - i_1 R_4 = 0.$$

Fig. 28-9 Sample Problem 28-2. (a) A multiloop circuit with an ideal battery of emf $\mathscr{E}$ and four resistances. (b) Assumed currents through the resistances. (c) A simplification of the circuit, with resistances R_2 and R_3 replaced with their equivalent resistance R_{23}. The current through R_{23} is equal to that through R_1 and R_4.

Unfortunately, this equation contains two unknowns, i_1 and i_2; we would need at least one more equation to find them.

A second Key Idea is that a much easier option is to simplify the circuit of Fig. 28-9b by finding equivalent resistances. Note carefully that R_1 and R_2 are *not* in series and thus cannot be replaced with an equivalent resistance. However, R_2 and R_3 are in parallel, so we can use either Eq. 28-21 or Eq. 28-22 to find their equivalent

resistance R_{23}. From the latter,

$$R_{23} = \frac{R_2 R_3}{R_2 + R_3} = \frac{(20\ \Omega)(30\ \Omega)}{50\ \Omega} = 12\ \Omega.$$

We can now redraw the circuit as in Fig. 28-9c; note that the current through R_{23} must be i_1 because charge that moves through R_1 and R_4 must also move through R_{23}. For this simple one-loop circuit, the loop rule (applied clockwise from point a) yields

$$+\mathscr{E} - i_1 R_1 - i_1 R_{23} - i_1 R_4 = 0.$$

Substituting the given data, we find

$$12\ \text{V} - i_1(20\ \Omega) - i_1(12\ \Omega) - i_1(8.0\ \Omega) = 0,$$

which gives us

$$i_1 = \frac{12\ \text{V}}{40\ \Omega} = 0.30\ \text{A}. \qquad \text{(Answer)}$$

(b) What is the current i_2 through R_2?

SOLUTION: One Key Idea here is that we must work backward from the equivalent circuit of Fig. 28-9c, where R_{23} has replaced the parallel resistances R_2 and R_3. A second Key Idea is that, because R_2 and R_3 are in parallel, they both have the same potential differ-ence across them as their equivalent R_{23}. We know the current through R_{23} is $i_1 = 0.30$ A. Thus, we can use Eq. 27-8 ($R = V/i$) to find the potential difference V_{23} across R_{23}:

$$V_{23} = i_1 R_{23} = (0.30\ \text{A})(12\ \Omega) = 3.6\ \text{V}.$$

The potential difference across R_2 is thus 3.6 V, so the current i_2 in R_2 must be, by Eq. 27-8,

$$i_2 = \frac{V_2}{R_2} = \frac{3.6\ \text{V}}{20\ \Omega} = 0.18\ \text{A}. \qquad \text{(Answer)}$$

(c) What is the current i_3 through R_3?

SOLUTION: We can answer by using the same technique as in (b), or we can use this Key Idea: The junction rule tells us that at point b in Fig. 28-9b, the incoming current i_1 and the outgoing currents i_2 and i_3 are related by

$$i_1 = i_2 + i_3.$$

This gives us

$$\begin{aligned} i_3 &= i_1 - i_2 = 0.30\ \text{A} - 0.18\ \text{A} \\ &= 0.12\ \text{A}. \qquad \text{(Answer)} \end{aligned}$$

Sample Problem 28-3

Figure 28-10 shows a circuit whose elements have the following values:

$$\mathscr{E}_1 = 3.0\ \text{V}, \quad \mathscr{E}_2 = 6.0\ \text{V},$$

$$R_1 = 2.0\ \Omega, \quad R_2 = 4.0\ \Omega.$$

The three batteries are ideal batteries. Find the magnitude and di-rection of the current in each of the three branches.

SOLUTION: It is not worthwhile to try to simplify this circuit, because no two resistors are in parallel, and the resistors that are in series (those in the right branch or those in the left branch) present no problem. So, our Key Idea is to apply the junction and loop rules.

Using arbitrarily chosen directions for the currents as shown in Fig. 28-10, we apply the junction rule at point a by writing

$$i_3 = i_1 + i_2. \qquad (28\text{-}23)$$

An application of the junction rule at junction b gives only the same equation, so we next apply the loop rule to any two of the three loops of the circuit. We first arbitrarily choose the left-hand loop, arbitrarily start at point a, and arbitrarily traverse the loop in the counterclockwise direction, obtaining

$$-i_1 R_1 - \mathscr{E}_1 - i_1 R_1 + \mathscr{E}_2 + i_2 R_2 = 0.$$

Substituting the given data and simplifying yield

$$i_1(4.0\ \Omega) - i_2(4.0\ \Omega) = 3.0\ \text{V}. \qquad (28\text{-}24)$$

For our second application of the loop rule, we arbitrarily choose to traverse the right-hand loop clockwise from point a, finding

$$+i_3 R_1 - \mathscr{E}_2 + i_3 R_1 + \mathscr{E}_2 + i_2 R_2 = 0.$$

Substituting the given data and simplifying yield

$$i_2(4.0\ \Omega) + i_3(4.0\ \Omega) = 0. \qquad (28\text{-}25)$$

Fig. 28-10 Sample Problem 28-3. A multiloop circuit with three ideal batteries and five resistances.

Using Eq. 28-23 to eliminate i_3 from Eq. 28-25 and simplifying give us

$$i_1(4.0\ \Omega) + i_2(8.0\ \Omega) = 0. \qquad (28\text{-}26)$$

We now have a system of two equations (Eqs. 28-24 and 28-26) in two unknowns (i_1 and i_2) to solve either "by hand" (which is easy enough here) or with a "math package." (One solution tech-nique is Cramer's rule, given in Appendix E.) We find

$$i_2 = -0.25\ \text{A}.$$

(The minus sign signals that our arbitrary choice of direction for i_2 in Fig. 28-10 is wrong; i_2 should point up through $\mathscr{E}_2$ and R_2.) Substituting $i_2 = -0.25$ A into Eq. 28-26 and solving for i_1 then give us

$$i_1 = 0.50\ \text{A}. \qquad \text{(Answer)}$$

With Eq. 28-23 we then find that

$$i_3 = i_1 + i_2 = 0.25\ \text{A}. \qquad \text{(Answer)}$$

The positive answers we obtained for i_1 and i_3 signal that our choices of directions for these currents are correct. We can now correct the direction for i_2 and write its magnitude as

$$i_2 = 0.25\ \text{A}. \qquad \text{(Answer)}$$

Sample Problem 28-4

Electric fish are able to generate current with biological cells called *electroplaques*, which are physiological emf devices. The electroplaques in the South American eel shown in the photograph that opens this chapter are arranged in 140 rows, each row stretching horizontally along the body and each containing 5000 electroplaques. The arrangement is suggested in Fig. 28-11a; each electroplaque has an emf $\mathscr{E}$ of 0.15 V and an internal resistance r of 0.25 Ω. The water surrounding the eel completes a circuit between the two ends of the electroplaque array, one end at the animal's head and the other near its tail.

(a) If the water surrounding the eel has resistance $R_w = 800$ Ω, how much current can the eel produce in the water?

SOLUTION: The Key Idea here is that we can simplify the circuit of Fig. 28-11a by replacing combinations of emfs and internal resistances with equivalent emfs and resistances. We first consider a single row. The total emf $\mathscr{E}_{\text{row}}$ along a row of 5000 electroplaques is the sum of the emfs:

$$\mathscr{E}_{\text{row}} = 5000\mathscr{E} = (5000)(0.15\text{ V}) = 750\text{ V}.$$

The total resistance R_{row} along a row is the sum of the internal resistances of the 5000 electroplaques:

$$R_{\text{row}} = 5000r = (5000)(0.25\text{ Ω}) = 1250\text{ Ω}.$$

We can now represent each of the 140 identical rows as having a single emf $\mathscr{E}_{\text{row}}$ and a single resistance R_{row}, as shown in Fig. 28-11b.

In Fig. 28-11b, the emf between point a and point b on any row is $\mathscr{E}_{\text{row}} = 750$ V. Because the rows are identical and because they are all connected together at the left in Fig. 28-11b, all points b in that figure are at the same electric potential. Thus, we can consider them to be connected so that there is only a single point b. The emf between point a and this single point b is $\mathscr{E}_{\text{row}} = 750$ V, so we can draw the circuit as shown in Fig. 28-11c.

Between points b and c in Fig. 28-11c are 140 resistances $R_{\text{row}} = 1250$ Ω, all in parallel. The equivalent resistance R_{eq} of this combination is given by Eq. 28-21 as

$$\frac{1}{R_{\text{eq}}} = \sum_{j=1}^{140} \frac{1}{R_j} = 140\,\frac{1}{R_{\text{row}}},$$

Fig. 28-11 Sample Problem 28-4. (*a*) A model of the electric circuit of an eel in water. Each electroplaque of the eel has an emf $\mathscr{E}$ and internal resistance r. Along each of 140 rows extending from the head to the tail of the eel, there are 5000 electroplaques. The surrounding water has resistance R_w. (*b*) The emf $\mathscr{E}_{\text{row}}$ and resistance R_{row} of each row. (*c*) The emf between points a and b is $\mathscr{E}_{\text{row}}$. Between points b and c are 140 parallel resistances R_{row}. (*d*) The simplified circuit, with R_{eq} replacing the parallel combination.

or $$R_{eq} = \frac{R_{row}}{140} = \frac{1250\ \Omega}{140} = 8.93\ \Omega.$$

Replacing the parallel combination with R_{eq}, we obtain the simplified circuit of Fig. 28-11d. Applying the loop rule to this circuit counterclockwise from point b, we have

$$\mathscr{E}_{row} - iR_w - iR_{eq} = 0.$$

Solving for i and substituting the known data, we find

$$i = \frac{\mathscr{E}_{row}}{R_w + R_{eq}} = \frac{750\ V}{800\ \Omega + 8.93\ \Omega}$$
$$= 0.927\ A \approx 0.93\ A. \qquad \text{(Answer)}$$

If the head or tail of the eel is near a fish, some of this current could pass along a narrow path through the fish, stunning or killing it.

(b) How much current i_{row} travels through each row of Fig. 28-11a?

SOLUTION: The Key Idea here is that since the rows are identical, the current into and out of the eel is evenly divided among them:

$$i_{row} = \frac{i}{140} = \frac{0.927\ A}{140} = 6.6 \times 10^{-3}\ A. \qquad \text{(Answer)}$$

Thus, the current through each row is small, about two orders of magnitude smaller than the current through the water. This tends to spread the current through the eel's body, so that it need not stun or kill itself when it stuns or kills a fish.

PROBLEM-SOLVING TACTICS

Tactic 2: *Solving Circuits of Batteries and Resistors*
Here are two general techniques for solving circuits for unknown currents or potential differences.

1. If a circuit can be simplified by replacing resistors in series or in parallel with their equivalents, do so. If you can reduce the circuit to a single loop, then you can find the current through the battery with that loop, as in Sample Problem 28-2a. You may then have to "work backward," undoing the resistor simplification process, to find the current or potential difference for any particular resistor, as in Sample Problem 28-2b.

2. If a circuit cannot be simplified to a single loop, use the junction rule and the loop rule to write a set of simultaneous equations, as in Sample Problem 28-3. You need have only as many independent equations as there are unknowns in those equations. If you have to find the current or potential difference for a particular resistor, you can ensure that its current or potential difference appears in the equations by having at least one of the loops pass through it.

Tactic 3: *Arbitrary Choices in Solving Circuit Problems*

In Sample Problem 28-3, we made several arbitrary choices. (1) We assumed directions for the currents in Fig. 28-10 arbitrarily. (2) We chose which of the three possible loops to write equations for arbitrarily. (3) We chose the direction in which to traverse each loop arbitrarily. (4) We chose the starting and ending point for each traversal arbitrarily.

Such arbitrariness often worries a beginning circuit solver, but an experienced circuit solver knows that it does not matter. Just keep two rules firmly in mind. First, make sure you traverse each chosen loop completely. Second, once you have chosen a direction for a current, stick with it until you get numerical values for all the currents. If you were wrong about a direction, the algebra will signal you with a minus sign. Then you can make a correction by simply erasing the minus sign and reversing the arrow representing that current in the circuit diagram. However, *you should not make this correction until you have completed all the required calculations for the circuit,* as we did in Sample Problem 28-3.

28-7 The Ammeter and the Voltmeter

Fig. 28-12 A single-loop circuit, showing how to connect an ammeter (A) and a voltmeter (V).

An instrument used to measure currents is called an *ammeter*. To measure the current in a wire, you usually have to break or cut the wire and insert the ammeter so that the current to be measured passes through the meter. (In Fig. 28-12, ammeter A is set up to measure current i.)

It is essential that the resistance R_A of the ammeter be very small compared to other resistances in the circuit. Otherwise, the very presence of the meter will change the current to be measured.

A meter used to measure potential differences is called a *voltmeter*. To find the potential difference between any two points in the circuit, the voltmeter terminals are connected between those points, without breaking or cutting the wire. (In Fig. 28-12, voltmeter V is set up to measure the voltage across R_1.)

It is essential that the resistance R_V of a voltmeter be very large compared to the resistance of any circuit element across which the voltmeter is connected. Otherwise, the meter itself becomes an important circuit element and alters the potential difference that is to be measured.

Often a single meter is packaged so that, by means of a switch, it can be made to serve as either an ammeter or a voltmeter—and usually also as an *ohmmeter,* designed to measure the resistance of any element connected between its terminals. Such a versatile unit is called a *multimeter.*

28-8 *RC* Circuits

In preceding sections we dealt only with circuits in which the currents did not vary with time. Here we begin a discussion of time-varying currents.

Charging a Capacitor

The capacitor of capacitance C in Fig. 28-13 is initially uncharged. To charge it, we close switch S on point a. This completes an *RC series circuit* consisting of the capacitor, an ideal battery of emf $\mathscr{E}$, and a resistance R.

From Section 26-2, we already know that as soon as the circuit is complete, charge begins to flow (current exists) between a capacitor plate and a battery terminal on each side of the capacitor. This current increases the charge q on the plates and the potential difference V_C $(= q/C)$ across the capacitor. When that potential difference equals the potential difference across the battery (which here is equal to the emf $\mathscr{E}$), the current is zero. From Eq. 26-1 $(q = CV)$, the *equilibrium* (final) *charge* on the then fully charged capacitor is equal to $C\mathscr{E}$.

Here we want to examine the charging process. In particular we want to know how the charge $q(t)$ on the capacitor plates, the potential difference $V_C(t)$ across the capacitor, and the current $i(t)$ in the circuit vary with time during the charging process. We begin by applying the loop rule to the circuit, traversing it clockwise from the negative terminal of the battery. We find

$$\mathscr{E} - iR - \frac{q}{C} = 0. \tag{28-27}$$

The last term on the left side represents the potential difference across the capacitor. The term is negative because the capacitor's top plate, which is connected to the battery's positive terminal, is at a higher potential than the lower plate. Thus, there is a drop in potential as we move down through the capacitor.

We cannot immediately solve Eq. 28-27 because it contains two variables, i and q. However, those variables are not independent but are related by

$$i = \frac{dq}{dt}. \tag{28-28}$$

Substituting this for i in Eq. 28-27 and rearranging, we find

$$R\frac{dq}{dt} + \frac{q}{C} = \mathscr{E} \qquad \text{(charging equation).} \tag{28-29}$$

This differential equation describes the time variation of the charge q on the capacitor in Fig. 28-13. To solve it, we need to find the function $q(t)$ that satisfies this equation and also satisfies the condition that the capacitor be initially uncharged; that is, $q = 0$ at $t = 0$.

We shall soon show that the solution to Eq. 28-29 is

$$q = C\mathscr{E}(1 - e^{-t/RC}) \qquad \text{(charging a capacitor).} \tag{28-30}$$

Fig. 28-13 When switch S is closed on a, the capacitor is *charged* through the resistor. When the switch is afterward closed on b, the capacitor *discharges* through the resistor.

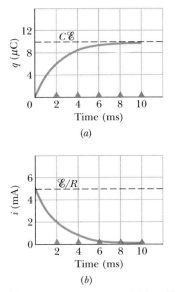

Fig. 28-14 (*a*) A plot of Eq. 28-30, which shows the buildup of charge on the capacitor of Fig. 28-13. (*b*) A plot of Eq. 28-31, which shows the decline of the charging current in the circuit of Fig. 28-13. The curves are plotted for $R = 2000\ \Omega$, $C = 1\ \mu\text{F}$, and $\mathscr{E} = 10$ V; the small triangles represent successive intervals of one time constant τ.

(Here e is the exponential base, 2.718 . . . , and not the elementary charge.) Note that Eq. 28-30 does indeed satisfy our required initial condition, because at $t = 0$ the term $e^{-t/RC}$ is unity; so the equation gives $q = 0$. Note also that as t goes to infinity (that is, a long time later), the term $e^{-t/RC}$ goes to zero; so the equation gives the proper value for the full (equilibrium) charge on the capacitor—namely, $q = C\mathscr{E}$. A plot of $q(t)$ for the charging process is given in Fig. 28-14*a*.

The derivative of $q(t)$ is the current $i(t)$ charging the capacitor:

$$i = \frac{dq}{dt} = \left(\frac{\mathscr{E}}{R}\right)e^{-t/RC} \qquad \text{(charging a capacitor)}. \qquad (28\text{-}31)$$

A plot of $i(t)$ for the charging process is given in Fig. 28-14*b*. Note that the current has the initial value $\mathscr{E}/R$ and that it decreases to zero as the capacitor becomes fully charged.

▶ A capacitor that is being charged initially acts like ordinary connecting wire relative to the charging current. A long time later, it acts like a broken wire.

By combining Eq. 26-1 ($q = CV$) and Eq. 28-30, we find that the potential difference $V_C(t)$ across the capacitor during the charging process is

$$V_C = \frac{q}{C} = \mathscr{E}(1 - e^{-t/RC}) \qquad \text{(charging a capacitor)}. \qquad (28\text{-}32)$$

This tells us that $V_C = 0$ at $t = 0$ and that $V_C = \mathscr{E}$ when the capacitor becomes fully charged as $t \rightarrow \infty$.

The Time Constant

The product RC that appears in Eqs. 28-30, 28-31, and 28-32 has the dimensions of time (both because the argument of an exponential must be dimensionless and because, in fact, $1.0\ \Omega \times 1.0$ F $= 1.0$ s). RC is called the **capacitive time constant** of the circuit and is represented with the symbol τ:

$$\tau = RC \qquad \text{(time constant)}. \qquad (28\text{-}33)$$

From Eq. 28-30, we can now see that at time $t = \tau\ (= RC)$, the charge on the initially uncharged capacitor of Fig. 28-13 has increased from zero to

$$q = C\mathscr{E}(1 - e^{-1}) = 0.63C\mathscr{E}. \qquad (28\text{-}34)$$

In words, during the first time constant τ the charge has increased from zero to 63% of its final value $C\mathscr{E}$. In Fig. 28-14, the small triangles along the time axes mark successive intervals of one time constant during the charging of the capacitor. The charging times for *RC* circuits are often stated in terms of τ; the greater τ is, the greater the charging time.

Discharging a Capacitor

Assume now that the capacitor of Fig. 28-13 is fully charged to a potential V_0 equal to the emf $\mathscr{E}$ of the battery. At a new time $t = 0$, switch S is thrown from *a* to *b* so that the capacitor can *discharge* through resistance R. How do the charge $q(t)$ on the capacitor and the current $i(t)$ through the discharge loop of capacitor and resistance now vary with time?

The differential equation describing $q(t)$ is like Eq. 28-29 except that now, with no battery in the discharge loop, $\mathscr{E} = 0$. Thus,

$$R \frac{dq}{dt} + \frac{q}{C} = 0 \qquad \text{(discharging equation)}. \qquad (28\text{-}35)$$

The solution to this differential equation is

$$q = q_0 e^{-t/RC} \qquad \text{(discharging a capacitor)}, \qquad (28\text{-}36)$$

where $q_0 \, (= CV_0)$ is the initial charge on the capacitor. You can verify by substitution that Eq. 28-36 is indeed a solution of Eq. 28-35.

Equation 28-36 tells us that q decreases exponentially with time, at a rate that is set by the capacitive time constant $\tau = RC$. At time $t = \tau$, the capacitor's charge has been reduced to $q_0 e^{-1}$, or about 37% of the initial value. Note that a greater τ means a greater discharge time.

Differentiating Eq. 28-36 gives us the current $i(t)$:

$$i = \frac{dq}{dt} = -\left(\frac{q_0}{RC}\right) e^{-t/RC} \qquad \text{(discharging a capacitor)}. \qquad (28\text{-}37)$$

This tells us that the current also decreases exponentially with time, at a rate set by τ. The initial current i_0 is equal to q_0/RC. Note that you can find i_0 by simply applying the loop rule to the circuit at $t = 0$; just then the capacitor's initial potential V_0 is connected across the resistance R, so the current must be $i_0 = V_0/R = (q_0/C)/R = q_0/RC$. The minus sign in Eq. 28-37 can be ignored; it merely means that the capacitor's charge q is decreasing.

Derivation of Eq. 28-30

To solve Eq. 28-29, we first rewrite it as

$$\frac{dq}{dt} + \frac{q}{RC} = \frac{\mathscr{E}}{R}. \qquad (28\text{-}38)$$

The general solution to this differential equation is of the form

$$q = q_p + Ke^{-at}, \qquad (28\text{-}39)$$

where q_p is a *particular solution* of the differential equation, K is a constant to be evaluated from the initial conditions, and $a = 1/RC$ is the coefficient of q in Eq. 28-38. To find q_p, we set $dq/dt = 0$ in Eq. 28-38 (corresponding to the final condition of no further charging), let $q = q_p$, and solve, obtaining

$$q_p = C\mathscr{E}. \qquad (28\text{-}40)$$

To evaluate K, we first substitute this into Eq. 28-39 to get

$$q = C\mathscr{E} + Ke^{-at}.$$

Then substituting the initial conditions $q = 0$ and $t = 0$ yields

$$0 = C\mathscr{E} + K,$$

or $K = -C\mathscr{E}$. Finally, with the values of q_p, a, and K inserted, Eq. 28-39 becomes

$$q = C\mathscr{E} - C\mathscr{E}e^{-t/RC},$$

which, with a slight modification, is Eq. 28-30.

✔CHECKPOINT 5:   The table gives four sets of values for the circuit elements in Fig. 28-13. Rank the sets according to (a) the initial current (as the switch is closed on a) and (b) the time required for the current to decrease to half its initial value, greatest first.

	1	2	3	4
$\mathscr{E}$ (V)	12	12	10	10
R (Ω)	2	3	10	5
C (μF)	3	2	0.5	2

Sample Problem 28-5

A capacitor of capacitance C is discharging through a resistor of resistance R.

(a) In terms of the time constant $\tau = RC$, when will the charge on the capacitor be half its initial value?

SOLUTION: The Key Idea here is that the charge on the capacitor varies according to Eq. 28-36,

$$q = q_0 e^{-t/RC},$$

in which q_0 is the initial charge. We are asked to find the time t at which $q = \frac{1}{2}q_0$, or at which

$$\tfrac{1}{2}q_0 = q_0 e^{-t/RC}. \qquad (28\text{-}41)$$

After canceling q_0, we realize that the time t we seek is "buried" inside an exponential function. To expose the symbol t in Eq. 28-41, we take the natural logarithms of both sides of the equation. (The natural logarithm is the inverse function of the exponential function.) We find

$$\ln \tfrac{1}{2} = \ln(e^{-t/RC}) = -\frac{t}{RC},$$

or $\qquad t = (-\ln \tfrac{1}{2})RC = 0.69RC = 0.69\tau.$ (Answer)

(b) When will the energy stored in the capacitor be half its initial value?

SOLUTION: There are two Key Ideas here. First, the energy U stored in a capacitor is related to the charge q on the capacitor according to Eq. 26-21 ($U = Q^2/2C$). Second, that charge is decreasing according to Eq. 28-36. Combining these two ideas gives us

$$U = \frac{q^2}{2C} = \frac{q_0^2}{2C}\,e^{-2t/RC} = U_0 e^{-2t/RC},$$

in which U_0 is the initial stored energy. We are asked to find the time at which $U = \frac{1}{2}U_0$, or at which

$$\tfrac{1}{2}U_0 = U_0 e^{-2t/RC}.$$

Canceling U_0 and taking the natural logarithms of both sides, we obtain

$$\ln \tfrac{1}{2} = -\frac{2t}{RC},$$

or $\qquad t = -RC\,\frac{\ln \tfrac{1}{2}}{2} = 0.35RC = 0.35\tau.$ (Answer)

It takes longer (0.69τ versus 0.35τ) for the *charge* to fall to half its initial value than for the *stored energy* to fall to half its initial value. Doesn't this result surprise you?

REVIEW & SUMMARY

Emf   An **emf device** does work on charges to maintain a potential difference between its output terminals. If dW is the work the device does to force positive charge dq from the negative to the positive terminal, then the **emf** (work per unit charge) of the device is

$$\mathscr{E} = \frac{dW}{dq} \qquad \text{(definition of }\mathscr{E}\text{)}. \qquad (28\text{-}1)$$

The volt is the SI unit of emf as well as of potential difference. An **ideal emf device** is one that lacks any internal resistance. The potential difference between its terminals is equal to the emf. A **real emf device** has internal resistance. The potential difference between its terminals is equal to the emf only if there is no current through the device.

Analyzing Circuits   The change in potential in traversing a resistance R in the direction of the current is $-iR$; in the opposite direction it is $+iR$. The change in potential in traversing an ideal emf device in the direction of the emf arrow is $+\mathscr{E}$; in the opposite direction it is $-\mathscr{E}$. Conservation of energy leads to the loop rule:

Loop Rule.   *The algebraic sum of the changes in potential encountered in a complete traversal of any loop of a circuit must be zero.*

Conservation of charge gives us the junction rule:

Junction Rule.   *The sum of the currents entering any junction must be equal to the sum of the currents leaving that junction.*

Single-Loop Circuits   The current in a single-loop circuit containing a single resistance R and an emf device with emf $\mathscr{E}$ and internal resistance r is

$$i = \frac{\mathscr{E}}{R + r}, \qquad (28\text{-}4)$$

which reduces to $i = \mathscr{E}/R$ for an ideal emf device with $r = 0$.

Power   When a real battery of emf $\mathscr{E}$ and internal resistance r does work on the charge carriers in a current i through it, the rate P of energy transfer to the charge carriers is

$$P = iV, \qquad (28\text{-}11)$$

where V is the potential across the terminals of the battery. The rate P_r of energy transfer to thermal energy within the battery is

$$P_r = i^2 r. \qquad (28\text{-}13)$$

The rate P_{emf} at which the chemical energy within the battery changes is

$$P_{\text{emf}} = i\mathscr{E}. \qquad (28\text{-}14)$$

Series Resistances   When resistances are in **series**, they have the same current. The equivalent resistance that can replace a series combination of resistances is

$$R_{\text{eq}} = \sum_{j=1}^{n} R_j \qquad (n \text{ resistances in series}). \qquad (28\text{-}7)$$

Parallel Resistances   When resistances are in **parallel**, they have the same potential difference. The equivalent resistance that can replace a parallel combination of resistances is given by

$$\frac{1}{R_{\text{eq}}} = \sum_{j=1}^{n} \frac{1}{R_j} \qquad (n \text{ resistances in parallel}). \qquad (28\text{-}21)$$

RC Circuits   When an emf $\mathscr{E}$ is applied to a resistance R and capacitance C in series, as in Fig. 28-13 with the switch at a, the charge on the capacitor increases according to

$$q = C\mathscr{E}(1 - e^{-t/RC}) \qquad (\text{charging a capacitor}), \qquad (28\text{-}30)$$

in which $C\mathscr{E} = q_0$ is the equilibrium (final) charge and $RC = \tau$ is the **capacitive time constant** of the circuit. During the charging, the current is

$$i = \frac{dq}{dt} = \left(\frac{\mathscr{E}}{R}\right) e^{-t/RC} \qquad (\text{charging a capacitor}). \qquad (28\text{-}31)$$

When a capacitor discharges through a resistance R, the charge on the capacitor decays according to

$$q = q_0 e^{-t/RC} \qquad (\text{discharging a capacitor}). \qquad (28\text{-}36)$$

During the discharging, the current is

$$i = \frac{dq}{dt} = -\left(\frac{q_0}{RC}\right) e^{-t/RC} \qquad (\text{discharging a capacitor}). \qquad (28\text{-}37)$$

QUESTIONS

1. Figure 28-15 shows current i passing through a battery. The following table gives four sets of values for i and the battery's emf $\mathscr{E}$ and internal resistance r; it also gives the *polarity* (orientation of the terminals) of the battery. Rank the sets according to the rate at which energy is transferred between the battery and the charge carriers, greatest transfer *to* the carriers first and greatest transfer *from* the carriers last.

	$\mathscr{E}$	r	i	Polarity
(1)	$15\mathscr{E}_1$	0	i_1	+ at left
(2)	$10\mathscr{E}_1$	0	$2i_1$	+ at left
(3)	$10\mathscr{E}_1$	0	$2i_1$	− at left
(4)	$10\mathscr{E}_1$	r_1	$2i_1$	− at left

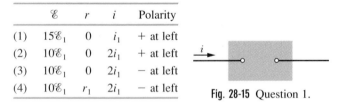

Fig. 28-15 Question 1.

2. For each circuit in Fig. 28-16, are the resistors connected in series, in parallel, or neither?

(a)   (b)   (c)

Fig. 28-16 Question 2.

3. (a) In Fig. 28-17a, are resistors R_1 and R_3 in series? (b) Are resistors R_1 and R_2 in parallel? (c) Rank the equivalent resistances of the four circuits shown in Fig. 28-17, greatest first.

4. (a) In Fig. 28-17a, with $R_1 > R_2$, is the potential difference across R_2 more than, less than, or equal to that across R_1? (b) Is the current through resistor R_2 more than, less than, or equal to that through resistor R_1?

(a)   (b)

(c)   (d)

Fig. 28-17 Questions 3 and 4.

5. You are to connect resistors R_1 and R_2, with $R_1 > R_2$, to a battery, first individually, then in series, and then in parallel. Rank those arrangements according to the amount of current through the battery, greatest first.

6. *Res-monster maze.* In Fig. 28-18, all the resistors have a resistance of 4.0 Ω and all the (ideal) batteries have an emf of 4.0 V. What is the current through resistor R? (If you can find the proper loop through this maze, you can answer the question with a few seconds of mental calculation.)

Fig. 28-18 Question 6.

7. Initially, a single resistor R_1 is wired to a battery. Then resistor R_2 is added in parallel. Are (a) the potential difference across R_1 and (b) the current i_1 through R_1 now more than, less than, or the same as previously? (c) Is the equivalent resistance R_{12} of R_1 and R_2 more than, less than, or equal to R_1? (d) Is the total current through R_1 and R_2 together more than, less than, or equal to the current through R_1 previously?

8. *Cap-monster maze.* In Fig. 28-19, all the capacitors have a capacitance of 6.0 μF, and all the batteries have an emf of 10 V. What is the charge on capacitor C? (If you can find the proper loop through this maze, you can answer the question with a few seconds of mental calculation.)

9. A resistor R_1 is wired to a battery, then resistor R_2 is added in series. Are (a) the potential difference across R_1 and (b) the current i_1 through R_1 now more than, less than, or the same as previously? (c) Is the equivalent resistance R_{12} of R_1 and R_2 more than, less than, or equal to R_1?

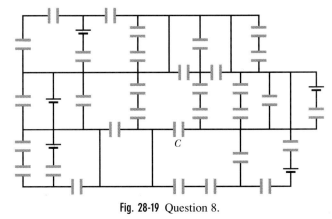

Fig. 28-19 Question 8.

10. Figure 28-20 shows three sections of circuit that are to be connected in turn to the same battery via a switch as in Fig. 28-13. The resistors are all identical, as are the capacitors. Rank the sections according to (a) the final (equilibrium) charge on the capacitor and (b) the time required for the capacitor to reach 50% of its final charge, greatest first.

Fig. 28-20 Question 10.

11. Figure 28-21 shows plots of $V(t)$ for three capacitors that discharge (separately) through the same resistor. Rank the plots according to the capacitances of the capacitors, greatest first.

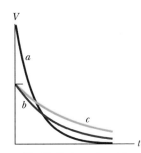

Fig. 28-21 Question 11.

EXERCISES & PROBLEMS

SEC. 28-5 Potential Differences

1E. A standard flashlight battery can deliver about 2.0 W · h of energy before it runs down. (a) If a battery costs 80¢, what is the cost of operating a 100 W lamp for 8.0 h using batteries? (b) What is the cost if energy is provided at 6¢ per kilowatt-hour? ssm

2E. A 5.0 A current is set up in a circuit for 6.0 min by a rechargeable battery with a 6.0 V emf. By how much is the chemical energy of the battery reduced?

3E. A certain car battery with a 12 V emf has an initial charge of 120 A · h. Assuming that the potential across the terminals stays constant until the battery is completely discharged, for how long can it deliver energy at the rate of 100 W? ssm

4E. In Fig. 28-22, $\mathcal{E}_1 = 12$ V and $\mathcal{E}_2 = 8$ V. (a) What is the direction of the current in the resistor? (b) Which battery is doing positive work? (c) Which point, A or B, is at the higher potential?

Fig. 28-22 Exercise 4.

5E. Assume that the batteries in Fig. 28-23 have negligible internal resistance. Find (a) the current in the circuit, (b) the power dissipated in each resistor, and (c) the power of each battery, stating whether energy is supplied by or absorbed by it. ssm

6E. A wire of resistance 5.0 Ω is connected to a battery whose emf $\mathcal{E}$ is 2.0 V and whose internal resistance is 1.0 Ω. In 2.0 min, (a) how much energy is transferred from chemical to electrical form? (b) How much energy appears in the wire as thermal energy? (c) Account for the difference between (a) and (b).

Fig. 28-23 Exercise 5.

7E. A car battery with a 12 V emf and an internal resistance of 0.040 Ω is being charged with a current of 50 A. (a) What is the potential difference across its terminals? (b) At what rate is energy being dissipated as thermal energy in the battery? (c) At what rate is electric energy being converted to chemical energy? (d) What are the answers to (a) and (b) when the battery is used to supply 50 A to the starter motor? ilw

8E. In Fig. 28-4a, put $\mathcal{E} = 2.0$ V and $r = 100$ Ω. Plot (a) the current and (b) the potential difference across R, as functions of R over the range 0 to 500 Ω. Make both plots on the same graph. (c) Make a third plot by multiplying together, for various values of R, the corresponding values on the two plotted curves. What is the physical significance of this third plot?

9E. In Fig. 28-24, circuit section AB absorbs energy at a rate of 50 W when a current $i = 1.0$ A passes through it in the indicated direction. (a) What is the potential difference between A and B? (b) Emf device X does not have internal resistance. What is its emf? (c) What is its *polarity* (the orientation of its positive and negative terminals)? ssm

Fig. 28-24 Exercise 9.

10E. In Fig. 28-25, if the potential at point P is 100 V, what is the potential at point Q?

Fig. 28-25 Exercise 10.

11E. In Fig. 28-6a, calculate the potential difference between a and c by considering a path that contains R, r_1, and $\mathcal{E}_1$.

12P. (a) In Fig. 28-26, what value must R have if the current in the circuit is to be 1.0 mA? Take $\mathcal{E}_1 = 2.0$ V, $\mathcal{E}_2 = 3.0$ V, and $r_1 = r_2 = 3.0$ Ω. (b) What is the rate at which thermal energy appears in R?

Fig. 28-26 Problem 12.

13P. The current in a single-loop circuit with one resistance R is 5.0 A. When an additional resistance of 2.0 Ω is inserted in series with R, the current drops to 4.0 A. What is R? ilw

14P. The starting motor of an automobile is turning too slowly, and the mechanic has to decide whether to replace the motor, the cable, or the battery. The manufacturer's manual says that the 12 V battery should have no more than 0.020 Ω internal resistance, the motor no more than 0.200 Ω resistance, and the cable no more than 0.040 Ω resistance. The mechanic turns on the motor and measures 11.4 V across the battery, 3.0 V across the cable, and a current of 50 A. Which part is defective?

15P. Two batteries having the same emf $\mathcal{E}$ but different internal resistances r_1 and r_2 $(r_1 > r_2)$ are connected in series to an external resistance R. (a) Find the value of R that makes the potential difference zero between the terminals of one battery. (b) Which battery is it? ssm

16P. A solar cell generates a potential difference of 0.10 V when a 500 Ω resistor is connected across it, and a potential difference of 0.15 V when a 1000 Ω resistor is substituted. What are (a) the internal resistance and (b) the emf of the solar cell? (c) The area of the cell is 5.0 cm², and the rate per unit area at which it receives energy from light is 2.0 mW/cm². What is the efficiency of the cell for converting light energy to thermal energy in the 1000 Ω external resistor?

17P. (a) In Fig. 28-4a, show that the rate at which energy is dissipated in R as thermal energy is a maximum when $R = r$. (b) Show that this maximum power is $P = \mathcal{E}^2/4r$. ssm www

SEC. 28-6 Multiloop Circuits

18E. By using only two resistors—singly, in series, or in parallel—you are able to obtain resistances of 3.0, 4.0, 12, and 16 Ω. What are the two resistances?

19E. Four 18.0 Ω resistors are connected in parallel across a 25.0 V ideal battery. What is the current through the battery? ssm

20E. In Fig. 28-27, find the equivalent resistance between points D and E. (*Hint:* Imagine that a battery is connected between points D and E.)

Fig. 28-27 Exercise 20.

21E. In Fig. 28-28 find the current in each resistor and the potential difference between points *a* and *b*. Put $\mathscr{E}_1 = 6.0$ V, $\mathscr{E}_2 = 5.0$ V, $\mathscr{E}_3 = 4.0$ V, $R_1 = 100$ Ω, and $R_2 = 50$ Ω. ssm

22E. Figure 28-29 shows a circuit containing three switches, labeled S_1, S_2, and S_3. Find the current at *a* for all possible combinations of switch settings. Put $\mathscr{E} = 120$ V, $R_1 = 20.0$ Ω, and $R_2 = 10.0$ Ω. Assume that the battery has no resistance.

23E. Two lightbulbs, one of resistance R_1 and the other of resistance R_2, where $R_1 > R_2$, are connected to a battery (a) in parallel and (b) in series. Which bulb is brighter (dissipates more energy) in each case? ssm

24E. In Fig. 28-7, calculate the potential difference between points *c* and *d* by as many paths as possible. Assume that $\mathscr{E}_1 = 4.0$ V, $\mathscr{E}_2 = 1.0$ V, $R_1 = R_2 = 10$ Ω, and $R_3 = 5.0$ Ω.

25E. Nine copper wires of length *l* and diameter *d* are connected in parallel to form a single composite conductor of resistance *R*. What must be the diameter *D* of a single copper wire of length *l* if it is to have the same resistance? ssm

26P. In Fig. 28-30, find the equivalent resistance between points (a) *F* and *H* and (b) *F* and *G*. (*Hint:* For each pair of points, imagine that a battery is connected across the pair.)

27P. You are given a number of 10 Ω resistors, each capable of dissipating only 1.0 W without being destroyed. What is the minimum number of such resistors that you need to combine in series or in parallel to make a 10 Ω resistance that is capable of dissipating at least 5.0 W? ssm

28P. (a) In Fig. 28-31, what is the equivalent resistance of the network shown? (b) What is the current in each resistor? Put $R_1 = 100$ Ω, $R_2 = R_3 = 50$ Ω, $R_4 = 75$ Ω, and $\mathscr{E} = 6.0$ V; assume the battery is ideal.

29P. Two batteries of emf $\mathscr{E}$ and internal resistance *r* are connected in parallel across a resistor *R*, as in Fig. 28-32*a*. (a) For what value of *R* is the rate of electrical energy dissipation by the resistor a maximum? (b) What is the maximum energy dissipation rate? ssm

30P. You are given two batteries of emf $\mathscr{E}$ and internal resistance *r*. They may be connected either in parallel (Fig. 28-32*a*) or in

Fig. 28-28 Exercise 21.

Fig. 28-29 Exercise 22.

Fig. 28-30 Problem 26.

Fig. 28-31 Problem 28.

Fig. 28-32 Problems 29 and 30.

series (Fig. 28-32*b*) and are to be used to establish a current in a resistor *R*. (a) Derive expressions for the current in *R* for both arrangements. Which will yield the larger current (b) when $R > r$ and (c) when $R < r$?

31P. In Fig. 28-33, $\mathscr{E}_1 = 3.00$ V, $\mathscr{E}_2 = 1.00$ V, $R_1 = 5.00$ Ω, $R_2 = 2.00$ Ω, $R_3 = 4.00$ Ω, and both batteries are ideal. What is the rate at which energy is dissipated in (a) R_1, (b) R_2, and (c) R_3? What is the power of (d) battery 1 and (e) battery 2? ssm www

32P. In the circuit of Fig. 28-34, for what value of *R* will the ideal battery transfer energy to the resistors (a) at a rate of 60.0 W, (b) at the maximum possible rate, and (c) at the minimum possible rate? (d) What are those rates?

33P. (a) Calculate the current through each ideal battery in Fig. 28-35. Assume that $R_1 = 1.0$ Ω, $R_2 = 2.0$ Ω, $\mathscr{E}_1 = 2.0$ V, and $\mathscr{E}_2 = \mathscr{E}_3 = 4.0$ V. (b) Calculate $V_a - V_b$. ilw

34P. In the circuit of Fig. 28-36, $\mathscr{E}$ has a constant value but *R* can be varied. Find the value of *R* that results in the maximum heating in that resistor. The battery is ideal.

35P. A copper wire of radius $a = 0.250$ mm has an aluminum jacket of outer radius $b = 0.380$ mm. (a) There is a current $i = 2.00$ A in the composite wire. Using Table 27-1, calculate the current in each material. (b) If a potential difference $V = 12.0$ V between the ends maintains the current, what is the length of the composite wire? ssm

Fig. 28-33 Problem 31.

Fig. 28-34 Problem 32.

Fig. 28-35 Problem 33.

Fig. 28-36 Problem 34.

SEC. 28-7   The Ammeter and the Voltmeter

36E. A simple ohmmeter is made by connecting a 1.50 V flashlight battery in series with a resistance R and an ammeter that reads from 0 to 1.00 mA, as shown in Fig. 28-37. Resistance R is adjusted so that when the clip leads are shorted together, the meter deflects to its full-scale value of 1.00 mA. What external resistance across the leads results in a deflection of (a) 10%, (b) 50%, and (c) 90% of full scale? (d) If the ammeter has a resistance of 20.0 Ω and the internal resistance of the battery is negligible, what is the value of R?

Fig. 28-37 Exercise 36.

37P. (a) In Fig. 28-38, determine what the ammeter will read, assuming $\mathcal{E} = 5.0$ V (for the ideal battery), $R_1 = 2.0\ \Omega$, $R_2 = 4.0\ \Omega$, and $R_3 = 6.0\ \Omega$. (b) The ammeter and the source of emf are now physically interchanged. Show that the ammeter reading remains unchanged. **itw**

Fig. 28-38 Problem 37.

38P. When the lights of an automobile are switched on, an ammeter in series with them reads 10 A and a voltmeter connected across them reads 12 V. See Fig. 28-39. When the electric starting motor is turned on, the ammeter reading drops to 8.0 A and the lights dim somewhat. If the internal resistance of the battery is 0.050 Ω and that of the ammeter is negligible, what are (a) the emf of the battery and (b) the current through the starting motor when the lights are on?

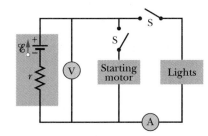

Fig. 28-39 Problem 38.

39P. In Fig. 28-12, assume that $\mathcal{E} = 3.0$ V, $r = 100\ \Omega$, $R_1 = 250\ \Omega$, and $R_2 = 300\ \Omega$. If the voltmeter resistance R_V is 5.0 kΩ, what percent error does it introduce into the measurement of the potential difference across R_1? Ignore the presence of the ammeter. **ssm**

40P. A voltmeter (of resistance R_V) and an ammeter (of resistance R_A) are connected to measure a resistance R, as in Fig. 28-40a. The resistance is given by $R = V/i$, where V is the voltmeter reading and i is the current in the resistance R. Some of the current i' registered by the ammeter goes through the voltmeter, so that the ratio of the meter readings ($= V/i'$) gives only an *apparent* resistance reading R'. Show that R and R' are related by

$$\frac{1}{R} = \frac{1}{R'} - \frac{1}{R_V}.$$

Note that as $R_V \to \infty$, $R' \to R$.

Fig. 28-40 Problems 40 to 42.

41P. (See Problem 40.) If an ammeter and a voltmeter are used to measure resistance, they may also be connected as in Fig. 28-40b. Again the ratio of the meter readings gives only an apparent resistance R'. Show that now R' is related to R by

$$R = R' - R_A,$$

in which R_A is the ammeter resistance. Note that as $R_A \to 0$, $R' \to R$.

42P. (See Problems 40 and 41.) In Fig. 28-40, the ammeter and voltmeter resistances are 3.00 and 300 Ω, respectively. Take $\mathcal{E} = 12.0$ V for the ideal battery and $R_0 = 100\ \Omega$. If $R = 85.0\ \Omega$, (a) what will the meters read for the two different connections (Figs. 28-40a and b)? (b) What apparent resistance R' will be computed in each case?

43P. In Fig. 28-41, R_s is to be adjusted in value by moving the sliding contact across it until points a and b are brought to the same potential. (One tests for this condition by momentarily connecting a sensitive ammeter between a and b; if these points are at the same potential, the ammeter will not deflect.) Show that when this adjustment is made, the following relation holds:

$$R_x = R_s\left(\frac{R_2}{R_1}\right).$$

Fig. 28-41 Problem 43.

An unknown resistance (R_x) can be measured in terms of a standard (R_s) using this device, which is called a Wheatstone bridge. **ssm**

SEC. 28-8   *RC* Circuits

44E. A capacitor with initial charge q_0 is discharged through a resistor. In terms of the time constant τ, how long is required for the capacitor to lose (a) the first one-third of its charge and (b) two-thirds of its charge?

45E. How many time constants must elapse for an initially uncharged capacitor in an *RC* series circuit to be charged to 99.0% of its equilibrium charge? **ssm**

46E. In an *RC* series circuit, $\mathcal{E} = 12.0$ V, $R = 1.40$ MΩ, and $C = 1.80\ \mu$F. (a) Calculate the time constant. (b) Find the maximum charge that will appear on the capacitor during charging. (c) How long does it take for the charge to build up to 16.0 μC?

47E. A 15.0 kΩ resistor and a capacitor are connected in series and then a 12.0 V potential difference is suddenly applied across them. The potential difference across the capacitor rises to 5.00 V in 1.30 μs. (a) Calculate the time constant of the circuit. (b) Find the capacitance of the capacitor. ilw

48P. The potential difference between the plates of a leaky (meaning that charge leaks from one plate to the other) 2.0 μF capacitor drops to one-fourth its initial value in 2.0 s. What is the equivalent resistance between the capacitor plates?

49P. A 3.00 MΩ resistor and a 1.00 μF capacitor are connected in series with an ideal battery of emf $\mathscr{E} = 4.00$ V. At 1.00 s after the connection is made, what are the rates at which (a) the charge of the capacitor is increasing, (b) energy is being stored in the capacitor, (c) thermal energy is appearing in the resistor, and (d) energy is being delivered by the battery? ssm

50P. An initially uncharged capacitor C is fully charged by a device of constant emf $\mathscr{E}$ connected in series with a resistor R. (a) Show that the final energy stored in the capacitor is half the energy supplied by the emf device. (b) By direct integration of i^2R over the charging time, show that the thermal energy dissipated by the resistor is also half the energy supplied by the emf device.

51P. A capacitor with an initial potential difference of 100 V is discharged through a resistor when a switch between them is closed at $t = 0$. At $t = 10.0$ s, the potential difference across the capacitor is 1.00 V. (a) What is the time constant of the circuit? (b) What is the potential difference across the capacitor at $t = 17.0$ s? ssm

52P. Figure 28-42 shows the circuit of a flashing lamp, like those attached to barrels at highway construction sites. The fluorescent lamp L (of negligible capacitance) is connected in parallel across the capacitor C of an RC circuit. There is a current through the lamp only when the potential difference across it reaches the breakdown voltage V_L; in this event, the capacitor discharges completely through the lamp and the lamp flashes briefly. Suppose that two flashes per second are needed. For a lamp with breakdown voltage $V_L = 72.0$ V, wired to a 95.0 V ideal battery and a 0.150 μF capacitor, what should be the resistance R?

Fig. 28-42 Problem 52.

53P. A 1.0 μF capacitor with an initial stored energy of 0.50 J is discharged through a 1.0 MΩ resistor. (a) What is the initial charge on the capacitor? (b) What is the current through the resistor when the discharge starts? (c) Determine V_C, the potential difference across the capacitor, and V_R, the potential difference across the resistor, as functions of time. (d) Express the production rate of thermal energy in the resistor as a function of time. ssm www

54P. A controller on an electronic arcade game consists of a variable resistor connected across the plates of a 0.220 μF capacitor. The capacitor is charged to 5.00 V, then discharged through the resistor. The time for the potential difference across the plates to decrease to 0.800 V is measured by a clock inside the game. If the range of discharge times that can be handled effectively is from 10.0 μs to 6.00 ms, what should be the resistance range of the resistor?

55P*. In the circuit of Fig. 28-43, $\mathscr{E} = 1.2$ kV, $C = 6.5$ μF, $R_1 = R_2 = R_3 = 0.73$ MΩ. With C completely uncharged, switch S is suddenly closed (at $t = 0$). (a) Determine the current through each resistor at $t = 0$ and as $t \to \infty$. (b) Draw qualitatively a graph of the potential difference V_2 across R_2 from $t = 0$ to $t = \infty$. (c) What are the numerical values of V_2 at $t = 0$ and as $t \to \infty$? (d) What is the physical meaning of "$t \to \infty$" in this case? ssm

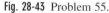

Fig. 28-43 Problem 55.

Additional Problem

56. *Heart attack or electrocution?* This story begins with Problem 45 in Chapter 27. Figure 28-44 shows the electrical pathway of the current up through one foot of the victim, across the torso (including the heart), and down through the other foot. (a) From the given data, find the potential difference between the man's feet, assuming that one foot was 0.50 m closer to the leaky rod than the other. (b) Assume that the resistance of a foot on wet soil is the typical value of 300 Ω and the resistance of the torso interior is the commonly accepted value of 1000 Ω. What then was the current across the victim's torso? (c) The human heart can be put into fibrillation by a current of 0.10 A to 1.0 A through the torso. Was the victim's fibrillation due to current leakage from the rod?

Fig. 28-44 Problem 56.

29 Magnetic Fields

If you are outside on a dark night in the middle to high latitudes, you might be able to see an aurora, a ghostly "curtain" of light that hangs down from the sky. This curtain is not only local; it may be several hundred kilometers high and several thousand kilometers long, stretching around Earth in an arc. However, it is less than 1 km thick.

What produces this huge display, and what makes it so thin?

The answer is in this chapter.

Fig. 29-1 Using an electromagnet to collect and transport scrap metal at a steel mill.

29-1 The Magnetic Field

We have discussed how a charged plastic rod produces a vector field—the electric field $\vec{E}$—at all points in the space around it. Similarly, a magnet produces a vector field—the **magnetic field $\vec{B}$**—at all points in the space around it. You get a hint of that magnetic field whenever you attach a note to a refrigerator door with a small magnet, or accidentally erase a computer disk by bringing it near a magnet. The magnet acts on the door or disk *by means of* its magnetic field.

In a familiar type of magnet, a wire coil is wound around an iron core and a current is sent through the coil; the strength of the magnetic field is determined by the size of the current. In industry, such **electromagnets** are used for sorting scrap metal (Fig. 29-1) among many other things. You are probably more familiar with **permanent magnets**—magnets, like the refrigerator-door type, that do not need current to have a magnetic field.

In Chapter 23 we saw that an *electric charge* sets up an electric field that can then affect other electric charges. Here, we might reasonably expect that a *magnetic charge* sets up a magnetic field that can then affect other magnetic charges. Although such magnetic charges, called *magnetic monopoles,* are predicted by certain theories, their existence has not been confirmed.

How then are magnetic fields set up? There are two ways. (1) Moving electrically charged particles, such as a current in a wire, create magnetic fields. (2) Elementary particles such as electrons have an *intrinsic* magnetic field around them; that is, this field is a basic characteristic of the particles, just as are their mass and electric charge (or lack of charge). As we shall discuss in Chapter 32, the magnetic fields of the electrons in certain materials add together to give a net magnetic field around the material. This is true for the material in permanent magnets (which is good, because they can then hold notes to a refrigerator door). In other materials, the magnetic fields of all the electrons cancel out, giving no net magnetic field surrounding the material. This is true for the material in your body (which is also good, because otherwise you might be slammed up against a refrigerator door every time you passed one).

Experimentally we find that when a charged particle (either alone or as part of a current) moves through a magnetic field, a force due to the field can act on the particle. In this chapter we focus on the relation between the magnetic field and this force.

29-2 The Definition of $\vec{B}$

We determined the electric field $\vec{E}$ at a point by putting a test particle of charge q at rest at that point and measuring the electric force $\vec{F}_E$ acting on the particle. We then defined $\vec{E}$ as

$$\vec{E} = \frac{\vec{F}_E}{q}. \tag{29-1}$$

If a magnetic monopole were available, we could define $\vec{B}$ in a similar way. Because such particles have not been found, we must define $\vec{B}$ in another way, in terms of the magnetic force $\vec{F}_B$ exerted on a moving electrically charged test particle.

In principle, we do this by firing a charged particle through the point at which $\vec{B}$ is to be defined, using various directions and speeds for the particle and determining the force $\vec{F}_B$ that acts on the particle at that point. After many such trials we would find that when the particle's velocity $\vec{v}$ is along a particular axis through the point, force $\vec{F}_B$ is zero. For all other directions of $\vec{v}$, the magnitude of $\vec{F}_B$ is always proportional to $v \sin \phi$, where ϕ is the angle between the zero-force axis and the direction of $\vec{v}$. Furthermore, the direction of $\vec{F}_B$ is always perpendicular to the direction of $\vec{v}$. (These results suggest that a cross product is involved.)

We can then define a magnetic field $\vec{B}$ to be a vector quantity that is directed along the zero-force axis. We can next measure the magnitude of $\vec{F}_B$ when $\vec{v}$ is directed perpendicular to that axis and then define the magnitude of $\vec{B}$ in terms of that force magnitude:

$$B = \frac{F_B}{|q|v},$$

where q is the charge of the particle.

We can summarize all these results with the following vector equation:

$$\vec{F}_B = q\vec{v} \times \vec{B}; \tag{29-2}$$

that is, the force $\vec{F}_B$ on the particle is equal to the charge q times the cross product of its velocity $\vec{v}$ and the field $\vec{B}$ (all measured in the same reference frame). Using Eq. 3-20 for the cross product, we can write the magnitude of $\vec{F}_B$ as

$$F_B = |q|vB \sin \phi, \tag{29-3}$$

where ϕ is the angle between the directions of velocity $\vec{v}$ and magnetic field $\vec{B}$.

Finding the Magnetic Force on a Particle

Equation 29-3 tells us that the magnitude of the force $\vec{F}_B$ acting on a particle in a magnetic field is proportional to the charge q and speed v of the particle. Thus, the force is equal to zero if the charge is zero or if the particle is stationary. Equation 29-3 also tells us that the magnitude of the force is zero if $\vec{v}$ and $\vec{B}$ are either parallel ($\phi = 0°$) or antiparallel ($\phi = 180°$), and the force is at its maximum when $\vec{v}$ and $\vec{B}$ are perpendicular to each other.

Equation 29-2 tells us all this plus the direction of $\vec{F}_B$. From Section 3-7, we know that the cross product $\vec{v} \times \vec{B}$ in Eq. 29-2 is a vector that is perpendicular to the two vectors $\vec{v}$ and $\vec{B}$. The right-hand rule (Fig. 29-2a) tells us that the thumb of the right hand points in the direction of $\vec{v} \times \vec{B}$ when the fingers sweep $\vec{v}$ into $\vec{B}$. If q is positive, then (by Eq. 29-2) the force $\vec{F}_B$ has the same sign as $\vec{v} \times \vec{B}$ and thus must be in the same direction; that is, for positive q, $\vec{F}_B$ is directed along the thumb (Fig. 29-2b). If q is negative, then the force $\vec{F}_B$ and cross product

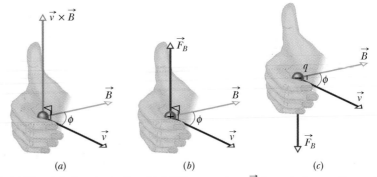

Fig. 29-2 (a) The right-hand rule (in which $\vec{v}$ is swept into $\vec{B}$ through the smaller angle ϕ between them) gives the direction of $\vec{v} \times \vec{B}$ as the direction of the thumb. (b) If q is positive, then the direction of $\vec{F}_B = q\vec{v} \times \vec{B}$ is in the direction of $\vec{v} \times \vec{B}$. (c) If q is negative, then the direction of $\vec{F}_B$ is opposite that of $\vec{v} \times \vec{B}$.

Fig. 29-3 The tracks of two electrons (e^-) and a positron (e^+) in a bubble chamber that is immersed in a uniform magnetic field that is directed out of the plane of the page.

$\vec{v} \times \vec{B}$ have opposite signs and thus must be in opposite directions. For negative q, $\vec{F}_B$ is directed opposite the thumb (Fig. 29-2c).

Regardless of the sign of the charge, however,

> The force $\vec{F}_B$ acting on a charged particle moving with velocity $\vec{v}$ through a magnetic field $\vec{B}$ is *always* perpendicular to $\vec{v}$ and $\vec{B}$.

Thus, $\vec{F}_B$ *never* has a component parallel to $\vec{v}$. This means that $\vec{F}_B$ cannot change the particle's speed v (and thus it cannot change the particle's kinetic energy). The force can change only the direction of $\vec{v}$ (and thus the direction of travel); only in this sense can $\vec{F}_B$ accelerate the particle.

To develop a feeling for Eq. 29-2, consider Fig. 29-3, which shows some tracks left by charged particles moving rapidly through a *bubble chamber* at the Lawrence Berkeley Laboratory. The chamber, which is filled with liquid hydrogen, is immersed in a strong uniform magnetic field that is directed out of the plane of the figure. An incoming gamma ray particle—which leaves no track because it is uncharged—transforms into an electron (spiral track marked e^-) and a positron (track marked e^+) while it knocks an electron out of a hydrogen atom (long track marked e^-). Check with Eq. 29-2 and Fig. 29-2 that the three tracks made by these two negative particles and one positive particle curve in the proper directions.

The SI unit for $\vec{B}$ that follows from Eqs. 29-2 and 29-3 is the newton per coulomb-meter per second. For convenience, this is called the **tesla** (T):

$$1 \text{ tesla} = 1 \text{ T} = 1 \frac{\text{newton}}{\text{(coulomb)(meter/second)}}.$$

Recalling that a coulomb per second is an ampere, we have

$$1 \text{ T} = 1 \frac{\text{newton}}{\text{(coulomb/second)(meter)}} = 1 \frac{\text{N}}{\text{A} \cdot \text{m}}. \qquad (29\text{-}4)$$

An earlier (non-SI) unit for $\vec{B}$, still in common use, is the *gauss* (G), and

$$1 \text{ tesla} = 10^4 \text{ gauss}. \qquad (29\text{-}5)$$

Table 29-1 lists the magnetic fields that occur in a few situations. Note that Earth's magnetic field near the planet's surface is about 10^{-4} T ($= 100 \ \mu$T or 1 gauss).

TABLE 29-1   Some Approximate Magnetic Fields

At the surface of a neutron star	10^8 T
Near a big electromagnet	1.5 T
Near a small bar magnet	10^{-2} T
At Earth's surface	10^{-4} T
In interstellar space	10^{-10} T
Smallest value in a magnetically shielded room	10^{-14} T

✔CHECKPOINT 1: The figure shows three situations in which a charged particle with velocity $\vec{v}$ travels through a uniform magnetic field $\vec{B}$. In each situation, what is the direction of the magnetic force $\vec{F}_B$ on the particle?

(a)   (b)   (c)

Magnetic Field Lines

We can represent magnetic fields with field lines, as we did for electric fields. Similar rules apply; that is, (1) the direction of the tangent to a magnetic field line at any point gives the direction of $\vec{B}$ at that point, and (2) the spacing of the lines represents the magnitude of $\vec{B}$—the magnetic field is stronger where the lines are closer together, and conversely.

Fig. 29-4 (*a*) The magnetic field lines for a bar magnet. (*b*) A "cow magnet"—a bar magnet that is intended to be slipped down into the rumen of a cow to prevent accidentally ingested bits of scrap iron from reaching the cow's intestines. The iron filings at its ends reveal the magnetic field lines.

(*a*)   (*b*)

(*a*)

(*b*)

Fig. 29-5 (*a*) A horseshoe magnet and (*b*) a **C**-shaped magnet. (Only some of the external field lines are shown.)

Figure 29-4*a* shows how the magnetic field near a *bar magnet* (a permanent magnet in the shape of a bar) can be represented by magnetic field lines. The lines all pass through the magnet, and they all form closed loops (even those that are not shown closed in the figure). The external magnetic effects of a bar magnet are strongest near its ends, where the field lines are most closely spaced. Thus, the magnetic field of the bar magnet in Fig. 29-4*b* collects the iron filings mainly near the two ends of the magnet.

The (closed) field lines enter one end of a magnet and exit the other end. The end of a magnet from which the field lines emerge is called the *north pole* of the magnet; the other end, where field lines enter the magnet, is called the *south pole*. The magnets we use to fix notes on refrigerators are short bar magnets. Figure 29-5 shows two other common shapes for magnets: a *horseshoe magnet* and a magnet that has been bent around into the shape of a **C** so that the *pole faces* are facing each other. (The magnetic field between the pole faces can then be approximately uniform.) Regardless of the shape of the magnets, if we place two of them near each other we find:

> Opposite magnetic poles attract each other, and like magnetic poles repel each other.

Earth has a magnetic field that is produced in its core by still unknown mechanisms. On Earth's surface, we can detect this magnetic field with a compass, which is essentially a slender bar magnet on a low-friction pivot. This bar magnet, or this needle, turns because its north-pole end is attracted toward the Arctic region of Earth. Thus, the *south* pole of Earth's magnetic field must be located toward the Arctic. Logically, we then should call the pole there a south pole. However, because we call that direction north, we are trapped into the statement that Earth has a *geomagnetic north pole* in that direction.

With more careful measurement we would find that in the northern hemisphere, the magnetic field lines of Earth generally point down into Earth and toward the Arctic. In the southern hemisphere, they generally point up out of Earth and away from the Antarctic—that is, away from Earth's *geomagnetic south pole*.

Sample Problem 29-1

A uniform magnetic field $\vec{B}$, with magnitude 1.2 mT, is directed vertically upward throughout the volume of a laboratory chamber. A proton with kinetic energy 5.3 MeV enters the chamber, moving horizontally from south to north. What magnetic deflecting force acts on the proton as it enters the chamber? The proton mass is 1.67×10^{-27} kg. (Neglect Earth's magnetic field.)

SOLUTION: Because the proton is charged and moving through a magnetic field, a magnetic force $\vec{F}_B$ can act on it. The Key Idea here is that, because the initial direction of the proton's velocity is not along a magnetic field line, $\vec{F}_B$ is not simply zero. To find the magnitude of $\vec{F}_B$, we can use Eq. 29-3 provided we first find the proton's speed v. We can find v from the given kinetic energy, since $K = \frac{1}{2}mv^2$. Solving for v, we obtain

$$v = \sqrt{\frac{2K}{m}} = \sqrt{\frac{(2)(5.3 \text{ MeV})(1.60 \times 10^{-13} \text{ J/MeV})}{1.67 \times 10^{-27} \text{ kg}}}$$

$$= 3.2 \times 10^7 \text{ m/s}.$$

Fig. 29-6 Sample Problem 29-1. An overhead view of a proton moving from south to north with velocity $\vec{v}$ in a chamber. A magnetic field is directed vertically upward in the chamber, as represented by the array of dots (which resemble the tips of arrows). The proton is deflected toward the east.

Equation 29-3 then yields

$$
\begin{aligned}
F_B &= |q|vB \sin \phi \\
&= (1.60 \times 10^{-19} \text{ C})(3.2 \times 10^7 \text{ m/s}) \\
&\quad \times (1.2 \times 10^{-3} \text{ T})(\sin 90°) \\
&= 6.1 \times 10^{-15} \text{ N}. \qquad \text{(Answer)}
\end{aligned}
$$

This may seem like a small force, but it acts on a particle of small mass, producing a large acceleration; namely,

$$a = \frac{F_B}{m} = \frac{6.1 \times 10^{-15} \text{ N}}{1.67 \times 10^{-27} \text{ kg}} = 3.7 \times 10^{12} \text{ m/s}^2.$$

To find the direction of $\vec{F}_B$, we use the Key Idea that $\vec{F}_B$ has the direction of the cross product $q\vec{v} \times \vec{B}$. Because the charge q is positive, $\vec{F}_B$ must have the same direction as $\vec{v} \times \vec{B}$, which can be determined with the right-hand rule for cross products (as in Fig. 29-2b). We know that $\vec{v}$ is directed horizontally from south to north and $\vec{B}$ is directed vertically up. The right-hand rule shows us that the deflecting force $\vec{F}_B$ must be directed horizontally from west to east, as Fig. 29-6 shows. (The array of dots in the figure represents a magnetic field directed out of the plane of the figure. An array of Xs would have represented a magnetic field directed into that plane.)

If the charge of the particle were negative, the magnetic deflecting force would be directed in the opposite direction—that is, horizontally from east to west. This is predicted automatically by Eq. 29-2 if we substitute a negative value for q.

29-3 Crossed Fields: Discovery of the Electron

Both an electric field $\vec{E}$ and a magnetic field $\vec{B}$ can produce a force on a charged particle. When the two fields are perpendicular to each other, they are said to be *crossed fields*. Here we shall examine what happens to charged particles—namely, electrons—as they move through crossed fields. We use as our example the experiment that led to the discovery of the electron in 1897 by J. J. Thomson at Cambridge University.

Figure 29-7 shows a modern, simplified version of Thomson's experimental apparatus—a *cathode ray tube* (which is like the picture tube in a standard television

Fig. 29-7 A modern version of J. J. Thomson's apparatus for measuring the ratio of mass to charge for the electron. An electric field $\vec{E}$ is established by connecting a battery across the deflecting-plate terminals. A magnetic field $\vec{B}$ is set up by means of a current in a system of coils (not shown). The magnetic field shown is into the plane of the figure, as represented by the array of Xs (which resemble the feathered ends of arrows).

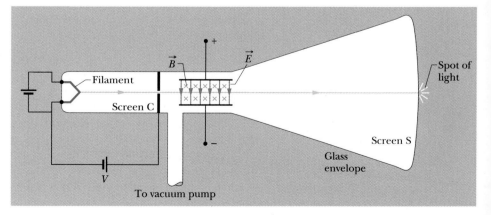

set). Charged particles (which we now know as electrons) are emitted by a hot filament at the rear of the evacuated tube and are accelerated by an applied potential difference V. After they pass through a slit in screen C, they form a narrow beam. They then pass through a region of crossed $\vec{E}$ and $\vec{B}$ fields, headed toward a fluorescent screen S, where they produce a spot of light (on a television screen the spot is part of the picture). The forces on the charged particles in the crossed-fields region can deflect them from the center of the screen. By controlling the magnitudes and directions of the fields, Thomson could thus control where the spot of light appeared on the screen. Recall that the force on a negatively charged particle due to an electric field is directed opposite the field. Thus, for the particular field arrangement of Fig. 29-7, electrons are forced up the page by the electric field $\vec{E}$ and down the page by the magnetic field $\vec{B}$; that is, the forces are in *opposition*. Thomson's procedure was equivalent to the following series of steps.

1. Set $E = 0$ and $B = 0$ and note the position of the spot on screen S due to the undeflected beam.

2. Turn on $\vec{E}$ and measure the resulting beam deflection.

3. Maintaining $\vec{E}$, now turn on $\vec{B}$ and adjust its value until the beam returns to the undeflected position. (With the forces in opposition, they can be made to cancel.)

We discussed the deflection of a charged particle moving through an electric field $\vec{E}$ between two plates (step 2 here) in Sample Problem 23-4. We found that the deflection of the particle at the far end of the plates is

$$y = \frac{qEL^2}{2mv^2},\qquad(29\text{-}6)$$

where v is the particle's speed, m its mass, and q its charge, and L is the length of the plates. We can apply this same equation to the beam of electrons in Fig. 29-7; if need be, we can calculate the deflection by measuring the deflection of the beam on screen S and then working back to calculate the deflection y at the end of the plates. (Because the direction of the deflection is set by the sign of the particle's charge, Thomson was able to show that the particles that were lighting up his screen were negatively charged.)

When the two fields in Fig. 29-7 are adjusted so that the two deflecting forces cancel (step 3), we have from Eqs. 29-1 and 29-3a

$$|q|E = |q|vB \sin(90°) = |q|vB$$

or
$$v = \frac{E}{B}.\qquad(29\text{-}7)$$

Thus, the crossed fields allow us to measure the speed of the charged particles passing through them. Substituting Eq. 29-7 for v in Eq. 29-6 and rearranging yield

$$\frac{m}{q} = \frac{B^2L^2}{2yE},\qquad(29\text{-}8)$$

in which all quantities on the right can be measured. Thus, the crossed fields allow us to measure the ratio m/q of the particles moving through Thomson's apparatus.

Thomson claimed that these particles are found in all matter. He also claimed that they are lighter than the lightest known atom (hydrogen) by a factor of more than 1000. (The exact ratio proved later to be 1836.15.) His m/q measurement, coupled with the boldness of his two claims, is considered to be the "discovery of the electron."

✔**CHECKPOINT 2:**   The figure shows four directions for the velocity vector $\vec{v}$ of a positively charged particle moving through a uniform electric field $\vec{E}$ (directed out of the page and represented with an encircled dot) and a uniform magnetic field $\vec{B}$. (a) Rank directions 1, 2, and 3 according to the magnitude of the net force on the particle, greatest first. (b) Of all four directions, which might result in a net force of zero?

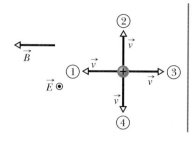

29-4 Crossed Fields: The Hall Effect

As we just discussed, a beam of electrons in a vacuum can be deflected by a magnetic field. Can the drifting conduction electrons in a copper wire also be deflected by a magnetic field? In 1879, Edwin H. Hall, then a 24-year-old graduate student at the Johns Hopkins University, showed that they can. This **Hall effect** allows us to find out whether the charge carriers in a conductor are positively or negatively charged. Beyond that, we can measure the number of such carriers per unit volume of the conductor.

Figure 29-8a shows a copper strip of width d, carrying a current i whose conventional direction is from the top of the figure to the bottom. The charge carriers are electrons and, as we know, they drift (with drift speed v_d) in the opposite direction, from bottom to top. At the instant shown in Fig. 29-8a, an external magnetic field $\vec{B}$, pointing into the plane of the figure, has just been turned on. From Eq. 29-2 we see that a magnetic deflecting force $\vec{F}_B$ will act on each drifting electron, pushing it toward the right edge of the strip.

As time goes on, electrons move to the right, mostly piling up on the right edge of the strip, leaving uncompensated positive charges in fixed positions at the left edge. The separation of positive and negative charges produces an electric field $\vec{E}$ within the strip, pointing from left to right in Fig. 29-8b. This field exerts an electric force $\vec{F}_E$ on each electron, tending to push it to the left.

An equilibrium quickly develops in which the electric force on each electron builds up until it just cancels the magnetic force. When this happens, as Fig. 29-8b shows, the force due to $\vec{B}$ and the force due to $\vec{E}$ are in balance. The drifting electrons then move along the strip toward the top of the page at velocity $\vec{v}_d$, with no further collection of electrons on the right edge of the strip and thus no further increase in the electric field $\vec{E}$.

A *Hall potential difference* V is associated with the electric field across strip width d. From Eq. 25-42, the magnitude of that potential difference is

$$V = Ed. \qquad (29\text{-}9)$$

By connecting a voltmeter across the width, we can measure the potential difference between the two edges of the strip. Moreover, the voltmeter can tell us which edge is at higher potential. For the situation of Fig. 29-8a, we would find that the left

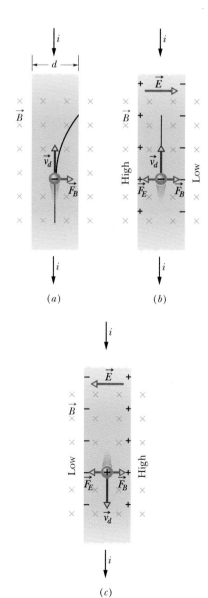

Fig. 29-8 A strip of copper carrying a current i is immersed in a magnetic field $\vec{B}$. (a) The situation immediately after the magnetic field is turned on. The curved path that will then be taken by an electron is shown. (b) The situation at equilibrium, which quickly follows. Note that negative charges pile up on the right side of the strip, leaving uncompensated positive charges on the left. Thus, the left side is at a higher potential than the right side. (c) For the same current direction, if the charge carriers were positively charged, *they* would pile up on the right side, and the right side would be at the higher potential.

edge is at higher potential, which is consistent with our assumption that the charge carriers are negatively charged.

For a moment, let us make the opposite assumption, that the charge carriers in current i are positively charged (Fig. 29-8c). Convince yourself that as these charge carriers move from top to bottom in the strip, they are pushed to the right edge by $\vec{F}_B$ and thus that the *right edge* is at higher potential. Because that last statement is contradicted by our voltmeter reading, the charge carriers must be negatively charged.

Now for the quantitative part. When the electric and magnetic forces are in balance (Fig. 29-8b), Eqs. 29-1 and 29-3 give us

$$eE = ev_dB. \tag{29-10}$$

From Eq. 27-7, the drift speed v_d is

$$v_d = \frac{J}{ne} = \frac{i}{neA}, \tag{29-11}$$

in which J ($= i/A$) is the current density in the strip, A is the cross-sectional area of the strip, and n is the *number density* of charge carriers (their number per unit volume).

In Eq. 29-10, substituting for E with Eq. 29-9 and substituting for v_d with Eq. 29-11, we obtain

$$n = \frac{Bi}{Vle}, \tag{29-12}$$

in which l ($= A/d$) is the thickness of the strip. With this equation we can find n from measurable quantities.

It is also possible to use the Hall effect to measure directly the drift speed v_d of the charge carriers, which you may recall is of the order of centimeters per hour. In this clever experiment, the metal strip is moved mechanically through the magnetic field in a direction opposite that of the drift velocity of the charge carriers. The speed of the moving strip is then adjusted until the Hall potential difference vanishes. At this condition, with no Hall effect, the velocity of the charge carriers *with respect to the laboratory frame* must be zero, so the velocity of the strip must be equal in magnitude but opposite the direction of the velocity of the negative charge carriers.

Sample Problem 29-2

Figure 29-9 shows a solid metal cube, of edge length $d = 1.5$ cm, moving in the positive y direction at a constant velocity $\vec{v}$ of magnitude 4.0 m/s. The cube moves through a uniform magnetic field $\vec{B}$ of magnitude 0.050 T directed toward positive z.

(a) Which cube face is at a lower electric potential and which is at a higher electric potential because of the motion through the field?

SOLUTION: One Key Idea here is that, because the cube is moving through a magnetic field $\vec{B}$, a magnetic force $\vec{F}_B$ acts on its charged particles, including its conduction electrons. A second Key Idea is how $\vec{F}_B$ causes an electric potential difference between certain faces of the cube. When the cube first begins to move through the magnetic field, its electrons do also. Because each electron has charge q and is moving through a magnetic field with velocity $\vec{v}$, the magnetic force $\vec{F}_B$ acting on it is given by Eq. 29-2. Because q is

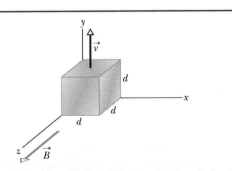

Fig. 29-9 Sample Problem 29-2. A solid metal cube of edge length d, at constant velocity $\vec{v}$ through a uniform magnetic field $\vec{B}$.

negative, the direction of $\vec{F}_B$ is opposite the cross product $\vec{v} \times \vec{B}$, which is in the positive direction of the x axis in Fig. 29-9. Thus, $\vec{F}_B$ acts in the negative direction of the x axis, toward the left face of the cube (which is hidden from view in Fig. 29-9).

Most of the electrons are fixed in place in the molecules of the cube. However, because the cube is a metal, it contains conduction electrons that are free to move. Some of those conduction electrons are deflected by $\vec{F}_B$ to the left cube face, making that face negatively charged and leaving the right face positively charged. This charge separation produces an electric field $\vec{E}$ directed from the positively charged right face to the negatively charged left face. Thus, the left face is at a lower electric potential, and the right face is at a higher electric potential.

(b) What is the potential difference between the faces of higher and lower electric potential?

SOLUTION: The **Key Ideas** here are these:

1. The electric field $\vec{E}$ created by the charge separation produces an electric force $\vec{F}_E = q\vec{E}$ on each electron. Because q is negative, this force is directed opposite the field $\vec{E}$—that is, toward the right. Thus on each electron, $\vec{F}_E$ acts toward the right and $\vec{F}_B$ acts toward the left.

2. When the cube had just begun to move through the magnetic field and the charge separation had just begun, the magnitude of $\vec{E}$ began to increase from zero. Thus, the magnitude of $\vec{F}_E$ also began to increase from zero and was initially smaller than the magnitude $\vec{F}_B$. During this early stage, the net force on any electron was dominated by $\vec{F}_B$, which continuously moved additional electrons to the left cube face, increasing the charge separation.

3. However, as the charge separation increased, eventually magnitude F_E became equal to magnitude F_B. The net force on any electron was then zero, and no additional electrons were moved to the left cube face. Thus, the magnitude of $\vec{F}_E$ could not increase further, and the electrons were then in equilibrium.

We seek the potential difference V between the left and right cube faces after equilibrium was reached (which occurred quickly).

We can obtain V with Eq. 29-9 ($V = Ed$) provided we first find the magnitude E of the electric field at equilibrium. We can do so with the equation for the balance of forces ($F_E = F_B$).

For F_E, we substitute $|q|E$. For F_B, we substitute $|q|vB \sin \phi$ from Eq. 29-3. From Fig. 29-9, we see that the angle ϕ between vectors $\vec{v}$ and $\vec{B}$ is 90°; so $\sin \phi = 1$ and $F_E = F_B$ yields

$$|q|E = |q|vB \sin 90° = |q|vB.$$

This gives us $E = vB$, so Eq. 29-9 ($V = Ed$) becomes

$$V = vBd. \qquad (29\text{-}13)$$

Substituting known values gives us

$$V = (4.0 \text{ m/s})(0.050 \text{ T})(0.015 \text{ m})$$
$$= 0.0030 \text{ V} = 3.0 \text{ mV}. \qquad \text{(Answer)}$$

✓**CHECKPOINT 3:** The figure shows a metallic, rectangular solid that is to move at a certain speed v through the uniform magnetic field $\vec{B}$. Its dimensions are multiples of d, as shown. You have six choices for the direction of the velocity of the solid: it can be parallel to x, y, or z, in either the positive or negative direction. **(a)** Rank the six choices according to the potential difference that would be set up across the solid, greatest first. **(b)** For which choice is the front face at lower potential?

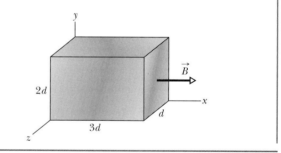

29-5 A Circulating Charged Particle

If a particle moves in a circle at constant speed, we can be sure that the net force acting on the particle is constant in magnitude and points toward the center of the circle, always perpendicular to the particle's velocity. Think of a stone tied to a string and whirled in a circle on a smooth horizontal surface, or of a satellite moving in a circular orbit around Earth. In the first case, the tension in the string provides the necessary force and centripetal acceleration. In the second case, Earth's gravitational attraction provides the force and acceleration.

Figure 29-10 shows another example: A beam of electrons is projected into a chamber by an *electron gun* G. The electrons enter in the plane of the page with speed v and then move in a region of uniform magnetic field $\vec{B}$ directed out of that plane. As a result, a magnetic force $\vec{F}_B = q\vec{v} \times \vec{B}$ continually deflects the electrons, and because $\vec{v}$ and $\vec{B}$ are always perpendicular to each other, this deflection causes the electrons to follow a circular path. The path is visible in the photo because atoms of gas in the chamber emit light when some of the circulating electrons collide with them.

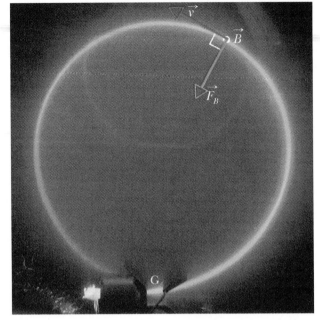

Fig. 29-10 Electrons circulating in a chamber containing gas at low pressure (their path is the glowing circle). A uniform magnetic field $\vec{B}$, pointing directly out of the plane of the page, fills the chamber. Note the radially directed magnetic force $\vec{F}_B$; for circular motion to occur, $\vec{F}_B$ *must* point toward the center of the circle. Use the right-hand rule for cross products to confirm that $\vec{F}_B = q\vec{v} \times \vec{B}$ gives $\vec{F}_B$ the proper direction. (Don't forget the sign of q.)

We would like to determine the parameters that characterize the circular motion of these electrons, or of any particle of charge magnitude q and mass m moving perpendicular to a uniform magnetic field $\vec{B}$ at speed v. From Eq. 29-3, the force acting on the particle has a magnitude of qvB. From Newton's second law ($\vec{F} = m\vec{a}$) applied to uniform circular motion (Eq. 6-18),

$$F = m\frac{v^2}{r},$$
(29-14)

we have

$$qvB = \frac{mv^2}{r}.$$
(29-15)

Solving for r, we find the radius of the circular path as

$$r = \frac{mv}{qB} \quad \text{(radius).}$$
(29-16)

The period T (the time for one full revolution) is equal to the circumference divided by the speed:

$$T = \frac{2\pi r}{v} = \frac{2\pi}{v}\frac{mv}{qB} = \frac{2\pi m}{qB} \quad \text{(period).}$$
(29-17)

The frequency f (the number of revolutions per unit time) is

$$f = \frac{1}{T} = \frac{qB}{2\pi m} \quad \text{(frequency).}$$
(29-18)

The angular frequency ω of the motion is then

$$\omega = 2\pi f = \frac{qB}{m} \quad \text{(angular frequency).}$$
(29-19)

The quantities T, f, and ω do not depend on the speed of the particle (provided that speed is much less than the speed of light). Fast particles move in large circles and slow ones in small circles, but all particles with the same charge-to-mass ratio q/m

take the same time T (the period) to complete one round trip. Using Eq. 29-2, you can show that if you are looking in the direction of $\vec{B}$, the direction of rotation for a positive particle is always counterclockwise, and the direction for a negative particle is always clockwise.

Helical Paths

If the velocity of a charged particle has a component parallel to the (uniform) magnetic field, the particle will move in a helical path about the direction of the field vector. Figure 29-11a, for example, shows the velocity vector $\vec{v}$ of such a particle resolved into two components, one parallel to $\vec{B}$ and one perpendicular to it:

$$v_{\parallel} = v \cos \phi \quad \text{and} \quad v_{\perp} = v \sin \phi. \quad (29\text{-}20)$$

The parallel component determines the *pitch p* of the helix—that is, the distance between adjacent turns (Fig. 29-11b). The perpendicular component determines the radius of the helix and is the quantity to be substituted for v in Eq. 29-16.

Figure 29-11c shows a charged particle spiraling in a nonuniform magnetic field. The more closely spaced field lines at the left and right sides indicate that the magnetic field is stronger there. When the field at an end is strong enough, the particle "reflects" from that end. If the particle reflects from both ends, it is said to be trapped in a *magnetic bottle*.

Electrons and protons are trapped in this way by the terrestrial magnetic field; the trapped particles form the *Van Allen radiation belts,* which loop well above Earth's atmosphere between Earth's north and south geomagnetic poles. These particles bounce back and forth, from one end of this magnetic bottle to the other, within a few seconds.

When a large solar flare shoots additional energetic electrons and protons into the radiation belts, an electric field is produced in the region where electrons normally reflect. This field eliminates the reflection and instead drives electrons down into the atmosphere, where they collide with atoms and molecules of air, causing

(a)

(b)

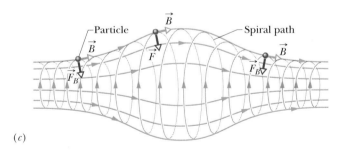

(c)

Fig. 29-11 (a) A charged particle moves in a uniform magnetic field $\vec{B}$, its velocity $\vec{v}$ making an angle ϕ with the field direction. (b) The particle follows a helical path, of radius r and pitch p. (c) A charged particle spiraling in a nonuniform magnetic field. (The particle can become trapped, spiraling back and forth between the strong field regions at either end.) Note that the magnetic force vectors at the left and right sides have a component pointing toward the center of the figure.

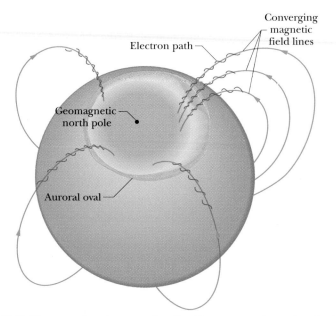

Fig. 29-12 The auroral oval surrounding Earth's geomagnetic north pole (in northwestern Greenland). Magnetic field lines converge toward that pole. Electrons moving toward Earth are "caught by" and spiral around these field lines, entering the terrestrial atmosphere at high latitudes and producing aurora within the oval.

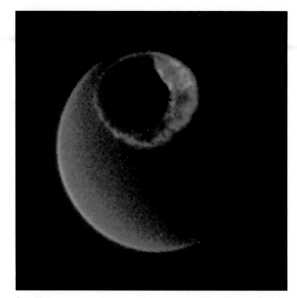

Fig. 29-13 A false-color image of aurora inside the north auroral oval, recorded by the satellite *Dynamic Explorer*, using ultraviolet light emitted by oxygen atoms excited in the aurora. The sun-lit portion of Earth is the crescent at the left.

that air to emit light. This light forms the aurora—a curtain of light that hangs down to an altitude of about 100 km. Green light is emitted by oxygen atoms, and pink light is emitted by nitrogen molecules, but often the light is so dim that we perceive only white light.

Aurora extend in arcs above Earth and can occur in a region called the *auroral oval* that is shown in Figs. 29-12 and 29-13 as seen from space. Although an aurora is long, it is less than 1 km thick (north to south) because the paths of the electrons producing it converge as the electrons spiral down the converging magnetic field lines (Fig. 29-12).

CHECKPOINT 4: The figure here shows the circular paths of two particles that travel at the same speed in a uniform magnetic field $\vec{B}$, which is directed into the page. One particle is a proton; the other is an electron (which is less massive). (a) Which particle follows the smaller circle, and (b) does that particle travel clockwise or counterclockwise?

Sample Problem 29-3

Figure 29-14 shows the essentials of a *mass spectrometer*, which can be used to measure the mass of an ion; an ion of mass m (to be measured) and charge q is produced in source S. The initially stationary ion is accelerated by the electric field due to a potential difference V. The ion leaves S and enters a separator chamber in which a uniform magnetic field $\vec{B}$ is perpendicular to the path of the ion. The magnetic field causes the ion to move in a semicircle, striking (and thus altering) a photographic plate at distance x from the entry slit. Suppose that in a certain trial $B = 80.000$ mT and $V = 1000.0$ V, and ions of charge $q = +1.6022 \times 10^{-19}$ C strike the plate at $x = 1.6254$ m. What is the mass m of the individual ions, in unified atomic mass units (1 u $= 1.6605 \times 10^{-27}$ kg)?

SOLUTION: One **Key Idea** here is that, because the (uniform) magnetic field causes the (charged) ion to follow a circular path, we can relate the ion's mass m to the path's radius r with Eq. 29-16 ($r = mv/qB$). From Fig. 29-14 we see that $r = x/2$, and we are given the magnitude B of the magnetic field. However, we lack the ion's speed v in the magnetic field, after it has been accelerated due to the potential difference V.

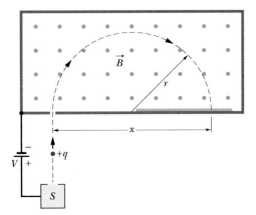

Fig. 29-14 Sample Problem 29-3. Essentials of a mass spectrometer. A positive ion, after being accelerated from its source S by potential difference V, enters a chamber of uniform magnetic field $\vec{B}$. There it travels through a semicircle of radius r and strikes a photographic plate at a distance x from where it entered the chamber.

To relate v and V, we use the Key Idea that mechanical energy ($E_{mec} = K + U$) is conserved during the acceleration. When the ion emerges from the source, its kinetic energy is approximately zero. At the end of the acceleration, its kinetic energy is $\frac{1}{2}mv^2$. Also,

during the acceleration, the positive ion moves through a change in potential of $-V$. Thus, because the ion has positive charge q, its potential energy changes by $-qV$. If we now write the conservation of mechanical energy as

$$\Delta K + \Delta U = 0,$$

we get

$$\tfrac{1}{2}mv^2 - qV = 0$$

or

$$v = \sqrt{\frac{2qV}{m}}. \tag{29-21}$$

Substituting this into Eq. 29-16 gives us

$$r = \frac{mv}{qB} = \frac{m}{qB}\sqrt{\frac{2qV}{m}} = \frac{1}{B}\sqrt{\frac{2mV}{q}}.$$

Thus,

$$x = 2r = \frac{2}{B}\sqrt{\frac{2mV}{q}}.$$

Solving this for m and substituting the given data yield

$$m = \frac{B^2qx^2}{8V}$$

$$= \frac{(0.080000\text{ T})^2(1.6022 \times 10^{-19}\text{ C})(1.6254\text{ m})^2}{8(1000.0\text{ V})}$$

$$= 3.3863 \times 10^{-25}\text{ kg} = 203.93\text{ u}. \qquad \text{(Answer)}$$

Sample Problem 29-4

An electron with a kinetic energy of 22.5 eV moves into a region of uniform magnetic field $\vec{B}$ of magnitude 4.55×10^{-4} T. The angle between the directions of $\vec{B}$ and the electron's velocity $\vec{v}$ is 65.5°. What is the pitch of the helical path taken by the electron?

SOLUTION: One Key Idea here is that the pitch p is the distance the electron travels parallel to the magnetic field $\vec{B}$ during one period T of circulation. A second Key Idea is that the period T is given by Eq. 29-17, regardless of the angle between the directions of $\vec{v}$ and $\vec{B}$ (provided the angle is not zero, for which there is no circulation of the electron). Thus, using Eqs. 29-20 and 29-17, we find

$$p = v_{\parallel}T = (v\cos\phi)\frac{2\pi m}{qB}. \tag{29-22}$$

We can calculate the electron's speed v from its kinetic energy as we did for the proton in Sample Problem 29-1. We find that $v = 2.81 \times 10^6$ m/s. Substituting this and known data in Eq. 29-22 gives us

$$p = (2.81 \times 10^6\text{ m/s})(\cos 65.5°)$$

$$\times \frac{2\pi(9.11 \times 10^{-31}\text{ kg})}{(1.60 \times 10^{-19}\text{ C})(4.55 \times 10^{-4}\text{ T})}$$

$$= 9.16\text{ cm}. \qquad \text{(Answer)}$$

29-6 Cyclotrons and Synchrotrons

What is the structure of matter on the smallest scale? This question has always intrigued physicists. One way of getting at the answer is to allow an energetic charged particle (a proton, for example) to slam into a solid target. Better yet, allow two such energetic protons to collide head-on. Then analyze the debris from many such collisions to learn the nature of the subatomic particles of matter. The Nobel Prizes in physics for 1976 and 1984 were awarded for just such studies.

How can we give a proton enough kinetic energy for such an experiment? The direct approach is to allow the proton to "fall" through a potential difference V, thereby increasing its kinetic energy by eV. As we want higher and higher energies, however, it becomes more and more difficult to establish the necessary potential difference.

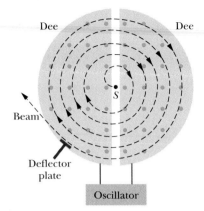

Fig. 29-15 The elements of a cyclotron, showing the particle source S and the dees. A uniform magnetic field is directed up from the plane of the page. Circulating protons spiral outward within the hollow dees, gaining energy every time they cross the gap between the dees.

A better way is to arrange for the proton to circulate in a magnetic field, and to give it a modest electrical "kick" once per revolution. For example, if a proton circulates 100 times in a magnetic field and receives an energy boost of 100 keV every time it completes an orbit, it will end up with a kinetic energy of (100)(100 keV) or 10 MeV. Two very useful accelerating devices are based on this principle.

The Cyclotron

Figure 29-15 is a top view of the region of a *cyclotron* in which the particles (protons, say) circulate. The two hollow **D**-shaped objects (open on their straight edges) are made of sheet copper. These *dees,* as they are called, are part of an electrical oscillator that alternates the electric potential difference across the gap between the dees. The electrical signs of the dees are alternated so that the electric field in the gap alternates in direction, first toward one dee and then the other dee, back and forth. The dees are immersed in a magnetic field ($B = 1.5$ T) whose direction is out of the plane of the page and that is set up by a large electromagnet.

Suppose that a proton, injected by source S at the center of the cyclotron in Fig. 29-15, initially moves toward a negatively charged dee. It will accelerate toward this dee and enter it. Once inside, it is shielded from electric fields by the copper walls of the dee; that is, the electric field does not enter the dee. The magnetic field, however, is not screened by the (nonmagnetic) copper dee, so the proton moves in a circular path whose radius, which depends on its speed, is given by Eq. 29-16 ($r = mv/qB$).

Let us assume that at the instant the proton emerges into the center gap from the first dee, the potential difference between the dees is reversed. Thus, the proton *again* faces a negatively charged dee and is *again* accelerated. This process continues, the circulating proton always being in step with the oscillations of the dee potential, until the proton has spiraled out to the edge of the dee system. There a deflector plate sends it out through a portal.

The key to the operation of the cyclotron is that the frequency f at which the proton circulates in the field (and that does *not* depend on its speed) must be equal to the fixed frequency f_{osc} of the electrical oscillator, or

$$f = f_{osc} \qquad \text{(resonance condition).} \qquad (29\text{-}23)$$

This *resonance condition* says that, if the energy of the circulating proton is to increase, energy must be fed to it at a frequency f_{osc} that is equal to the natural frequency f at which the proton circulates in the magnetic field.

Combining Eqs. 29-18 and 29-23 allows us to write the resonance condition as

$$qB = 2\pi m f_{osc}. \qquad (29\text{-}24)$$

For the proton, q and m are fixed. The oscillator (we assume) is designed to work at a single fixed frequency f_{osc}. We then "tune" the cyclotron by varying B until Eq. 29-24 is satisfied and then many protons circulate through the magnetic field, to emerge as a beam.

The Proton Synchrotron

At proton energies above 50 MeV, the conventional cyclotron begins to fail because one of the assumptions of its design—that the frequency of revolution of a charged particle circulating in a magnetic field is independent of the particle's speed—is true only for speeds that are much less than the speed of light. At greater proton speeds (above about 10% of the speed of light), we must treat the problem relativistically. According to relativity theory, as the speed of a circulating proton approaches that of light, the proton's frequency of revolution decreases steadily. Thus, the protons get out of step with the cyclotron's oscillator—whose frequency remains fixed at f_{osc}—and eventually the energy of the circulating proton stops increasing.

There is another problem. For a 500 GeV proton in a magnetic field of 1.5 T, the path radius is 1.1 km. The corresponding magnet for a conventional cyclotron of the proper size would be impossibly expensive, the area of its pole faces being about 4×10^6 m^2.

The *proton synchrotron* is designed to meet these two difficulties. The magnetic field B and the oscillator frequency f_{osc}, instead of having fixed values as in the conventional cyclotron, are made to vary with time during the accelerating cycle. When this is done properly, (1) the frequency of the circulating protons remains in step with the oscillator at all times, and (2) the protons follow a circular—not a spiral—path. Thus, the magnet need extend only along that circular path, not over some 4×10^6 m^2. The circular path, however, still must be large if high energies are to be achieved. The proton synchrotron at the Fermi National Accelerator Laboratory (Fermilab) in Illinois has a circumference of 6.3 km and can produce protons with energies of about 1 TeV ($= 10^{12}$ eV).

Sample Problem 29-5

Suppose a cyclotron is operated at an oscillator frequency of 12 MHz and has a dee radius $R = 53$ cm.

(a) What is the magnitude of the magnetic field needed for deuterons to be accelerated in the cyclotron? A deuteron is the nucleus of deuterium, an isotope of hydrogen. It consists of a proton and a neutron and thus has the same charge as a proton. Its mass is $m = 3.34 \times 10^{-27}$ kg.

SOLUTION: The Key Idea here is that, for a given oscillator frequency f_{osc}, the magnetic field magnitude B required to accelerate any particle in a cyclotron depends on the ratio m/q of mass to charge for the particle, according to Eq. 29-24. For deuterons and the oscillator frequency $f_{osc} = 12$ MHz, we find

$$B = \frac{2\pi m f_{osc}}{q} = \frac{(2\pi)(3.34 \times 10^{-27} \text{ kg})(12 \times 10^6 \text{ s}^{-1})}{1.60 \times 10^{-19} \text{ C}}$$

$$= 1.57 \text{ T} \approx 1.6 \text{ T}. \qquad \text{(Answer)}$$

Note that, to accelerate protons, B would have to be reduced by a factor of 2, provided that the oscillator frequency remained fixed at 12 MHz.

(b) What is the resulting kinetic energy of the deuterons?

SOLUTION: One Key Idea here is that the kinetic energy ($\frac{1}{2}mv^2$) of a deuteron exiting the cyclotron is equal to the kinetic energy it had just before exiting, when it was traveling in a circular path with a radius approximately equal to the radius R of the cyclotron dees. A second Key Idea is that we can find the speed v of the deuteron in that circular path with Eq. 29-16 ($r = mv/qB$). Solving that equation for v, substituting R for r, and then substituting known data, we find

$$v = \frac{RqB}{m} = \frac{(0.53 \text{ m})(1.60 \times 10^{-19} \text{ C})(1.57 \text{ T})}{3.34 \times 10^{-27} \text{ kg}}$$

$$= 3.99 \times 10^7 \text{ m/s}.$$

This speed corresponds to a kinetic energy of

$$K = \frac{1}{2}mv^2$$

$$= \frac{1}{2}(3.34 \times 10^{-27} \text{ kg})(3.99 \times 10^7 \text{ m/s})^2$$

$$= 2.7 \times 10^{-12} \text{ J}, \qquad \text{(Answer)}$$

or about 17 MeV.

29-7 Magnetic Force on a Current-Carrying Wire

We have already seen (in connection with the Hall effect) that a magnetic field exerts a sideways force on electrons moving in a wire. This force must then be transmitted to the wire itself, because the conduction electrons cannot escape sideways out of the wire.

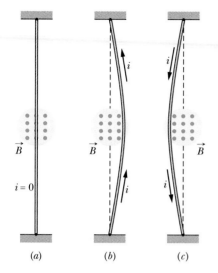

(a)   (b)   (c)

Fig. 29-16 A flexible wire passes between the pole faces of a magnet (only the farther pole face is shown). (a) Without current in the wire, the wire is straight. (b) With upward current, the wire is deflected rightward. (c) With downward current, the deflection is leftward. The connections for getting the current into the wire at one end and out of it at the other end are not shown.

In Fig. 29-16a, a vertical wire, carrying no current and fixed in place at both ends, extends through the gap between the vertical pole faces of a magnet. The magnetic field between the faces is directed outward from the page. In Fig. 29-16b, a current is sent upward through the wire; the wire deflects to the right. In Fig. 29-16c, we reverse the direction of the current and the wire deflects to the left.

Figure 29-17 shows what happens inside the wire of Fig. 29-16. We see one of the conduction electrons, drifting downward with an assumed drift speed v_d. Equation 29-3, in which we must put $\phi = 90°$, tells us that a force $\vec{F}_B$ of magnitude ev_dB must act on each such electron. From Eq. 29-2 we see that this force must be directed to the right. We expect then that the wire as a whole will experience a force to the right, in agreement with Fig. 29-16b.

If, in Fig. 29-17, we were to reverse *either* the direction of the magnetic field *or* the direction of the current, the force on the wire would reverse, being directed now to the left. Note too that it does not matter whether we consider negative charges drifting downward in the wire (the actual case) or positive charges drifting upward. The direction of the deflecting force on the wire is the same. We are safe then in dealing with a current of positive charge.

Consider a length L of the wire in Fig. 29-17. All the conduction electrons in this section of wire will drift past plane xx in Fig. 29-17 in a time $t = L/v_d$. Thus, in that time a charge given by

$$q = it = i\frac{L}{v_d}$$

will pass through that plane. Substituting this into Eq. 29-3 yields

$$F_B = qv_dB \sin \phi = \frac{iL}{v_d}v_dB \sin 90°$$

or $$F_B = iLB. \tag{29-25}$$

This equation gives the magnetic force that acts on a length L of straight wire carrying a current i and immersed in a magnetic field $\vec{B}$ that is perpendicular to the wire.

If the magnetic field is *not* perpendicular to the wire, as in Fig. 29-18, the magnetic force is given by a generalization of Eq. 29-25:

$$\vec{F}_B = i\vec{L} \times \vec{B} \quad \text{(force on a current).} \tag{29-26}$$

Here $\vec{L}$ is a *length vector* that has magnitude L and is directed along the wire segment in the direction of the (conventional) current. The force magnitude F_B is

$$F_B = iLB \sin \phi, \tag{29-27}$$

where ϕ is the angle between the directions of $\vec{L}$ and $\vec{B}$. The direction of $\vec{F}_B$ is that of the cross product $\vec{L} \times \vec{B}$, because we take current i to be a positive quantity. Equation 29-26 tells us that $\vec{F}_B$ is always perpendicular to the plane defined by vectors $\vec{L}$ and $\vec{B}$, as indicated in Fig. 29-18.

Equation 29-26 is equivalent to Eq. 29-2 in that either can be taken as the

Fig. 29-17 A close-up view of a section of the wire of Fig. 29-16b. The current direction is upward, which means that electrons drift downward. A magnetic field that emerges from the plane of the page causes the electrons and the wire to be deflected to the right.

Fig. 29-18 A wire carrying current i makes an angle ϕ with magnetic field $\vec{B}$. The wire has length L in the field and length vector $\vec{L}$ (in the direction of the current). A magnetic force $\vec{F}_B = i\vec{L} \times \vec{B}$ acts on the wire.

defining equation for $\vec{B}$. In practice, we define $\vec{B}$ from Eq. 29-26. It is much easier to measure the magnetic force acting on a wire than that on a single moving charge.

If a wire is not straight or the field is not uniform, we can imagine it broken up into small straight segments and apply Eq. 29-26 to each segment. The force on the wire as a whole is then the vector sum of all the forces on the segments that make it up. In the differential limit, we can write

$$d\vec{F}_B = i \, d\vec{L} \times \vec{B}, \qquad (29\text{-}28)$$

and we can find the resultant force on any given arrangement of currents by integrating Eq. 29-28 over that arrangement.

In using Eq. 29-28, bear in mind that there is no such thing as an isolated current-carrying wire segment of length dL. There must always be a way to introduce the current into the segment at one end and take it out at the other end.

✔**CHECKPOINT 5:**   The figure shows a current i through a wire in a uniform magnetic field $\vec{B}$, as well as the magnetic force $\vec{F}_B$ acting on the wire. The field is oriented so that the force is maximum. In what direction is the field?

Sample Problem 29-6

A straight, horizontal length of copper wire has a current $i = 28$ A through it. What are the magnitude and direction of the minimum magnetic field $\vec{B}$ needed to suspend the wire—that is, to balance the gravitational force on it? The linear density (mass per unit length) of the wire is 46.6 g/m.

SOLUTION: One **Key Idea** is that, because the wire carries a current, a magnetic force $\vec{F}_B$ can act on the wire if we place it in a magnetic field $\vec{B}$. To balance the downward gravitational force $\vec{F}_g$ on the wire, we want $\vec{F}_B$ to be directed upward (Fig. 29-19).

A second **Key Idea** is that the direction of $\vec{F}_B$ is related to the directions of $\vec{B}$ and the wire's length vector $\vec{L}$ by Eq. 29-26. Because $\vec{L}$ is directed horizontally (and the current is taken to be positive), Eq. 29-26 and the right-hand rule for cross products tell us that $\vec{B}$ must be horizontal and rightward (in Fig. 29-19) to give the required upward $\vec{F}_B$.

The magnitude of $\vec{F}_B$ is given by Eq. 29-27 ($F_B = iLB \sin \phi$). Because we want $\vec{F}_B$ to balance $\vec{F}_g$, we want

$$iLB \sin \phi = mg, \qquad (29\text{-}29)$$

where mg is the magnitude of $\vec{F}_g$ and m is the mass of the wire. We also want the minimal field magnitude B for $\vec{F}_B$ to balance $\vec{F}_g$. Thus, we need to maximize $\sin \phi$ in Eq. 29-29. To do so, we set $\phi = 90°$, thereby arranging for $\vec{B}$ to be perpendicular to the

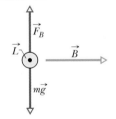

Fig. 29-19 Sample Problem 29-6. A current-carrying wire (shown in cross section) can be made to "float" in a magnetic field. The current in the wire emerges from the plane of the page, and the magnetic field is directed to the right.

wire. We then have $\sin \phi = 1$, so Eq. 29-29 yields

$$B = \frac{mg}{iL \sin \phi} = \frac{(m/L)g}{i}. \qquad (29\text{-}30)$$

We write the result this way because we know m/L, the linear density of the wire. Substituting known data then gives us

$$B = \frac{(46.6 \times 10^{-3} \text{ kg/m})(9.8 \text{ m/s}^2)}{28 \text{ A}}$$

$$= 1.6 \times 10^{-2} \text{ T.} \qquad \text{(Answer)}$$

This is about 160 times the strength of Earth's magnetic field.

Fig. 29-20 The elements of an electric motor. A rectangular loop of wire, carrying a current and free to rotate about a fixed axis, is placed in a magnetic field. Magnetic forces on the wire produce a torque that rotates it. A commutator (not shown) reverses the direction of the current every half-revolution so that the torque always acts in the same direction.

29-8 Torque on a Current Loop

Much of the world's work is done by electric motors. The forces behind this work are the magnetic forces that we studied in the preceding section—that is, the forces that a magnetic field exerts on a wire that carries a current.

Figure 29-20 shows a simple motor, consisting of a single current-carrying loop immersed in a magnetic field $\vec{B}$. The two magnetic forces $\vec{F}$ and $-\vec{F}$ produce a torque on the loop, tending to rotate it about its central axis. Although many essential details have been omitted, the figure does suggest how the action of a magnetic field on a current loop produces rotary motion. Let us analyze that action.

Figure 29-21a shows a rectangular loop of sides a and b, carrying current i through uniform magnetic field $\vec{B}$. We place it in the field so that its long sides, labeled 1 and 3, are perpendicular to the field direction (which is into the page), but its short sides, labeled 2 and 4, are not. Wires to lead the current into and out of the loop are needed but, for simplicity, they are not shown.

To define the orientation of the loop in the magnetic field, we use a normal vector $\vec{n}$ that is perpendicular to the plane of the loop. Figure 29-21b shows a right-hand rule for finding the direction of $\vec{n}$. Point or curl the fingers of your right hand in the direction of the current at any point on the loop. Your extended thumb then points in the direction of the normal vector $\vec{n}$.

In Fig. 29-21c, the normal vector of the loop is shown at an arbitrary angle θ to the direction of the magnetic field $\vec{B}$. We wish to find the net force and net torque acting on the loop in this orientation.

The net force on the loop is the vector sum of the forces acting on its four sides. For side 2 the vector $\vec{L}$ in Eq. 29-26 points in the direction of the current and has magnitude b. The angle between $\vec{L}$ and $\vec{B}$ for side 2 (see Fig. 29-21c) is $90° - \theta$. Thus, the magnitude of the force acting on this side is

$$F_2 = ibB \sin(90° - \theta) = ibB \cos \theta. \tag{29-31}$$

You can show that the force $\vec{F}_4$ acting on side 4 has the same magnitude as $\vec{F}_2$ but the opposite direction. Thus, $\vec{F}_2$ and $\vec{F}_4$ cancel out exactly. Their net force is zero and, because their common line of action is through the center of the loop, their net torque is also zero.

The situation is different for sides 1 and 3. For them, $\vec{L}$ is perpendicular to $\vec{B}$,

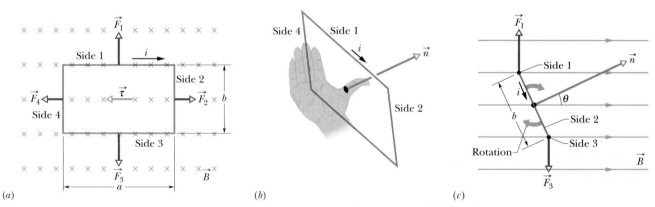

Fig. 29-21 A rectangular loop, of length a and width b and carrying a current i, is located in a uniform magnetic field. A torque $\vec{\tau}$ acts to align the normal vector $\vec{n}$ with the direction of the field. (a) The loop as seen by looking in the direction of the magnetic field. (b) A perspective of the loop showing how the right-hand rule gives the direction of $\vec{n}$, which is perpendicular to the plane of the loop. (c) A side view of the loop, from side 2. The loop rotates as indicated.

so the forces $\vec{F}_1$ and $\vec{F}_3$ have the common magnitude iaB. Because these two forces have opposite directions, they do not tend to move the loop up or down. However, as Fig. 29-21c shows, these two forces do *not* share the same line of action so they *do* produce a net torque. The torque tends to rotate the loop so as to align its normal vector $\vec{n}$ with the direction of the magnetic field $\vec{B}$. That torque has moment arm $(b/2) \sin \theta$ about the central axis of the loop. The magnitude τ' of the torque due to forces $\vec{F}_1$ and $\vec{F}_3$ is then (see Fig. 29-21c)

$$\tau' = \left(iaB \frac{b}{2} \sin \theta \right) + \left(iaB \frac{b}{2} \sin \theta \right) = iabB \sin \theta. \tag{29-32}$$

Suppose we replace the single loop of current with a *coil* of N loops, or *turns*. Further, suppose that the turns are wound tightly enough that they can be approximated as all having the same dimensions and lying in a plane. Then the turns form a *flat coil* and a torque τ' with the magnitude given in Eq. 29-32 acts on each of them. The total torque on the coil then has magnitude

$$\tau = N\tau' = NiabB \sin \theta = (NiA)B \sin \theta, \tag{29-33}$$

in which $A\ (= ab)$ is the area enclosed by the coil. The quantities in parentheses (NiA) are grouped together because they are all properties of the coil: its number of turns, its area, and the current it carries. Equation 29-33 holds for all flat coils, no matter what their shape, provided the magnetic field is uniform.

Instead of focusing on the motion of the coil, it is simpler to keep track of the vector $\vec{n}$, which is normal to the plane of the coil. Equation 29-33 tells us that a current-carrying flat coil placed in a magnetic field will tend to rotate so that $\vec{n}$ has the same direction as the field.

In a motor, the current in the coil is reversed as $\vec{n}$ begins to line up with the field direction, so that a torque continues to rotate the coil. This automatic reversal of the current is done via a commutator that electrically connects the rotating coil with the stationary contacts on the wires that supply the current from some source.

Sample Problem 29-7

Analog voltmeters and ammeters work by measuring the torque exerted by a magnetic field on a current-carrying coil. The reading is displayed by means of the deflection of a pointer over a scale. Figure 29-22 shows the essentials of a *galvanometer*, on which both analog ammeters and analog voltmeters are based. Assume the coil is 2.1 cm high and 1.2 cm wide, has 250 turns, and is mounted so that it can rotate about an axis (into the page) in a uniform *radial* magnetic field with $B = 0.23$ T. For any orientation of the coil, the net magnetic field through the coil is perpendicular to the normal vector of the coil (and thus parallel to the plane of the coil). A spring Sp provides a countertorque that balances the magnetic torque, so that a given steady current i in the coil results in a steady angular deflection ϕ. The greater the current is, the greater the deflection is, and thus the greater the torque required of the spring is. If a current of 100 μA produces an angular deflection of 28°, what must be the torsional constant κ of the spring, as used in Eq. 16-22 ($\tau = -\kappa\phi$)?

SOLUTION: The Key Idea here is that, with a constant current through the device, the resulting magnetic torque (Eq. 29-33) is balanced by the spring torque. Thus, the magnitudes of those torques are

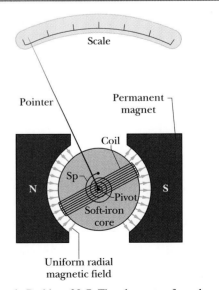

Fig. 29-22 Sample Problem 29-7. The elements of a galvanometer. Depending on the external circuit, this device can be wired up as either a voltmeter or an ammeter.

equal:

$$NiAB \sin \theta = \kappa\phi. \qquad (29\text{-}34)$$

Here ϕ is the angular deflection of the coil and pointer, and A ($= 2.52 \times 10^{-4}$ m^2) is the area encircled by the coil. Since the net magnetic field through the coil is always perpendicular to the normal vector of the coil, $\theta = 90°$ for any orientation of the pointer.

Solving Eq. 29-34 for κ, we find

$$
\begin{aligned}
\kappa &= \frac{NiAB \sin \theta}{\phi} \\
&= (250)(100 \times 10^{-6}\ \text{A})(2.52 \times 10^{-4}\ \text{m}^2) \\
&\quad \times \frac{(0.23\ \text{T})(\sin 90°)}{28°} \\
&= 5.2 \times 10^{-8}\ \text{N} \cdot \text{m/degree}. \qquad \text{(Answer)}
\end{aligned}
$$

Many modern ammeters and voltmeters are of the digital, direct-reading type and do not use a moving coil.

29-9 The Magnetic Dipole Moment

We can describe the current-carrying coil of the preceding section with a single vector $\vec{\mu}$, its **magnetic dipole moment**. We take the direction of $\vec{\mu}$ to be that of the normal vector $\vec{n}$ to the plane of the coil, as in Fig. 29-21c. We define the magnitude of $\vec{\mu}$ as

$$\mu = NiA \qquad \text{(magnetic moment)}, \qquad (29\text{-}35)$$

in which N is the number of turns in the coil, i is the current through the coil, and A is the area enclosed by each turn of the coil. (Equation 29-35 tells us that the unit of $\vec{\mu}$ is the ampere-square meter.) Using $\vec{\mu}$, we can rewrite Eq. 29-33 for the torque on the coil due to a magnetic field as

$$\tau = \mu B \sin \theta, \qquad (29\text{-}36)$$

in which θ is the angle between the vectors $\vec{\mu}$ and $\vec{B}$.

We can generalize this to the vector relation

$$\vec{\tau} = \vec{\mu} \times \vec{B}, \qquad (29\text{-}37)$$

which reminds us very much of the corresponding equation for the torque exerted by an *electric* field on an *electric* dipole—namely, Eq. 23-34:

$$\vec{\tau} = \vec{p} \times \vec{E}.$$

In each case the torque due to the field—either magnetic or electric—is equal to the vector product of the corresponding dipole moment and the field vector.

A magnetic dipole in an external magnetic field has a **magnetic potential energy** that depends on the dipole's orientation in the field. For electric dipoles we have shown (Eq. 23-38) that

$$U(\theta) = -\vec{p} \cdot \vec{E}.$$

In strict analogy, we can write for the magnetic case

$$U(\theta) = -\vec{\mu} \cdot \vec{B}. \qquad (29\text{-}38)$$

Fig. 29-23 The orientations of highest and lowest energy of a magnetic dipole (here a coil carrying current) in an external magnetic field $\vec{B}$. The direction of the current i gives the direction of the magnetic dipole moment $\vec{\mu}$ via the right-hand rule shown for $\vec{n}$ in Fig. 29-21b.

A magnetic dipole has its lowest energy ($= -\mu B \cos 0 = -\mu B$) when its dipole moment $\vec{\mu}$ is lined up with the magnetic field (Fig. 29-23). It has its highest energy ($= -\mu B \cos 180° = +\mu B$) when $\vec{\mu}$ is directed opposite the field.

When a magnetic dipole rotates from an initial orientation θ_i to another orientation θ_f, the work W done on the dipole by the magnetic field is

$$W = -\Delta U = -(U_f - U_i), \qquad (29\text{-}39)$$

where U_f and U_i are calculated with Eq. 29-38. If an applied torque (due to "an

TABLE 29-2   Some Magnetic Dipole Moments

A small bar magnet	5 J/T
Earth	8.0×10^{22} J/T
A proton	1.4×10^{-26} J/T
An electron	9.3×10^{-24} J/T

external agent") acts on the dipole during the change in its orientation, then work W_a is done on the dipole by the applied torque. *If the dipole is stationary* before and after the change in its orientation, then work W_a is the negative of the work done on the dipole by the field. Thus,

$$W_a = -W = U_f - U_i. \tag{29-40}$$

So far, we have identified only a current-carrying coil as a magnetic dipole. However, a simple bar magnet is also a magnetic dipole, as is a rotating sphere of charge. Earth itself is (approximately) a magnetic dipole. Finally, most subatomic particles, including the electron, the proton, and the neutron, have magnetic dipole moments. As you will see in Chapter 32, all these quantities can be viewed as current loops. For comparison, some approximate magnetic dipole moments are shown in Table 29-2.

✓**CHECKPOINT 6:**   The figure shows four orientations, at angle θ, of a magnetic dipole moment $\vec{\mu}$ in a magnetic field. Rank the orientations according to (a) the magnitude of the torque on the dipole and (b) the potential energy of the dipole, greatest first.

Sample Problem 29-8

Figure 29-24 shows a circular coil with 250 turns, an area A of 2.52×10^{-4} m², and a current of 100 μA. The coil is at rest in a uniform magnetic field of magnitude $B = 0.85$ T, with its magnetic dipole moment $\vec{\mu}$ initially aligned with $\vec{B}$.

(a) In Fig. 29-24, what is the direction of the current in the coil?

SOLUTION: The Key Idea here is to apply the following right-hand rule to the coil: Imagine cupping the coil with your right hand so that your right thumb is outstretched in the direction of $\vec{\mu}$. The direction in which your fingers curl around the coil is the direction of the current in the coil. Thus, in the wires on the near side of the coil—those we see in Fig. 29-24—the current is from top to bottom.

(b) How much work would the torque applied by an external agent have to do on the coil to rotate it 90° from its initial orientation, so that $\vec{\mu}$ is perpendicular to $\vec{B}$ and the coil is again at rest?

SOLUTION: The Key Idea here is that the work W_a done by the applied torque would be equal to the change in the coil's potential energy

Fig. 29-24 Sample Problem 29-8. A side view of a circular coil carrying a current and oriented so that its magnetic dipole moment $\vec{\mu}$ is aligned with magnetic field $\vec{B}$.

due to its change in orientation. From Eq. 29-40 ($W_a = U_f - U_i$), we find

$$W_a = U(90°) - U(0°)$$
$$= -\mu B \cos 90° - (-\mu B \cos 0°) = 0 + \mu B$$
$$= \mu B.$$

Substituting for μ from Eq. 29-35 ($\mu = NiA$), we find that

$$W_a = (NiA)B$$
$$= (250)(100 \times 10^{-6} \text{ A})(2.52 \times 10^{-4} \text{ m}^2)(0.85 \text{ T})$$
$$= 5.356 \times 10^{-6} \text{ J} \approx 5.4 \text{ μJ}. \qquad \text{(Answer)}$$

REVIEW & SUMMARY

Magnetic Field $\vec{B}$ A **magnetic field** $\vec{B}$ is defined in terms of the force $\vec{F}_B$ acting on a test particle with charge q moving through the field with velocity $\vec{v}$:

$$\vec{F}_B = q\vec{v} \times \vec{B}. \tag{29-2}$$

The SI unit for $\vec{B}$ is the **tesla** (T): 1 T = 1 N/(A · m) = 10^4 gauss.

The Hall Effect   When a conducting strip of thickness l carry-

ing a current i is placed in a uniform magnetic field $\vec{B}$, some charge carriers (with charge e) build up on the sides of the conductor, creating a potential difference V across the strip. The polarities of the sides indicate the sign of the charge carriers; the number density n of charge carriers can be calculated with

$$n = \frac{Bi}{Vle}. \tag{29-12}$$

A Charged Particle Circulating in a Magnetic Field A charged particle with mass m and charge magnitude q moving with velocity $\vec{v}$ perpendicular to a uniform magnetic field $\vec{B}$ will travel in a circle. Applying Newton's second law to the circular motion yields

$$qvB = \frac{mv^2}{r}, \qquad (29\text{-}15)$$

from which we find the radius r of the circle to be

$$r = \frac{mv}{qB}. \qquad (29\text{-}16)$$

The frequency of revolution f, the angular frequency ω, and the period of the motion T are given by

$$f = \frac{\omega}{2\pi} = \frac{1}{T} = \frac{qB}{2\pi m}. \qquad (29\text{-}19, 29\text{-}18, 29\text{-}17)$$

Cyclotrons and Synchrotrons A cyclotron is a particle accelerator that uses a magnetic field to hold a charged particle in a circular orbit of increasing radius so that a modest accelerating potential may act on the particle repeatedly, providing it with high energy. Because the moving particle gets out of step with the oscillator as its speed approaches that of light, there is an upper limit to the energy attainable with the cyclotron. A synchrotron avoids this difficulty. Here both B and the oscillator frequency f_{osc} are programmed to change cyclically so that the particle not only can go to high energies but can do so at a constant orbital radius.

Magnetic Force on a Current-Carrying Wire A straight wire carrying a current i in a uniform magnetic field experiences a sideways force

$$\vec{F}_B = i\vec{L} \times \vec{B}. \qquad (29\text{-}26)$$

The force acting on a current element $i\,d\vec{L}$ in a magnetic field is

$$d\vec{F}_B = i\,d\vec{L} \times \vec{B}. \qquad (29\text{-}28)$$

The direction of the length vector $\vec{L}$ or $d\vec{L}$ is that of the current i.

Torque on a Current-Carrying Coil A coil (of area A and carrying current i, with N turns) in a uniform magnetic field $\vec{B}$ will experience a torque $\vec{\tau}$ given by

$$\vec{\tau} = \vec{\mu} \times \vec{B}. \qquad (29\text{-}37)$$

Here $\vec{\mu}$ is the **magnetic dipole moment** of the coil, with magnitude $\mu = NiA$ and direction given by the right-hand rule.

Orientation Energy of a Magnetic Dipole The **magnetic potential energy** of a magnetic dipole in a magnetic field is

$$U(\theta) = -\vec{\mu} \cdot \vec{B}. \qquad (29\text{-}38)$$

If a magnetic dipole rotates from an initial orientation θ_i to another orientation θ_f, the work W done on the dipole by the magnetic field is

$$W = -\Delta U = -(U_f - U_i). \qquad (29\text{-}39)$$

QUESTIONS

1. Figure 29-25 shows three situations in which a positive particle of velocity $\vec{v}$ moves through a uniform magnetic field $\vec{B}$ and experiences a magnetic force $\vec{F}_B$. In each situation, determine whether the orientations of the vectors are physically reasonable.

Fig. 29-25 Question 1.

2. For four situations, here is the velocity $\vec{v}$ of a proton at a certain instant as it moves through a uniform magnetic field $\vec{B}$:

(a) $\vec{v} = 2\hat{i} - 3\hat{j}$ and $\vec{B} = 4\hat{k}$
(b) $\vec{v} = 3\hat{i} + 2\hat{j}$ and $\vec{B} = -4\hat{k}$
(c) $\vec{v} = 3\hat{j} - 2\hat{k}$ and $\vec{B} = 4\hat{i}$
(d) $\vec{v} = 20\hat{i}$ and $\vec{B} = -4\hat{i}$.

Without written calculation, rank the situations according to the magnitude of the magnetic force on the proton, greatest first.

3. In Section 29-3, we discussed a charged particle moving through crossed fields with the forces $\vec{F}_E$ and $\vec{F}_B$ in opposition. We found that the particle moves in a straight line (that is, neither force dominates the motion) if its speed is given by Eq. 29-7 ($v = E/B$).

Which of the two forces dominates if the speed of the particle is, instead, (a) $v < E/B$ and (b) $v > E/B$?

4. Figure 29-26 shows crossed and uniform electric and magnetic fields $\vec{E}$ and $\vec{B}$ and, at a certain instant, the velocity vectors of the 10 charged particles listed in Table 29-3. (The vectors are not drawn to scale.) The table gives the signs of the charges and the speeds of the particles; the speeds are given as either less than or greater than E/B (see Question 3). Which particles will move out of the page toward you after the instant of Fig. 29-26?

Fig. 29-26 Question 4.

TABLE 29-3 **Question 4**

Particle	Charge	Speed	Particle	Charge	Speed
1	+	Less	6	−	Greater
2	+	Greater	7	+	Less
3	+	Less	8	+	Greater
4	+	Greater	9	−	Less
5	−	Less	10	−	Greater

5. In Fig. 29-27, a charged particle enters a uniform magnetic field $\vec{B}$ with speed v_0, moves through a half-circle in time T_0, and then leaves the field. (a) Is the charge positive or negative? (b) Is the final speed of the particle greater than, less than, or equal to v_0? (c) If the initial speed had been $0.5v_0$, would the time spent in field $\vec{B}$ have been greater than, less than, or equal to T_0? (d) Would the path have been a half-circle, more than a half-circle, or less than a half-circle?

Fig. 29-27 Question 5.

6. Figure 29-28 shows the path of a particle through six regions of uniform magnetic field, where the path is either a half-circle or a quarter-circle. Upon leaving the last region, the particle travels between two charged, parallel plates and is deflected toward the plate of higher potential. What are the directions of the magnetic fields in the six regions?

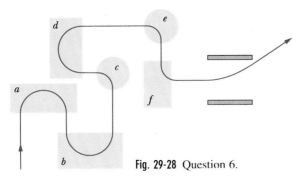

Fig. 29-28 Question 6.

7. Figure 29-29 shows the path of an electron that passes through two regions containing uniform magnetic fields of magnitudes B_1 and B_2. Its path in each region is a half-circle. (a) Which field is stronger? (b) What are the directions of the two fields? (c) Is the time spent by the electron in the $\vec{B}_1$ region greater than, less than, or the same as the time spent in the $\vec{B}_2$ region?

8. *Particle Roundabout.* Figure 29-30 shows 11 paths through a

Fig. 29-29 Question 7.

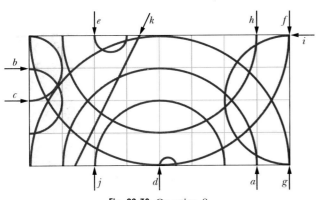

Fig. 29-30 Question 8.

region of uniform magnetic field. One path is a straight line; the rest are half-circles. Table 29-4 gives the masses, charges, and speeds of 11 particles that take these paths through the field in the directions shown. Which path in the figure corresponds to which particle in the table?

TABLE 29-4	Question 8		
Particle	Mass	Charge	Speed
1	$2m$	q	v
2	m	$2q$	v
3	$m/2$	q	$2v$
4	$3m$	$3q$	$3v$
5	$2m$	q	$2v$
6	m	$-q$	$2v$
7	m	$-4q$	v
8	m	$-q$	v
9	$2m$	$-2q$	$8v$
10	m	$-2q$	$8v$
11	$3m$	0	$3v$

9. Figure 29-31 shows eight wires that carry identical currents through the same uniform magnetic field (directed into the page) in eight separate experiments. Each wire consists of two straight sections (each of length L and either parallel or perpendicular to the x and y axes shown) and one curved section (with radius of curvature R). The directions of the currents through the wires are indicated by the arrows next to the wires. (a) Give the direction of the net magnetic force on each wire in terms of an angle measured counterclockwise from the positive direction of the x axis. (b) Rank wires 1 through 4 according to the magnitude of the net magnetic force on them, greatest first. (c) Do the same type of ranking for wires 5 through 8.

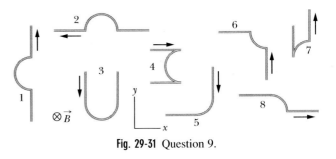

Fig. 29-31 Question 9.

10. (a) In Checkpoint 6, if the dipole moment $\vec{\mu}$ rotates from orientation 1 to orientation 2, is the work done on the dipole *by the magnetic field* positive, negative, or zero? (b) Rank the work done on the dipole by the magnetic field for rotations from orientation 1 to (1) orientation 2, (2) orientation 3, and (3) orientation 4, greatest first.

EXERCISES & PROBLEMS

SEC. 29-2   The Definition of $\vec{B}$

1E. An alpha particle travels at a velocity $\vec{v}$ of magnitude 550 m/s through a uniform magnetic field $\vec{B}$ of magnitude 0.045 T. (An alpha particle has a charge of $+3.2 \times 10^{-19}$ C and a mass of 6.6×10^{-27} kg.) The angle between $\vec{v}$ and $\vec{B}$ is 52°. What are the magnitudes of (a) the force $\vec{F}_B$ acting on the particle due to the field and (b) the acceleration of the particle due to $\vec{F}_B$? (c) Does the speed of the particle increase, decrease, or remain equal to 550 m/s?

2E. An electron in a TV camera tube is moving at 7.20×10^6 m/s in a magnetic field of strength 83.0 mT. (a) Without knowing the direction of the field, what can you say about the greatest and least magnitudes of the force acting on the electron due to the field? (b) At one point the electron has an acceleration of magnitude 4.90×10^{14} m/s². What is the angle between the electron's velocity and the magnetic field?

3E. A proton traveling at 23.0° with respect to the direction of a magnetic field of strength 2.60 mT experiences a magnetic force of 6.50×10^{-17} N. Calculate (a) the proton's speed and (b) its kinetic energy in electron-volts. ssm ilw

4P. An electron that has velocity

$$\vec{v} = (2.0 \times 10^6 \text{ m/s})\hat{i} + (3.0 \times 10^6 \text{ m/s})\hat{j}$$

moves through the magnetic field $\vec{B} = (0.030 \text{ T})\hat{i} - (0.15 \text{ T})\hat{j}$. (a) Find the force on the electron. (b) Repeat your calculation for a proton having the same velocity.

5P. Each of the electrons in the beam of a television tube has a kinetic energy of 12.0 keV. The tube is oriented so that the electrons move horizontally from geomagnetic south to geomagnetic north. The vertical component of Earth's magnetic field points down and has a magnitude of 55.0 μT. (a) In what direction will the beam deflect? (b) What is the acceleration of a single electron due to the magnetic field? (c) How far will the beam deflect in moving 20.0 cm through the television tube? ssm www

SEC. 29-3   Crossed Fields: Discovery of the Electron

6E. A proton travels through uniform magnetic and electric fields. The magnetic field is $\vec{B} = -2.5\hat{i}$ mT. At one instant the velocity of the proton is $\vec{v} = 2000\hat{j}$ m/s. At that instant, what is the magnitude of the net force acting on the proton if the electric field is (a) $4.0\hat{k}$ V/m, (b) $-4.0\hat{k}$ V/m, and (c) $4.0\hat{i}$ V/m?

7E. An electron with kinetic energy 2.5 keV moves horizontally into a region of space in which there is a downward-directed uniform electric field of magnitude 10 kV/m. (a) What are the magnitude and direction of the (smallest) uniform magnetic field that will cause the electron to continue to move horizontally? Ignore the gravitational force, which is rather small. (b) Is it possible for a proton to pass through this combination of fields undeflected? If so, under what circumstances? ssm

8E. An electric field of 1.50 kV/m and a magnetic field of 0.400 T act on a moving electron to produce no net force. (a) Calculate the minimum speed v of the electron. (b) Draw the vectors $\vec{E}$, $\vec{B}$, and $\vec{v}$.

9P. An electron is accelerated through a potential difference of 1.0 kV and directed into a region between two parallel plates separated by 20 mm with a potential difference of 100 V between them. The electron is moving perpendicular to the electric field of the plates when it enters the region between the plates. What uniform magnetic field, applied perpendicular to both the electron path and the electric field, will allow the electron to travel in a straight line? ilw

10P. An electron has an initial velocity of $(12.0\hat{j} + 15.0\hat{k})$ km/s and a constant acceleration of $(2.00 \times 10^{12} \text{ m/s}^2)\hat{i}$ in a region in which uniform electric and magnetic fields are present. If $\vec{B} = (400 \ \mu\text{T})\hat{i}$, find the electric field $\vec{E}$.

11P. An ion source is producing ions of ^{6}Li (mass = 6.0 u), each with a charge of $+e$. The ions are accelerated by a potential difference of 10 kV and pass horizontally into a region in which there is a uniform vertical magnetic field of magnitude $B = 1.2$ T. Calculate the strength of the smallest electric field, to be set up over the same region, that will allow the ^{6}Li ions to pass through undeflected. ssm

SEC. 29-4   Crossed Fields: The Hall Effect

12E. A strip of copper 150 μm wide is placed in a uniform magnetic field $\vec{B}$ of magnitude 0.65 T, with $\vec{B}$ perpendicular to the strip. A current $i = 23$ A is then sent through the strip such that a Hall potential difference V appears across the width of the strip. Calculate V. (The number of charge carriers per unit volume for copper is 8.47×10^{28} electrons/m³.)

13P. (a) In Fig. 29-8, show that the ratio of the Hall electric field E to the electric field E_C responsible for moving charge (the current) along the length of the strip is

$$\frac{E}{E_C} = \frac{B}{ne\rho},$$

where ρ is the resistivity of the material and n is the number density of the charge carriers. (b) Compute this ratio numerically for Exercise 12. (See Table 27-1.)

14P. A metal strip 6.50 cm long, 0.850 cm wide, and 0.760 mm thick moves with constant velocity $\vec{v}$ through a uniform magnetic field $B = 1.20$ mT directed perpendicular to the strip, as shown in Fig. 29-32. A potential difference of 3.90 μV is measured between points x and y across the strip. Calculate the speed v.

Fig. 29-32 Problem 14.

SEC. 29-5   A Circulating Charged Particle

15E. What uniform magnetic field, applied perpendicular to a beam of electrons moving at 1.3×10^6 m/s, is required to make the electrons travel in a circular arc of radius 0.35 m? ssm

16E. An electron is accelerated from rest by a potential difference of 350 V. It then enters a uniform magnetic field of magnitude 200 mT with its velocity perpendicular to the field. Calculate (a) the speed of the electron and (b) the radius of its path in the magnetic field.

17E. An electron with kinetic energy 1.20 keV circles in a plane perpendicular to a uniform magnetic field. The orbit radius is 25.0 cm. Find (a) the speed of the electron, (b) the magnetic field, (c) the frequency of circling, and (d) the period of the motion. ssm

18E. Physicist S. A. Goudsmit devised a method for measuring the masses of heavy ions by timing their periods of revolution in a known magnetic field. A singly charged ion of iodine makes 7.00 rev in a field of 45.0 mT in 1.29 ms. Calculate its mass, in unified atomic mass units. (Actually, the method allows mass measurements to be carried out to much greater accuracy than these approximate data suggest.)

19E. (a) Find the frequency of revolution of an electron with an energy of 100 eV in a uniform magnetic field of 35.0 μT. (b) Calculate the radius of the path of this electron if its velocity is perpendicular to the magnetic field. ilw

20E. An alpha particle ($q = +2e$, $m = 4.00$ u) travels in a circular path of radius 4.50 cm in a uniform magnetic field with $B = 1.20$ T. Calculate (a) its speed, (b) its period of revolution, (c) its kinetic energy in electron-volts, and (d) the potential difference through which it would have to be accelerated to achieve this energy.

21E. A beam of electrons whose kinetic energy is K emerges from a thin-foil "window" at the end of an accelerator tube. There is a metal plate a distance d from this window and perpendicular to the direction of the emerging beam (Fig. 29-33). Show that we can prevent the beam from hitting the plate if we apply a uniform magnetic field $\vec{B}$ such that

$$B \geq \sqrt{\frac{2mK}{e^2 d^2}},$$

in which m and e are the electron mass and charge. How should $\vec{B}$ be oriented? ssm

Foil window — Electron beam

Plate

Tube

d

Fig. 29-33 Exercise 21.

22P. A source injects an electron of speed $v = 1.5 \times 10^7$ m/s into a uniform magnetic field of magnitude $B = 1.0 \times 10^{-3}$ T. The velocity of the electron makes an angle $\theta = 10°$ with the direction of the magnetic field. Find the distance d from the point of injection at which the electron next crosses the field line that passes through the injection point.

23P. In a nuclear experiment a proton with kinetic energy 1.0 MeV moves in a circular path in a uniform magnetic field. What energy must (a) an alpha particle ($q = +2e$, $m = 4.0$ u) and (b) a deuteron ($q = +e$, $m = 2.0$ u) have if they are to circulate in the same circular path?

24P. A proton, a deuteron ($q = +e$, $m = 2.0$ u), and an alpha particle ($q = +2e$, $m = 4.0$ u) with the same kinetic energies enter a region of uniform magnetic field $\vec{B}$, moving perpendicular to $\vec{B}$. Compare the radii of their circular paths.

25P. A certain commercial mass spectrometer (see Sample Problem 29-3) is used to separate uranium ions of mass 3.92×10^{-25} kg and charge 3.20×10^{-19} C from related species. The ions are accelerated through a potential difference of 100 kV and then pass into a uniform magnetic field, where they are bent in a path of radius 1.00 m. After traveling through 180° and passing through a slit of width 1.00 mm and height 1.00 cm, they are collected in a cup. (a) What is the magnitude of the (perpendicular) magnetic field in the separator? If the machine is used to separate out 100 mg of material per hour, calculate (b) the current of the desired ions in the machine and (c) the thermal energy produced in the cup in 1.00 h. ssm

26P. A proton of charge $+e$ and mass m enters a uniform magnetic field $\vec{B} = B\hat{i}$ with an initial velocity $\vec{v} = v_{0x}\hat{i} + v_{0y}\hat{j}$. Find an expression in unit-vector notation for its velocity $\vec{v}$ at any later time t.

27P. A positron with kinetic energy 2.0 keV is projected into a uniform magnetic field $\vec{B}$ of magnitude 0.10 T, with its velocity vector making an angle of 89° with $\vec{B}$. Find (a) the period, (b) the pitch p, and (c) the radius r of its helical path. ssm www

28P. In Fig. 29-34, a charged particle moves into a region of uniform magnetic field $\vec{B}$, goes through half a circle, and then exits that region. The particle is either a proton or an electron (you must decide which). It spends 130 ns within the region. (a) What is the magnitude of $\vec{B}$? (b) If the particle is sent back through the magnetic field (along the same initial path) but with 2.00 times its previous kinetic energy, how much time does it spend within the field?

$\odot \vec{B}$

Fig. 29-34 Problem 28.

29P. A neutral particle is at rest in a uniform magnetic field $\vec{B}$. At time $t = 0$ it decays into two charged particles, each of mass m. (a) If the charge of one of the particles is $+q$, what is the charge of the other? (b) The two particles move off in separate paths, both of which lie in the plane perpendicular to $\vec{B}$. At a later time the particles collide. Express the time from decay until collision in terms of m, B, and q. ssm

SEC. 29-6   Cyclotrons and Synchrotrons

30E. In a certain cyclotron a proton moves in a circle of radius 0.50 m. The magnitude of the magnetic field is 1.2 T. (a) What is the oscillator frequency? (b) What is the kinetic energy of the proton, in electron-volts?

31P. Estimate the total path length traveled by a deuteron in the cyclotron of Sample Problem 29-5 during the (entire) acceleration process. Assume that the accelerating potential between the dees is 80 kV. ssm www

32P. The oscillator frequency of the cyclotron in Sample Problem 29-5 has been adjusted to accelerate deuterons ($q = +e$, $m = 2.0$ u). (a) If protons are injected instead of deuterons, to what kinetic energy can the protons be accelerated, using the same oscillator frequency? (b) What magnetic field would be required? (c) What kinetic energy could be produced for protons if the magnetic field were left at the value used for deuterons? (d) What oscillator frequency would then be required? (e) Answer the same questions for alpha particles ($q = +2e$, $m = 4.0$ u).

SEC. 29-7 Magnetic Force on a Current-Carrying Wire

33E. A horizontal conductor that is part of a power line carries a current of 5000 A from south to north. Earth's magnetic field (60.0 μT) is directed toward the north and is inclined downward at 70° to the horizontal. Find the magnitude and direction of the magnetic force on 100 m of the conductor due to Earth's field. ssm

34E. A wire 1.80 m long carries a current of 13.0 A and makes an angle of 35.0° with a uniform magnetic field $B = 1.50$ T. Calculate the magnetic force on the wire.

35E. A wire of 62.0 cm length and 13.0 g mass is suspended by a pair of flexible leads in a uniform magnetic field of magnitude 0.440 T (Fig. 29-35). What are the magnitude and direction of the current required to remove the tension in the supporting leads? ssm ilw

36P. A wire 50 cm long lying along the x axis carries a current of 0.50 A in the positive x direction, through a magnetic field $\vec{B} =$ (0.0030 T)$\hat{j}$ + (0.010 T)$\hat{k}$. Find the magnetic force on the wire.

Fig. 29-35 Exercise 35.

37P. A 1.0 kg copper rod rests on two horizontal rails 1.0 m apart and carries a current of 50 A from one rail to the other. The coefficient of static friction between rod and rails is 0.60. What is the smallest magnetic field (not necessarily vertical) that would cause the rod to slide? ssm

38P. Consider the possibility of a new design for an electric train. The engine is driven by the force on a conducting axle due to the vertical component of Earth's magnetic field. To produce the force, current is maintained down one rail, through a conducting wheel, through the axle, through another conducting wheel, and then back to the source via the other rail. (a) What current is needed to provide a modest 10 kN force? Take the vertical component of Earth's field to be 10 μT and the length of the axle to be 3.0 m. (b) At what rate would electric energy be lost for each ohm of resistance in the rails? (c) Is such a train totally or just marginally unrealistic?

SEC. 29-8 Torque on a Current Loop

39E. Figure 29-36 shows a rectangular 20-turn coil of wire, of dimensions 10 cm by 5.0 cm. It carries a current of 0.10 A and is hinged along one long side. It is mounted in the xy plane, at 30° to the direction of a uniform magnetic field of magnitude 0.50 T. Find the magnitude and direction of the torque acting on the coil about the hinge line. ssm

Fig. 29-36 Exercise 39.

40E. A single-turn current loop, carrying a current of 4.00 A, is in the shape of a right triangle with sides 50.0, 120, and 130 cm. The loop is in a uniform magnetic field of magnitude 75.0 mT whose direction is parallel to the current in the 130 cm side of the loop. (a) Find the magnitude of the magnetic force on each of the three sides of the loop. (b) Show that the total magnetic force on the loop is zero.

41E. A length L of wire carries a current i. Show that if the wire is formed into a circular coil, then the maximum torque in a given magnetic field is developed when the coil has one turn only, and that maximum torque has the magnitude $\tau = L^2iB/4\pi$. ssm ilw

42P. Prove that the relation $\tau = NiAB \sin \theta$ holds for closed loops of arbitrary shape and not only for rectangular loops as in Fig. 29-21. (*Hint:* Replace the loop of arbitrary shape with an assembly of adjacent long, thin, approximately rectangular loops that are nearly equivalent to the loop of arbitrary shape as far as the distribution of current is concerned.)

43P. Figure 29-37 shows a wire ring of radius a that is perpendicular to the general direction of a radially symmetric, diverging magnetic field. The magnetic field at the ring is everywhere of the same magnitude B, and its direction at the ring everywhere makes an angle θ with a normal to the plane of the ring. The twisted lead wires have no effect on the problem. Find the magnitude and direction of the force the field exerts on the ring if the ring carries a current i. ssm www

Fig. 29-37 Problem 43.

44P. A closed wire loop with current i is in a uniform magnetic field $\vec{B}$, with the plane of the loop at angle θ to the direction of $\vec{B}$. Show that the total magnetic force on the loop is zero. Does your proof also hold for a nonuniform magnetic field?

45P. The coil of a certain galvanometer (see Sample Problem 29-7) has a resistance of 75.3 Ω; its needle shows a full-scale deflection when a current of 1.62 mA passes through the coil. (a) Determine the value of the auxiliary resistance required to convert the galvanometer to a voltmeter that reads 1.00 V at full-scale deflection. How should this resistance be connected? (b) Determine the value of the auxiliary resistance required to convert the galvanometer to an ammeter that reads 50.0 mA at full-scale deflection. How should this resistance be connected? ssm

46P. A particle of charge q moves in a circle of radius a with speed v. Treating the circular path as a current loop with constant current equal to its average current, find the maximum torque exerted on the loop by a uniform magnetic field of magnitude B.

47P. Figure 29-38 shows a wood cylinder of mass $m = 0.250$ kg and length $L = 0.100$ m, with $N = 10.0$ turns of wire wrapped around it longitudinally, so that the plane of the wire coil contains the axis of the cylinder. What is the least current i through the coil

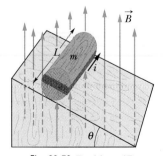

Fig. 29-38 Problem 47.

that will prevent the cylinder from rolling down a plane inclined at an angle θ to the horizontal, in the presence of a vertical, uniform magnetic field of magnitude 0.500 T, if the plane of the coil is parallel to the inclined plane? ssm

SEC. 29-9 The Magnetic Dipole Moment

48E. The magnetic dipole moment of Earth is 8.00×10^{22} J/T. Assume that this is produced by charges flowing in Earth's molten outer core. If the radius of their circular path is 3500 km, calculate the current they produce.

49E. A circular coil of 160 turns has a radius of 1.90 cm. (a) Calculate the current that results in a magnetic dipole moment of 2.30 A·m². (b) Find the maximum torque that the coil, carrying this current, can experience in a uniform 35.0 mT magnetic field. ssm

50E. A circular wire loop whose radius is 15.0 cm carries a current of 2.60 A. It is placed so that the normal to its plane makes an angle of 41.0° with a uniform magnetic field of 12.0 T. (a) Calculate the magnetic dipole moment of the loop. (b) What torque acts on the loop?

51E. A current loop, carrying a current of 5.0 A, is in the shape of a right triangle with sides 30, 40, and 50 cm. The loop is in a uniform magnetic field of magnitude 80 mT whose direction is parallel to the current in the 50 cm side of the loop. Find the magnitude of (a) the magnetic dipole moment of the loop and (b) the torque on the loop. ssm

52E. A stationary circular wall clock has a face with a radius of 15 cm. Six turns of wire are wound around its perimeter; the wire carries a current of 2.0 A in the clockwise direction. The clock is located where there is a constant, uniform external magnetic field of magnitude 70 mT (but the clock still keeps perfect time). At exactly 1:00 p.m., the hour hand of the clock points in the direction of the external magnetic field. (a) After how many minutes will the minute hand point in the direction of the torque on the winding due to the magnetic field? (b) Find the torque magnitude.

53E. Two concentric, circular wire loops, of radii 20.0 and 30.0 cm, are located in the xy plane; each carries a clockwise current of 7.00 A (Fig. 29-39). (a) Find the net magnetic dipole moment of this system. (b) Repeat for reversed current in the inner loop. ssm

54P. Figure 29-40 shows a current loop $ABCDEFA$ carrying a current $i = 5.00$ A. The sides of the loop are parallel to the coor-

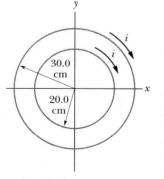

Fig. 29-39 Exercise 53.

dinate axes, with $AB = 20.0$ cm, $BC = 30.0$ cm, and $FA = 10.0$ cm. Calculate the magnitude and direction of the magnetic dipole moment of this loop. (*Hint:* Imagine equal and opposite currents i in the line segment AD; then treat the two rectangular loops $ABCDA$ and $ADEFA$.)

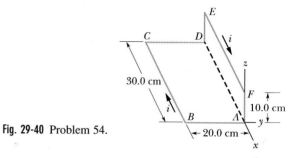

Fig. 29-40 Problem 54.

55P. A circular loop of wire having a radius of 8.0 cm carries a current of 0.20 A. A vector of unit length and parallel to the dipole moment $\vec{\mu}$ of the loop is given by $0.60\hat{i} - 0.80\hat{j}$. If the loop is located in a uniform magnetic field given by $\vec{B} = (0.25\ \text{T})\hat{i} + (0.30\ \text{T})\hat{k}$, find (a) the torque on the loop (in unit-vector notation) and (b) the magnetic potential energy of the loop. ssm

Additional Problems

56. A wire lying along a y axis from $y = 0$ to $y = 0.250$ m carries a current of 2.00 mA in the negative y direction. The wire lies in a nonuniform magnetic field given by

$$\vec{B} = (0.300\ \text{T/m})y\hat{i} + (0.400\ \text{T/m})y\hat{j}.$$

In unit-vector notation, what is the magnetic force on (a) an element dy of the wire at position y and (b) the entire wire?

57. A proton moves at a constant velocity of $+50$ m/s along an x axis, through crossed electric and magnetic fields. The magnetic field is $\vec{B} = (2.0\ \text{mT})\hat{j}$. What is the electric field?

58. A magnetic dipole with a dipole moment of magnitude 0.020 J/T is released from rest in a uniform magnetic field of magnitude 52 mT. The rotation of the dipole due to the magnetic force on it is unimpeded. When the dipole rotates through the orientation where its dipole moment is aligned with the magnetic field, its kinetic energy is 0.80 mJ. (a) What is the initial angle between the dipole moment and the magnetic field? (b) What is the angle when the dipole is next (momentarily) at rest?

59. An electron moves through a uniform magnetic field given by $\vec{B} = B_x\hat{i} + (3B_x)\hat{j}$. At a particular instant, the electron has the velocity $\vec{v} = (2.0\hat{i} + 4.0\hat{j})$ m/s and the magnetic force acting on it is $(6.4 \times 10^{-19}\ \text{N})\hat{k}$. Find B_x.

30 Magnetic Fields Due to Currents

This is the way we presently launch materials into space. However, when we begin mining the Moon and the asteroids, where we will not have a source of fuel for such conventional rockets, we shall need a more effective way. Electromagnetic launchers may be the answer. A small prototype, the *electromagnetic rail gun,* can presently accelerate a projectile from rest to a speed of 10 km/s (36 000 km/h) within 1 ms.

How can such rapid acceleration possibly be accomplished?

The answer is in this chapter.

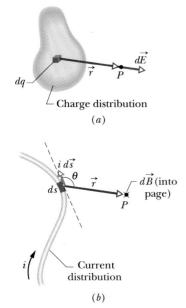

Charge distribution

(a)

Current distribution

(b)

Fig. 30-1 (a) A charge element dq produces a differential electric field $d\vec{E}$ at point P. (b) A current-length element $i\,d\vec{s}$ produces a differential magnetic field $d\vec{B}$ at point P. The green × (the tail of an arrow) at the dot for point P indicates that $d\vec{B}$ is directed *into* the page there.

30-1 Calculating the Magnetic Field Due to a Current

As we discussed in Section 29-1, one way to produce a magnetic field is with moving charges — that is, with a current. Our goal in this chapter is to calculate the magnetic field that is produced by a given distribution of currents. We shall use the same basic procedure we used in Chapter 23 to calculate the electric field produced by a given distribution of charged particles.

Let us quickly review that basic procedure. We first mentally divide the charge distribution into charge elements dq, as is done for a charge distribution of arbitrary shape in Fig. 30-1a. We then calculate the field $d\vec{E}$ produced at some point P by a typical charge element. Because the electric fields contributed by different elements can be superimposed, we calculate the net field $\vec{E}$ at P by summing, via integration, the contributions $d\vec{E}$ from all the elements.

Recall that we express the magnitude of $d\vec{E}$ as

$$dE = \frac{1}{4\pi\varepsilon_0}\frac{dq}{r^2}, \tag{30-1}$$

in which r is the distance from the charge element dq to point P. For a positively charged element, the direction of $d\vec{E}$ is that of $\vec{r}$, where $\vec{r}$ is the vector that extends from the charge element dq to the point P. Using $\vec{r}$, we can rewrite Eq. 30-1 in vector form as

$$d\vec{E} = \frac{1}{4\pi\varepsilon_0}\frac{dq}{r^3}\,\vec{r}, \tag{30-2}$$

which indicates that the direction of the vector $d\vec{E}$ produced by a positively charged element is the direction of the vector $\vec{r}$. Note that Eq. 30-2 is an inverse-square law ($d\vec{E}$ depends on inverse r^2) in spite of the exponent 3 in the denominator. That exponent is in the equation only because we added a factor of magnitude r in the numerator.

Now let us use the same basic procedure to calculate the magnetic field due to a current. Figure 30-1b shows a wire of arbitrary shape carrying a current i. We want to find the magnetic field $\vec{B}$ at a nearby point P. We first mentally divide the wire into differential elements ds and then define for each element a length vector $d\vec{s}$ that has length ds and whose direction is the direction of the current in ds. We can then define a differential *current-length element* to be $i\,d\vec{s}$; we wish to calculate the field $d\vec{B}$ produced at P by a typical current-length element. From experiment we find that magnetic fields, like electric fields, can be superimposed to find a net field. Thus, we can calculate the net field $\vec{B}$ at P by summing, via integration, the contributions $d\vec{B}$ from all the current-length elements. However, this summation is more challenging than the process associated with electric fields because of a complexity; whereas a charge element dq producing an electric field is a scalar, a current-length element $i\,d\vec{s}$ producing a magnetic field is the product of a scalar and a vector.

The magnitude of the field $d\vec{B}$ produced at point P by a current-length element $i\,d\vec{s}$ turns out to be

$$dB = \frac{\mu_0}{4\pi}\frac{i\,ds\,\sin\theta}{r^2}, \tag{30-3}$$

where θ is the angle between the directions of $d\vec{s}$ and $\vec{r}$, the vector that extends from ds to P. Symbol μ_0 is a constant, called the *permeability constant,* whose value is defined to be exactly

$$\mu_0 = 4\pi \times 10^{-7}\ \text{T} \cdot \text{m/A} \approx 1.26 \times 10^{-6}\ \text{T} \cdot \text{m/A}. \tag{30-4}$$

The direction of $d\vec{B}$, shown as being into the page in Fig. 30-1b, is that of the cross product $d\vec{s} \times \vec{r}$. We can therefore write Eq. 30-3 in vector form as

$$d\vec{B} = \frac{\mu_0}{4\pi} \frac{i \, d\vec{s} \times \vec{r}}{r^3} \qquad \text{(Biot–Savart law).} \qquad (30\text{-}5)$$

This vector equation and its scalar form, Eq. 30-3, are known as the **law of Biot and Savart** (rhymes with "Leo and bazaar"). The law, which is experimentally deduced, is an inverse-square law (the exponent in the denominator of Eq. 30-5 is 3 only because of the factor $\vec{r}$ in the numerator). We shall use this law to calculate the net magnetic field $\vec{B}$ produced at a point by various distributions of current.

Magnetic Field Due to a Current in a Long Straight Wire

Shortly we shall use the law of Biot and Savart to prove that the magnitude of the magnetic field at a perpendicular distance R from a long (infinite) straight wire carrying a current i is given by

$$B = \frac{\mu_0 i}{2\pi R} \qquad \text{(long straight wire).} \qquad (30\text{-}6)$$

The field magnitude B in Eq. 30-6 depends only on the current and the perpendicular distance R of the point from the wire. We shall show in our derivation that the field lines of $\vec{B}$ form concentric circles around the wire, as Fig. 30-2 shows and as the iron filings in Fig. 30-3 suggest. The increase in the spacing of the lines in Fig. 30-2 with increasing distance from the wire represents the $1/R$ decrease in the magnitude of $\vec{B}$ predicted by Eq. 30-6. The lengths of the two vectors $\vec{B}$ in the figure also show the $1/R$ decrease.

Here is a simple right-hand rule for finding the direction of the magnetic field set up by a current-length element, such as a section of a long wire:

> *Right-hand rule:* Grasp the element in your right hand with your extended thumb pointing in the direction of the current. Your fingers will then naturally curl around in the direction of the magnetic field lines due to that element.

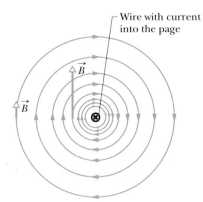

Fig. 30-2 The magnetic field lines produced by a current in a long straight wire form concentric circles around the wire. Here the current is into the page, as indicated by the ×.

Fig. 30-3 Iron filings that have been sprinkled onto cardboard collect in concentric circles when current is sent through the central wire. The alignment, which is along magnetic field lines, is caused by the magnetic field produced by the current.

(a)

(b)

Fig. 30-4 A right-hand rule gives the direction of the magnetic field due to a current in a wire. (*a*) The situation of Fig. 30-2, seen from the side. The magnetic field $\vec{B}$ at any point to the left of the wire is perpendicular to the dashed radial line and directed into the page, in the direction of the fingertips, as indicated by the ×. (*b*) If the current is reversed, $\vec{B}$ at any point to the left is still perpendicular to the dashed radial line but now is directed out of the page, as indicated by the dot.

The result of applying this right-hand rule to the current in the straight wire of Fig. 30-2 is shown in a side view in Fig. 30-4*a*. To determine the direction of the magnetic field $\vec{B}$ set up at any particular point by this current, mentally wrap your right hand around the wire with your thumb in the direction of the current. Let your fingertips pass through the point; their direction is then the direction of the magnetic field at that point. In the view of Fig. 30-2, $\vec{B}$ at any point is *tangent to a magnetic field line;* in the view of Fig. 30-4, it is *perpendicular to a dashed radial line connecting the point and the current.*

Proof of Equation 30-6

Figure 30-5, which is just like Fig. 30-1*b* except that now the wire is straight and of infinite length, illustrates the task at hand; we seek the field $\vec{B}$ at point *P*, a perpendicular distance *R* from the wire. The magnitude of the differential magnetic field produced at *P* by the current-length element $i\,d\vec{s}$ located a distance *r* from *P* is given by Eq. 30-3:

$$dB = \frac{\mu_0}{4\pi} \frac{i\,ds\,\sin\theta}{r^2}.$$

The direction of $d\vec{B}$ in Fig. 30-5 is that of the vector $d\vec{s} \times \vec{r}$—namely, directly into the page.

Note that $d\vec{B}$ at point *P* has this same direction for all the current-length elements into which the wire can be divided. Thus, we can find the magnitude of the magnetic field produced at *P* by the current-length elements in the upper half of the infinitely long wire by integrating *dB* in Eq. 30-3 from 0 to ∞.

Now consider a current-length element in the lower half of the wire, one that is as far below *P* as $d\vec{s}$ is above *P*. By Eq. 30-5, the magnetic field produced at *P* by this current-length element has the same magnitude and direction as that from element $i\,d\vec{s}$ in Fig. 30-5. Further, the magnetic field produced by the lower half of the wire is exactly the same as that produced by the upper half. To find the magnitude of the *total* magnetic field $\vec{B}$ at *P*, we need only multiply the result of our integration by 2. We get

$$B = 2\int_0^\infty dB = \frac{\mu_0 i}{2\pi} \int_0^\infty \frac{\sin\theta\,ds}{r^2}. \tag{30-7}$$

The variables θ, *s*, and *r* in this equation are not independent but (see Fig. 30-5) are related by

$$r = \sqrt{s^2 + R^2}$$

and

$$\sin\theta = \sin(\pi - \theta) = \frac{R}{\sqrt{s^2 + R^2}}.$$

Fig. 30-5 Calculating the magnetic field produced by a current *i* in a long straight wire. The field $d\vec{B}$ at *P* associated with the current-length element $i\,d\vec{s}$ is into the page, as shown.

With these substitutions and integral 19 in Appendix E, Eq. 30-7 becomes

$$B = \frac{\mu_0 i}{2\pi} \int_0^\infty \frac{R \, ds}{(s^2 + R^2)^{3/2}}$$

$$= \frac{\mu_0 i}{2\pi R} \left[\frac{s}{(s^2 + R^2)^{1/2}} \right]_0^\infty = \frac{\mu_0 i}{2\pi R}, \qquad (30\text{-}8)$$

which is the relation we set out to prove. Note that the magnetic field at P due to either the lower half or the upper half of the infinite wire in Fig. 30-5 is half this value; that is,

$$B = \frac{\mu_0 i}{4\pi R} \qquad \text{(semi-infinite straight wire).} \qquad (30\text{-}9)$$

Magnetic Field Due to a Current in a Circular Arc of Wire

To find the magnetic field produced at a point by a current in a curved wire, we would again use Eq. 30-3 to write the magnitude of the field produced by a single current-length element, and we would again integrate to find the net field produced by all the current-length elements. That integration can be difficult, depending on the shape of the wire; it is fairly straightforward, however, when the wire is a circular arc and the point is the center of curvature.

Figure 30-6a shows such an arc-shaped wire with central angle ϕ, radius R, and center C, carrying current i. At C, each current-length element $i \, d\vec{s}$ of the wire produces a magnetic field of magnitude dB given by Eq. 30-3. Moreover, as Fig. 30-6b shows, no matter where the element is located on the wire, the angle θ between the vectors $d\vec{s}$ and $\vec{r}$ is 90°; also, $r = R$. Thus, by substituting R for r and 90° for θ, we obtain from Eq. 30-3,

$$dB = \frac{\mu_0}{4\pi} \frac{i \, ds \sin 90°}{R^2} = \frac{\mu_0}{4\pi} \frac{i \, ds}{R^2}. \qquad (30\text{-}10)$$

The field at C due to each current-length element in the arc has this magnitude.

An application of the right-hand rule anywhere along the wire (as in Fig. 30-6c) will show that all the differential fields $d\vec{B}$ have the same direction at C—directly out of the page. Thus, the total field at C is simply the sum (via integration) of all the differential fields $d\vec{B}$. We use the identity $ds = R \, d\phi$ to change the variable of integration from ds to $d\phi$ and obtain, from Eq. 30-10,

$$B = \int dB = \int_0^\phi \frac{\mu_0}{4\pi} \frac{iR \, d\phi}{R^2} = \frac{\mu_0 i}{4\pi R} \int_0^\phi d\phi.$$

Integrating, we find that

$$B = \frac{\mu_0 i \phi}{4\pi R} \qquad \text{(at center of circular arc).} \qquad (30\text{-}11)$$

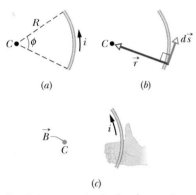

Fig. 30-6 (a) A wire in the shape of a circular arc with center C carries current i. (b) For any element of wire along the arc, the angle between the directions of $d\vec{s}$ and $\vec{r}$ is 90°. (c) Determining the direction of the magnetic field at the center C due to the current in the wire; the field is out of the page, in the direction of the fingertips, as indicated by the colored dot at C.

Note that this equation gives us the magnetic field *only* at the center of curvature of a circular arc of current. When you insert data into the equation, you must be careful to express ϕ in radians rather than degrees. For example, to find the magnitude of the magnetic field at the center of a full circle of current, you would substitute 2π rad for ϕ in Eq. 30-11, finding

$$B = \frac{\mu_0 i (2\pi)}{4\pi R} = \frac{\mu_0 i}{2R} \qquad \text{(at center of full circle).} \qquad (30\text{-}12)$$

Sample Problem 30-1

The wire in Fig. 30-7a carries a current i and consists of a circular arc of radius R and central angle $\pi/2$ rad, and two straight sections whose extensions intersect the center C of the arc. What magnetic field $\vec{B}$ does the current produce at C?

SOLUTION: One Key Idea here is that we can find the magnetic field $\vec{B}$ at point C by applying the Biot–Savart law of Eq. 30-5 to the wire. A second Key Idea is that the application of Eq. 30-5 can be simplified by evaluating $\vec{B}$ separately for the three distinguishable sections of the wire—namely, (1) the straight section at the left, (2) the straight section at the right, and (3) the circular arc.

Straight sections. For any current-length element in section 1, the angle θ between $d\vec{s}$ and $\vec{r}$ is zero (Fig. 30-7b), so Eq. 30-3 gives us

$$dB_1 = \frac{\mu_0}{4\pi} \frac{i\,ds\,\sin\theta}{r^2} = \frac{\mu_0}{4\pi} \frac{i\,ds\,\sin 0}{r^2} = 0.$$

Thus, the current along the entire length of wire in straight section 1 contributes no magnetic field at C:

$$B_1 = 0.$$

The same situation prevails in straight section 2, where the angle θ between $d\vec{s}$ and $\vec{r}$ for any current-length element is $180°$. Thus,

$$B_2 = 0.$$

Circular arc. The Key Idea here is that application of the Biot–Savart law to evaluate the magnetic field at the center of a circular arc leads to Eq. 30-11 ($B = \mu_0 i\phi/4\pi R$). Here the central angle ϕ of the arc is $\pi/2$ rad. Thus from Eq. 30-11, the magnitude of the magnetic field $\vec{B}_3$ at the arc's center C is

$$B_3 = \frac{\mu_0 i(\pi/2)}{4\pi R} = \frac{\mu_0 i}{8R}.$$

To find the direction of $\vec{B}_3$, we apply the right-hand rule displayed in Fig. 30-4. Mentally grasp the circular arc with your right hand as in Fig. 30-7c, with your thumb in the direction of the current. The direction in which your fingers curl around the wire indicates the direction of the magnetic field lines around the wire. In the region of point C (inside the circular arc), your fingertips point *into the plane* of the page. Thus, $\vec{B}_3$ is directed into that plane.

Net field. Generally, when we must combine two or more magnetic fields to find the net magnetic field, we must combine the fields as vectors and not simply add their magnitudes. Here, however, only the circular arc produces a magnetic field at point C.

(a)

(b)

(c)

Fig. 30-7 Sample Problem 30-1. (*a*) A wire consists of two straight sections (1 and 2) and a circular arc (3), and carries current i. (*b*) For a current-length element in section 1, the angle between $d\vec{s}$ and $\vec{r}$ is zero. (*c*) Determining the direction of magnetic field $\vec{B}_3$ at C due to the current in the circular arc; the field is into the page there.

Thus, we can write the magnitude of the net field $\vec{B}$ as

$$B = B_1 + B_2 + B_3 = 0 + 0 + \frac{\mu_0 i}{8R} = \frac{\mu_0 i}{8R}. \quad \text{(Answer)}$$

The direction of $\vec{B}$ is the direction of $\vec{B}_3$—namely, into the plane of Fig. 30-7.

✔ **CHECKPOINT 1:** The figure here shows three circuits consisting of concentric circular arcs (either half- or quarter-circles of radii r, $2r$, and $3r$) and radial lengths. The circuits carry the same current. Rank them according to the magnitude of the magnetic field produced at the center of curvature (the dot), greatest first.

(a)   (b)   (c)

Sample Problem 30-2

Figure 30-8a shows two long parallel wires carrying currents i_1 and i_2 in opposite directions. What are the magnitude and direction of the net magnetic field at point P? Assume the following values: $i_1 = 15$ A, $i_2 = 32$ A, and $d = 5.3$ cm.

SOLUTION: One Key Idea here is that the net magnetic field $\vec{B}$ at point P is the vector sum of the magnetic fields due to the currents in the two wires. A second Key Idea is that we can find the magnetic field due to any current by applying the Biot–Savart law to the

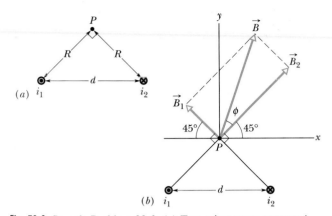

Fig. 30-8 Sample Problem 30-2. (*a*) Two wires carry currents i_1 and i_2 in opposite directions (out of and into the page). Note the right angle at P. (*b*) The separate fields $\vec{B}_1$ and $\vec{B}_2$ are combined vectorially to yield the net field $\vec{B}$.

current. For points near the current in a long straight wire, that law leads to Eq. 30-6.

In Fig. 30-8*a*, point P is distance R from both currents i_1 and i_2. Thus, Eq. 30-6 tells us that at point P those currents produce magnetic fields $\vec{B}_1$ and $\vec{B}_2$ with magnitudes

$$B_1 = \frac{\mu_0 i_1}{2\pi R} \quad \text{and} \quad B_2 = \frac{\mu_0 i_2}{2\pi R}.$$

In the right triangle of Fig. 30-8*a*, note that the base angles (between sides R and d) are both 45°. Thus, we may write cos 45° = R/d and replace R with d cos 45°. Then the field magnitudes B_1 and B_2 become

$$B_1 = \frac{\mu_0 i_1}{2\pi d \cos 45°} \quad \text{and} \quad B_2 = \frac{\mu_0 i_2}{2\pi d \cos 45°}.$$

We want to combine $\vec{B}_1$ and $\vec{B}_2$ to find their vector sum, which is the net field $\vec{B}$ at P. To find the directions of $\vec{B}_1$ and $\vec{B}_2$, we apply the right-hand rule of Fig. 30-4 to each current in Fig. 30-8*a*. For wire 1, with current out of the page, we mentally grasp the wire

with the right hand, with the thumb pointing out of the page. Then the curled fingers indicate that the field lines run counterclockwise. In particular, in the region of point P, they are directed upward to the left. Recall that the magnetic field at a point near a long, straight current-carrying wire must be directed perpendicular to a radial line between the point and the current. Thus, $\vec{B}_1$ must be directed upward to the left as drawn in Fig. 30-8*b*. (Note carefully the perpendicular symbol between vector $\vec{B}_1$ and the line connecting point P and wire 1.)

Repeating this analysis for the current in wire 2, we find that $\vec{B}_2$ is directed upward to the right as drawn in Fig. 30-8*b*. (Note the perpendicular symbol between vector $\vec{B}_2$ and the line connecting point P and wire 2.)

We can now vectorially add $\vec{B}_1$ and $\vec{B}_2$ to find the net magnetic field $\vec{B}$ at point P, either by using a vector-capable calculator or by resolving the vectors into components and then combining the components of $\vec{B}$. However, in Fig. 30-8*b*, there is a third method: Because $\vec{B}_1$ and $\vec{B}_2$ are perpendicular to each other, they form the legs of a right triangle, with $\vec{B}$ as the hypotenuse. The Pythagorean theorem then gives us

$$B = \sqrt{B_1^2 + B_2^2} = \frac{\mu_0}{2\pi d(\cos 45°)}\sqrt{i_1^2 + i_2^2}$$

$$= \frac{(4\pi \times 10^{-7} \text{ T}\cdot\text{m/A})\sqrt{(15 \text{ A})^2 + (32 \text{ A})^2}}{(2\pi)(5.3 \times 10^{-2} \text{ m})(\cos 45°)}$$

$$= 1.89 \times 10^{-4} \text{ T} \approx 190 \ \mu\text{T}. \qquad \text{(Answer)}$$

The angle ϕ between the directions of $\vec{B}$ and $\vec{B}_2$ in Fig. 30-8*b* follows from

$$\phi = \tan^{-1}\frac{B_1}{B_2},$$

which, with B_1 and B_2 as given above, yields

$$\phi = \tan^{-1}\frac{i_1}{i_2} = \tan^{-1}\frac{15 \text{ A}}{32 \text{ A}} = 25°.$$

The angle between the direction of $\vec{B}$ and the x axis shown in Fig. 30-8*b* is then

$$\phi + 45° = 25° + 45° = 70°. \qquad \text{(Answer)}$$

PROBLEM-SOLVING TACTICS

Tactic 1: *Right-Hand Rules*

To help you sort out the right-hand rules you have now seen (and the ones coming up), here is a review.

Right-Hand Rule for Cross Products. Introduced in Section 3-7, this is a way to determine the direction of the vector that results from a cross product. You point the fingers of your right hand so as to sweep the first vector expressed in the product into the second vector, through the smaller angle between the two vectors. Your outstretched thumb gives you the direction of the vector resulting from the cross product. In Chapter 12, we used this right-hand rule to find the directions of torque and angular momentum vectors; in Chapter 29, we used it to find the direction of the force on a current-carrying wire in a magnetic field.

Curled–Straight Right-Hand Rules for Magnetism. In many situations involving magnetism, you need to relate a "curled"

element and a "straight" element. You can do so with the (curled) fingers and the (straight) thumb on your right hand. You have already seen an example in Section 29-8, in which we related the current around a loop (curled element) to the normal vector $\vec{n}$ (straight element) of the loop: You curl the fingers of your right hand around in the direction of the current along the loop; your outstretched thumb then gives the direction of $\vec{n}$. This is also the direction of the magnetic dipole moment $\vec{\mu}$ of the loop.

In this section you were introduced to a second curled–straight right-hand rule. To determine the direction of the magnetic field lines around a current-length element, you point the outstretched thumb of your right hand in the direction of the current. The fingers then curl around the current-length element in the direction of the field lines.

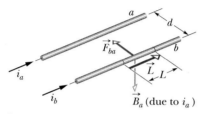

Fig. 30-9 Two parallel wires carrying currents in the same direction attract each other. $\vec{B}_a$ is the magnetic field at wire b produced by the current in wire a. $\vec{F}_{ba}$ is the resulting force acting on wire b because it carries current in $\vec{B}_a$.

30-2 Force Between Two Parallel Currents

Two long parallel wires carrying currents exert forces on each other. Figure 30-9 shows two such wires, separated by a distance d and carrying currents i_a and i_b. Let us analyze the forces on these wires due to each other.

We seek first the force on wire b in Fig. 30-9 due to the current in wire a. That current produces a magnetic field $\vec{B}_a$, and it is this magnetic field that actually causes the force we seek. To find the force, then, we need the magnitude and direction of the field $\vec{B}_a$ *at the site of wire b.* The magnitude of $\vec{B}_a$ at every point of wire b is, from Eq. 30-6,

$$B_a = \frac{\mu_0 i_a}{2\pi d}. \tag{30-13}$$

A (curled–straight) right-hand rule tells us that the direction of $\vec{B}_a$ at wire b is down, as Fig. 30-9 shows.

Now that we have the field, we can find the force it produces on wire b. Equation 29-27 tells us that the force $\vec{F}_{ba}$ on a length L of wire b due to the external magnetic field $\vec{B}_a$ is

$$\vec{F}_{ba} = i_b \vec{L} \times \vec{B}_a, \tag{30-14}$$

where $\vec{L}$ is the length vector of the wire. In Fig. 30-9, vectors $\vec{L}$ and $\vec{B}_a$ are perpendicular, so with Eq. 30-13, we can write

$$F_{ba} = i_b L B_a \sin 90° = \frac{\mu_0 L i_a i_b}{2\pi d}. \tag{30-15}$$

The direction of $\vec{F}_{ba}$ is the direction of the cross product $\vec{L} \times \vec{B}_a$. Applying the right-hand rule for cross products to $\vec{L}$ and $\vec{B}_a$ in Fig. 30-9, we see that $\vec{F}_{ba}$ is directly toward wire a, as shown.

The general procedure for finding the force on a current-carrying wire is this:

> To find the force on a current-carrying wire due to a second current-carrying wire, first find the field due to the second wire at the site of the first wire. Then find the force on the first wire due to that field.

We could now use this procedure to compute the force on wire a due to the current in wire b. We would find that the force is directly toward wire b; hence, the two wires with parallel currents attract each other. Similarly, if the two currents were antiparallel, we could show that the two wires repel each other. Thus,

> Parallel currents attract, and antiparallel currents repel.

The force acting between currents in parallel wires is the basis for the definition of the ampere, which is one of the seven SI base units. The definition, adopted in 1946, is this: The ampere is that constant current which, if maintained in two straight, parallel conductors of infinite length, of negligible circular cross section, and placed 1 m apart in vacuum, would produce on each of these conductors a force of magnitude 2×10^{-7} newton per meter of length.

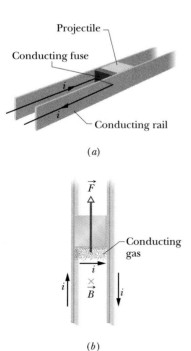

Fig. 30-10 (a) A rail gun, as a current i is set up in it. The current rapidly causes the conducting fuse to vaporize. (b) The current produces a magnetic field $\vec{B}$ between the rails, and the field causes a force $\vec{F}$ to act on the conducting gas, which is part of the current path. The gas propels the projectile along the rails, launching it.

Rail Gun

A rail gun is a device in which a magnetic force can accelerate a projectile to a high speed in a short time. The basics of a rail gun are shown in Fig. 30-10a. A large current is sent out along one of two parallel conducting rails, across a conducting "fuse" (such as a narrow piece of copper) between the rails, and then back to the current source along the second rail. The projectile to be fired lies on the far side of the fuse and fits loosely between the rails. Immediately after the current begins, the fuse element melts and vaporizes, creating a conducting gas between the rails where the fuse had been.

The curled–straight right-hand rule of Fig. 30-4 reveals that the currents in the rails of Fig. 30-10*a* produce magnetic fields that are directed downward between the rails. The net magnetic field $\vec{B}$ exerts a force $\vec{F}$ on the gas due to the current i through the gas (Fig. 30-10*b*). With Eq. 30-14 and the right-hand rule for cross products, we find that $\vec{F}$ points outward along the rails. As the gas is forced outward along the rails, it pushes the projectile, accelerating it by as much as $5 \times 10^{6}g$, and then launches it with a speed of 10 km/s, all within 1 ms.

✓**CHECKPOINT 2:**   The figure here shows three long, straight, parallel, equally spaced wires with identical currents either into or out of the page. Rank the wires according to the magnitude of the force on each due to the currents in the other two wires, greatest first.

$\odot \quad \odot \quad \otimes$
$a \quad\quad b \quad\quad c$

30-3 Ampere's Law

We can find the net electric field due to *any* distribution of charges with the inverse-square law for the differential field $d\vec{E}$ (Eq. 30-2), but if the distribution is complicated, we may have to use a computer. Recall, however, that if the distribution has planar, cylindrical, or spherical symmetry, we can apply Gauss' law to find the net electric field with considerably less effort.

Similarly, we can find the net magnetic field due to *any* distribution of currents with the inverse-square law for the differential field $d\vec{B}$ (Eq. 30-5), but again we may have to use a computer for a complicated distribution. However, if the distribution has some symmetry, we may be able to apply **Ampere's law** to find the magnetic field with considerably less effort. This law, which can be derived from the Biot–Savart law, has traditionally been credited to André Marie Ampère (1775–1836), for whom the SI unit of current is named. However, the law actually was advanced by English physicist James Clerk Maxwell.

Ampere's law is

$$\oint \vec{B} \cdot d\vec{s} = \mu_0 i_{\text{enc}} \qquad \text{(Ampere's law).} \qquad (30\text{-}16)$$

The circle on the integral sign means that the scalar (or dot) product $\vec{B} \cdot d\vec{s}$ is to be integrated around a *closed* loop, called an *Amperian loop*. The current i_{enc} on the right is the *net* current encircled by that loop.

To see the meaning of the scalar product $\vec{B} \cdot d\vec{s}$ and its integral, let us first apply Ampere's law to the general situation of Fig. 30-11. The figure shows cross sections of three long straight wires that carry currents i_1, i_2, and i_3 either directly into or directly out of the page. An arbitrary Amperian loop lying in the plane of the page encircles two of the currents but not the third. The counterclockwise direction marked on the loop indicates the arbitrarily chosen direction of integration for Eq. 30-16.

To apply Ampere's law, we mentally divide the loop into differential vector elements $d\vec{s}$ that are everywhere directed along the tangent to the loop in the direction of integration. Assume that at the location of the element $d\vec{s}$ shown in Fig. 30-11, the net magnetic field due to the three currents is $\vec{B}$. Because the wires are perpendicular to the page, we know that the magnetic field at $d\vec{s}$ due to each current

Fig. 30-11 Ampere's law applied to an arbitrary Amperian loop that encircles two long straight wires but excludes a third wire. Note the directions of the currents.

is in the plane of Fig. 30-11; thus, their net magnetic field $\vec{B}$ at $d\vec{s}$ must also be in that plane. However, we do not know the orientation of $\vec{B}$ within the plane. In Fig. 30-11, $\vec{B}$ is arbitrarily drawn at an angle θ to the direction of $d\vec{s}$.

The scalar product $\vec{B} \cdot d\vec{s}$ on the left side of Eq. 30-16 is then equal to $B \cos \theta \, ds$. Thus, Ampere's law can be written as

$$\oint \vec{B} \cdot d\vec{s} = \oint B \cos \theta \, ds = \mu_0 i_{\text{enc}}. \tag{30-17}$$

We can now interpret the scalar product $\vec{B} \cdot d\vec{s}$ as being the product of a length ds of the Amperian loop and the field component $B \cos \theta$ that is tangent to the loop. Then we can interpret the integration as being the summation of all such products around the entire loop.

When we can actually perform this integration, we do not need to know the direction of $\vec{B}$ before integrating. Instead, we arbitrarily assume $\vec{B}$ to be generally in the direction of integration (as in Fig. 30-11). Then we use the following curled–straight right-hand rule to assign a plus sign or a minus sign to each of the currents that make up the net encircled current i_{enc}:

> ▶ Curl your right hand around the Amperian loop, with the fingers pointing in the direction of integration. A current through the loop in the general direction of your outstretched thumb is assigned a plus sign, and a current generally in the opposite direction is assigned a minus sign.

Finally, we solve Eq. 30-17 for the magnitude of $\vec{B}$. If B turns out positive, then the direction we assumed for $\vec{B}$ is correct. If it turns out negative, we neglect the minus sign and redraw $\vec{B}$ in the opposite direction.

In Fig. 30-12 we apply the curled–straight rule for Ampere's law to the situation of Fig. 30-11. With the indicated counterclockwise direction of integration, the net current encircled by the loop is

$$i_{\text{enc}} = i_1 - i_2.$$

(Current i_3 is not encircled by the loop.) We can then rewrite Eq. 30-17 as

$$\oint B \cos \theta \, ds = \mu_0(i_1 - i_2). \tag{30-18}$$

You might wonder why, since current i_3 contributes to the magnetic-field magnitude B on the left side of Eq. 30-18, it is not needed on the right side. The answer is that the contributions of current i_3 to the magnetic field cancel out because the integration in Eq. 30-18 is made around the full loop. In contrast, the contributions of an encircled current to the magnetic field do not cancel out.

We cannot solve Eq. 30-18 for the magnitude B of the magnetic field, because for the situation of Fig. 30-11 we do not have enough information to simplify and solve the integral. However, we do know the outcome of the integration; it must be equal to the value of $\mu_0(i_1 - i_2)$, which is set by the net current passing through the loop.

We shall now apply Ampere's law to two situations in which symmetry does allow us to simplify and solve the integral, hence to find the magnetic field.

Fig. 30-12 A right-hand rule for Ampere's law, to determine the signs for currents encircled by an Amperian loop. The situation is that of Fig. 30-11.

The Magnetic Field Outside a Long Straight Wire with Current

Figure 30-13 shows a long straight wire that carries current i directly out of the page. Equation 30-6 tells us that the magnetic field $\vec{B}$ produced by the current has the same

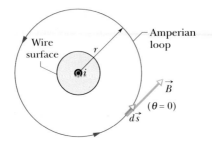

Fig. 30-13 Using Ampere's law to find the magnetic field produced by a current i in a long straight wire. The Amperian loop is a concentric circle that lies outside the wire.

magnitude at all points that are the same distance r from the wire; that is, the field $\vec{B}$ has cylindrical symmetry about the wire. We can take advantage of that symmetry to simplify the integral in Ampere's law (Eqs. 30-16 and 30-17) if we encircle the wire with a concentric circular Amperian loop of radius r, as in Fig. 30-13. The magnetic field $\vec{B}$ then has the same magnitude B at every point on the loop. We shall integrate counterclockwise, so that $d\vec{s}$ has the direction shown in Fig. 30-13.

We can further simplify the quantity $B \cos \theta$ in Eq. 30-17 by noting that $\vec{B}$ is tangent to the loop at every point along the loop, as is $d\vec{s}$. Thus, $\vec{B}$ and $d\vec{s}$ are either parallel or antiparallel at each point of the loop, and we shall arbitrarily assume the former. Then at every point the angle θ between $d\vec{s}$ and $\vec{B}$ is 0°, so $\cos \theta = \cos 0° = 1$. The integral in Eq. 30-17 then becomes

$$\oint \vec{B} \cdot d\vec{s} = \oint B \cos \theta \, ds = B \oint ds = B(2\pi r).$$

Note that $\oint ds$ above is the summation of all the line segment lengths ds around the circular loop; that is, it simply gives the circumference $2\pi r$ of the loop.

Our right-hand rule gives us a plus sign for the current of Fig. 30-13. The right side of Ampere's law becomes $+\mu_0 i$ and we then have

$$B(2\pi r) = \mu_0 i$$

or

$$B = \frac{\mu_0 i}{2\pi r}. \tag{30-19}$$

With a slight change in notation, this is Eq. 30-6, which we derived earlier—with considerably more effort—using the law of Biot and Savart. In addition, because the magnitude B turned out positive, we know that the correct direction of $\vec{B}$ must be the one shown in Fig. 30-13.

The Magnetic Field Inside a Long Straight Wire with Current

Figure 30-14 shows the cross section of a long straight wire of radius R that carries a uniformly distributed current i directly out of the page. Because the current is uniformly distributed over a cross section of the wire, the magnetic field $\vec{B}$ that it produces must be cylindrically symmetrical. Thus, to find the magnetic field at points inside the wire, we can again use an Amperian loop of radius r, as shown in Fig. 30-14, where now $r < R$. Symmetry again suggests that $\vec{B}$ is tangent to the loop, as shown, so the left side of Ampere's law again yields

$$\oint \vec{B} \cdot d\vec{s} = B \oint ds = B(2\pi r). \tag{30-20}$$

To find the right side of Ampere's law, we note that because the current is uniformly distributed, the current i_{enc} encircled by the loop is proportional to the area encircled by the loop; that is,

$$i_{enc} = i \frac{\pi r^2}{\pi R^2}. \tag{30-21}$$

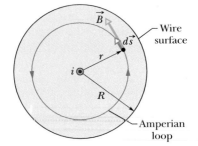

Fig. 30-14 Using Ampere's law to find the magnetic field that a current i produces inside a long straight wire of circular cross section. The current is uniformly distributed over the cross section of the wire and emerges from the page. An Amperian loop is drawn inside the wire.

Our right-hand rule tells us that i_{enc} gets a plus sign. Then Ampere's law gives us

$$B(2\pi r) = \mu_0 i \frac{\pi r^2}{\pi R^2}$$

or

$$B = \left(\frac{\mu_0 i}{2\pi R^2} \right) r. \tag{30-22}$$

Thus, inside the wire, the magnitude B of the magnetic field is proportional to r; that magnitude is zero at the center and a maximum at the surface, where $r = R$. Note that Eqs. 30-19 and 30-22 give the same value for B at $r = R$; that is, the expressions for the magnetic field outside the wire and inside the wire yield the same result at the surface of the wire.

✔CHECKPOINT 3:   The figure here shows three equal currents i (two parallel and one antiparallel) and four Amperian loops. Rank the loops according to the magnitude of $\oint \vec{B} \cdot d\vec{s}$ along each, greatest first.

Sample Problem 30-3

Figure 30-15a shows the cross section of a long conducting cylinder with inner radius $a = 2.0$ cm and outer radius $b = 4.0$ cm. The cylinder carries a current out of the page, and the current density in the cross section is given by $J = cr^2$, with $c = 3.0 \times 10^6$ A/m^4 and r in meters. What is the magnetic field $\vec{B}$ at a point that is 3.0 cm from the central axis of the cylinder?

SOLUTION: The point at which we want to evaluate $\vec{B}$ is inside the material of the conducting cylinder, between its inner and outer radii. We note that the current distribution has cylindrical symmetry (it is the same all around the cross section for any given radius). Thus, the Key Idea here is that the symmetry allows us to use Ampere's law to find $\vec{B}$ at the point. We first draw the Amperian loop shown in Fig. 30-15b. The loop is concentric with the cylinder and has radius $r = 3.0$ cm, because we want to evaluate $\vec{B}$ at that distance from the cylinder's central axis.

Next, we must compute the current i_{enc} that is encircled by the Amperian loop. However, a second Key Idea is that we *cannot* set up a proportionality as in Eq. 30-21, because here the current is not uniformly distributed. Instead, following the procedure of Sample Problem 27-2b, we must integrate the current density from the cylinder's inner radius a to the loop radius r:

$$i_{\text{enc}} = \int J \, dA = \int_a^r cr^2 (2\pi r \, dr)$$

$$= 2\pi c \int_a^r r^3 \, dr = 2\pi c \left[\frac{r^4}{4} \right]_a^r$$

$$= \frac{\pi c(r^4 - a^4)}{2}.$$

The direction of integration indicated in Fig. 30-15b is (arbitrarily) clockwise. Applying the right-hand rule for Ampere's law to that loop, we find that we should take i_{enc} as negative because the current is directed out of the page but our thumb is directed into the page.

We next evaluate the left side of Ampere's law exactly as we did in Fig. 30-14, and we again obtain Eq. 30-20. Then Ampere's law,

$$\oint \vec{B} \cdot d\vec{s} = \mu_0 i_{\text{enc}},$$

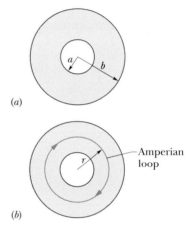

(a)

(b)

Fig. 30-15 Sample Problem 30-3. (a) Cross section of a conducting cylinder of inner radius a and outer radius b. (b) An Amperian loop of radius r is added to compute the magnetic field at points that are a distance r from the central axis.

gives us

$$B(2\pi r) = -\frac{\mu_0 \pi c}{2} (r^4 - a^4).$$

Solving for B and substituting known data yield

$$B = -\frac{\mu_0 c}{4r} (r^4 - a^4)$$

$$= -\frac{(4\pi \times 10^{-7} \text{ T} \cdot \text{m/A})(3.0 \times 10^6 \text{ A/m}^4)}{4(0.030 \text{ m})}$$

$$\times [(0.030 \text{ m})^4 - (0.020 \text{ m})^4]$$

$$= -2.0 \times 10^{-5} \text{ T}.$$

Thus, the magnetic field $\vec{B}$ at a point 3.0 cm from the central axis has the magnitude

$$B = 2.0 \times 10^{-5} \text{ T} \qquad \text{(Answer)}$$

and forms magnetic field lines that are directed opposite our direction of integration, hence counterclockwise in Fig. 30-15b.

30-4 Solenoids and Toroids

Magnetic Field of a Solenoid

Fig. 30-16 A solenoid carrying current i.

We now turn our attention to another situation in which Ampere's law proves useful. It concerns the magnetic field produced by the current in a long, tightly wound helical coil of wire. Such a coil is called a **solenoid** (Fig. 30-16). We assume that the length of the solenoid is much greater than the diameter.

Figure 30-17 shows a section through a portion of a "stretched-out" solenoid. The solenoid's magnetic field is the vector sum of the fields produced by the individual turns (loops) that make up the solenoid. For points very close to each turn, the wire behaves magnetically almost like a long straight wire, and the lines of $\vec{B}$ there are almost concentric circles. Figure 30-17 suggests that the field tends to cancel between adjacent turns. It also suggests that, at points inside the solenoid and reasonably far from the wire, $\vec{B}$ is approximately parallel to the (central) solenoid axis. In the limiting case of an *ideal solenoid,* which is infinitely long and consists of tightly packed (*close-packed*) turns of square wire, the field inside the coil is uniform and parallel to the solenoid axis.

At points above the solenoid, such as P in Fig. 30-17, the field set up by the upper parts of the solenoid turns (marked $\odot$) is directed to the left (as drawn near P) and tends to cancel the field set up by the lower parts of the turns (marked $\otimes$), which is directed to the right (not drawn). In the limiting case of an ideal solenoid, the magnetic field outside the solenoid is zero. Taking the external field to be zero is an excellent assumption for a real solenoid if its length is much greater than its diameter and if we consider external points such as point P that are not at either end of the solenoid. The direction of the magnetic field along the solenoid axis is given by a curled–straight right-hand rule: Grasp the solenoid with your right hand so that your fingers follow the direction of the current in the windings; your extended right thumb then points in the direction of the axial magnetic field.

Figure 30-18 shows the lines of $\vec{B}$ for a real solenoid. The spacing of the lines of $\vec{B}$ in the central region shows that the field inside the coil is fairly strong and uniform over the cross section of the coil. The external field, however, is relatively weak.

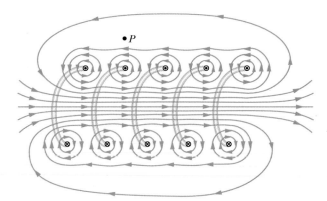

Fig. 30-17 A vertical cross section through the central axis of a "stretched-out" solenoid. The back portions of five turns are shown, as are the magnetic field lines due to a current through the solenoid. Each turn produces circular magnetic field lines near it. Near the solenoid's axis, the field lines combine into a net magnetic field that is directed along the axis. The closely spaced field lines there indicate a strong magnetic field. Outside the solenoid the field lines are widely spaced; the field there is very weak.

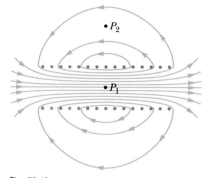

Fig. 30-18 Magnetic field lines for a real solenoid of finite length. The field is strong and uniform at interior points such as P_1 but relatively weak at external points such as P_2.

Let us now apply Ampere's law,

$$\oint \vec{B} \cdot d\vec{s} = \mu_0 i_{\text{enc}}, \qquad (30\text{-}23)$$

to the ideal solenoid of Fig. 30-19, where $\vec{B}$ is uniform within the solenoid and zero outside it, using the rectangular Amperian loop *abcda*. We write $\oint \vec{B} \cdot d\vec{s}$ as the sum of four integrals, one for each loop segment:

$$\oint \vec{B} \cdot d\vec{s} = \int_a^b \vec{B} \cdot d\vec{s} + \int_b^c \vec{B} \cdot d\vec{s}$$
$$+ \int_c^d \vec{B} \cdot d\vec{s} + \int_d^a \vec{B} \cdot d\vec{s}. \qquad (30\text{-}24)$$

The first integral on the right of Eq. 30-24 is Bh, where B is the magnitude of the uniform field $\vec{B}$ inside the solenoid and h is the (arbitrary) length of the segment from a to b. The second and fourth integrals are zero because for every element ds of these segments, $\vec{B}$ either is perpendicular to ds or is zero, and thus the product $\vec{B} \cdot d\vec{s}$ is zero. The third integral, which is taken along a segment that lies outside the solenoid, is zero because $B = 0$ at all external points. Thus, $\oint \vec{B} \cdot d\vec{s}$ for the entire rectangular loop has the value Bh.

The net current i_{enc} encircled by the rectangular Amperian loop in Fig. 30-19 is not the same as the current i in the solenoid windings because the windings pass more than once through this loop. Let n be the number of turns per unit length of the solenoid; then the loop encloses nh turns and

$$i_{\text{enc}} = i(nh).$$

Ampere's law then gives us

$$Bh = \mu_0 inh$$

or $$B = \mu_0 in \qquad \text{(ideal solenoid)}. \qquad (30\text{-}25)$$

Fig. 30-19 Application of Ampere's law to a section of a long ideal solenoid carrying a current i. The Amperian loop is the rectangle *abcd*.

Although we derived Eq. 30-25 for an infinitely long ideal solenoid, it holds quite well for actual solenoids if we apply it only at interior points, well away from the solenoid ends. Equation 30-25 is consistent with the experimental fact that the magnetic field magnitude B within a solenoid does not depend on the diameter or the length of the solenoid and that B is uniform over the solenoidal cross section. A solenoid thus provides a practical way to set up a known uniform magnetic field for experimentation, just as a parallel-plate capacitor provides a practical way to set up a known uniform electric field.

Magnetic Field of a Toroid

Figure 30-20a shows a **toroid**, which we may describe as a solenoid bent into the shape of a hollow doughnut. What magnetic field $\vec{B}$ is set up at its interior points (within the hollow of the doughnut)? We can find out from Ampere's law and the symmetry of the doughnut.

From the symmetry, we see that the lines of $\vec{B}$ form concentric circles inside the toroid, directed as shown in Fig. 30-20b. Let us choose a concentric circle of

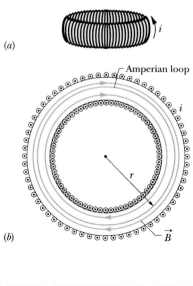

Fig. 30-20 (a) A toroid carrying a current i. (b) A horizontal cross section of the toroid. The interior magnetic field (inside the doughnut-shaped tube) can be found by applying Ampere's law with the Amperian loop shown.

radius r as an Amperian loop and traverse it in the clockwise direction. Ampere's law (Eq. 30-16) yields

$$(B)(2\pi r) = \mu_0 iN,$$

where i is the current in the toroid windings (and is positive for those windings enclosed by the Amperian loop) and N is the total number of turns. This gives

$$B = \frac{\mu_0 iN}{2\pi} \frac{1}{r} \qquad \text{(toroid).} \qquad (30\text{-}26)$$

In contrast to the situation for a solenoid, B is not constant over the cross section of a toroid. It is easy to show, with Ampere's law, that $B = 0$ for points outside an ideal toroid (as if the toroid were made from an ideal solenoid).

The direction of the magnetic field within a toroid follows from our curled–straight right-hand rule: Grasp the toroid with the fingers of your right hand curled in the direction of the current in the windings; your extended right thumb points in the direction of the magnetic field.

Sample Problem 30-4

A solenoid has length $L = 1.23$ m and inner diameter $d = 3.55$ cm, and it carries a current $i = 5.57$ A. It consists of five close-packed layers, each with 850 turns along length L. What is B at its center?

SOLUTION: One Key Idea here is that the magnitude B of the magnetic field along the solenoid's center is related to the solenoid's current i and number of turns per unit length n by Eq. 30-25. A second Key Idea is that B does not depend on the diameter of the windings, so the value of n for five identical layers is simply five times the value for each layer. Equation 30-25 then tells us

$$B = \mu_0 in = (4\pi \times 10^{-7} \text{ T} \cdot \text{m/A})(5.57 \text{ A}) \frac{5 \times 850 \text{ turns}}{1.23 \text{ m}}$$

$$= 2.42 \times 10^{-2} \text{ T} = 24.2 \text{ mT.} \qquad \text{(Answer)}$$

30-5 A Current-Carrying Coil as a Magnetic Dipole

So far we have examined the magnetic fields produced by current in a long straight wire, a solenoid, and a toroid. We turn our attention here to the field produced by a coil carrying a current. You saw in Section 29-9 that such a coil behaves as a magnetic dipole in that, if we place it in an external magnetic field $\vec{B}$, a torque $\vec{\tau}$ given by

$$\vec{\tau} = \vec{\mu} \times \vec{B} \qquad (30\text{-}27)$$

acts on it. Here $\vec{\mu}$ is the magnetic dipole moment of the coil and has the magnitude NiA, where N is the number of turns (or loops), i is the current in each turn, and A is the area enclosed by each turn.

Recall that the direction of $\vec{\mu}$ is given by a curled–straight right-hand rule: Grasp the coil so that the fingers of your right hand curl around it in the direction of the current; your extended thumb then points in the direction of the dipole moment $\vec{\mu}$.

Magnetic Field of a Coil

We turn now to the other aspect of a current-carrying coil as a magnetic dipole. What magnetic field does *it* produce at a point in the surrounding space? The problem does not have enough symmetry to make Ampere's law useful, so we must turn to

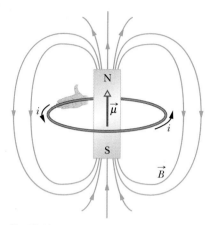

Fig. 30-21 A current loop produces a magnetic field like that of a bar magnet and thus has associated north and south poles. The magnetic dipole moment $\vec{\mu}$ of the loop, given by a curled–straight right-hand rule, points from the south pole to the north pole, in the direction of the field $\vec{B}$ within the loop.

the law of Biot and Savart. For simplicity, we first consider only a coil with a single circular loop and only points on its central axis, which we take to be a z axis. We shall show that the magnitude of the magnetic field at such points is

$$B(z) = \frac{\mu_0 iR^2}{2(R^2 + z^2)^{3/2}}, \tag{30-28}$$

in which R is the radius of the circular loop and z is the distance of the point in question from the center of the loop. Furthermore, the direction of the magnetic field $\vec{B}$ is the same as the direction of the magnetic dipole moment $\vec{\mu}$ of the loop.

For axial points far from the loop, we have $z \gg R$ in Eq. 30-28. With that approximation, the equation reduces to

$$B(z) \approx \frac{\mu_0 iR^2}{2z^3}.$$

Recalling that πR^2 is the area A of the loop and extending our result to include a coil of N turns, we can write this equation as

$$B(z) = \frac{\mu_0}{2\pi} \frac{NiA}{z^3}.$$

Further, since $\vec{B}$ and $\vec{\mu}$ have the same direction, we can write the equation in vector form, substituting from the identity $\mu = NiA$:

$$\vec{B}(z) = \frac{\mu_0}{2\pi} \frac{\vec{\mu}}{z^3} \quad \text{(current-carrying coil).} \tag{30-29}$$

Thus, we have two ways in which we can regard a current-carrying coil as a magnetic dipole: (1) it experiences a torque when we place it in an external magnetic field; (2) it generates its own intrinsic magnetic field, given, for distant points along its axis, by Eq. 30-29. Figure 30-21 shows the magnetic field of a current loop; one side of the loop acts as a north pole (in the direction of $\vec{\mu}$) and the other side as a south pole, as suggested by the lightly drawn magnet in the figure.

✔**CHECKPOINT 4:**   The figure here shows four arrangements of circular loops of radius r or $2r$, centered on vertical axes (perpendicular to the loops) and carrying identical currents in the directions indicated. Rank the arrangements according to the magnitude of the net magnetic field at the dot, midway between the loops on the central axis, greatest first.

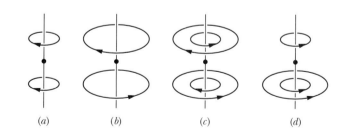

$(a) \qquad\qquad (b) \qquad\qquad (c) \qquad\qquad (d)$

Proof of Equation 30-28

Figure 30-22 shows the back half of a circular loop of radius R carrying a current i. Consider a point P on the axis of the loop, a distance z from its plane. Let us apply the law of Biot and Savart to a differential element ds of the loop, located at the left side of the loop. The length vector $\vec{ds}$ for this element points perpendicularly out of

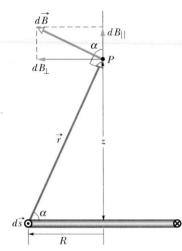

Fig. 30-22 A current loop of radius R. The plane of the loop is perpendicular to the page and only the back half of the loop is shown. We use the law of Biot and Savart to find the magnetic field at point P on the central axis of the loop.

the page. The angle θ between $d\vec{s}$ and $\vec{r}$ in Fig. 30-22 is 90°; the plane formed by these two vectors is perpendicular to the plane of the figure and contains both $\vec{r}$ and $d\vec{s}$. From the law of Biot and Savart (and the right-hand rule), the differential field $d\vec{B}$ produced at point P by the current in this element is perpendicular to this plane and thus is directed in the plane of the figure, perpendicular to $\vec{r}$, as indicated in Fig. 30-22.

Let us resolve $d\vec{B}$ into two components: $dB_{\parallel}$ along the axis of the loop and $dB_{\perp}$ perpendicular to this axis. From the symmetry, the vector sum of all the perpendicular components $dB_{\perp}$ due to all the loop elements ds is zero. This leaves only the axial components $dB_{\parallel}$ and we have

$$B = \int dB_{\parallel}.$$

For the element $d\vec{s}$ in Fig. 30-22, the law of Biot and Savart (Eq. 30-3) tells us that the magnetic field at distance r is

$$dB = \frac{\mu_0}{4\pi}\frac{i\,ds\,\sin 90°}{r^2}.$$

We also have

$$dB_{\parallel} = dB\cos\alpha.$$

Combining these two relations, we obtain

$$dB_{\parallel} = \frac{\mu_0 i \cos\alpha\, ds}{4\pi r^2}. \tag{30-30}$$

Figure 30-22 shows that r and α are not independent but are related to each other. Let us express each in terms of the variable z, the distance between point P and the center of the loop. The relations are

$$r = \sqrt{R^2 + z^2} \tag{30-31}$$

and

$$\cos\alpha = \frac{R}{r} = \frac{R}{\sqrt{R^2 + z^2}}. \tag{30-32}$$

Substituting Eqs. 30-31 and 30-32 into Eq. 30-30, we find

$$dB_{\parallel} = \frac{\mu_0 iR}{4\pi(R^2 + z^2)^{3/2}}\,ds.$$

Note that i, R, and z have the same values for all elements ds around the loop, so when we integrate this equation, we find that

$$B = \int dB_{\parallel}$$

$$= \frac{\mu_0 iR}{4\pi(R^2 + z^2)^{3/2}}\int ds$$

or, since $\int ds$ is simply the circumference $2\pi R$ of the loop,

$$B(z) = \frac{\mu_0 iR^2}{2(R^2 + z^2)^{3/2}}.$$

This is Eq. 30-28, the relation we sought to prove.

REVIEW & SUMMARY

The Biot–Savart Law   The magnetic field set up by a current-carrying conductor can be found from the *Biot–Savart law.* This law asserts that the contribution $d\vec{B}$ to the field produced by a current-length element $i\,d\vec{s}$ at a point P, a distance r from the current element, is

$$d\vec{B} = \frac{\mu_0}{4\pi}\frac{i\,d\vec{s} \times \vec{r}}{r^3} \qquad \text{(Biot–Savart law).} \qquad (30\text{-}5)$$

Here $\vec{r}$ is a vector that points from the element to P. The quantity μ_0, called the permeability constant, has the value $4\pi \times 10^{-7}$ T·m/A $\approx 1.26 \times 10^{-6}$ T·m/A.

Magnetic Field of a Long Straight Wire   For a long straight wire carrying a current i, the Biot–Savart law gives, for the magnitude of the magnetic field at a perpendicular distance R from the wire,

$$B = \frac{\mu_0 i}{2\pi R} \qquad \text{(long straight wire).} \qquad (30\text{-}6)$$

Magnetic Field of a Circular Arc   The magnitude of the magnetic field at the center of a circular arc, of radius R and central angle ϕ (in radians), carrying current i, is

$$B = \frac{\mu_0 i\phi}{4\pi R} \qquad \text{(at center of circular arc).} \qquad (30\text{-}11)$$

The Force Between Parallel Wires Carrying Currents   Parallel wires carrying currents in the same direction attract each other, whereas parallel wires carrying currents in opposite directions repel each other. The magnitude of the force on a length L of either wire is

$$F_{ba} = i_b L B_a \sin 90° = \frac{\mu_0 L i_a i_b}{2\pi d}, \qquad (30\text{-}15)$$

where d is the wire separation, and i_a and i_b are the currents in the wires.

Ampere's Law   **Ampere's law** states that

$$\oint \vec{B} \cdot d\vec{s} = \mu_0 i_{\text{enc}} \qquad \text{(Ampere's law).} \qquad (30\text{-}16)$$

The line integral in this equation is evaluated around a closed loop called an *Amperian loop.* The current i is the *net* current encircled by the loop. For some current distributions, Eq. 30-16 is easier to use than Eq. 30-5 to calculate the magnetic field due to the currents.

Fields of a Solenoid and a Toroid   Inside a *long solenoid* carrying current i, at points not near its ends, the magnitude B of the magnetic field is

$$B = \mu_0 i n \qquad \text{(ideal solenoid),} \qquad (30\text{-}25)$$

where n is the number of turns per unit length. At a point inside a *toroid,* the magnitude B of the magnetic field is

$$B = \frac{\mu_0 i N}{2\pi}\frac{1}{r} \qquad \text{(toroid),} \qquad (30\text{-}26)$$

where r is the distance from the center of the toroid to the point.

Field of a Magnetic Dipole   The magnetic field produced by a current-carrying coil, which is a *magnetic dipole,* at a point P located a distance z along the coil's central axis is parallel to the axis and is given by

$$\vec{B}(z) = \frac{\mu_0}{2\pi}\frac{\vec{\mu}}{z^3}, \qquad (30\text{-}29)$$

where $\vec{\mu}$ is the dipole moment of the coil. This equation applies only when z is much greater than the dimensions of the coil.

QUESTIONS

1. Figure 30-23 shows four arrangements in which long parallel wires carry equal currents directly into or out of the page at the corners of identical squares. Rank the arrangements according to the magnitude of the net magnetic field at the center of the square, greatest first.

(a)   *(b)*   *(c)*   *(d)*

Fig. 30-23 Question 1.

2. Figure 30-24 shows cross sections of two long straight wires; the left-hand wire carries current i_1 directly out of the page. If the net magnetic field due to the two currents is to be zero at point P, (a) should the direction of current i_2 in the right-hand wire be directly into or out of the page and (b) should i_2 be greater than, less than, or equal to i_1?

Fig. 30-24 Question 2.

3. Figure 30-25 shows three circuits, each consisting of two concentric circular arcs, one of radius r and the other of a larger radius

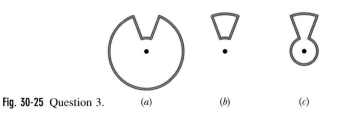

Fig. 30-25 Question 3.   *(a)*   *(b)*   *(c)*

R, and two radial lengths. The circuits have the same current through them, and the radial lengths have the same angle between them. Rank the circuits according to the magnitude of the net magnetic field at the center, greatest first.

4. Figure 30-26 shows four arrangements in which long, parallel, equally spaced wires carry equal currents directly into or out of the page. Rank the arrangements according to the magnitude of the net force on the central wire due to the currents in the other wires, greatest first.

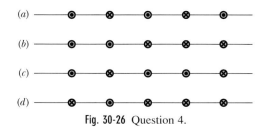

Fig. 30-26 Question 4.

5. Figure 30-27 shows three arrangements of three long straight wires, carrying equal currents directly into or out of the page. (a) Rank the arrangements according to the magnitude of the net force on the wire with the current directed out of the page due to the currents in the other wires, greatest first. (b) In arrangement 3, is the angle between the net force on that wire and the dashed line equal to, less than, or more than 45°?

Fig. 30-27 Question 5.

6. Figure 30-28 shows a uniform magnetic field $\vec{B}$ and four straight-line paths of equal lengths. Rank the paths according to the magnitude of $\int \vec{B} \cdot d\vec{s}$ taken along the paths, greatest first.

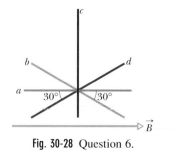

Fig. 30-28 Question 6.

7. Figure 30-29a shows four circular Amperian loops concentric with a wire whose current is directed out of the page. The current is uniform across the wire's circular cross section. Rank the loops according to the magnitude of $\oint \vec{B} \cdot d\vec{s}$ around each, greatest first.

8. Figure 30-29b shows four circular Amperian loops (red) and, in cross section, four long circular conductors (blue), all of which are concentric. Three of the conductors are hollow cylinders; the central conductor is a solid cylinder. The currents in the conductors are, from smallest radius to largest radius, 4 A out of the page, 9 A into the page, 5 A out of the page, and 3 A into the page. Rank the loops according to the magnitude of $\oint \vec{B} \cdot d\vec{s}$ around each, greatest first.

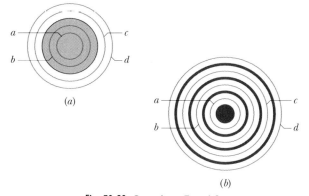

Fig. 30-29 Questions 7 and 8.

9. Figure 30-30 shows four identical currents i and five Amperian paths encircling them. Rank the paths according to the value of $\oint \vec{B} \cdot d\vec{s}$ taken in the directions shown, most positive first and most negative last.

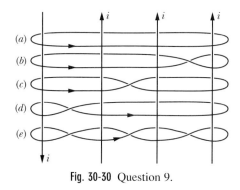

Fig. 30-30 Question 9.

10. The following table gives the number of turns per unit length n and the current i through six ideal solenoids of different radii. You want to combine several of them concentrically to produce a net magnetic field of zero along the central axis. Can this be done with (a) two of them, (b) three of them, (c) four of them, and (d) five of them? If so, answer by listing which solenoids are to be used and indicate the directions of the currents.

Solenoid:	1	2	3	4	5	6
n:	5	4	3	2	10	8
i:	5	3	7	6	2	3

EXERCISES & PROBLEMS

ssm   Solution is in the Student Solutions Manual.
www Solution is available on the World Wide Web at:
    http://www.wiley.com/college/hrw
ilw   Solution is available on the Interactive LearningWare.

SEC. 30-1   Calculating the Magnetic Field Due to a Current

1E. A surveyor is using a magnetic compass 6.1 m below a power line in which there is a steady current of 100 A. (a) What is the magnetic field at the site of the compass due to the power line? (b) Will this interfere seriously with the compass reading? The horizontal component of Earth's magnetic field at the site is 20 μT. ssm

2E. The electron gun in a traditional television tube fires electrons of kinetic energy 25 keV at the screen in a circular beam 0.22 mm in diameter; 5.6×10^{14} electrons arrive each second. Calculate the magnetic field produced by the beam at a point 1.5 mm from the beam axis.

3E. At a certain position in the Philippines, Earth's magnetic field of 39 μT is horizontal and directed due north. Suppose the net field is zero exactly 8.0 cm above a long, straight, horizontal wire that carries a constant current. What are (a) the magnitude and (b) the direction of the current? ssm

4E. A long wire carrying a current of 100 A is placed in a uniform external magnetic field of 5.0 mT. The wire is perpendicular to this magnetic field. Locate the points at which the net magnetic field is zero.

5E. A particle with positive charge q is a distance d from a long straight wire that carries a current i; the particle is traveling with speed v perpendicular to the wire. What are the direction and magnitude of the force on the particle if it is moving (a) toward and (b) away from the wire? ilw

6E. A straight conductor carrying a current i splits into identical semicircular arcs as shown in Fig. 30-31. What is the magnetic field at the center C of the resulting circular loop?

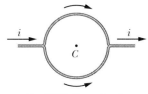

Fig. 30-31 Exercise 6.

7P. A wire carrying current i has the configuration shown in Fig. 30-32. Two semi-infinite straight sections, both tangent to the same circle, are connected by a circular arc, of central angle θ, along the circumference of the circle, with all sections lying in the same plane. What must θ be in order for B to be zero at the center of the circle? ssm

Fig. 30-32 Problem 7.

8P. Use the Biot–Savart law to calculate the magnetic field $\vec{B}$ at C, the common center of the semicircular arcs AD and HJ in Fig. 30-33a. The two arcs, of radii R_2 and R_1, respectively, form part of the circuit $ADJHA$ carrying current i.

9P. In the circuit of Fig. 30-33b, the curved segments are arcs of circles of radii a and b with common center P. The straight segments are along radii. Find the magnetic field $\vec{B}$ at point P, assuming a current i in the circuit. ssm

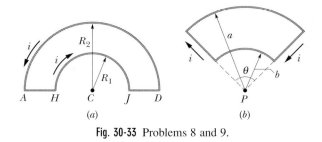

(a)   (b)

Fig. 30-33 Problems 8 and 9.

10P. The wire shown in Fig. 30-34 carries current i. What magnetic field $\vec{B}$ is produced at the center C of the semicircle by (a) each straight segment of length L, (b) the semicircular segment of radius R, and (c) the entire wire?

Fig. 30-34 Problem 10.

11P. In Fig. 30-35, a straight wire of length L carries current i. Show that the magnitude of the magnetic field $\vec{B}$ produced by this segment at P_1, a distance R from the segment along a perpendicular bisector, is

$$B = \frac{\mu_0 i}{2\pi R} \frac{L}{(L^2 + 4R^2)^{1/2}}.$$

Show that this expression for B reduces to an expected result as $L \to \infty$. ssm   www

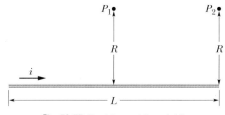

Fig. 30-35 Problems 11 and 13.

12P. A square loop of wire of edge length a carries current i. Using Problem 11, show that, at the center of the loop, the magnitude of the magnetic field produced by the current is

$$B = \frac{2\sqrt{2}\mu_0 i}{\pi a}.$$

13P. In Fig. 30-35, a straight wire of length L carries current i. Show that

$$B = \frac{\mu_0 i}{4\pi R} \frac{L}{(L^2 + R^2)^{1/2}}$$

gives the magnitude of the magnetic field $\vec{B}$ produced by the wire at P_2, a perpendicular distance R from one end of the wire. ssm

14P. Using Problem 11, show that the magnitude of the magnetic field produced at the center of a rectangular loop of wire of length l, and width W, carrying a current i, is

$$B = \frac{2\mu_0 i}{\pi} \frac{(L^2 + W^2)^{1/2}}{LW}.$$

15P. A square loop of wire of edge length a carries current i. Using Problem 11, show that the magnitude of the magnetic field produced at a point on the axis of the loop and a distance x from its center is

$$B(x) = \frac{4\mu_0 i a^2}{\pi(4x^2 + a^2)(4x^2 + 2a^2)^{1/2}}.$$

Prove that this result is consistent with the result of Problem 12. ssm

16P. In Fig. 30-36, a straight wire of length a carries a current i. Show that the magnitude of the magnetic field produced by the current at point P is $B = \sqrt{2}\mu_0 i/8\pi a$.

Fig. 30-36 Problem 16.

17P. Two wires, both of length L, are formed into a circle and a square, and each carries current i. Show that the square produces a greater magnetic field at its center than the circle produces at its center. (See Problem 12.) ssm

18P. Find the magnetic field $\vec{B}$ at point P in Fig. 30-37. (See Problem 16.)

Fig. 30-37 Problem 18.

19P. Figure 30-38 shows a cross section of a long thin ribbon of width w that is carrying a uniformly distributed total current i into the page. Calculate the magnitude and direction of the magnetic field $\vec{B}$ at a point P in the plane of the ribbon at a distance d from its edge. (*Hint:* Imagine the ribbon to be constructed from many long, thin, parallel wires.) ilw

Fig. 30-38 Problem 19.

20P. Find the magnetic field $\vec{B}$ at point P in Fig. 30-39 for $i = 10$ A and $a = 8.0$ cm. (See Problems 13 and 16.)

Fig. 30-39 Problem 20.

SEC. 30-2 Force Between Two Parallel Currents

21E. Two long parallel wires are 8.0 cm apart. What equal currents must be in the wires if the magnetic field halfway between them is to have a magnitude of 300 μT? Answer for both (a) parallel and (b) antiparallel currents. ssm

22E. Two long parallel wires a distance d apart carry currents of i and $3i$ in the same direction. Locate the point or points at which their magnetic fields cancel.

23E. Two long, straight, parallel wires, separated by 0.75 cm, are perpendicular to the plane of the page as shown in Fig. 30-40. Wire 1 carries a current of 6.5 A into the page. What must be the current (magnitude and direction) in wire 2 for the resultant magnetic field at point P to be zero?

Fig. 30-40 Exercise 23.

24E. Figure 30-41 shows five long parallel wires in the xy plane. Each wire carries a current $i = 3.00$ A in the positive x direction. The separation between adjacent wires is $d = 8.00$ cm. In unit-vector notation, what is the magnetic force per meter exerted on each of these five wires by the other wires?

Fig. 30-41 Exercise 24.

25P. Four long copper wires are parallel to each other, their cross sections forming the corners of a square with sides $a = 20$ cm. A 20 A current exists in each wire in the direction shown in Fig. 30-42. What are the magnitude and direction of $\vec{B}$ at the center of the square? ssm www

26P. Four identical parallel currents i are arranged to form a square of edge length a as in Fig. 30-42, *except* that they are *all* out of the page. What is the force per unit length (magnitude and direction) on any one wire?

Fig. 30-42 Problems 25, 26, and 27.

27P. In Fig. 30-42, what is the force per unit length acting on the lower left wire, in magnitude and direction, with the current directions as shown? The currents are i.

28P. Figure 30-43 is an idealized schematic drawing of a rail gun. Projectile P sits between two wide rails of circular cross section; a source of current sends current through the rails and through the (conducting) projectile itself (a fuse is not used). (a) Let w be the distance between the rails, R the radius of the rails, and i the current. Show that the force on the projectile is directed to the right along the rails and is given approximately by

$$F = \frac{i^2 \mu_0}{2\pi} \ln \frac{w + R}{R}.$$

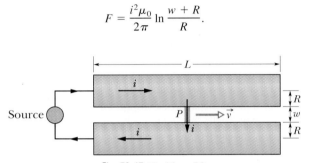

Fig. 30-43 Problem 28.

(b) If the projectile starts from the left end of the rails at rest, find the speed v at which it is expelled at the right. Assume that $i = 450$ kA, $w = 12$ mm, $R = 6.7$ cm, $L = 4.0$ m, and the mass of the projectile is $m = 10$ g.

29P. In Fig. 30-44, the long straight wire carries a current of 30 A and the rectangular loop carries a current of 20 A. Calculate the resultant force acting on the loop. Assume that $a = 1.0$ cm, $b = 8.0$ cm, and $L = 30$ cm. ilw

Fig. 30-44 Problem 29.

SEC. 30-3 Ampere's Law

30E. Eight wires cut the page perpendicularly at the points shown in Fig. 30-45. A wire labeled with the integer k ($k = 1, 2, \ldots, 8$) carries the current ki. For those with odd k, the current is out of the page; for those with even k, it is into the page. Evaluate $\oint \vec{B} \cdot d\vec{s}$ along the closed path in the direction shown.

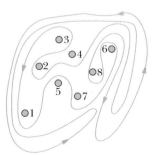

Fig. 30-45 Exercise 30.

31E. Each of the eight conductors in Fig. 30-46 carries 2.0 A of current into or out of the page. Two paths are indicated for the line integral $\oint \vec{B} \cdot d\vec{s}$. What is the value of the integral for the path (a) at the left and (b) at the right? ssm

Fig. 30-46 Exercise 31.

32E. Figure 30-47 shows a cross section of a long cylindrical conductor of radius a, carrying a uniformly distributed current i. Assume that $a = 2.0$ cm and $i = 100$ A, and plot $B(r)$ over the range $0 < r < 6.0$ cm.

Fig. 30-47 Exercise 32.

33P. Show that a uniform magnetic field $\vec{B}$ cannot drop abruptly to zero (as is suggested by the lack of field lines to the right of point a in Fig. 30-48) as one moves perpendicular to $\vec{B}$, say along the horizontal arrow in the figure. (*Hint:* Apply Ampere's law to the rectangular path shown by the dashed lines.) In actual magnets "fringing" of the magnetic field lines always occurs, which means that $\vec{B}$ approaches zero in a gradual manner. Modify the field lines in the figure to indicate a more realistic situation. ssm

Fig. 30-48 Problem 33.

34P. Two square conducting loops carry currents of 5.0 and 3.0 A as shown in Fig. 30-49. What is the value of $\oint \vec{B} \cdot d\vec{s}$ for each of the two closed paths shown?

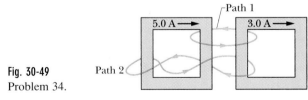

Fig. 30-49 Problem 34.

35P. The current density inside a long, solid, cylindrical wire of radius a is in the direction of the central axis and varies linearly with radial distance r from the axis according to $J = J_0 r/a$. Find the magnetic field inside the wire. ilw

36P. A long straight wire (radius = 3.0 mm) carries a constant current distributed uniformly over a cross section perpendicular to the axis of the wire. If the current density is 100 A/m², what are the magnitudes of the magnetic fields (a) 2.0 mm from the axis of the wire and (b) 4.0 mm from the axis of the wire?

37P. Figure 30-50 shows a cross section of a long cylindrical conductor of radius a containing a long cylindrical hole of radius b. The axes of the cylinder and hole are parallel and are a distance d apart; a current i is uniformly distributed over the tinted area.

Fig. 30-50 Problem 37.

(a) Use superposition to show that the magnetic field at the center of the hole is

$$B = \frac{\mu_0 i d}{2\pi(a^2 - b^2)}.$$

(b) Discuss the two special cases $b = 0$ and $d = 0$. (c) Use Ampere's law to show that the magnetic field in the hole is uniform. (*Hint:* Regard the cylindrical hole as resulting from the superposition of a complete cylinder (no hole) carrying a current in one direction and a cylinder of radius b carrying a current in the opposite direction, both cylinders having the same current density.) ssm www

38P. A long circular pipe with outside radius R carries a (uniformly distributed) current i into the page as shown in Fig. 30-51. A wire runs parallel to the pipe at a distance of $3R$ from center to center. Find the magnitude and direction of the current in the wire such that the net magnetic field at point P has the same magnitude as the net magnetic field at the center of the pipe but is in the opposite direction.

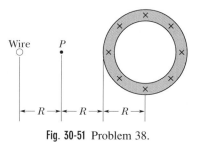

Fig. 30-51 Problem 38.

39P. Figure 30-52 shows a cross section of an infinite conducting sheet carrying a current per unit x-length of λ; the current emerges perpendicularly out of the page. (a) Use the Biot–Savart law and symmetry to show that for all points P above the sheet, and all points P' below it, the magnetic field $\vec{B}$ is parallel to the sheet and directed as shown. (b) Use Ampere's law to prove that $B = \frac{1}{2}\mu_0\lambda$ at all points P and P'. ssm

Fig. 30-52 Problems 39 and 44.

SEC. 30-4 Solenoids and Toroids

40E. A solenoid that is 95.0 cm long has a radius of 2.00 cm and a winding of 1200 turns; it carries a current of 3.60 A. Calculate the magnitude of the magnetic field inside the solenoid.

41E. A 200 turn solenoid having a length of 25 cm and a diameter of 10 cm carries a current of 0.30 A. Calculate the magnitude of the magnetic field $\vec{B}$ inside the solenoid. ssm

42E. A solenoid 1.30 m long and 2.60 cm in diameter carries a current of 18.0 A. The magnetic field inside the solenoid is 23.0 mT. Find the length of the wire forming the solenoid.

43E. A toroid having a square cross section, 5.00 cm on a side, and an inner radius of 15.0 cm has 500 turns and carries a current of 0.800 A. (It is made up of a square solenoid—instead of a round one as in Fig. 30-16—bent into a doughnut shape.) What is the magnetic field inside the toroid at (a) the inner radius and (b) the outer radius of the toroid? ssm

44P. Treat an ideal solenoid as a thin cylindrical conductor whose current per unit length, measured parallel to the cylinder axis, is λ. (a) By doing so, show that the magnitude of the magnetic field inside an ideal solenoid can be written as $B = \mu_0\lambda$. This is the value of the *change* in $\vec{B}$ that you encounter as you move from inside the solenoid to outside, through the solenoid wall. (b) Show that the same change occurs as you move through an infinite flat current sheet such as that of Fig. 30-52 (see Problem 39). Does this equality surprise you?

45P. In Section 30-4, we showed that the magnetic field at any radius r *inside* a toroid is given by

$$B = \frac{\mu_0 i N}{2\pi r}.$$

Show that as you move from any point just inside a toroid to a point just outside, the magnitude of the *change* in $\vec{B}$ that you encounter is just $\mu_0\lambda$. Here λ is the current per unit length along a circumference of radius r within the toroid. Compare this with the similar result found in Problem 44. Isn't the equality surprising? ssm

46P. A long solenoid has 100 turns/cm and carries current i. An electron moves within the solenoid in a circle of radius 2.30 cm perpendicular to the solenoid axis. The speed of the electron is $0.0460c$ (c = speed of light). Find the current i in the solenoid.

47P. A long solenoid with 10.0 turns/cm and a radius of 7.00 cm carries a current of 20.0 mA. A current of 6.00 A exists in a straight conductor located along the central axis of the solenoid. (a) At what radial distance from the axis will the direction of the resulting magnetic field be at 45.0° to the axial direction? (b) What is the magnitude of the magnetic field there? ssm ilw www

SEC. 30-5 A Current-Carrying Coil as a Magnetic Dipole

48E. Figure 30-53a shows a length of wire carrying a current i and bent into a circular coil of one turn. In Fig. 30-53b the same length of wire has been bent more sharply, to give a coil of two

Fig. 30-53 Exercise 48.

turns, each of half the original radius. (a) If B_a and B_b are the magnitudes of the magnetic fields at the centers of the two coils, what is the ratio B_b/B_a? (b) What is the ratio of the dipole moments, μ_b/μ_a, of the coils?

49E. What is the magnetic dipole moment $\vec{\mu}$ of the solenoid described in Exercise 41? ssm

50E. Figure 30-54 shows an arrangement known as a Helmholtz coil. It consists of two circular coaxial coils, each of N turns and radius R, separated by a distance R. The two coils carry equal currents i in the same direction. Find the magnitude of the net magnetic field at P, midway between the coils.

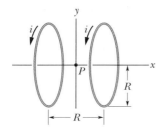

Fig. 30-54 Exercise 50; Problems 53 and 55.

51E. A student makes a short electromagnet by winding 300 turns of wire around a wooden cylinder of diameter $d = 5.0$ cm. The coil is connected to a battery producing a current of 4.0 A in the wire. (a) What is the magnetic moment of this device? (b) At what axial distance $z \gg d$ will the magnetic field of this dipole have the magnitude 5.0 μT (approximately one-tenth that of Earth's magnetic field)? ssm

52E. The magnitude $B(x)$ of the magnetic field at points on the axis of a square current loop of side a is given in Problem 15. (a) Show that the axial magnetic field of this loop, for $x \gg a$, is that of a magnetic dipole (see Eq. 30-29). (b) What is the magnetic dipole moment of this loop?

53P. Two 300-turn coils of radius R each carry a current i. They are arranged a distance R apart, as in Fig. 30-54. For $R = 5.0$ cm and $i = 50$ A, plot the magnitude B of the net magnetic field as a function of distance x along the common x axis over the range $x = -5$ cm to $x = +5$ cm, taking $x = 0$ at the midpoint P. (Such coils provide an especially uniform field $\vec{B}$ near point P.) (*Hint:* See Eq. 30-28.)

54P. A conductor carries a current of 6.0 A along the closed path *abcdefgha* involving 8 of the 12 edges of a cube of side 10 cm as shown in Fig. 30-55. (a) Why can one regard this as the superposition of three square loops: *bcfgb*, *abgha*, and *cdefc*? (*Hint:* Draw currents around those square loops.) (b) Use this superposition to find the magnetic dipole moment $\vec{\mu}$ (magnitude and direction) of the closed path. (c) Calculate $\vec{B}$ at the points $(x, y, z) = (0, 5.0$ m, $0)$ and $(5.0$ m, $0, 0)$.

55P. In Exercise 50 (Fig. 30-54), let the separation of the coils be a variable s (not necessarily equal to the coil radius R). (a) Show that the first derivative of the magnitude of the net magnetic field of the coils (dB/dx) vanishes at the midpoint P regardless of the value of s. Why would you expect this to be true from symmetry? (b) Show that the second derivative (d^2B/dx^2) also vanishes at P,

provided $s = R$. This accounts for the uniformity of B near P for this particular coil separation. ssm

Fig. 30-55 Problem 54.

56P. A length of wire is formed into a closed circuit with radii a and b, as shown in Fig. 30-56, and carries a current i. (a) What are the magnitude and direction of $\vec{B}$ at point P? (b) Find the magnetic dipole moment of the circuit.

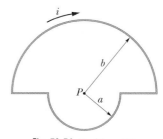

Fig. 30-56 Problem 56.

57P. A circular loop of radius 12 cm carries a current of 15 A. A flat coil of radius 0.82 cm, having 50 turns and a current of 1.3 A, is concentric with the loop. (a) What magnetic field $\vec{B}$ does the loop produce at its center? (b) What torque acts on the coil? Assume that the planes of the loop and coil are perpendicular and that the magnetic field due to the loop is essentially uniform throughout the volume occupied by the coil.

58P. (a) A long wire is bent into the shape shown in Fig. 30-57, without the wire actually touching itself at P. The radius of the circular section is R. Determine the magnitude and direction of $\vec{B}$ at the center C of the circular section when the current i is as indicated. (b) Suppose the circular section of the wire is rotated without distortion about the indicated diameter, until the plane of the circle is perpendicular to the straight sections of the wire. The magnetic dipole moment associated with the circular section is now in the direction of the current in the straight section of the wire. Determine $\vec{B}$ at C in this case.

Fig. 30-57 Problem 58.

31 Induction and Inductance

Soon after rock began in the mid-1950s, guitarists switched from acoustic guitars to electric guitars—but it was Jimi Hendrix who first understood the electric guitar as an electronic instrument. He exploded on the scene in the 1960s, ripping his pick along the strings, positioning himself and his guitar in front of a speaker to sustain feedback, and then laying down chords on top of the feedback. He shoved rock forward from the melodies of Buddy Holly into the psychedelia of the late 1960s and into the early heavy metal of Led Zeppelin and the raw energy of Joy Division in the 1970s, and his ideas continue to influence rock today.

What is it about an electric guitar that distinguishes it from an acoustic guitar and enabled Hendrix to make so much broader use of this electronic instrument?

The answer is in this chapter.

Fig. 31-1 A current meter registers a current in the wire loop when the magnet is moving with respect to the loop.

31-1 Two Symmetric Situations

In Section 29-8, we saw that if we put a closed conducting loop in a magnetic field and then send current through the loop, forces due to the magnetic field create a torque to turn the loop:

$$\text{current loop} + \text{magnetic field} \Rightarrow \text{torque}. \qquad (31\text{-}1)$$

Suppose that, instead, with the current off, we turn the loop by hand. Will the opposite of Eq. 31-1 occur? That is, will a current now appear in the loop:

$$\text{torque} + \text{magnetic field} \Rightarrow \text{current}? \qquad (31\text{-}2)$$

The answer is yes—a current does appear. The situations of Eqs. 31-1 and 31-2 are symmetric. The physical law on which Eq. 31-2 depends is called *Faraday's law of induction*. Whereas Eq. 31-1 is the basis for the electric motor, Eq. 31-2 and Faraday's law are the basis for the electric generator. This chapter is concerned with that law and the process it describes.

31-2 Two Experiments

Let us examine two simple experiments to prepare for our discussion of Faraday's law of induction.

First Experiment. Figure 31-1 shows a conducting loop connected to a sensitive current meter. Since there is no battery or other source of emf included, there is no current in the circuit. However, if we move a bar magnet toward the loop, a current suddenly appears in the circuit. The current disappears when the magnet stops. If we then move the magnet away, a current again suddenly appears, but now in the opposite direction. If we experimented for a while, we would discover the following:

1. A current appears only if there is relative motion between the loop and the magnet (one must move relative to the other); the current disappears when the relative motion between them ceases.

2. Faster motion produces a greater current.

3. If moving the magnet's north pole toward the loop causes, say, clockwise current, then moving the north pole away causes counterclockwise current. Moving the south pole toward or away from the loop also causes currents, but in the reversed directions.

The current produced in the loop is called an **induced current;** the work done per unit charge to produce that current (to move the conduction electrons that constitute the current) is called an **induced emf;** and the process of producing the current and emf is called **induction.**

Second Experiment. For this experiment we use the apparatus of Fig. 31-2, with the two conducting loops close to each other but not touching. If we close switch S, to turn on a current in the right-hand loop, the meter suddenly and briefly registers a current—an induced current—in the left-hand loop. If we then open the switch, another sudden and brief induced current appears in the left-hand loop, but in the opposite direction. We get an induced current (and thus an induced emf) only when the current in the right-hand loop is changing (either turning on or turning off) and not when it is constant (even if it is large).

The induced emf and induced current in these experiments are apparently caused when something changes—but what is that "something"? Faraday knew.

Fig. 31-2 The current meter registers a current in the left-hand wire loop just as switch S is closed (to turn on the current in the right-hand wire loop) or opened (to turn off the current in the right-hand loop). No motion of the coils is involved.

31-3   Faraday's Law of Induction

Faraday realized that an emf and a current can be induced in a loop, as in our two experiments, by changing the *amount of magnetic field* passing through the loop. He further realized that the "amount of magnetic field" can be visualized in terms of the magnetic field lines passing through the loop. **Faraday's law of induction,** stated in terms of our experiments, is this:

> ▶ An emf is induced in the loop at the left in Figs. 31-1 and 31-2 when the number of magnetic field lines that pass through the loop is changing.

The actual number of field lines passing through the loop does not matter; the values of the induced emf and induced current are determined by the *rate* at which that number changes.

In our first experiment (Fig. 31-1), the magnetic field lines spread out from the north pole of the magnet. Thus, as we move the north pole closer to the loop, the number of field lines passing through the loop increases. That increase apparently causes conduction electrons in the loop to move (the induced current) and provides energy (the induced emf) for their motion. When the magnet stops moving, the number of field lines through the loop no longer changes and the induced current and induced emf disappear.

In our second experiment (Fig. 31-2), when the switch is open (no current), there are no field lines. However, when we turn on the current in the right-hand loop, the increasing current builds up a magnetic field around that loop and at the left-hand loop. While the field builds, the number of magnetic field lines through the left-hand loop increases. As in the first experiment, the increase in field lines through that loop apparently induces a current and an emf there. When the current in the right-hand loop reaches a final, steady value, the number of field lines through the left-hand loop no longer changes, and the induced current and induced emf disappear.

Faraday's law does not explain *why* a current and an emf are induced in either experiment; it is just a statement that helps us visualize the induction.

A Quantitative Treatment

To put Faraday's law to work, we need a way to calculate the *amount of magnetic field* that passes through a loop. In Chapter 24, in a similar situation, we needed to calculate the amount of an electric field that passes through a surface. There we defined an electric flux $\Phi_E = \int \vec{E} \cdot d\vec{A}$. Here we define a *magnetic flux:* Suppose a loop enclosing an area A is placed in a magnetic field $\vec{B}$. Then the magnetic flux through the loop is

$$\Phi_B = \int \vec{B} \cdot d\vec{A} \qquad \text{(magnetic flux through area } A\text{).} \qquad (31\text{-}3)$$

As in Chapter 24, $d\vec{A}$ is a vector of magnitude dA that is perpendicular to a differential area dA.

As a special case of Eq. 31-3, suppose that the loop lies in a plane and that the magnetic field is perpendicular to the plane of the loop. Then we can write the dot product in Eq. 31-3 as $B \, dA \cos 0° = B \, dA$. If the magnetic field is also uniform, then B can be brought out in front of the integral sign. The remaining $\int dA$ then gives just the area A of the loop. Thus, Eq. 31-3 reduces to

$$\Phi_B = BA \qquad (\vec{B} \perp \text{area } A, \vec{B} \text{ uniform}). \qquad (31\text{-}4)$$

From Eqs. 31-3 and 31-4, we see that the SI unit for magnetic flux is the tesla–square meter, which is called the *weber* (abbreviated Wb):

$$1 \text{ weber} = 1 \text{ Wb} = 1 \text{ T} \cdot \text{m}^2. \qquad (31\text{-}5)$$

With the notion of magnetic flux, we can state Faraday's law in a more quantitative and useful way:

> ▶ The magnitude of the emf $\mathcal{E}$ induced in a conducting loop is equal to the rate at which the magnetic flux Φ_B through that loop changes with time.

As you will see in the next section, the induced emf $\mathcal{E}$ tends to oppose the flux change, so Faraday's law is formally written as

$$\mathcal{E} = -\frac{d\Phi_B}{dt} \qquad \text{(Faraday's law)}, \qquad (31\text{-}6)$$

with the minus sign indicating that opposition. We often neglect the minus sign in Eq. 31-6, seeking only the magnitude of the induced emf.

If we change the magnetic flux through a coil of N turns, an induced emf appears in every turn and the total emf induced in the coil is the sum of these individual induced emfs. If the coil is tightly wound (*closely packed*), so that the same magnetic flux Φ_B passes through all the turns, the total emf induced in the coil is

$$\mathcal{E} = -N\frac{d\Phi_B}{dt} \qquad \text{(coil of N turns)}. \qquad (31\text{-}7)$$

Here are the general means by which we can change the magnetic flux through a coil:

1. Change the magnitude B of the magnetic field within the coil.

2. Change the area of the coil, or the portion of that area that happens to lie within the magnetic field (for example, by expanding the coil or sliding it in or out of the field).

3. Change the angle between the direction of the magnetic field $\vec{B}$ and the area of the coil (for example, by rotating the coil so that field $\vec{B}$ is first perpendicular to the plane of the coil and then is along that plane).

✔**CHECKPOINT 1:**   The graph gives the magnitude $B(t)$ of a uniform magnetic field that exists throughout a conducting loop, perpendicular to the plane of the loop. Rank the five regions of the graph according to the magnitude of the emf induced in the loop, greatest first.

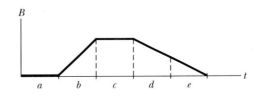

Sample Problem 31-1

The long solenoid S shown (in cross section) in Fig. 31-3 has 220 turns/cm and carries a current $i = 1.5$ A; its diameter D is 3.2 cm. At its center we place a 130-turn closely packed coil C of diameter $d = 2.1$ cm. The current in the solenoid is reduced to zero at a steady rate in 25 ms. What is the magnitude of the emf that is induced in coil C while the current in the solenoid is changing?

SOLUTION: The **Key Ideas** here are these:

1. Because coil C is located in the interior of the solenoid, it lies within the magnetic field produced by current i in the solenoid; thus, there is a magnetic flux Φ_B through coil C.

2. Because current i decreases, flux Φ_B also decreases.

3. As Φ_B decreases, emf $\mathcal{E}$ is induced in coil C, according to Faraday's law.

Because coil C consists of more than one turn, we apply Faraday's law in the form of Eq. 31-7 ($\mathcal{E} = -N \, d\Phi_B/dt$), where the number of turns N is 130 and $d\Phi_B/dt$ is the rate at which the flux in each turn changes.

Because the current in the solenoid decreases at a steady rate, flux Φ_B also decreases at a steady rate and we can write $d\Phi_B/dt$ as $\Delta\Phi_B/\Delta t$. Then, to evaluate $\Delta\Phi_B$, we need the final and initial flux. The final flux $\Phi_{B,f}$ is zero because the final current in the solenoid is zero. To find the initial flux $\Phi_{B,i}$, we need two more **Key Ideas**:

4. The flux through each turn of coil C depends on the area A and orientation of that turn in the solenoid's magnetic field $\vec{B}$. Because $\vec{B}$ is uniform and directed perpendicular to area A, the flux is given by Eq. 31-4 ($\Phi_B = BA$).

5. The magnitude B of magnetic field in the interior of a solenoid depends on the solenoid's current i and its number n of turns per unit length, according to Eq. 30-25 ($B = \mu_0 in$).

Fig. 31-3 Sample Problem 31-1. A coil C is located inside a solenoid S, which carries current i.

For the situation of Fig. 31-3, A is $\frac{1}{4}\pi d^2$ ($= 3.46 \times 10^{-4}$ m²) and n is 220 turns/cm, or 22 000 turns/m. Substituting Eq. 30-25 into Eq. 31-4 then leads to

$$\Phi_{B,i} = BA = (\mu_0 in)A$$
$$= (4\pi \times 10^{-7} \text{ T} \cdot \text{m/A})(1.5 \text{ A})(22\,000 \text{ turns/m})$$
$$\times (3.46 \times 10^{-4} \text{ m}^2)$$
$$= 1.44 \times 10^{-5} \text{ Wb}.$$

Now we can write

$$\frac{d\Phi_B}{dt} = \frac{\Delta\Phi_B}{\Delta t} = \frac{\Phi_{B,f} - \Phi_{B,i}}{\Delta t}$$
$$= \frac{(0 - 1.44 \times 10^{-5} \text{ Wb})}{25 \times 10^{-3} \text{ s}}$$
$$= -5.76 \times 10^{-4} \text{ Wb/s} = -5.76 \times 10^{-4} \text{ V}.$$

We are interested only in magnitudes, so we ignore the minus signs here and in Eq. 31-7, writing

$$\mathcal{E} = N\frac{d\Phi_B}{dt} = (130 \text{ turns})(5.76 \times 10^{-4} \text{ V})$$
$$= 7.5 \times 10^{-2} \text{ V} = 75 \text{ mV}. \qquad \text{(Answer)}$$

31-4 Lenz's Law

Soon after Faraday proposed his law of induction, Heinrich Friedrich Lenz devised a rule—now known as **Lenz's law**—for determining the direction of an induced current in a loop:

> ▶ An induced current has a direction such that the magnetic field due to *the current* opposes the change in the magnetic flux that induces the current.

Furthermore, the direction of an induced emf is that of the induced current. To get a feel for Lenz's law, let us apply it in two different but equivalent ways to Fig. 31-4, where the north pole of a magnet is being moved toward a conducting loop.

Fig. 31-4 Lenz's law at work. As the magnet is moved toward the loop, a current is induced in the loop. The current produces its own magnetic field, with magnetic dipole moment $\vec{\mu}$ oriented so as to oppose the motion of the magnet. Thus, the induced current must be counterclockwise as shown.

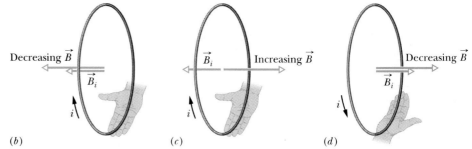

(a)   (b)   (c)   (d)

Fig. 31-5 The current i induced in a loop has the direction such that the current's magnetic field $\vec{B}_i$ opposes the *change* in the magnetic field $\vec{B}$ inducing i. The field $\vec{B}_i$ is always directed opposite an increasing field $\vec{B}$ (a, c) and in the same direction as a decreasing field $\vec{B}$ (b, d). The curled–straight right-hand rule gives the direction of the induced current based on the direction of the induced field.

1. ***Opposition to Pole Movement.*** The approach of the magnet's north pole in Fig. 31-4 increases the magnetic flux in the loop and thereby induces a current in the loop. From Fig. 30-21, we know that the loop then acts as a magnetic dipole with a south pole and a north pole, and that its magnetic dipole moment $\vec{\mu}$ is directed from south to north. To *oppose* the magnetic flux increase being caused by the approaching magnet, the loop's north pole (and thus $\vec{\mu}$) must face *toward* the approaching north pole so as to repel it (Fig. 31-4). Then the curled–straight right-hand rule for $\vec{\mu}$ (Fig. 30-21) tells us that the current induced in the loop must be counterclockwise in Fig. 31-4.

    If we next pull the magnet away from the loop, a current will again be induced in the loop. Now, however, the loop will have a south pole facing the retreating north pole of the magnet, so as to oppose the retreat. Thus, the induced current will be clockwise.

2. ***Opposition to Flux Change.*** In Fig. 31-4, with the magnet initially distant, no magnetic flux passes through the loop. As the north pole of the magnet then nears the loop with its magnetic field $\vec{B}$ directed *toward the left,* the flux through the loop increases. To oppose this increase in flux, the induced current i must set up its own field $\vec{B}_i$ *directed toward the right* inside the loop, as shown in Fig. 31-5a; then the rightward flux of field $\vec{B}_i$ opposes the increasing leftward flux of field $\vec{B}$. The curled–straight right-hand rule of Fig. 30-21 then tells us that i must be counterclockwise in Fig. 31-5a.

    Note carefully that the flux of $\vec{B}_i$ always opposes the *change* in the flux of $\vec{B}$, but that does not always mean that $\vec{B}_i$ points opposite $\vec{B}$. For example, if we next pull the magnet away from the loop, the flux Φ_B from the magnet is still directed to the left through the loop, but it is now decreasing. The flux of $\vec{B}_i$ must now be to the left inside the loop, to oppose the *decrease* in Φ_B, as shown in Fig. 31-5b. Thus, $\vec{B}_i$ and $\vec{B}$ are now in the same direction.

    Figures 31-5c and d show the situations in which the south pole of the magnet approaches and retreats from the loop, respectively.

Fig. 31-6 A Fender Stratocaster has three groups of six electric pickups each (within the wide part of the body). A toggle switch (at the bottom of the guitar) allows the musician to determine which group of pickups sends signals to an amplifier and thus to a speaker system.

Electric Guitars

Figure 31-6 shows a Fender Stratocaster, the type of electric guitar that was used by Jimi Hendrix and by many other musicians. Whereas an acoustic guitar depends for its sound on the acoustic resonance produced in the hollow body of the instrument

Fig. 31-7 A side view of an electric guitar pickup. When the metal string (which acts like a magnet) is made to oscillate, it causes a variation in magnetic flux that induces a current in the coil.

by the oscillations of the strings, an electric guitar is a solid instrument, so there is no body resonance. Instead, the oscillations of the metal strings are sensed by electric "pickups" that send signals to an amplifier and a set of speakers.

The basic construction of a pickup is shown in Fig. 31-7. Wire connecting the instrument to the amplifier is coiled around a small magnet. The magnetic field of the magnet produces a north and south pole in the section of the metal string just above the magnet. That section of string then has its own magnetic field. When the string is plucked and thus made to oscillate, its motion relative to the coil changes the flux of its magnetic field through the coil, inducing a current in the coil. As the string oscillates toward and away from the coil, the induced current changes direction at the same frequency as the string's oscillations, thus relaying the frequency of oscillation to the amplifier and speaker.

On a Stratocaster, there are three groups of pickups, placed at the near end of the strings (on the wide part of the body). The group closest to the near end better detects the high-frequency oscillations of the strings; the group farthest from the near end better detects the low-frequency oscillations. By throwing a toggle switch on the guitar, the musician can select which group or which pair of groups will send signals to the amplifier and speakers.

To gain further control over his music, Hendrix sometimes rewrapped the wire in the pickup coils of his guitar to change the number of turns. In this way, he altered the amount of emf induced in the coils and thus their relative sensitivity to string oscillations. Even without this additional measure, you can see that the electric guitar offers far more control over the sound that is produced than can be obtained with an acoustic guitar.

✓**CHECKPOINT 2:**   The figure shows three situations in which identical circular conducting loops are in uniform magnetic fields that are either increasing (Inc) or decreasing (Dec) in magnitude at identical rates. In each, the dashed line coincides with a diameter. Rank the situations according to the magnitude of the current induced in the loops, greatest first.

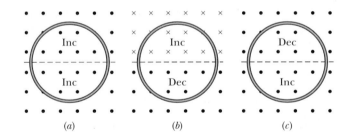

(a) (b) (c)

Sample Problem 31-2

Figure 31-8 shows a conducting loop consisting of a half-circle of radius $r = 0.20$ m and three straight sections. The half-circle lies in a uniform magnetic field $\vec{B}$ that is directed out of the page; the field magnitude is given by $B = 4.0t^2 + 2.0t + 3.0$, with B in teslas and t in seconds. An ideal battery with emf $\mathscr{E}_{bat} = 2.0$ V is connected to the loop. The resistance of the loop is 2.0 Ω.

(a) What are the magnitude and direction of the emf $\mathscr{E}_{ind}$ induced around the loop by field $\vec{B}$ at $t = 10$ s?

SOLUTION: One Key Idea here is that, according to Faraday's law, the magnitude of $\mathscr{E}_{ind}$ is equal to the rate $d\Phi_B/dt$ at which the magnetic

Fig. 31-8 Sample Problem 31-2. A battery is connected to a conducting loop consisting of a half-circle of radius r that lies in a uniform magnetic field. The field is directed out of the page; its magnitude is changing.

flux through the loop changes. A second Key Idea is that the flux through the loop depends on the loop's area A and its orientation in the magnetic field $\vec{B}$. Because $\vec{B}$ is uniform and is perpendicular to the plane of the loop, the flux is given by Eq. 31-4 ($\Phi_B = BA$). Using this equation and realizing that only the field magnitude B changes in time (not the area A), we rewrite Faraday's law, Eq. 31-6, as

$$\mathscr{E}_{ind} = \frac{d\Phi_B}{dt} = \frac{d(BA)}{dt} = A\,\frac{dB}{dt}.$$

A third Key Idea is that, because the flux penetrates the loop only within the half-circle, the area A in this equation is $\frac{1}{2}\pi r^2$. Substituting this and the given expression for B yields

$$\mathscr{E}_{ind} = A\,\frac{dB}{dt} = \frac{\pi r^2}{2}\,\frac{d}{dt}(4.0t^2 + 2.0t + 3.0)$$

$$= \frac{\pi r^2}{2}(8.0t + 2.0).$$

At $t = 10$ s, then,

$$\mathscr{E}_{ind} = \frac{\pi(0.20\ \text{m})^2}{2}[8.0(10) + 2.0]$$

$$= 5.152\ \text{V} \approx 5.2\ \text{V}. \qquad \text{(Answer)}$$

To find the direction of $\mathscr{E}_{ind}$, we first note that in Fig. 31-8 the flux through the loop is out of the page and increasing. Then the Key Idea here is that the induced field B_i (due to the induced current) must oppose that increase, and thus be *into* the page. Using the curled–straight right-hand rule (Fig. 30-7c), we find that the induced current must be clockwise around the loop. The induced emf $\mathscr{E}_{ind}$ must then also be clockwise.

(b) What is the current in the loop at $t = 10$ s?

SOLUTION: The Key Idea here is that two emfs tend to move charges around the loop. The induced emf $\mathscr{E}_{ind}$ tends to drive a current clockwise around the loop; the battery's emf $\mathscr{E}_{bat}$ tends to drive a current counterclockwise. Because $\mathscr{E}_{ind}$ is greater than $\mathscr{E}_{bat}$, the net emf $\mathscr{E}_{net}$ is clockwise, and thus so is the current. To find the current at $t = 10$ s, we use Eq. 28-2 ($i = \mathscr{E}/R$):

$$i = \frac{\mathscr{E}_{net}}{R} = \frac{\mathscr{E}_{ind} - \mathscr{E}_{bat}}{R}$$

$$= \frac{5.152\ \text{V} - 2.0\ \text{V}}{2.0\ \Omega} = 1.58\ \text{A} \approx 1.6\ \text{A}. \quad \text{(Answer)}$$

Sample Problem 31-3

Figure 31-9 shows a rectangular loop of wire immersed in a non-uniform and varying magnetic field $\vec{B}$ that is perpendicular to and directed into the page. The field's magnitude is given by $B = 4t^2x^2$, with B in teslas, t in seconds, and x in meters. The loop has width $W = 3.0$ m and height $H = 2.0$ m. What are the magnitude and direction of the induced emf $\mathscr{E}$ around the loop at $t = 0.10$ s?

SOLUTION: One Key Idea here is that because the magnitude of the magnetic field $\vec{B}$ is changing with time, the magnetic flux Φ_B through the loop is also changing. A second Key Idea is that the changing flux induces an emf $\mathscr{E}$ in the loop according to Faraday's law, which we can write as $\mathscr{E} = d\Phi_B/dt$.

To use that law, we need an expression for the flux Φ_B at any time t. However, a third Key Idea is that because B is *not* uniform over the area enclosed by the loop, we *cannot* use Eq. 31-4 ($\Phi_B = BA$) to find that expression; instead we must use Eq. 31-3 ($\Phi_B = \int \vec{B} \cdot d\vec{A}$).

In Fig. 31-9, $\vec{B}$ is perpendicular to the plane of the loop (and hence parallel to the differential area vector $d\vec{A}$), so the dot product in Eq. 31-3 gives $B\,dA$. Because the magnetic field varies with the coordinate x but not with the coordinate y, we can take the differential area dA to be the area of a vertical strip of height H and width dx (as shown in Fig. 31-9). Then $dA = H\,dx$, and the flux through the loop is

$$\Phi_B = \int \vec{B} \cdot d\vec{A} = \int B\,dA = \int BH\,dx = \int 4t^2x^2H\,dx.$$

Treating t as a constant for this integration and inserting the integration limits $x = 0$ and $x = 3.0$ m, we obtain

$$\Phi_B = 4t^2H \int_0^{3.0} x^2\,dx = 4t^2H \left[\frac{x^3}{3}\right]_0^{3.0} = 72t^2,$$

where we have substituted $H = 2.0$ m and Φ_B is in webers. Now we can use Faraday's law to find the magnitude of $\mathscr{E}$ at any time t:

$$\mathscr{E} = \frac{d\Phi_B}{dt} = \frac{d(72t^2)}{dt} = 144t,$$

in which $\mathscr{E}$ is in volts. At $t = 0.10$ s,

$$\mathscr{E} = (144\ \text{V/s})(0.10\ \text{s}) \approx 14\ \text{V}. \qquad \text{(Answer)}$$

The flux of $\vec{B}$ through the loop is into the page in Fig. 31-9 and is increasing in magnitude because B is increasing in magnitude with time. According to Lenz's law, the field B_i of the induced current must oppose this increase and so is directed out of the page. The curled–straight right-hand rule of Fig. 31-5 then tells us that the induced current is counterclockwise around the loop, and thus so is the induced emf $\mathscr{E}$.

Fig. 31-9 Sample Problem 31-3. A closed conducting loop, of width W and height H, lies in a nonuniform, varying magnetic field that points directly into the page. To apply Faraday's law, we use the vertical strip of height H, width dx, and area dA.

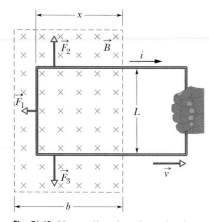

Fig. 31-10 You pull a closed conducting loop out of a magnetic field at constant velocity $\vec{v}$. While the loop is moving, a clockwise current i is induced in the loop, and the loop segments still within the magnetic field experience forces $\vec{F}_1$, $\vec{F}_2$, and $\vec{F}_3$.

31-5 Induction and Energy Transfers

By Lenz's law, whether you move the magnet toward or away from the loop in Fig. 31-1, a magnetic force resists the motion, requiring your applied force to do positive work. At the same time, thermal energy is produced in the material of the loop because of the material's electrical resistance to the current that is induced by the motion. The energy you transfer to the closed *loop + magnet* system via your applied force ends up in this thermal energy. (For now, we neglect energy that is radiated away from the loop as electromagnetic waves during the induction.) The faster you move the magnet, the more rapidly your applied force does work, and the greater the rate at which your energy is transferred to thermal energy in the loop; that is, the power of the transfer is greater.

Regardless of how current is induced in a loop, energy is always transferred to thermal energy during the process because of the electrical resistance of the loop (unless the loop is superconducting). For example, in Fig. 31-2, when switch S is closed and a current is briefly induced in the left-hand loop, energy is transferred from the battery to thermal energy in that loop.

Figure 31-10 shows another situation involving induced current. A rectangular loop of wire of width L has one end in a uniform external magnetic field that is directed perpendicularly into the plane of the loop. This field may be produced, for example, by a large electromagnet. The dashed lines in Fig. 31-10 show the assumed limits of the magnetic field; the fringing of the field at its edges is neglected. You are asked to pull this loop to the right at a constant velocity $\vec{v}$.

The situation of Fig. 31-10 does not differ in any essential way from that of Fig. 31-1. In each case a magnetic field and a conducting loop are in relative motion; in each case the flux of the field through the loop is changing with time. It is true that in Fig. 31-1 the flux is changing because $\vec{B}$ is changing and in Fig. 31-10 the flux is changing because the area of the loop still in the magnetic field is changing, but that difference is not important. The important difference between the two arrangements is that the arrangement of Fig. 31-10 makes calculations easier. Let us now calculate the rate at which you do mechanical work as you pull steadily on the loop in Fig. 31-10.

As you will see, to pull the loop at a constant velocity $\vec{v}$, you must apply a constant force $\vec{F}$ to the loop because a magnetic force of equal magnitude but opposite direction acts on the loop to oppose you. From Eq. 7-48, the rate at which you do work is then

$$P = Fv, \qquad (31\text{-}8)$$

where F is the magnitude of your force. We wish to find an expression for P in terms of the magnitude B of the magnetic field and the characteristics of the loop—namely, its resistance R to current and its dimension L.

As you move the loop to the right in Fig. 31-10, the portion of its area within the magnetic field decreases. Thus, the flux through the loop also decreases and, according to Lenz's law, a current is produced in the loop. It is the presence of this current that causes the force that opposes your pull.

To find the current, we first apply Faraday's law. When x is the length of the loop still in the magnetic field, the area of the loop still in the field is Lx. Then from Eq. 31-4, the magnitude of the flux through the loop is

$$\Phi_B = BA = BLx. \qquad (31\text{-}9)$$

As x decreases, the flux decreases. Faraday's law tells us that with this flux decrease, an emf is induced in the loop. Dropping the minus sign in Eq. 31-6 and using Eq.

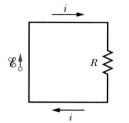

Fig. 31-11 A circuit diagram for the loop of Fig. 31-10 while the loop is moving.

31-9, we can write the magnitude of this emf as

$$\mathcal{E} = \frac{d\Phi_B}{dt} = \frac{d}{dt}BLx = BL\frac{dx}{dt} = BLv, \tag{31-10}$$

in which we have replaced dx/dt with v, the speed at which the loop moves.

Figure 31-11 shows the loop as a circuit: induced emf $\mathcal{E}$ is represented on the left, and the collective resistance R of the loop is represented on the right. The direction of the induced current i is obtained with a right-hand rule as in Fig. 31-5b; $\mathcal{E}$ must have the same direction.

To find the magnitude of the induced current, we cannot apply the loop rule for potential differences in a circuit because, as you will see in Section 31-6, we cannot define a potential difference for an induced emf. However, we can apply the equation $i = \mathcal{E}/R$, as we did in Sample Problem 31-2. With Eq. 31-10, this becomes

$$i = \frac{BLv}{R}. \tag{31-11}$$

Because three segments of the loop in Fig. 31-10 carry this current through the magnetic field, sideways deflecting forces act on those segments. From Eq. 29-26 we know that such a deflecting force is, in general notation,

$$\vec{F}_d = i\vec{L} \times \vec{B}. \tag{31-12}$$

In Fig. 31-10, the deflecting forces acting on the three segments of the loop are marked $\vec{F}_1$, $\vec{F}_2$, and $\vec{F}_3$. Note, however, that from the symmetry, forces $\vec{F}_2$ and $\vec{F}_3$ are equal in magnitude and cancel. This leaves only $\vec{F}_1$, which is directed opposite your force $\vec{F}$ on the loop and thus is the force opposing you. So, $\vec{F} = -\vec{F}_1$.

Using Eq. 31-12 to obtain the magnitude of $\vec{F}_1$ and noting that the angle between $\vec{B}$ and the length vector $\vec{L}$ for the left segment is 90°, we write

$$F = F_1 = iLB \sin 90° = iLB. \tag{31-13}$$

Substituting Eq. 31-11 for i in Eq. 31-13 then gives us

$$F = \frac{B^2L^2v}{R}. \tag{31-14}$$

Since B, L, and R are constants, the speed v at which you move the loop is constant if the magnitude F of the force you apply to the loop is also constant.

By substituting Eq. 31-14 into Eq. 31-8, we find the rate at which you do work on the loop as you pull it from the magnetic field:

$$P = Fv = \frac{B^2L^2v^2}{R} \qquad \text{(rate of doing work).} \tag{31-15}$$

To complete our analysis, let us find the rate at which thermal energy appears in the loop as you pull it along at constant speed. We calculate it from Eq. 27-22,

$$P = i^2R. \tag{31-16}$$

Substituting for i from Eq. 31-11, we find

$$P = \left(\frac{BLv}{R}\right)^2 R = \frac{B^2L^2v^2}{R} \qquad \text{(thermal energy rate),} \tag{31-17}$$

which is exactly equal to the rate at which you are doing work on the loop (Eq. 31-15). Thus, the work that you do in pulling the loop through the magnetic field appears as thermal energy in the loop.

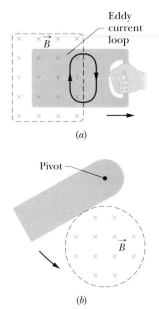

(a)

(b)

Fig. 31-12 (a) As you pull a solid conducting plate out of a magnetic field, *eddy currents* are induced in the plate. A typical loop of eddy current is shown. (b) A conducting plate is allowed to swing like a pendulum about a pivot and into a region of magnetic field. As it enters and leaves the field, eddy currents are induced in the plate.

Eddy Currents

Suppose we replace the conducting loop of Fig. 31-10 with a solid conducting plate. If we then move the plate out of the magnetic field as we did the loop (Fig. 31-12a), the relative motion of the field and the conductor again induces a current in the conductor. Thus, we again encounter an opposing force and must do work because of the induced current. With the plate, however, the conduction electrons making up the induced current do not follow one path as they do with the loop. Instead, the electrons swirl about within the plate as if they were caught in an eddy (or whirlpool) of water. Such a current is called an *eddy current* and can be represented as in Fig. 31-12a *as if* it followed a single path.

As with the conducting loop of Fig. 31-10, the current induced in the plate results in mechanical energy being dissipated as thermal energy. The dissipation is more apparent in the arrangement of Fig. 31-12b; a conducting plate, free to rotate about a pivot, is allowed to swing down through a magnetic field like a pendulum. Each time the plate enters and leaves the field, a portion of its mechanical energy is transferred to its thermal energy. After several swings, no mechanical energy remains and the warmed-up plate just hangs from its pivot.

CHECKPOINT 3:   The figure shows four wire loops, with edge lengths of either L or $2L$. All four loops will move through a region of uniform magnetic field $\vec{B}$ (directed out of the page) at the same constant velocity. Rank the four loops according to the maximum magnitude of the emf induced as they move through the field, greatest first.

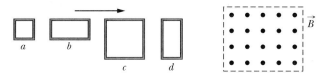

31-6 Induced Electric Fields

Let us place a copper ring of radius r in a uniform external magnetic field, as in Fig. 31-13a. The field—neglecting fringing—fills a cylindrical volume of radius R. Suppose that we increase the strength of this field at a steady rate, perhaps by increasing—in an appropriate way—the current in the windings of the electromagnet that produces the field. The magnetic flux through the ring will then change at a steady rate and—by Faraday's law—an induced emf and thus an induced current will appear in the ring. From Lenz's law we can deduce that the direction of the induced current is counterclockwise in Fig. 31-13a.

If there is a current in the copper ring, an electric field must be present along the ring; an electric field is needed to do the work of moving the conduction electrons. Moreover, the electric field must have been produced by the changing magnetic flux. This **induced electric field** $\vec{E}$ is just as real as an electric field produced by static charges; either field will exert a force $q_0\vec{E}$ on a particle of charge q_0.

By this line of reasoning, we are led to a useful and informative restatement of Faraday's law of induction:

▶ A changing magnetic field produces an electric field.

The striking feature of this statement is that the electric field is induced even if there is no copper ring.

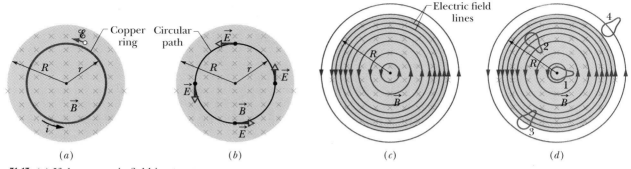

Fig. 31-13 (*a*) If the magnetic field increases at a steady rate, a constant induced current appears, as shown, in the copper ring of radius *r*. (*b*) An induced electric field exists even when the ring is removed; the electric field is shown at four points. (*c*) The complete picture of the induced electric field, displayed as field lines. (*d*) Four similar closed paths that enclose identical areas. Equal emfs are induced around paths 1 and 2, which lie entirely within the region of changing magnetic field. A smaller emf is induced around path 3, which only partially lies in that region. No emf is induced around path 4, which lies entirely outside the magnetic field.

To fix these ideas, consider Fig. 31-13*b*, which is just like Fig. 31-13*a* except the copper ring has been replaced by a hypothetical circular path of radius *r*. We assume, as previously, that the magnetic field $\vec{B}$ is increasing in magnitude at a constant rate dB/dt. The electric field induced at various points around the circular path must—from the symmetry—be tangent to the circle, as Fig. 31-13*b* shows.* Hence, the circular path is an electric field line. There is nothing special about the circle of radius *r*, so the electric field lines produced by the changing magnetic field must be a set of concentric circles, as in Fig. 31-13*c*.

As long as the magnetic field is *increasing* with time, the electric field represented by the circular field lines in Fig. 31-13*c* will be present. If the magnetic field remains *constant* with time, there will be no induced electric field and thus no electric field lines. If the magnetic field is *decreasing* with time (at a constant rate), the electric field lines will still be concentric circles as in Fig. 31-13*c*, but they will now have the opposite direction. All this is what we have in mind when we say: "A changing magnetic field produces an electric field."

A Reformulation of Faraday's Law

Consider a particle of charge q_0 moving around the circular path of Fig. 31-13*b*. The work W done on it in one revolution by the induced electric field is $\mathscr{E}q_0$, where $\mathscr{E}$ is the induced emf—that is, the work done per unit charge in moving the test charge around the path. From another point of view, the work is

$$\int \vec{F} \cdot d\vec{s} = (q_0 E)(2\pi r), \tag{31-18}$$

where $q_0 E$ is the magnitude of the force acting on the test charge and $2\pi r$ is the distance over which that force acts. Setting these two expressions for W equal to each other and canceling q_0, we find that

$$\mathscr{E} = 2\pi r E. \tag{31-19}$$

More generally, we can rewrite Eq. 31-18 to give the work done on a particle of charge q_0 moving along any closed path:

$$W = \oint \vec{F} \cdot d\vec{s} = q_0 \oint \vec{E} \cdot d\vec{s}. \tag{31-20}$$

---

*Arguments of symmetry would also permit the lines of $\vec{E}$ around the circular path to be *radial*, rather than tangential. However, such radial lines would imply that there are free charges, distributed symmetrically about the axis of symmetry, on which the electric field lines could begin or end; there are no such charges.

(The circle indicates that the integral is to be taken around the closed path.) Substituting $\mathscr{E}q_0$ for W, we find that

$$\mathscr{E} = \oint \vec{E} \cdot d\vec{s}. \tag{31-21}$$

This integral reduces at once to Eq. 31-19 if we evaluate it for the special case of Fig. 31-13b.

With Eq. 31-21, we can expand the meaning of induced emf. Previously, induced emf has meant the work per unit charge done in maintaining current due to a changing magnetic flux, or it has meant the work done per unit charge on a charged particle that moves around a closed path in a changing magnetic flux. However, with Fig. 31-13b and Eq. 31-21, an induced emf can exist without the need of a current or particle: An induced emf is the sum—via integration—of quantities $\vec{E} \cdot d\vec{s}$ around a closed path, where $\vec{E}$ is the electric field induced by a changing magnetic flux and $d\vec{s}$ is a differential length vector along the closed path.

If we combine Eq. 31-21 with Faraday's law in Eq. 31-6 ($\mathscr{E} = -d\Phi_B/dt$), we can rewrite Faraday's law as

$$\oint \vec{E} \cdot d\vec{s} = -\frac{d\Phi_B}{dt} \qquad \text{(Faraday's law)}. \tag{31-22}$$

This equation says simply that a changing magnetic field induces an electric field. The changing magnetic field appears on the right side of this equation, the electric field on the left.

Faraday's law in the form of Eq. 31-22 can be applied to *any* closed path that can be drawn in a changing magnetic field. Figure 31-13d, for example, shows four such paths, all having the same shape and area but located in different positions in the changing field. For paths 1 and 2, the induced emfs $\mathscr{E}$ ($-\oint \vec{E} \cdot d\vec{s}$) are equal because these paths lie entirely in the magnetic field and thus have the same value of $d\Phi_B/dt$. This is true even though the electric field vectors at points along these paths are different, as indicated by the patterns of electric field lines in the figure. For path 3 the induced emf is smaller because the enclosed flux Φ_B (hence $d\Phi_B/dt$) is smaller, and for path 4 the induced emf is zero, even though the electric field is not zero at any point on the path.

A New Look at Electric Potential

Induced electric fields are produced not by static charges but by a changing magnetic flux. Although electric fields produced in either way exert forces on charged particles, there is an important difference between them. The simplest evidence of this difference is that the field lines of induced electric fields form closed loops, as in Fig. 31-13c. Field lines produced by static charges never do so but must start on positive charges and end on negative charges.

In a more formal sense, we can state the difference between electric fields produced by induction and those produced by static charges in these words:

▶ Electric potential has meaning only for electric fields that are produced by static charges; it has no meaning for electric fields that are produced by induction.

You can understand this statement qualitatively by considering what happens to a charged particle that makes a single journey around the circular path in Fig. 31-13b. It starts at a certain point and, on its return to that same point, has experienced an emf $\mathscr{E}$ of, let us say, 5 V; that is, work of 5 J/C has been done on the particle, and thus the particle should then be at a point that is 5 V greater in potential. However, that is impossible because the particle is back at the same point, which cannot have two different values of potential. We must conclude that potential has no meaning for electric fields that are set up by changing magnetic fields.

We can take a more formal look by recalling Eq. 25-18, which defines the potential difference between two points i and f in an electric field $\vec{E}$:

$$V_f - V_i = -\int_i^f \vec{E} \cdot d\vec{s}. \tag{31-23}$$

In Chapter 25 we had not yet encountered Faraday's law of induction, so the electric fields involved in the derivation of Eq. 25-18 were those due to static charges. If i and f in Eq. 31-23 are the same point, the path connecting them is a closed loop, V_i and V_f are identical, and Eq. 31-23 reduces to

$$\oint \vec{E} \cdot d\vec{s} = 0. \tag{31-24}$$

However, when a changing magnetic flux is present, this integral is *not* zero but is $-d\Phi_B/dt$, as Eq. 31-22 asserts. Thus, assigning electric potential to an induced electric field leads us to a contradiction. We must conclude that electric potential has no meaning for electric fields associated with induction.

✓**CHECKPOINT 4:** The figure shows five lettered regions in which a uniform magnetic field extends either directly out of the page (as in region a) or into the page. The field is increasing in magnitude at the same steady rate in all five regions; the regions are identical in area. Also shown are four numbered paths along which $\oint \vec{E} \cdot d\vec{s}$ has the magnitudes given below in terms of a quantity mag. Determine whether the magnetic fields in regions b through e are directed into or out of the page.

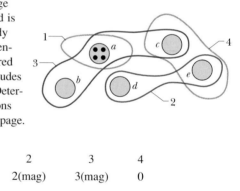

Path:	1	2	3	4
$\oint \vec{E} \cdot d\vec{s}$:	mag	2(mag)	3(mag)	0

Sample Problem 31-4

In Fig. 31-13b, take $R = 8.5$ cm and $dB/dt = 0.13$ T/s.

(a) Find an expression for the magnitude E of the induced electric field at points within the magnetic field, at radius r from the center of the magnetic field. Evaluate the expression for $r = 5.2$ cm.

SOLUTION: The Key Idea here is that an electric field is induced by the changing magnetic field, according to Faraday's law. To calculate the field magnitude E, we apply Faraday's law in the form of Eq. 31-22. We use a circular path of integration with radius $r \leq R$ because we want E for points within the magnetic field. We assume from the symmetry that $\vec{E}$ in Fig. 31-13b is tangent to the circular path at all points. The path vector $d\vec{s}$ is also always tangent to the circular path, so the dot product $\vec{E} \cdot d\vec{s}$ in Eq. 31-22 must have the magnitude $E\,ds$ at all points on the path. We can also assume from the symmetry that E has the same value at all points along the circular path. Then the left side of Eq. 31-22 becomes

$$\oint \vec{E} \cdot d\vec{s} = \oint E\,ds = E \oint ds = E(2\pi r). \tag{31-25}$$

(The integral $\oint ds$ is the circumference $2\pi r$ of the circular path.)

Next, we need to evaluate the right side of Eq. 31-22. Because $\vec{B}$ is uniform over the area A encircled by the path of integration and is directed perpendicular to that area, the magnetic flux is given by Eq. 31-4:

$$\Phi_B = BA = B(\pi r^2). \tag{31-26}$$

Substituting this and Eq. 31-25 into Eq. 31-22 and dropping the minus sign, we find that

$$E(2\pi r) = (\pi r^2) \frac{dB}{dt}$$

or

$$E = \frac{r}{2} \frac{dB}{dt}. \tag{Answer} \tag{31-27}$$

Equation 31-27 gives the magnitude of the electric field at any point for which $r \leq R$ (that is, within the magnetic field). Substituting given values yields, for the magnitude of $\vec{E}$ at $r = 5.2$ cm,

$$E = \frac{(5.2 \times 10^{-2} \text{ m})}{2}(0.13 \text{ T/s})$$

$$= 0.0034 \text{ V/m} = 3.4 \text{ mV/m}. \qquad \text{(Answer)}$$

(b) Find an expression for the magnitude E of the induced electric field at points that are outside the magnetic field, at radius r. Evaluate the expression for $r = 12.5$ cm.

SOLUTION: The Key Idea of part (a) applies here also, except that we use a circular path of integration with radius $r \geq R$, because we want to evaluate E for points outside the magnetic field. Proceeding as in (a), we again obtain Eq. 31-25. However, we do not then obtain Eq. 31-26, because the new path of integration is now outside the magnetic field, and we need this Key Idea: The magnetic flux encircled by the new path is only that in the area πR^2 of the magnetic field region. Therefore,

$$\Phi_B = BA = B(\pi R^2). \qquad (31\text{-}28)$$

Substituting this and Eq. 31-25 into Eq. 31-22 (without the minus sign) and solving for E yield

$$E = \frac{R^2}{2r}\frac{dB}{dt}. \qquad \text{(Answer)} \quad (31\text{-}29)$$

Since E is not zero here, we know that an electric field is induced even at points that are outside the changing magnetic field, an important result that (as you shall see in Section 33-11) makes transformers possible. With the given data, Eq. 31-29 yields the magnitude of $\vec{E}$ at $r = 12.5$ cm:

$$E = \frac{(8.5 \times 10^{-2} \text{ m})^2}{(2)(12.5 \times 10^{-2} \text{ m})}(0.13 \text{ T/s})$$

$$= 3.8 \times 10^{-3} \text{ V/m} = 3.8 \text{ mV/m}. \qquad \text{(Answer)}$$

Equations 31-27 and 31-29 give the same result, as they must, for $r = R$. Figure 31-14 shows a plot of $E(r)$ based on these two equations.

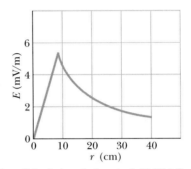

Fig. 31-14 A plot of the induced electric field $E(r)$ for the conditions of Sample Problem 31-4.

31-7 Inductors and Inductance

We found in Chapter 26 that a capacitor can be used to produce a desired electric field. We considered the parallel-plate arrangement as a basic type of capacitor. Similarly, an **inductor** (symbol ⟨⟨⟨⟨) can be used to produce a desired magnetic field. We shall consider a long solenoid (more specifically, a short length near the middle of a long solenoid) as our basic type of inductor.

If we establish a current i in the windings (or turns) of an inductor (a solenoid), the current produces a magnetic flux Φ_B through the central region of the inductor. The **inductance** of the inductor is then

$$L = \frac{N\Phi_B}{i} \qquad \text{(inductance defined)}, \qquad (31\text{-}30)$$

in which N is the number of turns. The windings of the inductor are said to be *linked* by the shared flux, and the product $N\Phi_B$ is called the *magnetic flux linkage*. The inductance L is thus a measure of the flux linkage produced by the inductor per unit of current.

Because the SI unit of magnetic flux is the tesla–square meter, the SI unit of inductance is the tesla–square meter per ampere ($\text{T} \cdot \text{m}^2/\text{A}$). We call this the **henry** (H), after American physicist Joseph Henry, the codiscoverer of the law of induction and a contemporary of Faraday. Thus,

$$1 \text{ henry} = 1 \text{ H} = 1 \text{ T} \cdot \text{m}^2/\text{A}. \qquad (31\text{-}31)$$

Through the rest of this chapter we assume that all inductors, no matter what their geometric arrangement, have no magnetic materials such as iron in their vicinity. Such materials would distort the magnetic field of an inductor.

The crude inductors with which Michael Faraday discovered the law of induction. In those days amenities such as insulated wire were not commercially available. It is said that Faraday insulated his wires by wrapping them with strips cut from one of his wife's petticoats.

Inductance of a Solenoid

Consider a long solenoid of cross-sectional area A. What is the inductance per unit length near its middle?

To use the defining equation for inductance (Eq. 31-30), we must calculate the flux linkage set up by a given current in the solenoid windings. Consider a length l near the middle of this solenoid. The flux linkage for this section of the solenoid is

$$N\Phi_B = (nl)(BA)$$

in which n is the number of turns per unit length of the solenoid and B is the magnitude of the magnetic field within the solenoid.

The magnitude B is given by Eq. 30-25,

$$B = \mu_0 in,$$

so from Eq. 31-30,

$$L = \frac{N\Phi_B}{i} = \frac{(nl)(BA)}{i} = \frac{(nl)(\mu_0 in)(A)}{i}$$

$$= \mu_0 n^2 l A. \tag{31-32}$$

Thus, the inductance per unit length for a long solenoid near its center is

$$\frac{L}{l} = \mu_0 n^2 A \quad \text{(solenoid).} \tag{31-33}$$

Inductance—like capacitance—depends only on the geometry of the device. The dependence on the square of the number of turns per unit length is to be expected. If you, say, triple n, you not only triple the number of turns (N) but you also triple the flux ($\Phi_B = BA = \mu_0 inA$) through each turn, multiplying the flux linkage $N\Phi_B$ and thus the inductance L by a factor of 9.

If the solenoid is very much longer than its radius, then Eq. 31-32 gives its inductance to a good approximation. This approximation neglects the spreading of the magnetic field lines near the ends of the solenoid, just as the parallel-plate capacitor formula ($C = \varepsilon_0 A/d$) neglects the fringing of the electric field lines near the edges of the capacitor plates.

From Eq. 31-32, and recalling that n is a number per unit length, we can see that an inductance can be written as a product of the permeability constant μ_0 and a quantity with the dimensions of a length. This means that μ_0 can be expressed in the unit henry per meter:

$$\mu_0 = 4\pi \times 10^{-7} \text{ T} \cdot \text{m/A}$$

$$= 4\pi \times 10^{-7} \text{ H/m.} \tag{31-34}$$

31-8 Self-Induction

If two coils—which we can now call inductors—are near each other, a current i in one coil produces a magnetic flux Φ_B through the second coil. We have seen that if we change this flux by changing the current, an induced emf appears in the second coil according to Faraday's law. An induced emf appears in the first coil as well.

▶ An induced emf $\mathcal{E}_L$ appears in any coil in which the current is changing.

This process (see Fig. 31-15) is called **self-induction,** and the emf that appears is called a **self-induced emf.** It obeys Faraday's law of induction just as other induced emfs do.

Fig. 31-15 If the current in a coil is changed by varying the contact position on a variable resistor, a self-induced emf $\mathcal{E}_L$ will appear in the coil *while the current is changing.*

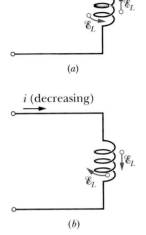

Fig. 31-16 (a) The current i is increasing and the self-induced emf $\mathscr{E}_L$ appears along the coil in a direction such that it opposes the increase. The arrow representing $\mathscr{E}_L$ can be drawn along a turn of the coil or alongside the coil. Both are shown. (b) The current i is decreasing and the self-induced emf appears in a direction such that it opposes the decrease.

For any inductor, Eq. 31-30 tells us that

$$N\Phi_B = Li. \qquad (31\text{-}35)$$

Faraday's law tells us that

$$\mathscr{E}_L = -\frac{d(N\Phi_B)}{dt}. \qquad (31\text{-}36)$$

By combining Eqs. 31-35 and 31-36 we can write:

$$\mathscr{E}_L = -L\frac{di}{dt} \qquad \text{(self-induced emf).} \qquad (31\text{-}37)$$

Thus, in any inductor (such as a coil, a solenoid, or a toroid) a self-induced emf appears whenever the current changes with time. The magnitude of the current has no influence on the magnitude of the induced emf; only the rate of change of the current counts.

You can find the *direction* of a self-induced emf from Lenz's law. The minus sign in Eq. 31-37 indicates that—as the law states—the self-induced emf $\mathscr{E}_L$ has the orientation such that it opposes the change in current i. We can drop the minus sign when we want only the magnitude of $\mathscr{E}_L$.

Suppose that, as in Fig. 31-16a, you set up a current i in a coil and arrange to have it increase with time at a rate di/dt. In the language of Lenz's law, this increase in the current is the "change" that the self-induction must oppose. For such opposition to occur, a self-induced emf must appear in the coil, pointing—as the figure shows—so as to oppose the increase in the current. If you cause the current to decrease with time, as in Fig. 31-16b, the self-induced emf must point in a direction that tends to oppose the decrease in the current, as the figure shows.

In Section 31-6 we saw that we cannot define an electric potential for an electric field (and thus for an emf) that is induced by a changing magnetic flux. This means that when a self-induced emf is produced in the inductor of Fig. 31-15, we cannot define an electric potential within the inductor itself, where the flux is changing. However, potentials can still be defined at points of the circuit that are not within the inductor—points where the electric fields are due to charge distributions and their associated electric potentials.

Moreover, we can define a self-induced potential difference V_L *across an inductor* (between its terminals, which we assume to be outside the region of changing flux). If the inductor is an *ideal inductor* (its wire has negligible resistance), the magnitude of V_L is equal to the magnitude of the self-induced emf $\mathscr{E}_L$.

If, instead, the wire in the inductor has resistance r, we mentally separate the inductor into a resistance r (which we take to be outside the region of changing flux) and an ideal inductor of self-induced emf $\mathscr{E}_L$. As with a real battery of emf $\mathscr{E}$ and internal resistance r, the potential difference across the terminals of a real inductor then differs from the emf. Unless otherwise indicated, we assume here that inductors are ideal.

✔ CHECKPOINT 5:   The figure shows an emf $\mathscr{E}_L$ induced in a coil. Which of the following can describe the current through the coil: (a) constant and rightward, (b) constant and leftward, (c) increasing and rightward, (d) decreasing and rightward, (e) increasing and leftward, (f) decreasing and leftward?

Fig. 31-17 An *RL* circuit. When switch S is closed on *a*, the current rises and approaches a limiting value of $\mathscr{E}/R$.

31-9 *RL* Circuits

In Section 28-8 we saw that if we suddenly introduce an emf $\mathscr{E}$ into a single-loop circuit containing a resistor R and a capacitor C, the charge on the capacitor does not build up immediately to its final equilibrium value $C\mathscr{E}$ but approaches it in an exponential fashion:

$$q = C\mathscr{E}(1 - e^{-t/\tau_C}). \tag{31-38}$$

The rate at which the charge builds up is determined by the capacitive time constant τ_C, defined in Eq. 28-33 as

$$\tau_C = RC. \tag{31-39}$$

If we suddenly remove the emf from this same circuit, the charge does not immediately fall to zero but approaches zero in an exponential fashion:

$$q = q_0 e^{-t/\tau_C}. \tag{31-40}$$

The time constant τ_C describes the fall of the charge as well as its rise.

An analogous slowing of the rise (or fall) of the current occurs if we introduce an emf $\mathscr{E}$ into (or remove it from) a single-loop circuit containing a resistor R and an inductor L. When the switch S in Fig. 31-17 is closed on *a*, for example, the current in the resistor starts to rise. If the inductor were not present, the current would rise rapidly to a steady value $\mathscr{E}/R$. Because of the inductor, however, a self-induced emf $\mathscr{E}_L$ appears in the circuit; from Lenz's law, this emf opposes the rise of the current, which means that it opposes the battery emf $\mathscr{E}$ in polarity. Thus, the current in the resistor responds to the difference between two emfs, a constant one $\mathscr{E}$ due to the battery and a variable one $\mathscr{E}_L$ ($= -L\,di/dt$) due to self-induction. As long as $\mathscr{E}_L$ is present, the current in the resistor will be less than $\mathscr{E}/R$.

As time goes on, the rate at which the current increases becomes less rapid and the magnitude of the self-induced emf, which is proportional to di/dt, becomes smaller. Thus, the current in the circuit approaches $\mathscr{E}/R$ asymptotically.

We can generalize these results as follows:

> Initially, an inductor acts to oppose changes in the current through it. A long time later, it acts like ordinary connecting wire.

Now let us analyze the situation quantitatively. With the switch S in Fig. 31-17 thrown to *a*, the circuit is equivalent to that of Fig. 31-18. Let us apply the loop rule, starting at point *x* in this figure and moving clockwise around the loop along with current *i*.

1. *Resistor.* Because we move through the resistor in the direction of current *i*, the electric potential decreases by iR. Thus, as we move from point *x* to point *y*, we encounter a potential change of $-iR$.

2. *Inductor.* Because current *i* is changing, there is a self-induced emf $\mathscr{E}_L$ in the inductor. The magnitude of $\mathscr{E}_L$ is given by Eq. 31-37 as $L\,di/dt$. The direction of $\mathscr{E}_L$ is upward in Fig. 31-18 because current *i* is downward through the inductor *and* increasing. Thus, as we move from point *y* to point *z*, opposite the direction of $\mathscr{E}_L$, we encounter a potential change of $-L\,di/dt$.

3. *Battery.* As we move from point *z* back to starting point *x*, we encounter a potential change of $+\mathscr{E}$ due to the battery's emf.

Fig. 31-18 The circuit of Fig. 31-17 with the switch closed on *a*. We apply the loop rule for circuits clockwise, starting at *x*.

Thus, the loop rule gives us

$$-iR - L\frac{di}{dt} + \mathscr{E} = 0$$

or $\qquad L\dfrac{di}{dt} + Ri = \mathscr{E}$ *(RL circuit)*.   (31-41)

Equation 31-41 is a differential equation involving the variable i and its first derivative di/dt. To solve it, we seek the function $i(t)$ such that when $i(t)$ and its first derivative are substituted in Eq. 31-41, the equation is satisfied and the initial condition $i(0) = 0$ is satisfied.

Equation 31-41 and its initial condition are of exactly the form of Eq. 28-29 for an RC circuit, with i replacing q, L replacing R, and R replacing $1/C$. The solution of Eq. 31-41 must then be of exactly the form of Eq. 28-30 with the same replacements. That solution is

$$i = \frac{\mathscr{E}}{R}(1 - e^{-Rt/L}),\qquad(31\text{-}42)$$

which we can rewrite as

$$i = \frac{\mathscr{E}}{R}(1 - e^{-t/\tau_L})\qquad\text{(rise of current).}\qquad(31\text{-}43)$$

Here τ_L, the **inductive time constant,** is given by

$$\tau_L = \frac{L}{R}\qquad\text{(time constant).}\qquad(31\text{-}44)$$

Let's examine Eq. 31-43 for when the switch is closed (at time $t = 0$) and for a time long after the switch is closed ($t \to \infty$). If we substitute $t = 0$ into Eq. 31-43, the exponential becomes $e^{-0} = 1$. Thus, Eq. 31-43 tells us that the current is initially $i = 0$, as we expected. Next, if we let t go to ∞, then the exponential goes to $e^{-\infty} = 0$. Thus, Eq. 31-43 tells us that the current goes to its equilibrium value of $\mathscr{E}/R$.

We can also examine the potential differences in the circuit. Figure 31-19 shows how the potential differences V_R ($= iR$) across the resistor and V_L ($= L\,di/dt$) across the inductor vary with time for particular values of $\mathscr{E}$, L, and R. Compare this figure carefully with the corresponding figure for an RC circuit (Fig. 28-14).

To show that the quantity τ_L ($= L/R$) has the dimension of time, we convert from Henries per ohm as follows:

$$1\,\frac{\text{H}}{\Omega} = 1\,\frac{\text{H}}{\Omega}\left(\frac{1\,\text{V} \cdot \text{s}}{1\,\text{H} \cdot \text{A}}\right)\left(\frac{1\,\Omega \cdot \text{A}}{1\,\text{V}}\right) = 1\,\text{s}.$$

The first quantity in parentheses is a conversion factor based on Eq. 31-37, and the second one is a conversion factor based on the relation $V = iR$.

The physical significance of the time constant follows from Eq. 31-43. If we put $t = \tau_L = L/R$ in this equation, it reduces to

$$i = \frac{\mathscr{E}}{R}(1 - e^{-1}) = 0.63\,\frac{\mathscr{E}}{R}.\qquad(31\text{-}45)$$

Thus, the time constant τ_L is the time it takes the current in the circuit to reach about 63% of its final equilibrium value $\mathscr{E}/R$. Since the potential difference V_R across the resistor is proportional to the current i, a graph of the increasing current versus time has the same shape as that of V_R in Fig. 31-19a.

If the switch S in Fig. 31-17 is closed on a long enough for the equilibrium current $\mathscr{E}/R$ to be established, and then is thrown to b, the effect will be to remove the battery from the circuit. (The connection to b must actually be made an instant

Fig. 31-19 The variation with time of (a) V_R, the potential difference across the resistor in the circuit of Fig. 31-18, and (b) V_L, the potential difference across the inductor in that circuit. The small triangles represent successive intervals of one inductive time constant $\tau_L = L/R$. The figure is plotted for $R = 2000\ \Omega$, $L = 4.0$ H, and $\mathscr{E} = 10$ V.

before the connection to *a* is broken. A switch that does this is called a *make-before-break* switch.)

With the battery gone, the current through the resistor will decrease. However, it cannot drop immediately to zero but must decay to zero over time. The differential equation that governs the decay can be found by putting $\mathscr{E} = 0$ in Eq. 31-41:

$$L\frac{di}{dt} + iR = 0. \tag{31-46}$$

By analogy with Eqs. 28-35 and 28-36, the solution of this differential equation that satisfies the initial condition $i(0) = i_0 = \mathscr{E}/R$ is

$$i = \frac{\mathscr{E}}{R}e^{-t/\tau_L} = i_0 e^{-t/\tau_L} \qquad \text{(decay of current).} \tag{31-47}$$

We see that both current rise (Eq. 31-43) and current decay (Eq. 31-47) in an *RL* circuit are governed by the same inductive time constant, τ_L.

We have used i_0 in Eq. 31-47 to represent the current at time $t = 0$. In our case that happened to be $\mathscr{E}/R$, but it could be any other initial value.

Sample Problem 31-5

Figure 31-20*a* shows a circuit that contains three identical resistors with resistance $R = 9.0\ \Omega$, two identical inductors with inductance $L = 2.0$ mH, and an ideal battery with emf $\mathscr{E} = 18$ V.

(a) What is the current *i* through the battery just after the switch is closed?

SOLUTION: The Key Idea here is that just after the switch is closed, the inductor acts to oppose a change in the current through it. Because the current through each inductor is zero before the switch is closed, it will also be zero just afterward. Thus, immediately after the switch is closed, the inductors act as broken wires, as indicated in Fig. 31-20*b*. We then have a single-loop circuit for which the loop rule gives us

$$\mathscr{E} - iR = 0.$$

Substituting given data, we find that

$$i = \frac{\mathscr{E}}{R} = \frac{18\ \text{V}}{9.0\ \Omega} = 2.0\ \text{A.} \qquad \text{(Answer)}$$

(b) What is the current *i* through the battery long after the switch has been closed?

SOLUTION: The Key Idea here is that long after the switch has been closed, the currents in the circuit have reached their equilibrium values, and the inductors act as simple connecting wires, as indicated in Fig. 31-20*c*. We then have a circuit with three identical resistors in parallel; from Eq. 28-20, their equivalent resistance is $R_{\text{eq}} = R/3 = (9.0\ \Omega)/3 = 3.0\ \Omega$. The equivalent circuit shown in Fig. 31-20*d* then yields the loop equation $\mathscr{E} - iR_{\text{eq}} = 0$, or

$$i = \frac{\mathscr{E}}{R_{\text{eq}}} = \frac{18\ \text{V}}{3.0\ \Omega} = 6.0\ \text{A.} \qquad \text{(Answer)}$$

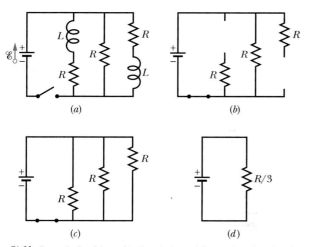

(a)   (b)

(c)   (d)

Fig. 31-20 Sample Problem 31-5. (*a*) A multiloop *RL* circuit with an open switch. (*b*) The equivalent circuit just after the switch has been closed. (*c*) The equivalent circuit a long time later. (*d*) The single-loop circuit that is equivalent to circuit (*c*).

✓**CHECKPOINT 6:** The figure shows three circuits with identical batteries, inductors, and resistors. Rank the circuits according to the current through the battery (a) just after the switch is closed and (b) a long time later, greatest first.

(1)   (2)   (3)

Sample Problem 31-6

A solenoid has an inductance of 53 mH and a resistance of 0.37 Ω. If it is connected to a battery, how long will the current take to reach half its final equilibrium value?

SOLUTION: One Key Idea here is that we can mentally separate the solenoid into a resistance and an inductance that are wired in series with a battery, as in Fig. 31-18. Then application of the loop rule leads to Eq. 31-41, which has the solution of Eq. 31-43 for the current i in the circuit.

The second Key Idea is that, according to that solution, current i increases exponentially from zero to its final equilibrium value of $\mathscr{E}/R$. Let t_0 be the time that current i takes to reach half its equilibrium value. Then Eq. 31-43 gives us

$$\frac{1}{2}\frac{\mathscr{E}}{R} = \frac{\mathscr{E}}{R}(1 - e^{-t_0/\tau_L}).$$

We solve for t_0 by canceling $\mathscr{E}/R$, isolating the exponential, and taking the natural logarithm of each side. We find

$$t_0 = \tau_L \ln 2 = \frac{L}{R}\ln 2 = \frac{53 \times 10^{-3}\text{ H}}{0.37\ \Omega}\ln 2$$

$$= 0.10\text{ s.} \qquad \text{(Answer)}$$

31-10 Energy Stored in a Magnetic Field

When we pull two particles with opposite signs of charge away from each other, we say that the resulting electric potential energy is stored in the electric field of the particles. We get it back from the field by letting the particles move closer together again. In the same way we can consider energy to be stored in a magnetic field.

To derive a quantitative expression for that stored energy, consider again Fig. 31-18, which shows a source of emf $\mathscr{E}$ connected to a resistor R and an inductor L. Equation 31-41, restated here for convenience,

$$\mathscr{E} = L\frac{di}{dt} + iR, \qquad (31\text{-}48)$$

is the differential equation that describes the growth of current in this circuit. Recall that this equation follows immediately from the loop rule and that the loop rule in turn is an expression of the principle of conservation of energy for single-loop circuits. If we multiply each side of Eq. 31-48 by i, we obtain

$$\mathscr{E}i = Li\frac{di}{dt} + i^2R, \qquad (31\text{-}49)$$

which has the following physical interpretation in terms of work and energy:

1. If a charge dq passes through the battery of emf $\mathscr{E}$ in Fig. 31-18 in time dt, the battery does work on it in the amount $\mathscr{E}\,dq$. The rate at which the battery does work is $(\mathscr{E}\,dq)/dt$, or $\mathscr{E}i$. Thus, the left side of Eq. 31-49 represents the rate at which the emf device delivers energy to the rest of the circuit.

2. The rightmost term in Eq. 31-49 represents the rate at which energy appears as thermal energy in the resistor.

3. Energy that is delivered to the circuit but does not appear as thermal energy must, by the conservation-of-energy hypothesis, be stored in the magnetic field of the inductor. Since Eq. 31-49 represents the principle of conservation of energy for RL circuits, the middle term must represent the rate dU_B/dt at which energy is stored in the magnetic field.

Thus

$$\frac{dU_B}{dt} = Li\frac{di}{dt}. \qquad (31\text{-}50)$$

We can write this as

$$dU_B = Li\,di.$$

Integrating yields

$$\int_0^{U_B} dU_B = \int_0^i Li \, di$$

or

$$U_B = \tfrac{1}{2}Li^2 \qquad \text{(magnetic energy)}, \qquad (31\text{-}51)$$

which represents the total energy stored by an inductor L carrying a current i. Note the similarity in form between this expression and the expression for the energy stored by a capacitor with capacitance C and charge q; namely,

$$U_E = \frac{q^2}{2C}. \qquad (31\text{-}52)$$

(The variable i^2 corresponds to q^2, and the constant L corresponds to $1/C$.)

Sample Problem 31-7

A coil has an inductance of 53 mH and a resistance of 0.35 Ω.

(a) If a 12 V emf is applied across the coil, how much energy is stored in the magnetic field after the current has built up to its equilibrium value?

SOLUTION: The Key Idea here is that the energy stored in the magnetic field of a coil at any time depends on the current through the coil at that time, according to Eq. 31-51 ($U_B = \tfrac{1}{2}Li^2$). Thus, to find the energy $U_{B\infty}$ stored at equilibrium, we must first find the equilibrium current. From Eq. 31-43, the equilibrium current is

$$i_\infty = \frac{\mathscr{E}}{R} = \frac{12 \text{ V}}{0.35 \ \Omega} = 34.3 \text{ A}. \qquad (31\text{-}53)$$

Then substitution yields

$$U_{B\infty} = \tfrac{1}{2}Li_\infty^2 = (\tfrac{1}{2})(53 \times 10^{-3} \text{ H})(34.3 \text{ A})^2$$
$$= 31 \text{ J}. \qquad \text{(Answer)}$$

(b) After how many time constants will half this equilibrium energy be stored in the magnetic field?

SOLUTION: The Key Idea of part (a) applies here also. Now we are being asked: At what time t will the relation

$$U_B = \tfrac{1}{2}U_{B\infty}$$

be satisfied? Using Eq. 31-51 twice allows us to rewrite this energy condition as

$$\tfrac{1}{2}Li^2 = (\tfrac{1}{2})\tfrac{1}{2}Li_\infty^2$$

or

$$i = \left(\frac{1}{\sqrt{2}}\right)i_\infty. \qquad (31\text{-}54)$$

However, i is given by Eq. 31-43 and i_∞ (see Eq. 31-53) is $\mathscr{E}/R$, so Eq. 31-54 becomes

$$\frac{\mathscr{E}}{R}(1 - e^{-t/\tau_L}) = \frac{\mathscr{E}}{\sqrt{2}R}.$$

By canceling $\mathscr{E}/R$ and rearranging, this can be written as

$$e^{-t/\tau_L} = 1 - \frac{1}{\sqrt{2}} = 0.293,$$

which yields

$$\frac{t}{\tau_L} = -\ln 0.293 = 1.23$$

or

$$t \approx 1.2\tau_L. \qquad \text{(Answer)}$$

Thus, the stored energy will reach half its equilibrium value 1.2 time constants after the emf is applied.

31-11 Energy Density of a Magnetic Field

Consider a length l near the middle of a long solenoid of cross-sectional area A carrying current i; the volume associated with this length is Al. The energy U_B stored by the length l of the solenoid must lie entirely within this volume because the magnetic field outside such a solenoid is approximately zero. Moreover, the stored energy must be uniformly distributed within the solenoid because the magnetic field is (approximately) uniform everywhere inside.

Thus, the energy stored per unit volume of the field is

$$u_B = \frac{U_B}{Al}$$

or, since

$$U_B = \tfrac{1}{2}Li^2,$$

we have

$$u_B = \frac{Li^2}{2Al} = \frac{L}{l}\frac{i^2}{2A}.$$

Here L is the inductance of length l of the solenoid.

Substituting for L/l from Eq. 31-33, we find

$$u_B = \tfrac{1}{2}\mu_0 n^2 i^2, \tag{31-55}$$

where n is the number of turns per unit length. From Eq. 30-25 ($B = \mu_0 in$) we can write this *energy density* as

$$u_B = \frac{B^2}{2\mu_0} \qquad \text{(magnetic energy density).} \tag{31-56}$$

This equation gives the density of stored energy at any point where the magnetic field is B. Even though we derived it by considering the special case of a solenoid, Eq. 31-56 holds for all magnetic fields, no matter how they are generated. The equation is comparable to Eq. 26-23; namely,

$$u_E = \tfrac{1}{2}\varepsilon_0 E^2, \tag{31-57}$$

which gives the energy density (in a vacuum) at any point in an electric field. Note that both u_B and u_E are proportional to the square of the appropriate field magnitude, B or E.

✔ **CHECKPOINT 7:**   The table lists the number of turns per unit length, current, and cross-sectional area for three solenoids. Rank the solenoids according to the magnetic energy density within them, greatest first.

Solenoid	Turns per Unit Length	Current	Area
a	$2n_1$	i_1	$2A_1$
b	n_1	$2i_1$	A_1
c	n_1	i_1	$6A_1$

Sample Problem 31-8

A long coaxial cable (Fig. 31-21) consists of two thin-walled concentric conducting cylinders with radii a and b. The inner cylinder carries a steady current i, the outer cylinder providing the return path for that current. The current sets up a magnetic field between the two cylinders.

(a) Calculate the energy stored in the magnetic field for a length ℓ of the cable.

SOLUTION: The Key Ideas for this challenging problem are these:

1. We can calculate the (total) energy U_B stored in the magnetic field from the energy density u_B of the field.

2. That energy density depends on the magnitude B of the magnetic field according to Eq. 31-56 ($u_B = B^2/2\mu_0$).

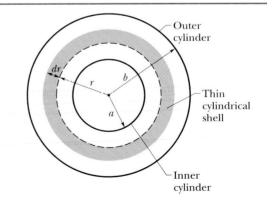

Fig. 31-21 Sample Problem 31-8. A cross section of a long coaxial cable consisting of two thin-walled conducting cylinders, the inner cylinder of radius a and the outer cylinder of radius b.

3. Because of the circular symmetry of the cable, we can find B by using Ampere's law with the given current i.

Finding B: To apply these ideas, we begin with Ampere's law, using a circular path of integration with radius r such that $a < r < b$ (between the two cylinders, as indicated by the dashed line in Fig. 31-21). The only current enclosed by this path is current i on the inner cylinder. Thus, we can write Ampere's law as

$$\oint \vec{B} \cdot d\vec{s} = \mu_0 i. \qquad (31\text{-}58)$$

Next, we simplify the integral: Because of the circular symmetry, at all points along the circular path, $\vec{B}$ is tangent to the path and has the same magnitude B. Let us take the direction of integration along the path as the direction of the magnetic field around the path. Then we can replace $\vec{B} \cdot d\vec{s}$ with $B\,ds\cos 0 = B\,ds$, and then move magnitude B in front of the integration. The integral that remains is $\oint ds$, which just gives the circumference $2\pi r$ of the path. Thus, Eq. 31-58 simplifies to

$$B(2\pi r) = \mu_0 i$$

or

$$B = \frac{\mu_0 i}{2\pi r}. \qquad (31\text{-}59)$$

Finding u_B: Next, to obtain the energy density, we substitute Eq. 31-59 into Eq. 31-56:

$$u_B = \frac{B^2}{2\mu_0} = \frac{\mu_0 i^2}{8\pi^2 r^2}. \qquad (31\text{-}60)$$

Finding U_B: Note that u_B is not uniform in the volume between the two cylinders, but instead depends on the radial distance r. Thus, to find the total energy U_B stored between the cylinders, we must integrate u_B over that volume.

Because the volume between the cylinders has circular symmetry about the cable's central axis, we consider the volume dV of a cylindrical shell located between the cylinders; the shell has inner radius r, outer radius $r + dr$ (Fig. 31-21), and length ℓ. The shell's cross-sectional area (or face area) is the product of its circumference $2\pi r$ and thickness dr. Thus, the shell's volume dV is $(2\pi r)(dr)(\ell)$; that is, $dV = 2\pi r\ell\,dr$.

Because points within this shell are all at approximately the same radial distance r, they all have approximately the same energy density u_B. Thus, the total energy dU_B contained in the shell of volume dV is given by

$$\text{energy} = \left(\begin{array}{c}\text{energy per}\\\text{unit volume}\end{array}\right)(\text{volume})$$

or

$$dU_B = u_B\,dV.$$

Substituting Eq. 31-60 for u_B and $2\pi r\ell\,dr$ for dV, we obtain

$$dU_B = \frac{\mu_0 i^2}{8\pi^2 r^2}(2\pi r\ell)\,dr = \frac{\mu_0 i^2 \ell}{4\pi}\frac{dr}{r}.$$

To find the total energy contained between the two cylinders, we integrate this equation over the volume between the two cylinders:

$$U_B = \int dU_B = \frac{\mu_0 i^2 \ell}{4\pi}\int_a^b \frac{dr}{r}$$

$$= \frac{\mu_0 i^2 \ell}{4\pi}\ln\frac{b}{a}. \qquad (\text{Answer}) \quad (31\text{-}61)$$

No energy is stored outside the outer cylinder or inside the inner cylinder because the magnetic field is zero in both locations, as you can show with Ampere's law.

(b) What is the stored energy per unit length of the cable if $a = 1.2$ mm, $b = 3.5$ mm, and $i = 2.7$ A?

SOLUTION: From Eq. 31-61 we have

$$\frac{U_B}{\ell} = \frac{\mu_0 i^2}{4\pi}\ln\frac{b}{a}$$

$$= \frac{(4\pi \times 10^{-7}\ \text{H/m})(2.7\ \text{A})^2}{4\pi}\ln\frac{3.5\ \text{mm}}{1.2\ \text{mm}}$$

$$= 7.8 \times 10^{-7}\ \text{J/m} = 780\ \text{nJ/m}. \qquad (\text{Answer})$$

31-12 Mutual Induction

In this section we return to the case of two interacting coils, which we first discussed in Section 31-2, and we treat it in a somewhat more formal manner. We saw earlier that if two coils are close together as in Fig. 31-2, a steady current i in one coil will set up a magnetic flux Φ through the other coil (*linking* the other coil). If we change i with time, an emf $\mathscr{E}$ given by Faraday's law appears in the second coil; we called this process *induction*. We could better have called it **mutual induction,** to suggest the mutual interaction of the two coils and to distinguish it from *self-induction,* in which only one coil is involved.

Let us look a little more quantitatively at mutual induction. Figure 31-22a shows two circular close-packed coils near each other and sharing a common central axis. There is a steady current i_1 in coil 1, produced by the battery in the external circuit. This current creates a magnetic field represented by the lines of $\vec{B}_1$ in the figure. Coil 2 is connected to a sensitive meter but contains no battery; a magnetic flux Φ_{21} (the flux through coil 2 associated with the current in coil 1) links the N_2 turns of coil 2.

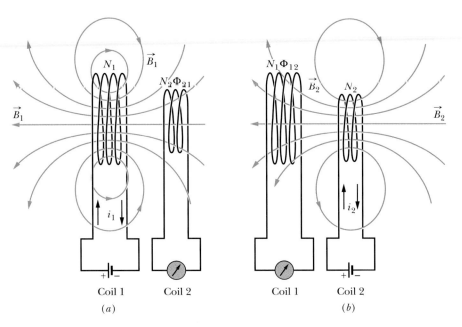

Fig. 31-22 Mutual induction. (*a*) If the current in coil 1 changes, an emf will be induced in coil 2. (*b*) If the current in coil 2 changes, an emf will be induced in coil 1.

We define the mutual inductance M_{21} of coil 2 with respect to coil 1 as

$$M_{21} = \frac{N_2 \Phi_{21}}{i_1}, \tag{31-62}$$

which has the same form as Eq. 31-30 ($L = N\Phi/i$), the definition of (self) inductance. We can recast Eq. 31-62 as

$$M_{21} i_1 = N_2 \Phi_{21}.$$

If, by external means, we cause i_1 to vary with time, we have

$$M_{21} \frac{di_1}{dt} = N_2 \frac{d\Phi_{21}}{dt}.$$

The right side of this equation is, according to Faraday's law, just the magnitude of the emf $\mathcal{E}_2$ appearing in coil 2 due to the changing current in coil 1. Thus, with a minus sign to indicate direction,

$$\mathcal{E}_2 = -M_{21} \frac{di_1}{dt}, \tag{31-63}$$

which you should compare with Eq. 31-37 for self-induction ($\mathcal{E} = -L\, di/dt$).

Let us now interchange the roles of coils 1 and 2, as in Fig. 31-22*b*; that is, we set up a current i_2 in coil 2 by means of a battery, and this produces a magnetic flux Φ_{12} that links coil 1. If we change i_2 with time, we have, by the argument given above,

$$\mathcal{E}_1 = -M_{12} \frac{di_2}{dt}. \tag{31-64}$$

Thus, we see that the emf induced in either coil is proportional to the rate of change of current in the other coil. The proportionality constants M_{21} and M_{12} seem to be different. We assert, without proof, that they are in fact the same so that no subscripts are needed. (This conclusion is true but is in no way obvious.) Thus, we have

$$M_{21} = M_{12} = M, \tag{31-65}$$

and we can rewrite Eqs. 31-63 and 31-64 as

$$\mathcal{E}_2 = -M \frac{di_1}{dt} \qquad (31\text{-}66)$$

and

$$\mathcal{E}_1 = -M \frac{di_2}{dt}. \qquad (31\text{-}67)$$

The induction is indeed mutual. The SI unit for M (as for L) is the henry.

Sample Problem 31-9

Figure 31-23 shows two circular close-packed coils, the smaller (radius R_2, with N_2 turns) being coaxial with the larger (radius R_1, with N_1 turns) and in the same plane.

(a) Derive an expression for the mutual inductance M for this arrangement of these two coils, assuming that $R_1 \gg R_2$.

SOLUTION: The Key Idea here is that the mutual inductance M for these coils is the ratio of the flux linkage ($N\Phi$) through one coil to the current i in the other coil, which produces that flux linkage. Thus, we need to assume that currents exist in the coils; then we need to calculate the flux linkage in one of the coils.

The magnetic field through the larger coil due to the smaller coil is nonuniform in both magnitude and direction, so the flux through the larger coil due to the smaller coil is nonuniform and difficult to calculate. However, the smaller coil is small enough for us to assume that the magnetic field through it due to the larger coil is approximately uniform. Thus, the flux through it due to the larger coil is also approximately uniform. Hence, to find M we shall assume a current i_1 in the larger coil and calculate the flux linkage $N_2\Phi_{21}$ in the smaller coil:

$$M = \frac{N_2\Phi_{21}}{i_1}. \qquad (31\text{-}68)$$

A second Key Idea is that the flux Φ_{21} through each turn of the smaller coil is, from Eq. 31-4,

$$\Phi_{21} = B_1 A_2,$$

where B_1 is the magnitude of the magnetic field at points within the small coil due to the larger coil, and A_2 ($= \pi R_2^2$) is the area enclosed by the turn. Thus, the flux linkage in the smaller coil (with its N_2 turns) is

$$N_2\Phi_{21} = N_2 B_1 A_2. \qquad (31\text{-}69)$$

A third Key Idea is that to find B_1 at points within the smaller coil, we can use Eq. 30-28, with z set to 0 because the smaller coil is in the plane of the larger coil. That equation tells us that each turn of the larger coil produces a magnetic field of magnitude $\mu_0 i_1 / 2R_1$ at points within the smaller coil. Thus, the larger coil (with its N_1 turns) produces a total magnetic field of magnitude

$$B_1 = N_1 \frac{\mu_0 i_1}{2R_1} \qquad (31\text{-}70)$$

at points within the smaller coil.

Substituting Eq. 31-70 for B_1 and πR_2^2 for A_2 in Eq. 31-69 yields

$$N_2\Phi_{21} = \frac{\pi \mu_0 N_1 N_2 R_2^2 i_1}{2R_1}.$$

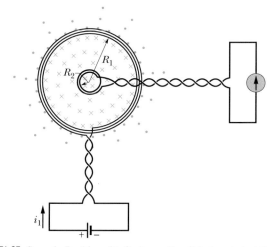

Fig. 31-23 Sample Problem 31-9. A small coil is located at the center of a large coil. The mutual inductance of the coils can be determined by sending current i_1 through the large coil.

Substituting this result into Eq. 31-68, we find

$$M = \frac{N_2\Phi_{21}}{i_1} = \frac{\pi \mu_0 N_1 N_2 R_2^2}{2R_1}. \quad \text{(Answer)} \quad (31\text{-}71)$$

(b) What is the value of M for $N_1 = N_2 = 1200$ turns, $R_2 = 1.1$ cm, and $R_1 = 15$ cm?

SOLUTION: Equation 31-71 yields

$$M = \frac{(\pi)(4\pi \times 10^{-7} \text{ H/m})(1200)(1200)(0.011 \text{ m})^2}{(2)(0.15 \text{ m})}$$
$$= 2.29 \times 10^{-3} \text{ H} \approx 2.3 \text{ mH}. \quad \text{(Answer)}$$

Consider the situation if we reverse the roles of the two coils—that is, if we produce a current i_2 in the smaller coil and try to calculate M from Eq. 31-62 in the form

$$M = \frac{N_1\Phi_{12}}{i_2}.$$

The calculation of Φ_{12} (the nonuniform flux of the smaller coil's magnetic field encompassed by the larger coil) is not simple. If we were to do the calculation numerically using a computer, we would find M to be 2.3 mH, as above! This emphasizes that Eq. 31-65 ($M_{21} = M_{12} = M$) is not obvious.

REVIEW & SUMMARY

Magnetic Flux   The *magnetic flux* Φ_B through an area A in a magnetic field $\vec{B}$ is defined as

$$\Phi_B = \int \vec{B} \cdot d\vec{A}, \tag{31-3}$$

where the integral is taken over the area. The SI unit of magnetic flux is the weber, where 1 Wb = 1 T · m². If $\vec{B}$ is perpendicular to the area and uniform over it, Eq. 31-3 becomes

$$\Phi_B = BA \qquad (\vec{B} \perp A, \vec{B} \text{ uniform}). \tag{31-4}$$

Faraday's Law of Induction   If the magnetic flux Φ_B through an area bounded by a closed conducting loop changes with time, a current and an emf are produced in the loop; this process is called *induction*. The induced emf is

$$\mathcal{E} = -\frac{d\Phi_B}{dt} \qquad \text{(Faraday's law)}. \tag{31-6}$$

If the loop is replaced by a closely packed coil of N turns, the induced emf is

$$\mathcal{E} = -N\frac{d\Phi_B}{dt}. \tag{31-7}$$

Lenz's Law   An induced current has a direction such that the magnetic field *of the current* opposes the change in the magnetic flux that produces the current. The induced emf has the same direction as the induced current.

Emf and the Induced Electric Field   An emf is induced by a changing magnetic flux even if the loop through which the flux is changing is not a physical conductor but an imaginary line. The changing magnetic field induces an electric field $\vec{E}$ at every point of such a loop; the induced emf is related to $\vec{E}$ by

$$\mathcal{E} = \oint \vec{E} \cdot d\vec{s}, \tag{31-21}$$

where the integration is taken around the loop. From Eq. 31-21 we can write Faraday's law in its most general form,

$$\oint \vec{E} \cdot d\vec{s} = -\frac{d\Phi_B}{dt} \qquad \text{(Faraday's law)}. \tag{31-22}$$

The essence of this law is that *a changing magnetic field induces an electric field $\vec{E}$.*

Inductors   An **inductor** is a device that can be used to produce a known magnetic field in a specified region. If a current i is established through each of the N windings of an inductor, a magnetic flux Φ_B links those windings. The **inductance** L of the inductor is

$$L = \frac{N\Phi_B}{i} \qquad \text{(inductance defined)}. \tag{31-30}$$

The SI unit of inductance is the **henry** (H), with

$$1 \text{ henry} = 1 \text{ H} = 1 \text{ T} \cdot \text{m}^2/\text{A}. \tag{31-31}$$

The inductance per unit length near the middle of a long solenoid of cross-sectional area A and n turns per unit length is

$$\frac{L}{l} = \mu_0 n^2 A \qquad \text{(solenoid)}. \tag{31-33}$$

Self-Induction   If a current i in a coil changes with time, an emf is induced in the coil. This self-induced emf is

$$\mathcal{E}_L = -L\frac{di}{dt}. \tag{31-37}$$

The direction of $\mathcal{E}_L$ is found from Lenz's law: The self-induced emf acts to oppose the change that produces it.

Series RL Circuits   If a constant emf $\mathcal{E}$ is introduced into a single-loop circuit containing a resistance R and an inductance L, the current rises to an equilibrium value of $\mathcal{E}/R$ according to

$$i = \frac{\mathcal{E}}{R}(1 - e^{-t/\tau_L}) \qquad \text{(rise of current)}. \tag{31-43}$$

Here $\tau_L\,(=L/R)$ governs the rate of rise of the current and is called the **inductive time constant** of the circuit. When the source of constant emf is removed, the current decays from a value i_0 according to

$$i = i_0 e^{-t/\tau_L} \qquad \text{(decay of current)}. \tag{31-47}$$

Magnetic Energy   If an inductor L carries a current i, the inductor's magnetic field stores an energy given by

$$U_B = \tfrac{1}{2}Li^2 \qquad \text{(magnetic energy)}. \tag{31-51}$$

If B is the magnitude of a magnetic field at any point (in an inductor or anywhere else), the density of stored magnetic energy at that point is

$$u_B = \frac{B^2}{2\mu_0} \qquad \text{(magnetic energy density)}. \tag{31-56}$$

Mutual Induction   If two coils (labeled 1 and 2) are near each other, a changing current in either coil can induce an emf in the other. This mutual induction is described by

$$\mathcal{E}_2 = -M\frac{di_1}{dt} \tag{31-66}$$

and

$$\mathcal{E}_1 = -M\frac{di_2}{dt}, \tag{31-67}$$

where M (measured in henries) is the mutual inductance for the coil arrangement.

QUESTIONS

1. In Fig. 31-24, a long straight wire with current i passes (without touching) three rectangular wire loops with edge lengths L, $1.5L$, and $2L$. The loops are widely spaced (so as to not affect one another). Loops 1 and 3 are symmetric about the long wire. Rank the loops according to the size of the current induced in them if current i is (a) constant and (b) increasing, greatest first.

Fig. 31-24 Question 1.

2. If the circular conductor in Fig. 31-25 undergoes thermal expansion while it is in a uniform magnetic field, a current will be induced clockwise around it. Is the magnetic field directed into the page or out of it?

Fig. 31-25 Question 2.

3. Figure 31-26 shows two circuits in which a conducting bar is slid at the same speed v through the same uniform magnetic field and along a **U**-shaped wire. The parallel lengths of the wire are separated by $2L$ in circuit 1 and by L in circuit 2. The current induced in circuit 1 is counterclockwise. (a) Is the direction of the magnetic field into or out of the page? (b) Is the direction of the current induced in circuit 2 clockwise or counterclockwise? (c) Is the emf induced in circuit 1 larger than, smaller than, or the same as that in circuit 2?

Fig. 31-26 Question 3.   (1)   (2)

4. Figure 31-27 shows two coils wrapped around nonconducting rods. Coil X is connected to a battery and a variable resistance. What is the direction of the induced current through the current meter connected to coil Y (a) when coil Y is moved toward coil X and (b) when the current in coil X is decreased without any change in the relative positions of the coils?

Fig. 31-27 Question 4.

5. Figure 31-28a shows a circular region in which an increasing uniform magnetic field is directed out of the page, as well as a concentric circular path along which $\oint \vec{E} \cdot d\vec{s}$ is to be evaluated. The table gives the initial magnitude of the magnetic field, the increase in that magnitude, and the time interval for the increase, in three situations. Rank the situations according to the magnitude of the electric field induced along the path, greatest first.

Situation	Initial Field	Increase	Time
a	B_1	ΔB_1	Δt_1
b	$2B_1$	$\Delta B_1/2$	Δt_1
c	$B_1/4$	ΔB_1	$\Delta t_1/2$

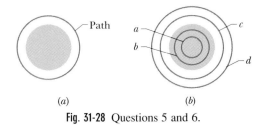

Fig. 31-28 Questions 5 and 6.

6. Figure 31-28b shows a circular region in which a decreasing uniform magnetic field is directed out of the page, as well as four concentric circular paths. Rank the paths according to the magnitude of $\oint \vec{E} \cdot d\vec{s}$ evaluated along them, greatest first.

7. Figure 31-29 gives the variation with time of the potential difference V_R across a resistor in three circuits wired as shown in Fig. 31-18. The circuits contain the same resistance R and emf $\mathscr{E}$ but differ in the inductance L. Rank the circuits according to the value of L, greatest first.

Fig. 31-29 Question 7.

8. Figure 31-30 shows three circuits with identical batteries, inductors, and resistors. Rank the circuits according to the time for the current to reach 50% of its equilibrium value after the switches are closed, greatest first.

Fig. 31-30 Question 8.

9. Figure 31-31 shows a circuit with two identical resistors and an ideal inductor. Is the current through the central resistor more than, less than, or the same as that through the other resistor (a) just after the closing of switch S, (b) a long time after the closing of S, (c) just after S is reopened, a long time later, and (d) a long time after the reopening of S?

Fig. 31-31 Question 9.

10. The switch in the circuit of Fig. 31-17 has been closed on *a* for a very long time when it is then thrown to *b*. The resulting current through the inductor is indicated in Fig. 31-32 for four sets of values for the resistance *R* and inductance *L*: (1) R_0 and L_0, (2) $2R_0$ and L_0, (3) R_0 and $2L_0$, (4) $2R_0$ and $2L_0$. Which set goes with which curve?

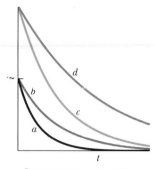

Fig. 31-32 Question 10.

EXERCISES & PROBLEMS

SEC. 31-4   Lenz's Law

1E. A UHF television loop antenna has a diameter of 11 cm. The magnetic field of a TV signal is normal to the plane of the loop and, at one instant of time, its magnitude is changing at the rate 0.16 T/s. The magnetic field is uniform. What emf is induced in the antenna?   ssm

2E. A small loop of area *A* is inside of, and has its axis in the same direction as, a long solenoid of *n* turns per unit length and current *i*. If $i = i_0 \sin \omega t$, find the emf induced in the loop.

3E. The magnetic flux through the loop shown in Fig. 31-33 increases according to the relation $\Phi_B = 6.0t^2 + 7.0t$, where Φ_B is in milliwebers and *t* is in seconds. (a) What is the magnitude of the emf induced in the loop when *t* = 2.0 s? (b) What is the direction of the current through *R*?

Fig. 31-33 Exercise 3 and Problem 11.

4E. The magnetic field through a single loop of wire, 12 cm in radius and of 8.5 Ω resistance, changes with time as shown in Fig. 31-34. Calculate the emf in the loop as a function of time. Consider the time intervals (a) *t* = 0 to *t* = 2.0 s, (b) *t* = 2.0 s to *t* = 4.0 s, (c) *t* = 4.0 s to *t* = 6.0 s. The (uniform) magnetic field is perpendicular to the plane of the loop.

Fig. 31-34 Exercise 4.

5E. A uniform magnetic field is normal to the plane of a circular loop 10 cm in diameter and made of copper wire (of diameter 2.5 mm). (a) Calculate the resistance of the wire. (See Table 27-1.) (b) At what rate must the magnetic field change with time if an induced current of 10 A is to appear in the loop?

6P. The current in the solenoid of Sample Problem 31-1 changes, not as stated there, but according to $i = 3.0t + 1.0t^2$, where *i* is in amperes and *t* is in seconds. (a) Plot the induced emf in the coil from *t* = 0 to *t* = 4.0 s. (b) The resistance of the coil is 0.15 Ω. What is the current in the coil at *t* = 2.0 s?

7P. In Fig. 31-35 a 120-turn coil of radius 1.8 cm and resistance 5.3 Ω is placed *outside* a solenoid like that of Sample Problem 31-1. If the current in the solenoid is changed as in that sample problem, what current appears in the coil while the solenoid current is being changed?   ssm

Fig. 31-35 Problem 7.

8P. An elastic conducting material is stretched into a circular loop of 12.0 cm radius. It is placed with its plane perpendicular to a uniform 0.800 T magnetic field. When released, the radius of the loop starts to shrink at an instantaneous rate of 75.0 cm/s. What emf is induced in the loop at that instant?

9P. Figure 31-36 shows two parallel loops of wire having a common axis. The smaller loop (radius *r*) is above the larger loop (radius *R*) by a distance $x \gg R$. Consequently, the magnetic field due to the current *i* in the larger loop is nearly constant throughout the smaller loop. Suppose that *x* is increasing at the constant rate of $dx/dt = v$. (a) Determine the magnetic flux through the area bounded by the smaller loop as a function of *x*. (*Hint:* See Eq. 30-29.) In the smaller loop, find (b) the induced emf and (c) the direction of the induced current.   ssm   www

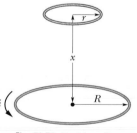

Fig. 31-36 Problem 9.

10P. In Fig. 31-37, a circular loop of wire 10 cm in diameter (seen edge-on) is placed with its normal $\vec{N}$ at an angle $\theta = 30°$ with the direction of a uniform magnetic field $\vec{B}$ of magnitude 0.50 T. The loop is then rotated such that $\vec{N}$ rotates in a cone about the field direction at the constant rate of 100 rev/min; the angle θ remains unchanged during the process. What is the emf induced in the loop?

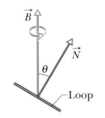

Fig. 31-37 Problem 10.

11P. In Fig. 31-33 let the flux through the loop be $\Phi_B(0)$ at time $t = 0$. Then let the magnetic field $\vec{B}$ vary in a continuous but unspecified way, in both magnitude and direction, so that at time t the flux is represented by $\Phi_B(t)$. (a) Show that the net charge $q(t)$ that has passed through resistor R in time t is

$$q(t) = \frac{1}{R}[\Phi_B(0) - \Phi_B(t)]$$

and is independent of the way $\vec{B}$ has changed. (b) If $\Phi_B(t) = \Phi_B(0)$ in a particular case, we have $q(t) = 0$. Is the induced current necessarily zero throughout the interval from 0 to t? ssm

12P. A small circular loop of area 2.00 cm^2 is placed in the plane of, and concentric with, a large circular loop of radius 1.00 m. The current in the large loop is changed uniformly from 200 A to -200 A (a change in direction) in a time of 1.00 s, beginning at $t = 0$. (a) What is the magnetic field at the center of the small circular loop due to the current in the large loop at $t = 0$, $t = 0.500$ s, and $t = 1.00$ s? (b) What emf is induced in the small loop at $t = 0.500$ s? (Since the inner loop is small, assume the field $\vec{B}$ due to the outer loop is uniform over the area of the smaller loop.)

13P. One hundred turns of insulated copper wire are wrapped around a wooden cylindrical core of cross-sectional area 1.20×10^{-3} m^2. The two terminals are connected to a resistor. The total resistance in the circuit is 13.0 Ω. If an externally applied uniform longitudinal magnetic field in the core changes from 1.60 T in one direction to 1.60 T in the opposite direction, how much charge flows through the circuit? (*Hint:* See Problem 11.) ilw

14P. At a certain place, Earth's magnetic field has magnitude $B = 0.590$ gauss and is inclined downward at an angle of 70.0° to the horizontal. A flat horizontal circular coil of wire with a radius of 10.0 cm has 1000 turns and a total resistance of 85.0 Ω. It is connected to a meter with 140 Ω resistance. The coil is flipped through a half-revolution about a diameter, so that it is again horizontal. How much charge flows through the meter during the flip? (*Hint:* See Problem 11.)

15P. A square wire loop with 2.00 m sides is perpendicular to a uniform magnetic field, with half the area of the loop in the field as shown in Fig. 31-38. The loop contains a 20.0 V battery with negligible internal resistance. If the magnitude of the field varies

Fig. 31-38 Problem 15.

with time according to $B = 0.0420 - 0.870t$, with B in teslas and t in seconds, what are (a) the net emf in the circuit and (b) the direction of the current through the battery? ssm

16P. A wire is bent into three circular segments, each of radius $r = 10$ cm, as shown in Fig. 31-39. Each segment is a quadrant of a circle, ab lying in the xy plane, bc lying in the yz plane, and ca lying in the zx plane. (a) If a uniform magnetic field $\vec{B}$ points in the positive x direction, what is the magnitude of the emf developed in the wire when B increases at the rate of 3.0 mT/s? (b) What is the direction of the current in segment bc?

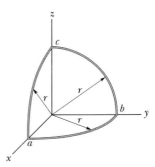

Fig. 31-39 Problem 16.

17P. A rectangular coil of N turns and of length a and width b is rotated at frequency f in a uniform magnetic field $\vec{B}$, as indicated in Fig. 31-40. The coil is connected to co-rotating cylinders, against which metal brushes slide to make contact. (a) Show that the emf induced in the coil is given (as a function of time t) by

$$\mathcal{E} = 2\pi fNabB \sin(2\pi ft) = \mathcal{E}_0 \sin(2\pi ft).$$

This is the principle of the commercial alternating-current generator. (b) Design a loop that will produce an emf with $\mathcal{E}_0 = 150$ V when rotated at 60.0 rev/s in a uniform magnetic field of 0.500 T. ssm www

Fig. 31-40 Problem 17.

18P. A stiff wire bent into a semicircle of radius a is rotated with frequency f in a uniform magnetic field, as suggested in Fig. 31-41. What are (a) the frequency and (b) the amplitude of the varying emf induced in the loop?

Fig. 31-41 Problem 18

19P. An electric generator consists of 100 turns of wire formed into a rectangular loop 50.0 cm by 30.0 cm, placed entirely in a uniform magnetic field with magnitude $B = 3.50$ T. What is the maximum value of the emf produced when the loop is spun at 1000 rev/min about an axis perpendicular to $\vec{B}$? ilw

20P. In Fig. 31-42, a wire forms a closed circular loop, with radius $R = 2.0$ m and resistance 4.0 Ω. The circle is centered on a long straight wire; at time $t = 0$, the current in the long straight wire is 5.0 A rightward. Thereafter, the current changes according to $i = 5.0$ A $- (2.0$ A/s$^2)t^2$. (The straight wire is insulated, so there is no electrical contact between it and the wire of the loop.) What are the magnitude and direction of the current induced in the loop at times $t > 0$?

Fig. 31-42 Problem 20.

21P. In Fig. 31-43, the square loop of wire has sides of length 2.0 cm. A magnetic field is directed out of the page; its magnitude is given by $B = 4.0t^2y$, where B is in teslas, t is in seconds, and y is in meters. Determine the emf around the square at $t = 2.5$ s and give its direction. ssm ilw

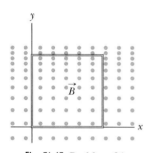

Fig. 31-43 Problem 21.

22P. For the situation shown in Fig. 31-44, $a = 12.0$ cm and $b = 16.0$ cm. The current in the long straight wire is given by $i = 4.50t^2 - 10.0t$, where i is in amperes and t is in seconds. (a) Find the emf in the square loop at $t = 3.00$ s. (b) What is the direction of the induced current in the loop?

Fig. 31-44 Problem 22.

23P*. Two long, parallel copper wires of diameter 2.5 mm carry currents of 10 A in opposite directions. (a) Assuming that their central axes are 20 mm apart, calculate the magnetic flux per meter of wire that exists in the space between those axes. (b) What fraction of this flux lies inside the wires? (c) Repeat part (a) for parallel currents. ssm

24P. A rectangular loop of wire with length a, width b, and resistance R is placed near an infinitely long wire carrying current i, as shown in Fig. 31-45. The distance from the long wire to the center of the loop is r. Find (a) the magnitude of the magnetic flux through the loop and (b) the current in the loop as it moves away from the long wire with speed v.

Fig. 31-45 Problem 24.

SEC. 31-5   Induction and Energy Transfers

25E. If 50.0 cm of copper wire (diameter = 1.00 mm) is formed into a circular loop and placed perpendicular to a uniform magnetic field that is increasing at the constant rate of 10.0 mT/s, at what rate is thermal energy generated in the loop? ssm ilw

26E. A loop antenna of area A and resistance R is perpendicular to a uniform magnetic field $\vec{B}$. The field drops linearly to zero in a time interval Δt. Find an expression for the total thermal energy dissipated in the loop.

27E. A metal rod is forced to move with constant velocity $\vec{v}$ along two parallel metal rails, connected with a strip of metal at one end, as shown in Fig. 31-46. A magnetic field $B = 0.350$ T points out of the page. (a) If the rails are separated by 25.0 cm and the speed of the rod is 55.0 cm/s, what emf is generated? (b) If the rod has a resistance of 18.0 Ω and the rails and connector have negligible resistance, what is the current in the rod? (c) At what rate is energy being transferred to thermal energy? ssm

Fig. 31-46 Exercise 27 and Problem 29.

28P. In Fig. 31-47, a long rectangular conducting loop, of width L, resistance R, and mass m, is hung in a horizontal, uniform magnetic field $\vec{B}$ that is directed into the page and that exists only above line aa. The loop is then dropped; during its fall, it accelerates until it reaches a certain terminal speed v_t. Ignoring air drag, find that terminal speed.

29P. The conducting rod shown in Fig. 31-46 has length L and is being pulled along horizontal, frictionless conducting rails at a constant velocity $\vec{v}$. The rails are connected at one end with a metal strip. A uniform magnetic field $\vec{B}$, directed out of the page, fills the region in which the rod moves. Assume that $L = 10$ cm, $v = 5.0$ m/s, and $B = 1.2$ T. (a) What are the magnitude and direction of the emf induced in the rod? (b) What is the current in the conducting loop? Assume that the resistance of the rod is 0.40 Ω and that the resistance of the rails and metal strip is negligibly small. (c) At what rate is thermal energy being generated in the rod? (d) What force must be applied to the rod by an external agent to maintain its motion? (e) At what rate does this external agent do work on the rod? Compare this answer with the answer to (c). ssm

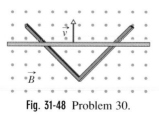

Fig. 31-47 Problem 28.

30P. Two straight conducting rails form a right angle where their ends are joined. A conducting bar in contact with the rails starts at the vertex at time $t = 0$ and moves with a constant velocity of 5.20 m/s along them, as shown in Fig. 31-48. A magnetic field with $B = 0.350$ T is directed out of the page. Calculate (a) the flux through the triangle formed by the rails and bar at $t = 3.00$ s and (b) the emf around the triangle at that time. (c) If we write the emf as $\mathscr{E} = at^n$, where a and n are constants, what is the value of n?

Fig. 31-48 Problem 30.

31P. Figure 31-49 shows a rod of length L caused to move at constant speed v along horizontal conducting rails. The magnetic field in which the rod moves is *not uniform* but is provided by a current

i in a long wire parallel to the rails. Assume that $v = 5.00$ m/s, $a = 10.0$ mm, $L = 10.0$ cm, and $i = 100$ A. (a) Calculate the emf induced in the rod. (b) What is the current in the conducting loop? Assume that the resistance of the rod is 0.400 Ω and that the resistance of the rails and the strip that connects them at the right is negligible. (c) At what rate is thermal energy being generated in the rod? (d) What force must be applied to the rod by an external agent to maintain its motion? (e) At what rate does this external agent do work on the rod? Compare this answer to that for (c). ssm

Fig. 31-49 Problem 31.

SEC. 31-6 Induced Electric Fields

32E. Figure 31-50 shows two circular regions R_1 and R_2 with radii $r_1 = 20.0$ cm and $r_2 = 30.0$ cm. In R_1 there is a uniform magnetic field $B_1 = 50.0$ mT into the page, and in R_2 there is a uniform magnetic field $B_2 = 75.0$ mT out of the page (ignore any fringing of these fields). Both fields are decreasing at the rate of 8.50 mT/s. Calculate the integral $\oint \vec{E} \cdot d\vec{s}$ for each of the three dashed paths.

Fig. 31-50 Exercise 32.

33E. A long solenoid has a diameter of 12.0 cm. When a current i exists in its windings, a uniform magnetic field $B = 30.0$ mT is produced in its interior. By decreasing i, the field is caused to decrease at the rate of 6.50 mT/s. Calculate the magnitude of the induced electric field (a) 2.20 cm and (b) 8.20 cm from the axis of the solenoid. ssm ilw

34P. Early in 1981 the Francis Bitter National Magnet Laboratory at M.I.T. commenced operation of a 3.3-cm-diameter cylindrical magnet, which produces a 30 T field, then the world's largest steady-state field. The field can be varied sinusoidally between the limits of 29.6 and 30.0 T at a frequency of 15 Hz. When this is done, what is the maximum value of the induced electric field at a radial distance of 1.6 cm from the axis? (*Hint:* See Sample Problem 31-4.)

35P. Prove that the electric field $\vec{E}$ in a charged parallel-plate capacitor cannot drop abruptly to zero (as is suggested at point a in Fig. 31-51), as one moves perpendicular to the field, say, along the horizontal arrow in the figure. Fringing of the field lines always occurs in actual capacitors, which means that $\vec{E}$ approaches zero in a continuous and gradual way (see Problem 33 in Chapter 30). (*Hint:* Apply Faraday's law to the rectangular path shown by the dashed lines.) ssm

Fig. 31-51 Problem 35.

SEC. 31-7 Inductors and Inductance

36E. A circular coil has a 10.0 cm radius and consists of 30.0 closely wound turns of wire. An externally produced magnetic field of 2.60 mT is perpendicular to the coil. (a) If no current is in the coil, what magnetic flux links its turns? (b) When the current in the coil is 3.80 A in a certain direction, the net flux through the coil is found to vanish. What is the inductance of the coil?

37E. The inductance of a close-packed coil of 400 turns is 8.0 mH. Calculate the magnetic flux through the coil when the current is 5.0 mA. ssm

38P. A wide copper strip of width W is bent to form a tube of radius R with two parallel planar extensions, as shown in Fig. 31-52. There is a current i through the strip, distributed uniformly over its width. In this way a "one-turn solenoid" is formed. (a) Derive an expression for the magnitude of the magnetic field $\vec{B}$ in the tubular part (far away from the edges). (*Hint:* Assume that the magnetic field outside this one-turn solenoid is negligibly small.) (b) Find the inductance of this one-turn solenoid, neglecting the two planar extensions.

Fig. 31-52 Problem 38.

39P. Two long parallel wires, both of radius a and whose centers are a distance d apart, carry equal currents in opposite directions. Show that, neglecting the flux within the wires, the inductance of a length l of such a pair of wires is given by

$$L = \frac{\mu_0 l}{\pi} \ln \frac{d - a}{a}.$$

(*Hint:* Calculate the flux through a rectangle of which the wires form two opposite sides.) ssm www

SEC. 31-8 Self-Induction

40E. At a given instant the current and self-induced emf in an inductor are directed as indicated in Fig. 31-53. (a) Is the current increasing or decreasing? (b) The induced emf is 17 V and the rate of change of the current is 25 kA/s; find the inductance.

Fig. 31-53 Exercise 40.

41E. A 12 H inductor carries a steady current of 2.0 A. How can a 60 V self-induced emf be made to appear in the inductor? ssm

42P. The current i through a 4.6 H inductor varies with time t as shown by the graph of Fig. 31-54. The inductor has a resistance of

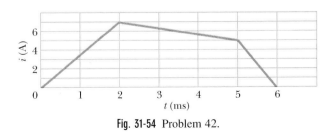

Fig. 31-54 Problem 42.

12 Ω. Find the magnitude of the induced emf $\mathscr{E}$ during the time intervals (a) $t = 0$ to $t = 2$ ms, (b) $t = 2$ ms to $t = 5$ ms, (c) $t = 5$ ms to $t = 6$ ms. (Ignore the behavior at the ends of the intervals.)

43P. *Inductors in series.* Two inductors L_1 and L_2 are connected in series and are separated by a large distance. (a) Show that the equivalent inductance is given by

$$L_{eq} = L_1 + L_2.$$

(*Hint:* Review the derivations for resistors in series and capacitors in series. Which is similar here?) (b) Why must their separation be large for this relationship to hold? (c) What is the generalization of (a) for N inductors in series?

44P. *Inductors in parallel.* Two inductors L_1 and L_2 are connected in parallel and separated by a large distance. (a) Show that the equivalent inductance is given by

$$\frac{1}{L_{eq}} = \frac{1}{L_1} + \frac{1}{L_2}.$$

(*Hint:* Review the derivations for resistors in parallel and capacitors in parallel. Which is similar here?) (b) Why must their separation be large for this relationship to hold? (c) What is the generalization of (a) for N inductors in parallel?

SEC. 31-9 *RL* Circuits

45E. In terms of τ_L, how long must we wait for the current in an *RL* circuit to build up to within 0.100% of its equilibrium value? ssm

46E. The current in an *RL* circuit builds up to one-third of its steady-state value in 5.00 s. Find the inductive time constant.

47E. The current in an *RL* circuit drops from 1.0 A to 10 mA in the first second following removal of the battery from the circuit. If L is 10 H, find the resistance R in the circuit. ilw

48E. Consider the *RL* circuit of Fig. 31-17. In terms of the battery emf $\mathscr{E}$, (a) what is the self-induced emf $\mathscr{E}_L$ when the switch has just been closed on a, and (b) what is $\mathscr{E}_L$ when $t = 2.0\tau_L$? (c) In terms of τ_L, when will $\mathscr{E}_L$ be just one-half the battery emf $\mathscr{E}$?

49E. A solenoid having an inductance of 6.30 μH is connected in series with a 1.20 kΩ resistor. (a) If a 14.0 V battery is switched across the pair, how long will it take for the current through the resistor to reach 80.0% of its final value? (b) What is the current through the resistor at time $t = 1.0\tau_L$? ssm

50P. Suppose the emf of the battery in the circuit of Fig. 31-18 varies with time t so that the current is given by $i(t) = 3.0 + 5.0t$, where i is in amperes and t is in seconds. Take $R = 4.0\ \Omega$ and $L = 6.0$ H, and find an expression for the battery emf as a function of time. (*Hint:* Apply the loop rule.)

51P. At time $t = 0$, a 45.0 V potential difference is suddenly applied to a coil with $L = 50.0$ mH and $R = 180\ \Omega$. At what rate is the current increasing at $t = 1.20$ ms? ilw

52P. A wooden toroidal core with a square cross section has an inner radius of 10 cm and an outer radius of 12 cm. It is wound with one layer of wire (of diameter 1.0 mm and resistance per meter 0.020 Ω/m). What are (a) the inductance and (b) the inductive time constant of the resulting toroid? Ignore the thickness of the insulation on the wire.

53P. In Fig. 31-55, $\mathscr{E} = 100$ V, $R_1 = 10.0\ \Omega$, $R_2 = 20.0\ \Omega$, $R_3 = 30.0\ \Omega$, and $L = 2.00$ H. Find the values of i_1 and i_2 (a) immediately after the closing of switch S, (b) a long time later, (c) immediately after the reopening of switch S, and (d) a long time after the reopening. ssm

Fig. 31-55 Problem 53.

54P. In the circuit of Fig. 31-56, $\mathscr{E} = 10$ V, $R_1 = 5.0\ \Omega$, $R_2 = 10\ \Omega$, and $L = 5.0$ H. For the two separate conditions (I) switch S just closed and (II) switch S closed for a long time, calculate (a) the current i_1 through R_1, (b) the current i_2 through R_2, (c) the current i through the switch, (d) the potential difference across R_2, (e) the potential difference across L, and (f) the rate of change di_2/dt.

Fig. 31-56 Problem 54.

55P*. In the circuit shown in Fig. 31-57, switch S is closed at time $t = 0$. Thereafter, the constant current source, by varying its emf, maintains a constant current i out of its upper terminal. (a) Derive an expression for the current through the inductor as a function of time. (b) Show that the current through the resistor equals the current through the inductor at time $t = (L/R) \ln 2$. ssm www

Fig. 31-57 Problem 55.

SEC. 31-10 Energy Stored in a Magnetic Field

56E. Consider the circuit of Fig. 31-18. In terms of the inductive time constant, at what instant after the battery is connected will the energy stored in the magnetic field of the inductor be half its steady-state value?

57E. Suppose that the inductive time constant for the circuit of Fig. 31-18 is 37.0 ms and the current in the circuit is zero at time $t = 0$. At what time does the rate at which energy is dissipated in the resistor equal the rate at which energy is being stored in the inductor? ilw

58E. A coil with an inductance of 2.0 H and a resistance of 10 Ω is suddenly connected to a resistanceless battery with $\mathscr{E} = 100$ V. At 0.10 s after the connection is made, what are the rates at which (a) energy is being stored in the magnetic field, (b) thermal energy is appearing in the resistance, and (c) energy is being delivered by the battery?

59P. A coil is connected in series with a 10.0 kΩ resistor. A 50.0 V battery is applied across the two devices, and the current reaches a value of 2.00 mA after 5.00 ms. (a) Find the inductance of the coil. (b) How much energy is stored in the coil at this same moment? ssm

60P. For the circuit of Fig. 31-18, assume that $\mathscr{E} = 10.0$ V, $R =$

6.70 Ω, and $L = 5.50$ H. The battery is connected at time $t = 0$. (a) How much energy is delivered by the battery during the first 2.00 s? (b) How much of this energy is stored in the magnetic field of the inductor? (c) How much of this energy is dissipated in the resistor?

61P. Prove that, after switch S in Fig. 31-17 has been thrown from a to b, all the energy stored in the inductor will ultimately appear as thermal energy in the resistor. ssm

SEC. 31-11 Energy Density of a Magnetic Field

62E. A toroidal inductor with an inductance of 90.0 mH encloses a volume of 0.0200 m³. If the average energy density in the toroid is 70.0 J/m³, what is the current through the inductor?

63E. A solenoid that is 85.0 cm long has a cross-sectional area of 17.0 cm². There are 950 turns of wire carrying a current of 6.60 A. (a) Calculate the energy density of the magnetic field inside the solenoid. (b) Find the total energy stored in the magnetic field there (neglect end effects). ssm

64E. The magnetic field in the interstellar space of our galaxy has a magnitude of about 10^{-10} T. How much energy is stored in this field in a cube 10 light-years on edge? (For scale, note that the nearest star is 4.3 light-years distant and the radius of our galaxy is about 8×10^4 light-years.)

65E. What must be the magnitude of a uniform electric field if it is to have the same energy density as that possessed by a 0.50 T magnetic field? ilw

66E. A circular loop of wire 50 mm in radius carries a current of 100 A. (a) Find the magnetic field strength at the center of the loop. (b) Calculate the energy density at the center of the loop.

67P. A length of copper wire carries a current of 10 A, uniformly distributed through its cross section. Calculate the energy density of (a) the magnetic field and (b) the electric field at the surface of the wire. The wire diameter is 2.5 mm, and its resistance per unit length is 3.3 Ω/km. ssm www

SEC. 31-12 Mutual Induction

68E. Coil 1 has $L_1 = 25$ mH and $N_1 = 100$ turns. Coil 2 has $L_2 = 40$ mH and $N_2 = 200$ turns. The coils are rigidly positioned with respect to each other; their mutual inductance M is 3.0 mH. A 6.0 mA current in coil 1 is changing at the rate of 4.0 A/s. (a) What magnetic flux Φ_{12} links coil 1, and what self-induced emf appears there? (b) What magnetic flux Φ_{21} links coil 2, and what mutually induced emf appears there?

69E. Two coils are at fixed locations. When coil 1 has no current and the current in coil 2 increases at the rate 15.0 A/s, the emf in coil 1 is 25.0 mV. (a) What is their mutual inductance? (b) When coil 2 has no current and coil 1 has a current of 3.60 A, what is the flux linkage in coil 2? ssm

70E. Two solenoids are part of the spark coil of an automobile. When the current in one solenoid falls from 6.0 A to zero in 2.5 ms, an emf of 30 kV is induced in the other solenoid. What is the mutual inductance M of the solenoids?

71P. Two coils, connected as shown in Fig. 31-58, separately have inductances L_1 and L_2. Their mutual inductance is M.

(a) Show that this combination can be replaced by a single coil of equivalent inductance given by

$$L_{eq} = L_1 + L_2 + 2M.$$

(b) How could the coils in Fig. 31-58 be reconnected to yield an equivalent inductance of

$$L_{eq} = L_1 + L_2 - 2M?$$

(This problem is an extension of Problem 43, but the requirement that the coils be far apart has been removed.) ssm

Fig. 31-58 Problem 71.

72P. A coil C of N turns is placed around a long solenoid S of radius R and n turns per unit length, as in Fig. 31-59. Show that the mutual inductance for the coil–solenoid combination is given by $M = \mu_0 \pi R^2 nN$. Explain why M does not depend on the shape, size, or possible lack of close-packing of the coil.

Fig. 31-59 Problem 72.

73P. Figure 31-60 shows, in cross section, two coaxial solenoids. Show that the mutual inductance M for a length l of this solenoid–solenoid combination is given by $M = \pi R_1^2 l \mu_0 n_1 n_2$, in which n_1 and n_2 are the respective numbers of turns per unit length and R_1 is the radius of the inner solenoid. Why does M depend on R_1 and not on R_2? ssm

Fig. 31-60
Problem 73.

74P. Figure 31-61 shows a coil of N_2 turns wound as shown around part of a toroid of N_1 turns. The toroid's inner radius is a, its outer radius is b, and its height is h. Show that the mutual inductance M for the toroid–coil combination is

$$M = \frac{\mu_0 N_1 N_2 h}{2\pi} \ln \frac{b}{a}.$$

75P. A rectangular loop of N close-packed turns is positioned near a long straight wire as shown in Fig. 31-62. (a) What is the mutual inductance M for the loop–wire combination? (b) Evaluate M for $N = 100$, $a = 1.0$ cm, $b = 8.0$ cm, and $l = 30$ cm. ilw

Fig. 31-61 Problem 74.

Fig. 31-62 Problem 75.

32 Magnetism of Matter: Maxwell's Equations

This is an overhead view of a frog that is being levitated in a magnetic field produced by current in a vertical solenoid below the frog. The solenoid's upward magnetic force on the frog balances the downward gravitational force on the frog. (The frog is not in discomfort; the sensation is like floating in water, which frogs like very much.) However, a frog is not magnetic (it would not, for example, stick to a refrigerator door).

How, then, can there be a magnetic force on the frog?

The answer is in this chapter.

32-1 Magnets

Fig. 32-1 A bar magnet is a magnetic dipole. The iron filings suggest the magnetic field lines. (The background is illuminated with colored light.)

The first known magnets were *lodestones*, which are stones that have been *magnetized* (made magnetic) naturally. When the ancient Greeks and ancient Chinese discovered these rare stones, they were amused by the stones' ability to attract metal over a short distance, as if by magic. Only much later did they learn to use lodestones (and artificially magnetized pieces of iron) in compasses to determine direction.

Today, magnets and magnetic materials are ubiquitous. We find them in VCRs, audio cassettes, ATM and credit cards, audio headsets, and even in the inks for paper money. In fact, some breakfast cereals that are "iron fortified" contain small bits of magnetic materials (you can collect them from a slurry of cereal and water with a magnet). More important, the modern electronics industry as we know it (including the music and information sectors) would not exist without magnetic materials.

The magnetic properties of materials can be traced back to their atoms and electrons. We begin here, however, with the bar magnet in Fig. 32-1. As you have seen, iron filings sprinkled around such a magnet tend to align with the magnetic field of the magnet, and their pattern reveals the magnetic field lines. The clustering of the lines at the ends of the magnet suggests that one end is a *source* of the lines (the field diverges from it) and the other end is a *sink* of the lines (the field converges toward it). By convention, we call the source the *north pole* of the magnet and the opposite end the *south pole,* and we say that the magnet, with its two poles, is an example of a **magnetic dipole.**

Suppose we break apart a bar magnet the way we break a piece of chalk (Fig. 32-2). We should, it seems, be able to isolate a single pole, or *monopole.* However, we cannot—not even if we break the magnet down to its individual atoms and then to its electrons and nuclei. Each fragment has a north pole and a south pole. Thus:

> The simplest magnetic structure that can exist is a magnetic dipole. Magnetic monopoles do not exist (as far as we know).

32-2 Gauss' Law for Magnetic Fields

Gauss' law for magnetic fields is a formal way of saying that magnetic monopoles do not exist. The law asserts that the net magnetic flux Φ_B through any closed Gaussian surface is zero:

$$\Phi_B = \oint \vec{B} \cdot d\vec{A} = 0 \qquad \text{(Gauss' law for magnetic fields).} \qquad (32\text{-}1)$$

Contrast this with Gauss' law for electric fields,

$$\Phi_E = \oint \vec{E} \cdot d\vec{A} = \frac{q_{\text{enc}}}{\varepsilon_0} \qquad \text{(Gauss' law for electric fields).}$$

Fig. 32-2 If you break a magnet, each fragment becomes a separate magnet, with its own north and south poles.

In both equations, the integral is taken over a *closed* Gaussian surface. Gauss' law for electric fields says that this integral (the net electric flux through the surface) is proportional to the net electric charge q_{enc} enclosed by the surface. Gauss' law for magnetic fields says that there can be no net magnetic flux through the surface because there can be no net "magnetic charge" (individual magnetic poles) enclosed by the surface. The simplest magnetic structure that can exist and thus be enclosed by a Gaussian surface is a dipole, which consists of both a source and a sink for the field lines. Thus, there must always be as much magnetic flux into the surface as out of it, and the net magnetic flux must always be zero.

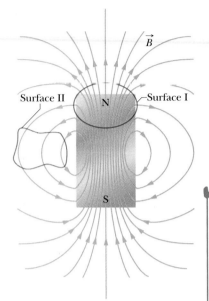

Fig. 32-3 The field lines for the magnetic field $\vec{B}$ of a short bar magnet. The red curves represent cross sections of closed, three-dimensional Gaussian surfaces.

Gauss' law for magnetic fields holds for more complicated structures than a magnetic dipole, and it holds even if the Gaussian surface does not enclose the entire structure. Gaussian surface II near the bar magnet of Fig. 32-3 encloses no poles, and we can easily conclude that the net magnetic flux through it is zero. Gaussian surface I is more difficult. It may seem to enclose only the north pole of the magnet because it encloses the label N and not the label S. However, a south pole must be associated with the lower boundary of the surface, because magnetic field lines enter the surface there. (The enclosed section is like one piece of the broken bar magnet in Fig. 32-2.) Thus, Gaussian surface I encloses a magnetic dipole and the net flux through the surface is zero.

✔**CHECKPOINT 1:**   The figure here shows four closed surfaces with flat top and bottom faces and curved sides. The table gives the areas A of the faces and the magnitudes B of the uniform and perpendicular magnetic fields through those faces; the units of A and B are arbitrary but consistent. Rank the surfaces according to the magnitudes of the magnetic flux through their curved sides, greatest first.

Surface	A_{top}	B_{top}	A_{bot}	B_{bot}
a	2	6, outward	4	3, inward
b	2	1, inward	4	2, inward
c	2	6, inward	2	8, outward
d	2	3, outward	3	2, outward

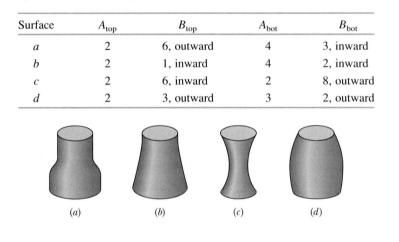

(a)   (b)   (c)   (d)

32-3 The Magnetism of Earth

Earth is a huge magnet; for points near Earth's surface, its magnetic field can be approximated as the field of a huge bar magnet—a magnetic dipole—that straddles the center of the planet. Figure 32-4 is an idealized symmetric depiction of the dipole field, without the distortion caused by passing charged particles from the Sun.

Because Earth's magnetic field is that of a magnetic dipole, a magnetic dipole moment $\vec{\mu}$ is associated with the field. For the idealized field of Fig. 32-4, the magnitude of $\vec{\mu}$ is 8.0×10^{22} J/T and the direction of $\vec{\mu}$ makes an angle of $11.5°$ with the rotation axis (*RR*) of Earth. The *dipole axis* (*MM* in Fig. 32-4) lies along $\vec{\mu}$ and intersects Earth's surface at the *geomagnetic north pole* in northwest Greenland and the *geomagnetic south pole* in Antarctica. The lines of the magnetic field $\vec{B}$ generally emerge in the southern hemisphere and reenter Earth in the northern hemisphere. Thus, the magnetic pole that is in Earth's northern hemisphere and known as a "north magnetic pole" *is really the south pole of Earth's magnetic dipole.*

The direction of the magnetic field at any location on Earth's surface is commonly specified in terms of two angles. The **field declination** is the angle (left or right) between geographic north (which is toward $90°$ latitude) and the horizontal component of the field. The **field inclination** is the angle (up or down) between a horizontal plane and the field's direction.

Magnetometers measure these angles and determine the field with much precision. However, you can do reasonably well with just a *compass* and a *dip meter*. A

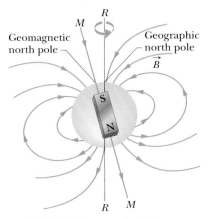

Fig. 32-4 Earth's magnetic field represented as a dipole field. The dipole axis *MM* makes an angle of 11.5° with Earth's rotational axis *RR*. The south pole of the dipole is in Earth's northern hemisphere.

compass is simply a needle-shaped magnet that is mounted so it can rotate freely about a vertical axis. When it is held in a horizontal plane, the north-pole end of the needle points, generally, toward the geomagnetic north pole (really a south magnetic pole, remember). The angle between the needle and geographic north is the field declination. A dip meter is a similar magnet that can rotate freely about a horizontal axis. When its vertical plane of rotation is aligned with the direction of the compass, the angle between the meter's needle and the horizontal is the field inclination.

At any point on Earth's surface, the measured magnetic field may differ appreciably, in both magnitude and direction, from the idealized dipole field of Fig. 32-4. In fact, the point where the field is actually perpendicular to Earth's surface and inward is not located at the geomagnetic north pole in Greenland as we would expect; instead, this so-called *dip north pole* is located in the Queen Elizabeth Islands in northern Canada, far from Greenland.

In addition, the field observed at any location on the surface of Earth varies with time, by measurable amounts over a period of a few years and by substantial amounts over, say, 100 years. For example, between 1580 and 1820 the direction indicated by compass needles in London changed by 35°.

In spite of these local variations, the average dipole field changes only slowly over such relatively short time periods. Variations over longer periods can be studied by measuring the weak magnetism of the ocean floor on either side of the Mid-Atlantic Ridge (Fig. 32-5). This floor has been formed by molten magma that oozed up through the ridge from Earth's interior, solidified, and was pulled away from the ridge (by the drift of tectonic plates) at the rate of a few centimeters per year. As the magma solidified, it became weakly magnetized with its magnetic field in the direction of Earth's magnetic field at the time of solidification. Study of this solidified magma across the ocean floor reveals that Earth's field has reversed its *polarity* (directions of the north pole and south pole) about every million years. The reason for the reversals is not known. In fact, the mechanism that produces Earth's magnetic field is only vaguely understood.

32-4 Magnetism and Electrons

Magnetic materials, from lodestones to videotapes, are magnetic because of the electrons within them. We have already seen one way in which electrons can generate a magnetic field: Send them through a wire as an electric current, and their motion produces a magnetic field around the wire. There are two more ways, each involving a magnetic dipole moment that produces a magnetic field in the surrounding space. However, their explanation requires quantum physics that is beyond the physics presented in this book, so here we shall only outline the results.

Fig. 32-5 A magnetic profile of the seafloor on either side of the Mid-Atlantic Ridge. The seafloor, extruded through the ridge and spreading out as part of the tectonic drift system, displays a record of the past magnetic history of Earth's core. The direction of the magnetic field produced by the core reverses about every million years.

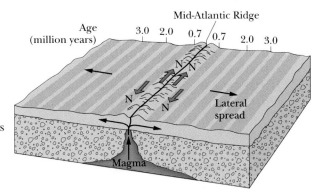

Spin Magnetic Dipole Moment

An electron has an intrinsic angular momentum called its **spin angular momentum** (or just **spin**) $\vec{S}$; associated with this spin is an intrinsic **spin magnetic dipole moment** $\vec{\mu}_s$. (By *intrinsic*, we mean that $\vec{S}$ and $\vec{\mu}_s$ are basic characteristics of an electron, like its mass and electric charge.) $\vec{S}$ and $\vec{\mu}_s$ are related by

$$\vec{\mu}_s = -\frac{e}{m}\,\vec{S}, \qquad (32\text{-}2)$$

in which e is the elementary charge (1.60×10^{-19} C) and m is the mass of an electron (9.11×10^{-31} kg). The minus sign means that $\vec{\mu}_s$ and $\vec{S}$ are oppositely directed.

Spin $\vec{S}$ is different from the angular momenta of Chapter 12 in two respects:

1. $\vec{S}$ itself cannot be measured. However, its component along any axis can be measured.

2. A measured component of $\vec{S}$ is *quantized*, which is a general term that means it is restricted to certain values. A measured component of $\vec{S}$ can have only two values, which differ only in sign.

Let us assume that the component of spin $\vec{S}$ is measured along the z axis of a coordinate system. Then the measured component S_z can have only the two values given by

$$S_z = m_s\,\frac{h}{2\pi}, \qquad \text{for } m_s = \pm\tfrac{1}{2}, \qquad (32\text{-}3)$$

where m_s is called the *spin magnetic quantum number* and h ($= 6.63 \times 10^{-34}$ J · s) is the Planck constant, the ubiquitous constant of quantum physics. The signs given in Eq. 32-3 have to do with the direction of S_z along the z axis. When S_z is parallel to the z axis, m_s is $+\tfrac{1}{2}$ and the electron is said to be *spin up*. When S_z is antiparallel to the z axis, m_s is $-\tfrac{1}{2}$ and the electron is said to be *spin down*.

The spin magnetic dipole moment $\vec{\mu}_s$ of an electron also cannot be measured; only its component along any axis can be measured, and that component too is quantized, with two possible values of the same magnitude but different signs. We can relate the component $\mu_{s,z}$ measured on the z axis to S_z by rewriting Eq. 32-2 in component form for the z axis as

$$\mu_{s,z} = -\frac{e}{m}\,S_z.$$

Substituting for S_z from Eq. 32-3 then gives us

$$\mu_{s,z} = \pm\,\frac{eh}{4\pi m}, \qquad (32\text{-}4)$$

where the plus and minus signs correspond to $\mu_{s,z}$ being parallel and antiparallel to the z axis, respectively.

The quantity on the right side of Eq. 32-4 is called the *Bohr magneton* μ_B:

$$\mu_B = \frac{eh}{4\pi m} = 9.27 \times 10^{-24} \text{ J/T} \qquad \text{(Bohr magneton).} \qquad (32\text{-}5)$$

Spin magnetic dipole moments of electrons and other elementary particles can be expressed in terms of μ_B. For an electron, the magnitude of the measured z component of $\vec{\mu}_s$ is

$$\mu_{s,z} = 1\mu_B. \qquad (32\text{-}6)$$

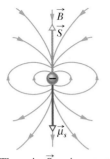

Fig. 32-6 The spin $\vec{S}$, spin magnetic dipole moment $\vec{\mu}_s$, and magnetic dipole field $\vec{B}$ of an electron represented as a microscopic sphere.

(The quantum physics of the electron, called *quantum electrodynamics,* or QED, reveals that $\mu_{s,z}$ is actually slightly greater than $1\mu_B$, but we shall neglect that fact.)

When an electron is placed in an external magnetic field $\vec{B}_{ext}$, a potential energy U can be associated with the orientation of the electron's spin magnetic dipole moment $\vec{\mu}_s$ just as a potential energy can be associated with the orientation of the magnetic dipole moment $\vec{\mu}$ of a current loop placed in $\vec{B}_{ext}$. From Eq. 29-38, the potential energy for the electron is

$$U = -\vec{\mu}_s \cdot \vec{B}_{ext} = -\mu_{s,z}B_{ext}, \qquad (32\text{-}7)$$

where the z axis is taken to be in the direction of $\vec{B}_{ext}$.

If we imagine an electron to be a microscopic sphere (which it is not), we can represent the spin $\vec{S}$, the spin magnetic dipole moment $\vec{\mu}_s$, and the associated magnetic dipole field as in Fig. 32-6. Although we use the word "spin" here, electrons do not spin like tops. How, then, can something have angular momentum without actually rotating? Again, we would need quantum physics to provide the answer.

Protons and neutrons also have an intrinsic angular momentum called spin and an associated intrinsic spin magnetic dipole moment. For a proton those two vectors have the same direction, and for a neutron they have opposite directions. We shall not examine the contributions of these dipole moments to the magnetic fields of atoms because they are about a thousand times smaller than that due to an electron.

✓**CHECKPOINT 2:** The figure here shows the spin orientations of two particles in an external magnetic field $\vec{B}_{ext}$. (a) If the particles are electrons, which spin orientation is at lower potential energy? (b) If, instead, the particles are protons, which spin orientation is at lower potential energy?

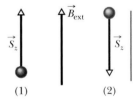

Orbital Magnetic Dipole Moment

When it is in an atom, an electron has an additional angular momentum called its **orbital angular momentum** $\vec{L}_{orb}$. Associated with $\vec{L}_{orb}$ is an **orbital magnetic dipole moment** $\vec{\mu}_{orb}$; the two are related by

$$\vec{\mu}_{orb} = -\frac{e}{2m}\vec{L}_{orb}. \qquad (32\text{-}8)$$

The minus sign means that $\vec{\mu}_{orb}$ and $\vec{L}_{orb}$ have opposite directions.

Orbital angular momentum $\vec{L}_{orb}$ cannot be measured; only its component along any axis can be measured, and that component is quantized. The component along, say, a z axis can have only the values given by

$$L_{orb,z} = m_l \frac{h}{2\pi}, \qquad \text{for } m_l = 0, \pm 1, \pm 2, \ldots, \pm(\text{limit}), \qquad (32\text{-}9)$$

in which m_l is called the *orbital magnetic quantum number* and "limit" refers to some largest allowed integer value for m_l. The signs in Eq. 32-9 have to do with the direction of $L_{orb,z}$ along the z axis.

The orbital magnetic dipole moment $\vec{\mu}_{orb}$ of an electron also cannot itself be measured; only its component along an axis can be measured, and that component is again quantized. By writing Eq. 32-8 for a component along the same z axis as above and then substituting for $L_{orb,z}$ from Eq. 32-9, we can write the z component $\mu_{orb,z}$ of the orbital magnetic dipole moment as

$$\mu_{orb,z} = -m_l \frac{eh}{4\pi m} \qquad (32\text{-}10)$$

and, in terms of the Bohr magneton, as

$$\mu_{\text{orb},z} = -m_l \mu_B. \tag{32-11}$$

When an atom is placed in an external magnetic field $\vec{B}_{\text{ext}}$, a potential energy U can be associated with the orientation of the orbital magnetic dipole moment of each electron in the atom. Its value is

$$U = -\vec{\mu}_{\text{orb}} \cdot \vec{B}_{\text{ext}} = -\mu_{\text{orb},z} B_{\text{ext}}, \tag{32-12}$$

where the z axis is taken in the direction of $\vec{B}_{\text{ext}}$.

Although we have used the words "orbit" and "orbital" here, electrons do not orbit the nucleus of an atom like planets orbiting the Sun. How can an electron have an orbital angular momentum without orbiting in the common meaning of the term? Once again, this can be explained only with quantum physics.

Loop Model for Electron Orbits

We can obtain Eq. 32-8 with the nonquantum derivation that follows, in which we assume that an electron moves along a circular path with a radius that is much larger than an atomic radius (hence the name "loop model"). However, the derivation does not apply to an electron within an atom (for which we need quantum physics).

We imagine an electron moving at constant speed v in a circular path of radius r, counterclockwise as shown in Fig. 32-7. The motion of the negative charge of the electron is equivalent to a conventional current i (of positive charge) that is clockwise, as also shown in Fig. 32-7. The magnitude of the orbital magnetic dipole moment of such a *current loop* is obtained from Eq. 29-35 with $N = 1$:

$$\mu_{\text{orb}} = iA, \tag{32-13}$$

where A is the area enclosed by the loop. The direction of this magnetic dipole moment is, from the right-hand rule of Fig. 30-21, downward in Fig. 32-7.

To evaluate Eq. 32-13, we need the current i. Current is, generally, the rate at which charge passes some point in a circuit. Here, the charge of magnitude e takes a time $T = 2\pi r/v$ to circle from any point back through that point, so

$$i = \frac{\text{charge}}{\text{time}} = \frac{e}{2\pi r/v}. \tag{32-14}$$

Substituting this and the area $A = \pi r^2$ of the loop into Eq. 32-13 gives us

$$\mu_{\text{orb}} = \frac{e}{2\pi r/v}\, \pi r^2 = \frac{evr}{2}. \tag{32-15}$$

To find the electron's orbital angular momentum $\vec{L}_{\text{orb}}$, we use Eq. 12-18, $\vec{\ell} = m(\vec{r} \times \vec{v})$. Because $\vec{r}$ and $\vec{v}$ are perpendicular, $\vec{L}_{\text{orb}}$ has the magnitude

$$L_{\text{orb}} = mrv \sin 90° = mrv. \tag{32-16}$$

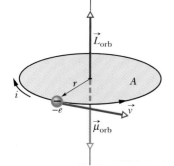

Fig. 32-7 An electron moving at constant speed v in a circular path of radius r that encloses an area A. The electron has an orbital angular momentum $\vec{L}_{\text{orb}}$ and an associated orbital magnetic dipole moment $\vec{\mu}_{\text{orb}}$. A clockwise current i (of positive charge) is equivalent to the counterclockwise circulation of the negatively charged electron.

$\vec{L}_{\text{orb}}$ is directed upward in Fig. 32-7 (see Fig. 12-11). Combining Eqs. 32-15 and 32-16, generalizing to a vector formulation, and indicating the opposite directions of the vectors with a minus sign yield

$$\vec{\mu}_{\text{orb}} = -\frac{e}{2m}\, \vec{L}_{\text{orb}},$$

which is Eq. 32-8. Thus, by "classical" (nonquantum) analysis we have obtained the same result, in both magnitude and direction, given by quantum physics. You might

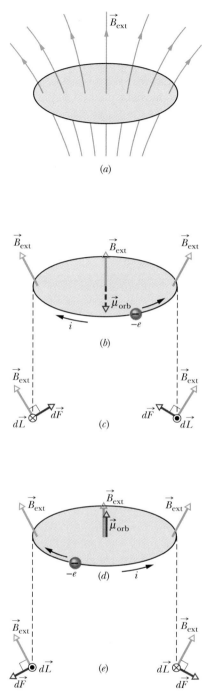

Fig. 32-8 (a) A loop model for an electron orbiting in an atom while in a nonuniform magnetic field $\vec{B}_{\text{ext}}$. (b) Charge $-e$ moves counterclockwise; the associated conventional current i is clockwise. (c) The magnetic forces $\vec{dF}$ on the left and right sides of the loop, as seen from the plane of the loop. The net force on the loop is upward. (d) Charge $-e$ now moves clockwise. (e) The net force on the loop is now downward.

wonder, since this derivation gives the correct result for an electron within an atom, why the derivation is invalid for that situation. The answer is that this line of reasoning yields other results that are contradicted by experiments.

Loop Model in a Nonuniform Field

We continue to consider an electron orbit as a current loop, as we did in Fig. 32-7. Now, however, we draw the loop in a nonuniform magnetic field $\vec{B}_{\text{ext}}$ as shown in Fig. 32-8a. (This field could be the diverging field near the north pole of the magnet in Fig. 32-3.) We make this change to prepare for the next several sections, in which we shall discuss the forces that act on magnetic materials when the materials are placed in a nonuniform magnetic field. We shall discuss these forces by assuming that the electron orbits in the materials are tiny current loops like that in Fig. 32-8a.

Here we assume that the magnetic field vectors all around the electron's circular path have the same magnitude and form the same angle with the vertical, as shown in Figs. 32-8b and d. We also assume that all the electrons in an atom move either counterclockwise (Fig. 32-8b) or clockwise (Fig. 32-8d). The associated conventional current i around the current loop and the orbital magnetic dipole moment $\vec{\mu}_{\text{orb}}$ produced by i are shown for each of these directions of motion.

Figures 32-8c and e show diametrically opposite views of a length element $\vec{dL}$ of the loop with the same direction as i, as seen from the plane of the orbit. Also shown are the field $\vec{B}_{\text{ext}}$ and the resulting magnetic force $\vec{dF}$ on $\vec{dL}$. Recall that a current along an element $\vec{dL}$ in a magnetic field $\vec{B}_{\text{ext}}$ experiences a magnetic force $\vec{dF}$ as given by Eq. 29-28:

$$\vec{dF} = i\,\vec{dL} \times \vec{B}_{\text{ext}}. \qquad (32\text{-}17)$$

On the left side of Fig. 32-8c, Eq. 32-17 tells us that the force $\vec{dF}$ is directed upward and rightward. On the right side, the force $\vec{dF}$ is just as large and is directed upward and leftward. Because their angles are the same, the horizontal components of these two forces cancel and the vertical components add. The same is true at any other two symmetric points on the loop. Thus, the net force on the current loop of Fig. 32-8b must be upward. The same reasoning leads to a downward net force on the loop in Fig. 32-8d. We shall use these two results shortly when we examine the behavior of magnetic materials in nonuniform magnetic fields.

32-5 Magnetic Materials

Each electron in an atom has an orbital magnetic dipole moment and a spin magnetic dipole moment that combine vectorially. The resultant of these two vector quantities combines vectorially with similar resultants for all other electrons in the atom, and the resultant for each atom combines with those for all the other atoms in a sample of a material. If the combination of all these magnetic dipole moments produces a magnetic field, then the material is magnetic. There are three general types of magnetism: diamagnetism, paramagnetism, and ferromagnetism.

1. *Diamagnetism* is exhibited by all common materials but is so feeble that it is masked if the material also exhibits magnetism of either of the other two types. In diamagnetism, weak magnetic dipole moments are produced in the atoms of the material when the material is placed in an external magnetic field $\vec{B}_{\text{ext}}$; the combination of all those induced dipole moments gives the material as a whole only a feeble net magnetic field. The dipole moments and thus their net field disappear when $\vec{B}_{\text{ext}}$ is removed. The term *diamagnetic material* usually refers to materials that exhibit only diamagnetism.

2. *Paramagnetism* is exhibited by materials containing transition elements, rare earth elements, and actinide elements (see Appendix G). Each atom of such a material has a permanent resultant magnetic dipole moment, but the moments are randomly oriented in the material and the material as a whole lacks a net magnetic field. However, an external magnetic field $\vec{B}_{ext}$ can partially align the atomic magnetic dipole moments to give the material a net magnetic field. The alignment and thus its field disappear when $\vec{B}_{ext}$ is removed. The term *paramagnetic material* usually refers to materials that exhibit primarily paramagnetism.

3. *Ferromagnetism* is a property of iron, nickel, and certain other elements (and of compounds and alloys of these elements). Some of the electrons in these materials have their resultant magnetic dipole moments aligned, which produces regions with strong magnetic dipole moments. An external field $\vec{B}_{ext}$ can then align the magnetic moments of such regions, producing a strong magnetic field for a sample of the material; the field partially persists when $\vec{B}_{ext}$ is removed. We usually use the term *ferromagnetic material*, and even the common term *magnetic material*, to refer to materials that exhibit primarily ferromagnetism.

The next three sections explore these three types of magnetism.

32-6 Diamagnetism

We cannot yet discuss the quantum physical explanation of diamagnetism, but we can provide a classical explanation with the loop model of Figs. 32-7 and 32-8. To begin, we assume that in an atom of a diamagnetic material each electron can orbit only clockwise as in Fig. 32-8d or counterclockwise as in Fig. 32-8b. To account for the lack of magnetism in the absence of an external magnetic field $\vec{B}_{ext}$, we assume the atom lacks a net magnetic dipole moment. This implies that before $\vec{B}_{ext}$ is applied, as many electrons orbit one way as orbit the other, with the result that the net upward magnetic dipole moment of the atom equals the net downward magnetic dipole moment.

Now let's turn on the nonuniform field $\vec{B}_{ext}$ of Fig. 32-8a, in which $\vec{B}_{ext}$ is directed upward but is diverging (the magnetic field lines are diverging). We could do this by increasing the current through an electromagnet or by moving the north pole of a bar magnet closer to, and below, the orbits. As the magnitude of $\vec{B}_{ext}$ increases from zero to its final maximum, steady-state value, a clockwise electric field is induced around each electron's orbital loop according to Faraday's law and Lenz's law. Let us see how this induced electric field affects the orbiting electrons in Figs. 32-8b and d.

In Fig. 32-8b, the counterclockwise electron is accelerated by the clockwise electric field. Thus, as the magnetic field $\vec{B}_{ext}$ increases to its maximum value, the electron speed increases to a maximum value. This means that the associated conventional current i and the downward magnetic dipole moment $\vec{\mu}$ due to i also *increase*.

In Fig. 32-8d, the clockwise electron is decelerated by the clockwise electric field. Thus, here, the electron speed, the associated current i, and the upward magnetic dipole moment $\vec{\mu}$ due to i all *decrease*. By turning on field $\vec{B}_{ext}$, we have given the atom a *net* magnetic dipole moment that is upward. This would also be so if the magnetic field were uniform.

The nonuniformity of field $\vec{B}_{ext}$ also affects the atom. Because the current i in Fig. 32-8b increases, the upward magnetic forces $d\vec{F}$ in Fig. 32-8c also increase, as does the net upward force on the current loop. Because current i in Fig. 32-8d decreases, the downward magnetic forces $d\vec{F}$ in Fig. 32-8e also decrease, as does the net downward force on the current loop. Thus, by turning on the *nonuniform* field $\vec{B}_{ext}$, we have produced a net force on the atom; moreover, that force is directed *away* from the region of greater magnetic field.

We have argued with fictitious electron orbits (current loops), but we have ended up with exactly what happens to a diamagnetic material: If we apply the magnetic field of Fig. 32-8, the material develops a downward magnetic dipole moment and experiences an upward force. When the field is removed, both the dipole moment and the force disappear. The external field need not be positioned as shown; similar arguments can be made for other orientations of $\vec{B}_{ext}$. In general,

> ▶ A diamagnetic material placed in an external magnetic field $\vec{B}_{ext}$ develops a magnetic dipole moment directed opposite $\vec{B}_{ext}$. If the field is nonuniform, the diamagnetic material is repelled *from* a region of greater magnetic field *toward* a region of lesser field.

The frog in the photograph opening this chapter is diamagnetic (as is any other animal). When the frog was placed in the diverging magnetic field near the top end of a vertical current-carrying solenoid, every atom in the frog was repelled upward, away from the region of stronger magnetic field at that end of the solenoid. The frog moved upward into weaker and weaker magnetic field until the upward magnetic force balanced the gravitational force on it, and there it hung in midair. If we built a solenoid that was large enough, we could similarly levitate a person in midair owing to the person's diamagnetism.

✔ **CHECKPOINT 3:** The figure shows two diamagnetic spheres located near the south pole of a bar magnet. Are (a) the magnetic forces on the spheres and (b) the magnetic dipole moments of the spheres directed toward or away from the bar magnet? (c) Is the magnetic force on sphere 1 greater than, less than, or equal to that on sphere 2?

32-7 Paramagnetism

In paramagnetic materials, the spin and orbital magnetic dipole moments of the electrons in each atom do not cancel but add vectorially to give the atom a net (and permanent) magnetic dipole moment $\vec{\mu}$. In the absence of an external magnetic field, these atomic dipole moments are randomly oriented, and the net magnetic dipole moment of the material is zero. However, if a sample of the material is placed in an external magnetic field $\vec{B}_{ext}$, the magnetic dipole moments tend to line up with the field, which gives the sample a net magnetic dipole moment. This alignment with the external field is the opposite of what we saw with diamagnetic materials.

> ▶ A paramagnetic material placed in an external magnetic field $\vec{B}_{ext}$ develops a magnetic dipole moment in the direction of $\vec{B}_{ext}$. If the field is nonuniform, the paramagnetic material is attracted *toward* a region of greater magnetic field *from* a region of lesser field.

A paramagnetic sample with N atoms would have a magnetic dipole moment of magnitude $N\mu$ if alignment of its atomic dipoles were complete. However, random collisions of atoms due to thermal agitation transfer energy among them, disrupting their alignment and thus reducing the sample's magnetic dipole moment.

The importance of thermal agitation may be measured by comparing two energies. One, from Eq. 20-24, is the mean translational kinetic energy $K \; (= \frac{3}{2}kT)$ of an atom at temperature T, where k is the Boltzmann constant (1.38×10^{-23} J/K) and T is in kelvins (not Celsius degrees). The other, from Eq. 29-38, is the difference in energy $\Delta U_B \; (= 2\mu B_{ext})$ between parallel alignment and antiparallel alignment of

Liquid oxygen is suspended between the two pole faces of a magnet because the liquid is paramagnetic and is magnetically attracted to the magnet.

Fig 32-9 A *magnetization curve* for potassium chromium sulfate, a paramagnetic salt. The ratio of magnetization M of the salt to the maximum possible magnetization M_{max} is plotted versus the ratio of the applied magnetic field B_{ext} to the temperature T. Curie's law fits the data at the left; quantum theory fits all the data. After W. E. Henry.

the magnetic dipole moment of an atom and the external field. As we shall show below, $K \gg \Delta U_B$, even for ordinary temperatures and field magnitudes. Thus, energy transfers during collisions among atoms can significantly disrupt the alignment of the atomic dipole moments, keeping the magnetic dipole moment of a sample much less than $N\mu$.

We can express the extent to which a given paramagnetic sample is magnetized by finding the ratio of its magnetic dipole moment to its volume V. This vector quantity, the magnetic dipole moment per unit volume, is the **magnetization** $\vec{M}$ of the sample, and its magnitude is

$$M = \frac{\text{measured magnetic moment}}{V}. \quad (32\text{-}18)$$

The unit of $\vec{M}$ is the ampere–square meter per cubic meter, or ampere per meter (A/m) Complete alignment of the atomic dipole moments, called *saturation* of the sample, corresponds to the maximum value $M_{max} = N\mu/V$.

In 1895 Pierre Curie discovered experimentally that the magnetization of a paramagnetic sample is directly proportional to the external magnetic field $\vec{B}_{ext}$ and inversely proportional to the temperature T in kelvins; that is,

$$M = C\frac{B_{ext}}{T}. \quad (32\text{-}19)$$

Equation 32-19 is known as *Curie's law,* and C is called the *Curie constant.* Curie's law is reasonable in that increasing B_{ext} tends to align the atomic dipole moments in a sample and thus to increase M, whereas increasing T tends to disrupt the alignment via thermal agitation and thus to decrease M. However, the law is actually an approximation that is valid only when the ratio B_{ext}/T is not too large.

Figure 32-9 shows the ratio M/M_{max} as a function of B_{ext}/T for a sample of the salt potassium chromium sulfate, in which chromium ions are the paramagnetic substance. The plot is called a *magnetization curve*. The straight line for Curie's law fits the experimental data at the left, for B_{ext}/T below about 0.5 T/K. The curve that fits all the data points is based on quantum physics. The data on the right side, near saturation, are very difficult to obtain because they require very strong magnetic fields (about 100 000 times Earth's field), even at very low temperatures.

✓**CHECKPOINT 4:** The figure here shows two paramagnetic spheres located near the south pole of a bar magnet. Are (a) the magnetic forces on the spheres and (b) the magnetic dipole moments of the spheres directed toward or away from the bar magnet? (c) Is the magnetic force on sphere 1 greater than, less than, or equal to that on sphere 2?

Sample Problem 32-1

A paramagnetic gas at room temperature ($T = 300$ K) is placed in an external uniform magnetic field of magnitude $B = 1.5$ T; the atoms of the gas have magnetic dipole moment $\mu = 1.0\mu_B$. Calculate the mean translational kinetic energy K of an atom of the gas and the energy difference ΔU_B between parallel alignment and antiparallel alignment of the atom's magnetic dipole moment with the external field.

SOLUTION: The first Key Idea here is that the mean translational kinetic energy K of an atom in a gas depends on the temperature of the gas. From Eq. 20-24, we have

$$K = \tfrac{3}{2}kT = \tfrac{3}{2}(1.38 \times 10^{-23} \text{ J/K})(300 \text{ K})$$
$$= 6.2 \times 10^{-21} \text{ J} = 0.039 \text{ eV}. \qquad \text{(Answer)}$$

The second Key Idea is that the potential energy U_B of a magnetic dipole $\vec{\mu}$ in an external magnetic field $\vec{B}$ depends on the angle θ between the directions of $\vec{\mu}$ and $\vec{B}$. From Eq. 29-38 ($U_B = -\vec{\mu} \cdot \vec{B}$), we can write the difference ΔU_B between parallel alignment ($\theta = 0°$) and antiparallel alignment ($\theta = 180°$) as

$$\Delta U_B = -\mu B \cos 180° - (-\mu B \cos 0°) = 2\mu B$$
$$= 2\mu_B B = 2(9.27 \times 10^{-24} \text{ J/T})(1.5 \text{ T})$$
$$= 2.8 \times 10^{-23} \text{ J} = 0.000\ 17 \text{ eV}. \qquad \text{(Answer)}$$

Here K is about 230 times ΔU_B, so energy exchanges among the atoms during their collisions with one another can easily reorient any magnetic dipole moments that might be aligned with the external magnetic field. The magnetic dipole moment exhibited by the paramagnetic gas must then be due to fleeting partial alignments of the atomic dipole moments.

32-8 Ferromagnetism

When we speak of magnetism in everyday conversation, we almost always have a mental picture of a bar magnet or a disk magnet (probably clinging to a refrigerator door). That is, we picture a ferromagnetic material having strong, permanent magnetism, and not a diamagnetic or paramagnetic material having weak, temporary magnetism.

Iron, cobalt, nickel, gadolinium, dysprosium, and alloys containing these elements exhibit ferromagnetism because of a quantum physical effect called *exchange coupling* in which the electron spins of one atom interact with those of neighboring atoms. The result is alignment of the magnetic dipole moments of the atoms, in spite of the randomizing tendency of atomic collisions. This persistent alignment is what gives ferromagnetic materials their permanent magnetism.

If the temperature of a ferromagnetic material is raised above a certain critical value, called the *Curie temperature,* the exchange coupling ceases to be effective. Most such materials then become simply paramagnetic; that is, the dipoles still tend to align with an external field but much more weakly, and thermal agitation can now more easily disrupt the alignment. The Curie temperature for iron is 1043 K ($= 770°$C).

The magnetization of a ferromagnetic material such as iron can be studied with an arrangement called a *Rowland ring* (Fig. 32-10). The material is formed into a thin toroidal core of circular cross section. A primary coil P having n turns per unit length is wrapped around the core and carries current i_P. (The coil is essentially a long solenoid bent into a circle.) If the iron core were not present, the magnitude of the magnetic field inside the coil would be, from Eq. 30-25,

$$B_0 = \mu_0 i_P n. \qquad (32\text{-}20)$$

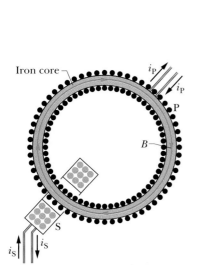

Fig. 32-10 A Rowland ring. Current i_P is sent through a primary coil P whose core is the ferromagnetic material to be studied (here iron) and that is magnetized by the current. (The turns of the coil are represented by dots.) The extent of magnetization of the core determines the total magnetic field $\vec{B}$ within coil P. Field $\vec{B}$ can be measured by means of a secondary coil S.

However, with the iron core present, the magnetic field $\vec{B}$ inside the coil is greater than $\vec{B}_0$, usually by a large amount. We can write the magnitude of this field as

$$B = B_0 + B_M, \qquad (32\text{-}21)$$

where B_M is the magnitude of the magnetic field contributed by the iron core. This contribution results from the alignment of the atomic dipole moments within the iron, due to exchange coupling and to the applied magnetic field B_0, and is proportional to the magnetization M of the iron. That is, the contribution B_M is proportional

Fig. 32-11 A magnetization curve for a ferromagnetic core material in the Rowland ring of Fig. 32-10. On the vertical axis, 1.0 corresponds to complete alignment (saturation) of the atomic dipoles within the material.

to the magnetic dipole moment per unit volume of the iron. To determine B_M we use a secondary coil S to measure B, compute B_0 with Eq. 32-20, and subtract as suggested by Eq. 32-21.

Figure 32-11 shows a magnetization curve for a ferromagnetic material in a Rowland ring; the ratio $B_M/B_{M\,max}$, where $B_{M,max}$ is the maximum possible value of B_M, corresponding to saturation, is plotted versus B_0. The curve is like Fig. 32-9, the magnetization curve for a paramagnetic substance: both curves show the extent to which an applied magnetic field can align the atomic dipole moments of a material.

For the ferromagnetic core yielding Fig. 32-11, the alignment of the dipole moments is about 70% complete for $B_0 \approx 1 \times 10^{-3}$ T. If B_0 were increased to 1 T, the alignment would be almost complete (but $B_0 = 1$ T, and thus almost complete saturation, is quite difficult to obtain).

Magnetic Domains

Exchange coupling produces strong alignment of adjacent atomic dipoles in a ferromagnetic material at a temperature below the Curie temperature. Why, then, isn't the material naturally at saturation even when there is no applied magnetic field B_0? That is, why isn't every piece of iron, such as an iron nail, a naturally strong magnet?

To understand this, consider a specimen of a ferromagnetic material such as iron that is in the form of a single crystal; that is, the arrangement of the atoms that make it up—its crystal lattice—extends with unbroken regularity throughout the volume of the specimen. Such a crystal will, in its normal state, be made up of a number of *magnetic domains*. These are regions of the crystal throughout which the alignment of the atomic dipoles is essentially perfect. The domains, however, are not all aligned. For the crystal as a whole, the domains are so oriented that they largely cancel each other as far as their external magnetic effects are concerned.

Figure 32-12 is a magnified photograph of such an assembly of domains in a single crystal of nickel. It was made by sprinkling a colloidal suspension of finely powdered iron oxide on the surface of the crystal. The domain boundaries, which are thin regions in which the alignment of the elementary dipoles changes from a certain orientation in one domain to a different orientation in the other, are the sites of intense, but highly localized and nonuniform, magnetic fields. The suspended colloidal particles are attracted to these boundaries and show up as the white lines (not all the domain boundaries are apparent in Fig. 32-12). Although the atomic dipoles in each domain are completely aligned as shown by the arrows, the crystal as a whole may have a very small resultant magnetic moment.

Actually, a piece of iron as we ordinarily find it is not a single crystal but an assembly of many tiny crystals, randomly arranged; we call it a *polycrystalline solid*. Each tiny crystal, however, has its array of variously oriented domains, just as in Fig. 32-12. If we magnetize such a specimen by placing it in an external magnetic field of gradually increasing strength, we produce two effects; together they produce a magnetization curve of the shape shown in Fig. 32-11. One effect is a growth in size of the domains that are oriented along the external field at the expense of those that are not. The second effect is a shift of the orientation of the dipoles within a domain, as a unit, to become closer to the field direction.

Exchange coupling and domain shifting give us the following result:

Fig. 32-12 A photograph of domain patterns within a single crystal of nickel; white lines reveal the boundaries of the domains. The white arrows superimposed on the photograph show the orientations of the magnetic dipoles within the domains and thus the orientations of the net magnetic dipoles of the domains. The crystal as a whole is unmagnetized if the net magnetic field (the vector sum over all the domains) is zero.

▶ A ferromagnetic material placed in an external magnetic field $\vec{B}_{ext}$ develops a strong magnetic dipole moment in the direction of $\vec{B}_{ext}$. If the field is nonuniform, the ferromagnetic material is attracted *toward* a region of greater magnetic field *from* a region of lesser field.

You can actually hear sound produced by shifting domains: Put an audio cassette player into its play mode without a cassette in place (or with a blank cassette) and turn the volume control to maximum. Then bring a strong magnet up to the play head (which is ferromagnetic). The magnetic field causes the domains in the play head to shift abruptly, which abruptly alters the magnetic field through a coil wrapped around the play head. The resulting suddenly induced currents in the coil are amplified and fed to the speaker, producing a fizzing sound.

Sample Problem 32-2

A compass needle made of pure iron (with density 7900 kg/m³) has a length L of 3.0 cm, a width of 1.0 mm, and a thickness of 0.50 mm. The magnitude of the magnetic dipole moment of an iron atom is $\mu_{Fe} = 2.1 \times 10^{-23}$ J/T. If the magnetization of the needle is equivalent to the alignment of 10% of the atoms in the needle, what is the magnitude of the needle's magnetic dipole moment $\vec{\mu}$?

SOLUTION: One Key Idea here is that alignment of all N atoms in the needle would give a magnitude of $N\mu_{Fe}$ for the needle's magnetic dipole moment $\vec{\mu}$. However, the needle has only 10% alignment (the random orientation of the rest does not give any net contribution to $\vec{\mu}$). Thus,

$$\mu = 0.10 N \mu_{Fe}. \tag{32-22}$$

A second Key Idea is that we can find the number of atoms N in the needle from the needle's mass:

$$N = \frac{\text{needle's mass}}{\text{iron's atomic mass}}. \tag{32-23}$$

Iron's atomic mass is not listed in Appendix F, but its molar mass M is. Thus, we write

$$\text{iron's atomic mass} = \frac{\text{iron's molar mass } M}{\text{Avogadro's number } N_A}. \tag{32-24}$$

Equation 32-23 then becomes

$$N = \frac{mN_A}{M}. \tag{32-25}$$

The needle's mass m is the product of its density and its volume. The volume works out to be 1.5×10^{-8} m³, so we can write

$$\begin{aligned}\text{needle's mass } m &= (\text{needle's density})(\text{needle's volume}) \\ &= (7900 \text{ kg/m}^3)(1.5 \times 10^{-8} \text{ m}^3) \\ &= 1.185 \times 10^{-4} \text{ kg}.\end{aligned}$$

Substituting into Eq. 32-25 with this value for m, and also 55.847 g/mole (= 0.055 847 kg/mole) for M and 6.02×10^{23} for N_A, we find

$$\begin{aligned} N &= \frac{(1.185 \times 10^{-4} \text{ kg})(6.02 \times 10^{23})}{0.055\,847 \text{ kg/mole}} \\ &= 1.2774 \times 10^{21}.\end{aligned}$$

Substituting this and the value of μ_{Fe} into Eq. 32-22 then yields

$$\begin{aligned} \mu &= (0.10)(1.2774 \times 10^{21})(2.1 \times 10^{-23} \text{ J/T}) \\ &= 2.682 \times 10^{-3} \text{ J/T} \approx 2.7 \times 10^{-3} \text{ J/T}. \quad \text{(Answer)}\end{aligned}$$

Hysteresis

Magnetization curves for ferromagnetic materials are not retraced as we increase and then decrease the external magnetic field B_0. Figure 32-13 is a plot of B_M versus B_0 during the following operations with a Rowland ring: (1) Starting with the iron unmagnetized (point a), increase the current in the toroid until B_0 (= $\mu_0 in$) has the value corresponding to point b; (2) reduce the current in the toroid winding (and thus B_0) back to zero (point c); (3) reverse the toroid current and increase it in magnitude until B_0 has the value corresponding to point d; (4) reduce the current to zero again (point e); (5) reverse the current once more until point b is reached again.

The lack of retraceability shown in Fig. 32-13 is called **hysteresis,** and the curve $bcdeb$ is called a *hysteresis loop.* Note that at points c and e the iron core is magnetized, even though there is no current in the toroid windings; this is the familiar phenomenon of permanent magnetism.

Hysteresis can be understood through the concept of magnetic domains. Evidently the motions of the domain boundaries and the reorientations of the domain directions are not totally reversible. When the applied magnetic field B_0 is increased and then decreased back to its initial value, the domains do not return completely to

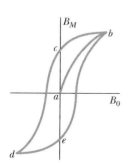

Fig. 32-13 A magnetization curve (*ab*) for a ferromagnetic specimen and an associated hysteresis loop (*bcdeb*).

their original configuration but retain some "memory" of their alignment after the initial increase. This memory of magnetic materials is essential for the magnetic storage of information, as on cassette tapes and computer disks.

This memory of the alignment of domains can also occur naturally. When lightning sends currents along multiple tortuous paths through the ground, the currents produce intense magnetic fields that can suddenly magnetize any ferromagnetic material in nearby rock. Because of hysteresis, such rock material retains some of that magnetization after the lightning strike (after the currents disappear). Pieces of the rock, later exposed, broken, and loosened by weathering, are then lodestones.

32-9 Induced Magnetic Fields

In Chapter 31 you saw that a changing magnetic flux induces an electric field, and we ended up with Faraday's law of induction in the form

$$\oint \vec{E} \cdot d\vec{s} = -\frac{d\Phi_B}{dt} \qquad \text{(Faraday's law of induction).} \qquad (32\text{-}26)$$

Here $\vec{E}$ is the electric field induced along a closed loop by the changing magnetic flux Φ_B encircled by that loop. Because symmetry is often so powerful in physics, we should be tempted to ask whether induction can occur in the opposite sense; that is, can a changing electric flux induce a magnetic field?

The answer is that it can; furthermore, the equation governing the induction of a magnetic field is almost symmetric with Eq. 32-26. We often call it Maxwell's law of induction after James Clerk Maxwell, and we write it as

$$\oint \vec{B} \cdot d\vec{s} = \mu_0 \varepsilon_0 \frac{d\Phi_E}{dt} \qquad \text{(Maxwell's law of induction).} \qquad (32\text{-}27)$$

Here $\vec{B}$ is the magnetic field induced along a closed loop by the changing electric flux Φ_E in the region encircled by that loop.

As an example of this sort of induction, we consider the charging of a parallel-plate capacitor with circular plates (Fig. 32-14a). (Although we shall focus on this particular arrangement, a changing electric flux will always induce a magnetic field whenever it occurs.) We assume that the charge on the capacitor is being increased at a steady rate by a constant current i in the connecting wires. Then the electric field magnitude between the plates must also be increasing at a steady rate.

Figure 32-14b is a view of the right-hand plate of Fig. 32-14a from between the plates. The electric field is directed into the page. Let us consider a circular loop through point 1 in Figs. 32-14a and b, concentric with the capacitor plates and with a radius smaller than that of the plates. Because the electric field through the loop is changing, the electric flux through the loop must also be changing. According to Eq. 32-27, this changing electric flux induces a magnetic field around the loop.

Experiment proves that a magnetic field $\vec{B}$ is indeed induced around such a loop, directed as shown. This magnetic field has the same magnitude at every point around the loop and thus has circular symmetry about the central axis of the capacitor plates.

If we now consider a larger loop, say, through point 2 outside the plates in Figs. 32-14a and b, we find that a magnetic field is induced around that loop as well. Thus, while the electric field is changing, magnetic fields are induced between the plates, both inside and outside the gap. When the electric field stops changing, these induced magnetic fields disappear.

Although Eq. 32-27 is similar to Eq. 32-26, the equations differ in two ways. First, Eq. 32-27 has the two extra symbols, μ_0 and ε_0, but they appear only because

Fig. 32-14 (a) A circular parallel-plate capacitor, shown in side view, is being charged by a constant current i. (b) A view from within the capacitor, toward the plate at the right. The electric field $\vec{E}$ is uniform, is directed into the page (toward the plate), and grows in magnitude as the charge on the capacitor increases. The magnetic field $\vec{B}$ induced by this changing electric field is shown at four points on a circle with a radius r less than the plate radius R.

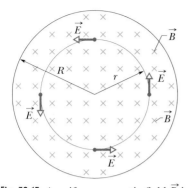

Fig. 32-15 A uniform magnetic field $\vec{B}$ in a circular region. The field, directed into the page, is increasing in magnitude. The electric field $\vec{E}$ induced by the changing magnetic field is shown at four points on a circle concentric with the circular region. Compare this situation with that of Fig. 32-14b.

we employ SI units. Second, Eq. 32-27 lacks the minus sign of Eq. 32-26, meaning that the induced electric field $\vec{E}$ and the induced magnetic field $\vec{B}$ have opposite directions when they are produced in otherwise similar situations. To see this opposition, examine Fig. 32-15, in which an increasing magnetic field $\vec{B}$, directed into the page, induces an electric field $\vec{E}$. The induced field $\vec{E}$ is counterclockwise, opposite the induced magnetic field $\vec{B}$ in Fig. 32-14b.

Ampere–Maxwell Law

Now recall that the left side of Eq. 32-27, the integral of the dot product $\vec{B} \cdot d\vec{s}$ around a closed loop, appears in another equation—namely, Ampere's law:

$$\oint \vec{B} \cdot d\vec{s} = \mu_0 i_{\text{enc}} \qquad \text{(Ampere's law)}, \qquad (32\text{-}28)$$

where i_{enc} is the current encircled by the closed loop. Thus, our two equations that specify the magnetic field $\vec{B}$ produced by means other than a magnetic material (that is, by a current and by a changing electric field) give the field in exactly the same form. We can combine the two equations into the single equation

$$\oint \vec{B} \cdot d\vec{s} = \mu_0 \varepsilon_0 \frac{d\Phi_E}{dt} + \mu_0 i_{\text{enc}} \qquad \text{(Ampere–Maxwell law)}. \qquad (32\text{-}29)$$

When there is a current but no change in electric flux (such as with a wire carrying a constant current), the first term on the right side of Eq. 32-29 is zero, and so Eq. 32-29 reduces to Eq. 32-28, Ampere's law. When there is a change in electric flux but no current (such as inside or outside the gap of a charging capacitor), the second term on the right side of Eq. 32-29 is zero, and so Eq. 32-29 reduces to Eq. 32-27, Maxwell's law of induction.

Sample Problem 32-3

A parallel-plate capacitor with circular plates of radius R is being charged as in Fig. 32-14a.

(a) Derive an expression for the magnetic field at radii r for the case $r \leq R$.

SOLUTION: The Key Idea here is that a magnetic field can be set up by a current and by induction due to a changing electric flux; both effects are included in Eq. 32-29. There is no current between the capacitor plates of Fig. 32-14, but the electric flux there is changing. Thus, Eq. 32-29 reduces to

$$\oint \vec{B} \cdot d\vec{s} = \mu_0 \varepsilon_0 \frac{d\Phi_E}{dt}. \qquad (32\text{-}30)$$

We shall separately evaluate the left and right sides of this equation.

Left side of Eq. 32-30: We choose a circular Amperian loop with a radius $r \leq R$ as shown in Fig. 32-14, because we want to evaluate the magnetic field for $r \leq R$—that is, inside the capacitor. The magnetic field $\vec{B}$ at all points along the loop is tangent to the loop, as is the path element $d\vec{s}$. Thus, $\vec{B}$ and $d\vec{s}$ are either parallel or antiparallel at each point of the loop. For simplicity, assume they are parallel (the choice does not alter our outcome here). Then

$$\oint \vec{B} \cdot d\vec{s} = \oint B \, ds \cos 0° = \oint B \, ds.$$

Due to the circular symmetry of the plates, we can also assume that $\vec{B}$ has the same magnitude at every point around the loop. Thus, B can be taken outside the integral on the right side of the above equation. The integral that remains is $\oint ds$, which simply gives the circumference $2\pi r$ of the loop. The left side of Eq. 32-30 is then $(B)(2\pi r)$.

Right side of Eq. 32-30: We assume that the electric field $\vec{E}$ is uniform between the capacitor plates and directed perpendicular to the plates. Then the electric flux Φ_E through the Amperian loop is EA, where A is the area encircled by the loop within the electric field. Thus, the right side of Eq. 32-30 is $\mu_0 \varepsilon_0 \, d(EA)/dt$.

Substituting our results for the left and right sides into Eq. 32-30, we get

$$(B)(2\pi r) = \mu_0 \varepsilon_0 \frac{d(EA)}{dt}.$$

Because A is a constant, we write $d(EA)$ as $A \, dE$, so we have

$$(B)(2\pi r) = \mu_0 \varepsilon_0 A \frac{dE}{dt}. \qquad (32\text{-}31)$$

We next use this Key Idea: The area A that is encircled by the Amperian loop within the electric field is the full area πr^2 of the loop, because the loop's radius r is less than (or equal to) the plate radius R. Substituting πr^2 for A in Eq. 32-31 and solving the result for B give us, for $r \leq R$,

$$B = \frac{\mu_0 \varepsilon_0 r}{2} \frac{dE}{dt}. \qquad \text{(Answer)} \quad \text{(32-32)}$$

This equation tells us that, inside the capacitor, B increases linearly with increased radial distance r, from zero at the center of the plates to a maximum value at the plate edges (where $r = R$).

(b) Evaluate the field magnitude B for $r = R/5 = 11.0$ mm and $dE/dt = 1.50 \times 10^{12}$ V/m · s.

SOLUTION: From the answer to (a), we have

$$B = \frac{1}{2} \mu_0 \varepsilon_0 r \frac{dE}{dt}$$

$$= \frac{1}{2}(4\pi \times 10^{-7}\ \text{T} \cdot \text{m/A})(8.85 \times 10^{-12}\ \text{C}^2/\text{N} \cdot \text{m}^2)$$

$$\times (11.0 \times 10^{-3}\ \text{m})(1.50 \times 10^{12}\ \text{V/m} \cdot \text{s})$$

$$= 9.18 \times 10^{-8}\ \text{T}. \qquad \text{(Answer)}$$

(c) Derive an expression for the induced magnetic field for the case $r \geq R$.

SOLUTION: Our procedure is the same as in (a) except we now use an Amperian loop with a radius r that is greater than the plate radius R, to evaluate B outside the capacitor. Evaluating the left and right sides of Eq. 32-30 again leads to Eq. 32-31. However, we then need this subtle Key Idea: The electric field exists only between the plates, not outside the plates. Thus, the area A that is encircled by the Amperian loop in the electric field is *not* the full area πr^2 of the loop. Rather, A is only the plate area πR^2.

Substituting πR^2 for A in Eq. 32-31 and solving the result for B give us, for $r \geq R$,

$$B = \frac{\mu_0 \varepsilon_0 R^2}{2r} \frac{dE}{dt}. \qquad \text{(Answer)} \quad \text{(32-33)}$$

This equation tells us that, outside the capacitor, B decreases with increased radial distance r, from a maximum value at the plate edges (where $r = R$). By substituting $r = R$ into Eqs. 32-32 and 32-33, you can show that these equations are consistent; that is, they give the same maximum value of B at the plate radius.

The magnitude of the induced magnetic field calculated in (b) is so small that it can scarcely be measured with simple apparatus. This is in sharp contrast to the magnitudes of induced electric fields (Faraday's law), which can be measured easily. This experimental difference exists partly because induced emfs can easily be multiplied by using a coil of many turns. No technique of comparable simplicity exists for multiplying induced magnetic fields. In any case, the experiment suggested by this sample problem has been done, and the presence of the induced magnetic fields has been verified quantitatively.

✔CHECKPOINT 5: The figure shows graphs of the electric field magnitude E versus time t for four uniform electric fields, all contained within identical circular regions as in Fig. 32-14b. Rank the fields according to the magnitudes of the magnetic fields they induce at the edge of the region, greatest first.

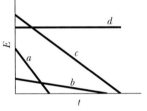

32-10 Displacement Current

If you compare the two terms on the right side of Eq. 32-29, you will see that the product $\varepsilon_0 (d\Phi_E/dt)$ must have the dimension of a current. In fact, that product has been treated as being a fictitious current called the **displacement current** i_d:

$$i_d = \varepsilon_0 \frac{d\Phi_E}{dt} \qquad \text{(displacement current).} \quad \text{(32-34)}$$

"Displacement" is poorly chosen in that nothing is being displaced, but we are stuck with the word. Nevertheless, we can now rewrite Eq. 32-29 as

$$\oint \vec{B} \cdot d\vec{s} = \mu_0 i_{d,\text{enc}} + \mu_0 i_{\text{enc}} \qquad \text{(Ampere–Maxwell law),} \quad \text{(32-35)}$$

in which $i_{d,\text{enc}}$ is the displacement current that is encircled by the integration loop.

Let us again focus on a charging capacitor with circular plates, as in Fig. 32-16a. The real current i that is charging the plates changes the electric field $\vec{E}$ between the plates. The fictitious displacement current i_d between the plates is associated with that changing field $\vec{E}$. Let us relate these two currents.

The charge q on the plates at any time is related to the magnitude E of the field between the plates at that time by Eq. 26-4:

$$q = \varepsilon_0 A E, \qquad \text{(32-36)}$$

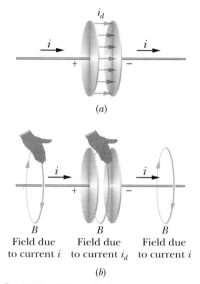

Fig. 32-16 (a) The displacement current i_d between the plates of a capacitor that is being charged by a current i. (b) The right-hand rule for finding the direction of the magnetic field around a wire with a real current (as at the left) also gives the direction of the magnetic field around a displacement current (as in the center).

in which A is the plate area. To get the real current i, we differentiate Eq. 32-36 with respect to time, finding

$$\frac{dq}{dt} = i = \varepsilon_0 A \frac{dE}{dt}. \qquad (32\text{-}37)$$

To get the displacement current i_d, we can use Eq. 32-34. Assuming that the electric field $\vec{E}$ between the two plates is uniform (we neglect any fringing), we can replace the electric flux Φ_E in that equation with EA. Then Eq. 32-34 becomes

$$i_d = \varepsilon_0 \frac{d\Phi_E}{dt} = \varepsilon_0 \frac{d(EA)}{dt} = \varepsilon_0 A \frac{dE}{dt}. \qquad (32\text{-}38)$$

Comparing Eqs. 32-37 and 32-38, we see that the real current i charging the capacitor and the fictitious displacement current i_d between the plates have the same magnitude:

$$i_d = i \qquad \text{(displacement current in a capacitor)}. \qquad (32\text{-}39)$$

Thus, we can consider the fictitious displacement current i_d to be simply a continuation of the real current i from one plate, across the capacitor gap, to the other plate. Because the electric field is uniformly spread over the plates, the same is true of this fictitious displacement current i_d, as suggested by the spread of current arrows in Fig. 32-16a. Although no charge actually moves across the gap between the plates, the idea of the fictitious current i_d can help us to quickly find the direction and magnitude of an induced magnetic field, as follows.

Finding the Induced Magnetic Field

In Chapter 30 we found the direction of the magnetic field produced by a real current i by using the right-hand rule of Fig. 30-4. We can apply the same rule to find the direction of an induced magnetic field produced by a fictitious displacement current i_d, as is shown in the center of Fig. 32-16b for a capacitor.

We can also use i_d to find the magnitude of the magnetic field induced by a charging capacitor with parallel circular plates of radius R. We simply consider the space between the plates to be an imaginary circular wire of radius R carrying the imaginary current i_d. Then, from Eq. 30-22, the magnitude of the magnetic field at a point inside the capacitor at radius r from the center is

$$B = \left(\frac{\mu_0 i_d}{2\pi R^2}\right) r \qquad \text{(inside a circular capacitor)}. \qquad (32\text{-}40)$$

Similarly, from Eq. 30-19, the magnitude of the magnetic field at a point outside the capacitor at radius r is

$$B = \frac{\mu_0 i_d}{2\pi r} \qquad \text{(outside a circular capacitor)}. \qquad (32\text{-}41)$$

Sample Problem 32-4

The circular parallel-plate capacitor in Sample Problem 32-3 is being charged with a current i.

(a) Between the plates, what is the magnitude of $\oint \vec{B} \cdot d\vec{s}$, in terms of μ_0 and i, at a radius $r = R/5$ from their center?

SOLUTION: The first **Key Idea** of Sample Problem 32-3a holds here too. However, now we can replace the product $\varepsilon_0 \, d\Phi_E/dt$ in Eq.

32-29 with a fictitious displacement current i_d. Then integral $\oint \vec{B} \cdot d\vec{s}$ is given by Eq. 32-35, but because there is no real current i between the capacitor plates, the equation reduces to

$$\oint \vec{B} \cdot d\vec{s} = \mu_0 i_{d,\text{enc}}. \qquad (32\text{-}42)$$

Because we want to evaluate $\oint \vec{B} \cdot d\vec{s}$ at radius $r = R/5$ (within the capacitor), the integration loop encircles only a portion $i_{d,\text{enc}}$ of the total displacement current i_d. A second Key Idea is to assume that i_d is uniformly spread over the full plate area. Then the portion of the displacement current encircled by the loop is proportional to the area encircled by the loop:

$$\frac{\left(\begin{array}{c}\text{encircled displacement}\\ \text{current } i_{d,enc}\end{array}\right)}{\left(\begin{array}{c}\text{total displacement}\\ \text{current } i_d\end{array}\right)} = \frac{\text{encircled area } \pi r^2}{\text{full plate area } \pi R^2}.$$

This gives us

$$i_{d,enc} = i_d \frac{\pi r^2}{\pi R^2}.$$

Substituting this into Eq. 32-42, we obtain

$$\oint \vec{B} \cdot d\vec{s} = \mu_0 i_d \frac{\pi r^2}{\pi R^2}. \qquad (32\text{-}43)$$

Now substituting $i_d = i$ (from Eq. 32-39) and $r = R/5$ into Eq. 32-43 leads to

$$\oint \vec{B} \cdot d\vec{s} = \mu_0 i \frac{(R/5)^2}{R^2} = \frac{\mu_0 i}{25}. \qquad \text{(Answer)}$$

(b) In terms of the maximum induced magnetic field, what is the magnitude of the magnetic field induced at $r = R/5$, inside the capacitor?

SOLUTION: The Key Idea here is that, because the capacitor has parallel circular plates, we can treat the space between the plates as an imaginary wire of radius R carrying the imaginary current i_d. Then

we can use Eq. 32-40 to find the induced magnetic field magnitude B at any point inside the capacitor. At $r = R/5$, that equation yields

$$B = \left(\frac{\mu_0 i_d}{2\pi R^2}\right) r = \frac{\mu_0 i_d (R/5)}{2\pi R^2} = \frac{\mu_0 i_d}{10\pi R}. \qquad (32\text{-}44)$$

The maximum field magnitude B_{max} within the capacitor occurs at $r = R$. It is

$$B_{max} = \left(\frac{\mu_0 i_d}{2\pi R^2}\right) R = \frac{\mu_0 i_d}{2\pi R}. \qquad (32\text{-}45)$$

Dividing Eq. 32-44 by Eq. 32-45 and rearranging the result, we find

$$B = \frac{B_{max}}{5}. \qquad \text{(Answer)}$$

We should be able to obtain this result with a little reasoning and less work. Equation 32-40 tells us that inside the capacitor, B increases linearly with r. Therefore, a point $\frac{1}{5}$ the distance out to the full radius R of the plates, where B_{max} occurs, should have a field B that is $\frac{1}{5}B_{max}$.

✓**CHECKPOINT 6:** The figure is a view of one plate of a parallel-plate capacitor from within the capacitor. The dashed lines show four integration paths (path b follows the edge of the plate). Rank the paths according to the magnitude of $\oint \vec{B} \cdot d\vec{s}$ along the paths during the discharging of the capacitor, greatest first.

32-11 Maxwell's Equations

Equation 32-29 is the last of the four fundamental equations of electromagnetism, called *Maxwell's equations* and displayed in Table 32-1. These four equations explain a diverse range of phenomena, from why a compass needle points north to why a car starts when you turn the ignition key. They are the basis for the functioning of such electromagnetic devices as electric motors, cyclotrons, television transmitters and receivers, telephones, fax machines, radar, and microwave ovens.

Maxwell's equations are the basis from which many of the equations you have seen since Chapter 22 can be derived. They are also the basis of many of the equations you will see in Chapters 34 through 37, which introduce you to optics.

TABLE 32-1 Maxwell's Equations[a]

Name	Equation	
Gauss' law for electricity	$\oint \vec{E} \cdot d\vec{A} = q_{enc}/\varepsilon_0$	Relates net electric flux to net enclosed electric charge
Gauss' law for magnetism	$\oint \vec{B} \cdot d\vec{A} = 0$	Relates net magnetic flux to net enclosed magnetic charge
Faraday's law	$\oint \vec{E} \cdot d\vec{s} = -\dfrac{d\Phi_B}{dt}$	Relates induced electric field to changing magnetic flux
Ampere–Maxwell law	$\oint \vec{B} \cdot d\vec{s} = \mu_0\varepsilon_0 \dfrac{d\Phi_E}{dt} + \mu_0 i_{enc}$	Relates induced magnetic field to changing electric flux and to current

[a]Written on the assumption that no dielectric or magnetic materials are present.

REVIEW & SUMMARY

Gauss' Law for Magnetic Fields The simplest magnetic structures are magnetic dipoles. Magnetic monopoles do not exist (as far as we know). **Gauss' law** for magnetic fields,

$$\Phi_B = \oint \vec{B} \cdot d\vec{A} = 0, \qquad (32\text{-}1)$$

states that the net magnetic flux through any (closed) Gaussian surface is zero. It implies that magnetic monopoles don't exist.

Earth's Magnetic Field Earth's magnetic field can be approximated as being that of a magnetic dipole whose dipole moment makes an angle of 11.5° with Earth's rotation axis, and with the south pole of the dipole in the northern hemisphere. The direction of the local magnetic field at any point on Earth's surface is given by the *field declination* (the angle left or right from geographic north) and the *field inclination* (the angle up or down from the horizontal).

Spin Magnetic Dipole Moment An electron has an intrinsic angular momentum called *spin angular momentum* (or *spin*) $\vec{S}$, with which an intrinsic *spin magnetic dipole moment* $\vec{\mu}_s$ is associated:

$$\vec{\mu}_s = -\frac{e}{m}\,\vec{S}. \qquad (32\text{-}2)$$

Spin $\vec{S}$ cannot itself be measured, but any component can be measured. Assuming that the measurement is along the z axis of a coordinate system, the component S_z can have only the values given by

$$S_z = m_s\frac{h}{2\pi}, \qquad \text{for } m_s = \pm\tfrac{1}{2}, \qquad (32\text{-}3)$$

where $h\ (= 6.63 \times 10^{-34}\ \mathrm{J \cdot s})$ is the Planck constant. Similarly, the electron's spin magnetic dipole moment $\vec{\mu}_s$ cannot itself be measured but its component can be measured. Along a z axis, the component is

$$\mu_{s,z} = \pm\frac{eh}{4\pi m} = \pm\mu_B, \qquad (32\text{-}4,\ 32\text{-}6)$$

where μ_B is the *Bohr magneton:*

$$\mu_B = \frac{eh}{4\pi m} = 9.27 \times 10^{-24}\ \mathrm{J/T}. \qquad (32\text{-}5)$$

The potential energy U associated with the orientation of the spin magnetic dipole moment in an external magnetic field $\vec{B}_{ext}$ is

$$U = -\vec{\mu}_s \cdot \vec{B}_{ext} = -\mu_{s,z}B. \qquad (32\text{-}7)$$

Orbital Magnetic Dipole Moment An electron in an atom has an additional angular momentum called its *orbital angular momentum* $\vec{L}_{orb}$, with which an *orbital magnetic dipole moment* $\vec{\mu}_{orb}$ is associated:

$$\vec{\mu}_{orb} = -\frac{e}{2m}\,\vec{L}_{orb}. \qquad (32\text{-}8)$$

Orbital angular momentum is quantized and can have only values given by

$$L_{orb,z} = m_l\frac{h}{2\pi}, \qquad \text{for } m_l = 0,\ \pm1,\ \pm2,\ \ldots,\ \pm(\text{limit}). \qquad (32\text{-}9)$$

Thus, the magnitude of the orbital angular momentum is

$$\mu_{orb,z} = -m_l\frac{eh}{4\pi m} = -m_l\mu_B. \qquad (32\text{-}10,\ 32\text{-}11)$$

The potential energy U associated with the orientation of the orbital magnetic dipole moment in an external magnetic field $\vec{B}_{ext}$ is

$$U = -\vec{\mu}_{orb} \cdot \vec{B}_{ext} = -\mu_{orb,z}B_{ext}. \qquad (32\text{-}12)$$

Diamagnetism *Diamagnetic materials* do not exhibit magnetism until they are placed in an external magnetic field $\vec{B}_{ext}$. They then develop a magnetic dipole moment directed opposite $\vec{B}_{ext}$. If the field is nonuniform, the diamagnetic material is repelled from regions of greater magnetic field. This property is called *diamagnetism.*

Paramagnetism In a *paramagnetic material,* each atom has a permanent magnetic dipole moment $\vec{\mu}$, but the dipole moments are randomly oriented and the material as a whole lacks a magnetic field. However, an external magnetic field $\vec{B}_{ext}$ can partially align the atomic dipole moments to give the material a net magnetic dipole moment in the direction of $\vec{B}_{ext}$. If $\vec{B}_{ext}$ is nonuniform, the material is attracted to regions of greater magnetic field. These properties are called *paramagnetism.*

The alignment of the atomic dipole moments increases with an increase in $\vec{B}_{ext}$ and decreases with an increase in temperature T. The extent to which a sample of volume V is magnetized is given by its *magnetization $\vec{M}$,* whose magnitude is

$$M = \frac{\text{measured magnetic moment}}{V}. \qquad (32\text{-}18)$$

Complete alignment of all N atomic magnetic dipoles in a sample, called *saturation* of the sample, corresponds to the maximum magnetization value $M_{max} = N\mu/V$. For low values of the ratio B_{ext}/T, we have the approximation

$$M = C\frac{B_{ext}}{T} \qquad (\text{Curie's law}), \qquad (32\text{-}19)$$

where C is called the *Curie constant.*

Ferromagnetism In the absence of an external magnetic field, some of the electrons in a ferromagnetic material have their magnetic dipole moments aligned by means of a quantum physical interaction called *exchange coupling,* producing regions (domains) within the material with strong magnetic dipole moments. An external field $\vec{B}_{ext}$ can align the magnetic dipole moments of those regions, producing a strong net magnetic dipole moment for the material as a whole, in the direction of $\vec{B}_{ext}$. This net magnetic dipole moment can partially persist when field $\vec{B}_{ext}$ is removed. If $\vec{B}_{ext}$ is nonuniform, the ferromagnetic material is attracted to regions of greater magnetic field. These properties are called *ferromagnetism.* Exchange coupling disappears when a sample's temperature exceeds its *Curie temperature,* and then the sample has only paramagnetism.

Maxwell's Extension of Ampere's Law   A changing electric flux induces a magnetic field $\vec{B}$. Maxwell's law,

$$\oint \vec{B} \cdot d\vec{s} = \mu_0 \varepsilon_0 \frac{d\Phi_E}{dt} \qquad \text{(Maxwell's law of induction)}, \qquad (32\text{-}27)$$

relates the magnetic field induced along a closed loop to the changing electric flux Φ_E through the loop. Ampere's law, $\oint \vec{B} \cdot d\vec{s} = \mu_0 i_{enc}$ (Eq. 32-28), gives the magnetic field generated by a current i_{enc} encircled by a closed loop. Maxwell's law and Ampere's law can be written as the single equation

$$\oint \vec{B} \cdot d\vec{s} = \mu_0 \varepsilon_0 \frac{d\Phi_E}{dt} + \mu_0 i_{enc} \qquad \text{(Ampere–Maxwell law)}. \quad (32\text{-}29)$$

Displacement Current   We define the fictitious *displacement current* due to a changing electric field as

$$i_d = \varepsilon_0 \frac{d\Phi_E}{dt}. \qquad (32\text{-}34)$$

Equation 32-29 then becomes

$$\oint \vec{B} \cdot d\vec{s} = \mu_0 i_{d,enc} + \mu_0 i_{enc} \qquad \text{(Ampere–Maxwell law)}, \quad (32\text{-}35)$$

where $i_{d,enc}$ is the displacement current encircled by the integration loop. The idea of a displacement current allows us to retain the notion of continuity of current through a capacitor. However, displacement current is *not* a transfer of charge.

Maxwell's Equations   Maxwell's equations, displayed in Table 32-1, summarize electromagnetism and form its foundation.

QUESTIONS

1. An electron in an external magnetic field $\vec{B}_{ext}$ has its spin angular momentum S_z antiparallel to $\vec{B}_{ext}$. If the electron undergoes a *spin-flip* so that S_z is then parallel with $\vec{B}_{ext}$, must energy be supplied to or lost by the electron?

2. Figure 32-17a shows a pair of opposite spin orientations for an electron in an external magnetic field $\vec{B}_{ext}$. Figure 32-17b gives three choices for the graph of the potential energies associated with those orientations as a function of the magnitude of $\vec{B}_{ext}$. Choices *b* and *c* consist of intersecting lines, choice *a* of parallel lines. Which is the correct choice?

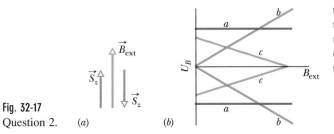

Fig. 32-17
Question 2.   (*a*)   (*b*)

3. Figure 32-18 shows three loop models of an electron orbiting counterclockwise within a magnetic field. The fields are nonuniform for models 1 and 2 and uniform for model 3. For each model, are (a) the magnetic dipole moment of the loop and (b) the magnetic force on the loop directed up, directed down, or zero?

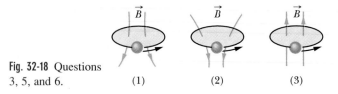

Fig. 32-18 Questions
3, 5, and 6.   (1)   (2)   (3)

4. Does the magnitude of the net force on the loop of Figs. 32-8a and *b* increase, decrease, or remain the same if we increase (a) the magnitude of $\vec{B}_{ext}$ and (b) the divergence of $\vec{B}_{ext}$?

5. Replace the current loops of Question 3 and Fig. 32-18 with diamagnetic spheres. For each field, are (a) the magnetic dipole moment of the sphere and (b) the magnetic force on the sphere directed up, directed down, or zero?

6. Replace the current loops of Question 3 and Fig. 32-18 with paramagnetic spheres. For each field, are (a) the magnetic dipole moment of the sphere and (b) the magnetic force on the sphere up, down, or zero?

7. The magnetic dipoles in a diamagnetic material are represented, for three situations, in Fig. 32-19. (For simplicity, the dipole moments are assumed to be directed only up or down the page.) The three situations differ in the magnitude of a magnetic field applied to the material. (a) For each situation, is the applied field directed up or down the page? Rank the three situations according to (b) the magnitude of the applied field and (c) the magnetization of the material, greatest first.

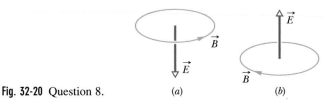

Fig. 32-19
Question 7.   (1)   (2)   (3)

8. Figure 32-20 shows, in two situations, an electric field vector and an induced magnetic field line. In each, is the magnitude of $\vec{E}$ increasing or decreasing?

Fig. 32-20 Question 8.   (*a*)   (*b*)

9. Figure 32-21 shows a parallel-plate capacitor and the current in the connecting wires that is discharging the capacitor. Are the directions of (a) electric field $\vec{E}$ and (b) displacement current i_d leftward or rightward between the plates? (c) Is the magnetic field at point P into the page or out of the page?

Fig. 32-21 Question 9.

10. A parallel-plate capacitor with rectangular plates is being discharged. A rectangular loop, centered on the plates and between them, measures L by $2L$; the plates measure $2L$ by $4L$. What fraction of the displacement current is encircled by the loop if that current is uniform?

11. Figure 32-22a shows a capacitor, with circular plates, that is being charged. Point a (near one of the connecting wires) and point

b (inside the capacitor gap) are equidistant from the central axis, as are point c (not so near the wire) and point d (between the plates but outside the gap). In Fig. 32-22b, one curve gives the variation with distance r of the magnitude of the magnetic field inside and outside the wire. The other curve gives the variation with distance r of the magnitude of the magnetic field inside and outside the gap. The two curves partially overlap. Which of the three points on the curves correspond to which of the four points of Fig. 32-22a?

(a)   (b)

Fig. 32-22 Question 11.

EXERCISES & PROBLEMS

SEC. 32-2   Gauss' Law for Magnetic Fields

1E. Imagine rolling a sheet of paper into a cylinder and placing a bar magnet near its end as shown in Fig. 32-23. (a) Sketch the magnetic field lines that pass through the surface of the cylinder. (b) What can you say about the sign of $\vec{B} \cdot d\vec{A}$ for every area $d\vec{A}$ on the surface? (c) Does this contradict Gauss' law for magnetism? Explain.

Fig. 32-23 Exercise 1.

2E. The magnetic flux through each of five faces of a die (singular of "dice") is given by $\Phi_B = \pm N$ Wb, where N (= 1 to 5) is the number of spots on the face. The flux is positive (outward) for N even and negative (inward) for N odd. What is the flux through the sixth face of the die?

3P. A Gaussian surface in the shape of a right circular cylinder with end caps has a radius of 12.0 cm and a length of 80.0 cm. Through one end there is an inward magnetic flux of 25.0 μWb. At the other end there is a uniform magnetic field of 1.60 mT, normal to the surface and directed outward. What is the net magnetic flux through the curved surface? ssm ilw www

SEC. 32-3   The Magnetism of Earth

4E. Assume the average value of the vertical component of Earth's magnetic field is 43 μT (downward) for all of Arizona, which has an area of 2.95×10^5 km^2, and calculate the net magnetic flux through the rest of Earth's surface (the entire surface excluding Arizona). Is that net magnetic flux outward or inward?

5E. In New Hampshire the average horizontal component of Earth's magnetic field in 1912 was 16 μT and the average inclination or

"dip" was 73°. What was the corresponding magnitude of Earth's magnetic field? ssm

6P. The magnetic field of Earth can be approximated as the magnetic field of a dipole, with horizontal and vertical components, at a point a distance r from Earth's center, given by

$$B_h = \frac{\mu_0 \mu}{4\pi r^3} \cos \lambda_m, \qquad B_v = \frac{\mu_0 \mu}{2\pi r^3} \sin \lambda_m,$$

where λ_m is the *magnetic latitude* (this type of latitude is measured from the geomagnetic equator toward the north or south geomagnetic pole). Assume that Earth's magnetic dipole moment is $\mu = 8.00 \times 10^{22}$ A · m^2. (a) Show that the magnitude of Earth's field at latitude λ_m is given by

$$B = \frac{\mu_0 \mu}{4\pi r^3} \sqrt{1 + 3 \sin^2 \lambda_m}.$$

(b) Show that the inclination ϕ_i of the magnetic field is related to the magnetic latitude λ_m by

$$\tan \phi_i = 2 \tan \lambda_m.$$

7P. Use the results displayed in Problem 6 to predict Earth's magnetic field (both magnitude and inclination) at (a) the geomagnetic equator, (b) a point at geomagnetic latitude 60°, and (c) the north geomagnetic pole. ssm

8P. Using the approximations given in Problem 6, find (a) the altitude above Earth's surface where the magnitude of its magnetic field is 50% of the surface value at the same latitude; (b) the maximum magnitude of the magnetic field at the core–mantle boundary, 2900 km below Earth's surface; and (c) the magnitude and inclination of Earth's magnetic field at the north geographic pole. Suggest why the values you calculated for (c) differ from measured values.

SEC. 32-4   Magnetism and Electrons

9E. What is the measured component of the orbital magnetic dipole moment of an electron with (a) $m_l = 1$ and (b) $m_l = -2$? ssm

10E. What is the energy difference between parallel and antiparallel alignment of the z component of an electron's spin magnetic dipole moment with an external magnetic field of magnitude 0.25 T, directed parallel to the z axis?

11E. If an electron in an atom has an orbital angular momentum with $m_l = 0$, what are the components (a) $L_{orb,z}$ and (b) $\mu_{orb,z}$? If the atom is in an external magnetic field $\vec{B}$ of magnitude 35 mT and directed along the z axis, what are the potential energies associated with the orientations of (c) the electron's orbital magnetic dipole moment and (d) the electron's spin magnetic dipole moment? (e) Repeat (a) through (d) for $m_l = -3$.   ssm

SEC. 32-6   Diamagnetism

12E. Figure 32-24 shows a loop model (loop L) for a diamagnetic material. (a) Sketch the magnetic field lines through and about the material due to the bar magnet. (b) What are the directions of the loop's net magnetic dipole moment $\vec{\mu}$ and the conventional current i in the loop? (c) What is the direction of the magnetic force on the loop?

Fig. 32-24 Exercises 12 and 16.

13P*. Assume that an electron of mass m and charge magnitude e moves in a circular orbit of radius r about a nucleus. A uniform magnetic field $\vec{B}$ is then established perpendicular to the plane of the orbit. Assuming also that the radius of the orbit does not change and that the change in the speed of the electron due to field $\vec{B}$ is small, find an expression for the change in the orbital magnetic dipole moment of the electron due to the field.   ssm

SEC. 32-7   Paramagnetism

14E. A 0.50 T magnetic field is applied to a paramagnetic gas whose atoms have an intrinsic magnetic dipole moment of 1.0×10^{-23} J/T. At what temperature will the mean kinetic energy of translation of the gas atoms be equal to the energy required to reverse such a dipole end for end in this magnetic field?

15E. A magnet in the form of a cylindrical rod has a length of 5.00 cm and a diameter of 1.00 cm. It has a uniform magnetization of 5.30×10^3 A/m. What is its magnetic dipole moment?   ssm   ilw

16E. Repeat Exercise 12 for the case in which loop L is the model for a paramagnetic material.

17E. A sample of the paramagnetic salt to which the magnetization curve of Fig. 32-9 applies is to be tested to see whether it obeys Curie's law. The sample is placed in a uniform 0.50 T magnetic field that remains constant throughout the experiment. The magnetization M is then measured at temperatures ranging from 10 to 300 K. Will it be found that Curie's law is valid under these conditions?   ssm

18E. A sample of the paramagnetic salt to which the magnetization curve of Fig. 32-9 applies is held at room temperature (300 K). At what applied magnetic field will the degree of magnetic saturation of the sample be (a) 50% and (b) 90%? (c) Are these fields attainable in the laboratory?

19P. An electron with kinetic energy K_e travels in a circular path that is perpendicular to a uniform magnetic field, the electron's motion subject only to the force due to the field. (a) Show that the magnetic dipole moment of the electron due to its orbital motion has magnitude $\mu = K_e/B$ and that it is in the direction opposite that of $\vec{B}$. (b) What are the magnitude and direction of the magnetic dipole moment of a positive ion with kinetic energy K_i under the same circumstances? (c) An ionized gas consists of 5.3×10^{21} electrons/m³ and the same number density of ions. Take the average electron kinetic energy to be 6.2×10^{-20} J and the average ion kinetic energy to be 7.6×10^{-21} J. Calculate the magnetization of the gas when it is in a magnetic field of 1.2 T.   ssm   www

SEC. 32-8   Ferromagnetism

20E. Measurements in mines and boreholes indicate that Earth's interior temperature increases with depth at the average rate of 30 C°/km. Assuming a surface temperature of 10°C, at what depth does iron cease to be ferromagnetic? (The Curie temperature of iron varies very little with pressure.)

21E. The exchange coupling mentioned in Section 32-8 as being responsible for ferromagnetism is *not* the mutual magnetic interaction between two elementary magnetic dipoles. To show this, calculate (a) the magnitude of the magnetic field a distance of 10 nm away, along the dipole axis, from an atom with magnetic dipole moment 1.5×10^{-23} J/T (cobalt), and (b) the minimum energy required to turn a second identical dipole end for end in this field. By comparing the latter with the results of Sample Problem 32-1, what can you conclude?   ssm

22E. The dipole moment associated with an atom of iron in an iron bar is 2.1×10^{-23} J/T. Assume that all the atoms in the bar, which is 5.0 cm long and has a cross-sectional area of 1.0 cm², have their dipole moments aligned. (a) What is the dipole moment of the bar? (b) What torque must be exerted to hold this magnet perpendicular to an external field of 1.5 T? (The density of iron is 7.9 g/cm³.)

23E. The saturation magnetization M_{max} of the ferromagnetic metal nickel is 4.70×10^5 A/m. Calculate the magnetic moment of a single nickel atom. (The density of nickel is 8.90 g/cm³ and its molar mass is 58.71 g/mol.)   ssm

24P. Figure 32-25 shows the apparatus used in a lecture demonstration of para- and diamagnetism. A sample of the magnetic material is suspended by a long string in the nonuniform field ($d = 2$ cm) between the poles of a powerful electromagnet. Pole P_1 is sharply pointed and pole P_2 is rounded as indicated. Any deflection of the string from the vertical is visible to the audience by means of an optical projection system (not shown). (a) First a bismuth (highly diamagnetic) sample is used. When the electromagnet is turned on, the sample is observed to deflect slightly (about 1 mm) toward one of the poles. What is the direction of this deflection? (b) Next an aluminum (paramagnetic, conducting) sample is used. When the electromagnet is turned on, the sample is observed to de-

Fig. 32-25 Problem 24.

flect strongly (about 1 cm) toward one pole for about a second and then deflect moderately (a few millimeters) toward the other pole. Explain and indicate the direction of these deflections. (*Hint:* The aluminum sample is a conductor, for which Lenz's law applies.) (c) What would happen if a ferromagnetic sample were used?

25P. The magnetic dipole moment of Earth is 8.0×10^{22} J/T. (a) If the origin of this magnetism were a magnetized iron sphere at the center of Earth, what would be its radius? (b) What fraction of the volume of Earth would such a sphere occupy? Assume complete alignment of the dipoles. The density of Earth's inner core is 14 g/cm^3. The magnetic dipole moment of an iron atom is 2.1×10^{-23} J/T. (*Note:* Earth's inner core is in fact thought to be in both liquid and solid forms and partly iron, but a permanent magnet as the source of Earth's magnetism has been ruled out by several considerations. For one, the temperature is certainly above the Curie point.) ssm ilw

SEC. 32-9   Induced Magnetic Fields

26E. Sample Problem 32-3 describes the charging of a parallel-plate capacitor with circular plates of radius 55.0 mm. At what two radii r from the central axis of the capacitor is the magnitude of the induced magnetic field equal to 50% of its maximum value?

27E. The induced magnetic field 6.0 mm from the central axis of a circular parallel-plate capacitor and between the plates is 2.0×10^{-7} T. The plates have radius 3.0 mm. At what rate $d\vec{E}/dt$ is the electric field between the plates changing? ssm

28P. Suppose that a parallel-plate capacitor has circular plates with radius $R = 30$ mm and a plate separation of 5.0 mm. Suppose also that a sinusoidal potential difference with a maximum value of 150 V and a frequency of 60 Hz is applied across the plates; that is,

$$V = (150 \text{ V}) \sin[2\pi(60 \text{ Hz})t].$$

(a) Find $B_{max}(R)$, the maximum value of the induced magnetic field that occurs at $r = R$. (b) Plot $B_{max}(r)$ for $0 < r < 10$ cm.

SEC. 32-10   Displacement Current

29E. Prove that the displacement current in a parallel-plate capacitor of capacitance C can be written as $i_d = C(dV/dt)$, where V is the potential difference between the plates. ssm

30E. At what rate must the potential difference between the plates of a parallel-plate capacitor with a 2.0 μF capacitance be changed to produce a displacement current of 1.5 A?

31E. For the situation of Sample Problem 32-3, show that the current density of the displacement current is $J_d = \varepsilon_0(dE/dt)$ for $r \le R$. ssm

32E. A parallel-plate capacitor with circular plates of radius 0.10 m is being discharged. A circular loop of radius 0.20 m is concentric with the capacitor and halfway between the plates. The displacement current through the loop is 2.0 A. At what rate is the electric field between the plates changing?

33P. As a parallel-plate capacitor with circular plates 20 cm in diameter is being charged, the current density of the displacement current in the region between the plates is uniform and has a mag-

nitude of 20 A/m^2. (a) Calculate the magnitude B of the magnetic field at a distance $r = 50$ mm from the axis of symmetry of this region. (b) Calculate dE/dt in this region. ssm ilw

34P. The magnitude of the electric field between the two circular parallel plates in Fig. 32-26 is $E = (4.0 \times 10^5) - (6.0 \times 10^4 t)$, with E in volts per meter and t in seconds. At $t = 0$, the field is upward as shown. The plate area is 4.0×10^{-2} m^2. For $t \ge 0$,

Fig. 32-26 Problem 34.

(a) what are the magnitude and direction of the displacement current between the plates and (b) is the direction of the induced magnetic field clockwise or counterclockwise around the plates?

35P. A uniform electric field collapses to zero from an initial strength of 6.0×10^5 N/C in a time of 15 μs in the manner shown in Fig. 32-27. Calculate the magnitude of the displacement current, through a 1.6 m^2 area perpendicular to the field, during each of the time intervals a, b, and c shown on the graph. (Ignore the behavior at the ends of the intervals.) ssm ilw

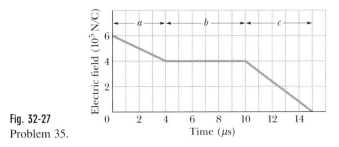

Fig. 32-27 Problem 35.

36P. A parallel-plate capacitor with circular plates is being charged. Consider a circular loop centered on the central axis between the plates. The loop radius is 0.20 m; the plate radius is 0.10 m; and the displacement current through the loop is 2.0 A. What is the rate at which the electric field between the plates is changing?

37P. A parallel-plate capacitor has square plates 1.0 m on a side as shown in Fig. 32-28. A current of 2.0 A charges the capacitor, producing a uniform electric field $\vec{E}$ between the plates, with $\vec{E}$ perpendicular to the plates. (a) What is the displacement current i_d through the region between the plates? (b) What is dE/dt in this region? (c) What is the displacement current through the square dashed path between the plates? (d) What is $\oint \vec{B} \cdot d\vec{s}$ around this square dashed path? ssm ilw www

Fig. 32-28 Problem 37.

38P. A capacitor with parallel circular plates of radius R is discharging via a current of 12.0 A. Consider a loop of radius $R/3$ that is centered on the central axis between the plates. (a) How much displacement current is encircled by the loop? The maximum induced magnetic field has a magnitude of 12.0 mT. (b) At what radial distance from the central axis of the plate is the magnitude of the induced magnetic field 3.00 mT?

33 Electromagnetic Oscillations and Alternating Current

When a high-voltage power transmission line requires repair, a utility company cannot just shut it down, perhaps blacking out an entire city. Repairs must be made while the lines are electrically "hot." The man outside the helicopter in this photograph has just replaced a spacer between 500 kV lines *by hand*, a procedure that requires considerable expertise.

How does he manage this repair without being electrocuted?

The answer is in this chapter.

33-1 New Physics—Old Mathematics

In this chapter you will see how the electric charge q varies with time in a circuit made up of an inductor L, a capacitor C, and a resistor R. From another point of view, we shall discuss how energy shuttles back and forth between the magnetic field of the inductor and the electric field of the capacitor, while it is being gradually dissipated as thermal energy in the resistor.

We have discussed oscillations before, in another context. In Chapter 16 we saw how displacement x varies with time in a mechanical oscillating system made up of a block of mass m, a spring of spring constant k, and a viscous or frictional element such as oil; Fig. 16-15 shows such a system. We also saw how energy shuttles back and forth between the kinetic energy of the oscillating mass and the potential energy of the spring, being gradually dissipated as thermal energy.

The parallel between these two idealized systems is exact, and the controlling differential equations are identical. Thus, there is no new mathematics to be learned; we can simply change the symbols and give our full attention to the physics of the situation.

33-2 *LC* Oscillations, Qualitatively

Of the three circuit elements, resistance R, capacitance C, and inductance L, we have so far discussed the series combinations RC (in Section 28-8) and RL (in Section 31-9). In these two kinds of circuit we found that the charge, current, and potential difference grow and decay exponentially. The time scale of the growth or decay is given by a *time constant* τ, which is either capacitive or inductive.

We now examine the remaining two-element circuit combination LC. You will see that in this case the charge, current, and potential difference do not decay exponentially with time but vary sinusoidally (with period T and angular frequency ω). The resulting oscillations of the capacitor's electric field and the inductor's magnetic field are said to be **electromagnetic oscillations.** Such a circuit is said to oscillate.

Parts *a* through *h* of Fig. 33-1 show succeeding stages of the oscillations in a simple *LC* circuit. From Eq. 26-21, the energy stored in the electric field of the capacitor at any time is

$$U_E = \frac{q^2}{2C},\tag{33-1}$$

where q is the charge on the capacitor at that time. From Eq. 31-51, the energy stored in the magnetic field of the inductor at any time is

$$U_B = \frac{Li^2}{2},\tag{33-2}$$

where i is the current through the inductor at that time.

The method of repairing high-voltage lines shown in the opening photograph is patented by Scott H. Yenzer and is licensed exclusively to Haverfield Corporation of Gettysburg, PA. As the lineman approaches a hot line, the electric field surrounding the line brings his body to nearly the potential of the line. To match the two potentials, he then extends a conducting "wand" to the line. To avoid being electrocuted, he must be isolated from anything electrically connected to the ground. To ensure that his body is always at a single potential—that of the line he is working on—he wears a conducting suit, hood, and gloves, all of which are electrically connected to the line via the wand.

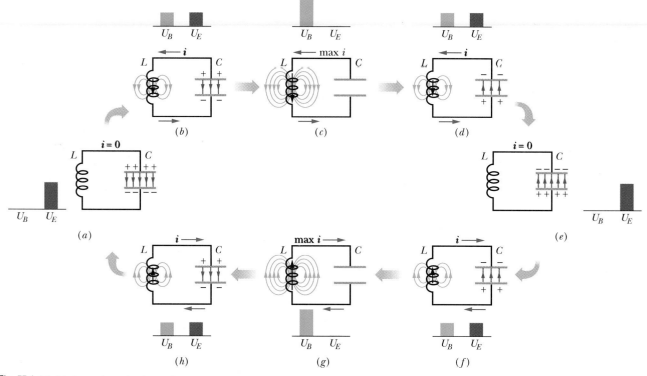

Fig. 33-1 Eight stages in a single cycle of oscillation of a resistanceless *LC* circuit. The bar graphs by each figure show the stored magnetic and electric energies. The magnetic field lines of the inductor and the electric field lines of the capacitor are shown. (*a*) Capacitor with maximum charge, no current. (*b*) Capacitor discharging, current increasing. (*c*) Capacitor fully discharged, current maximum. (*d*) Capacitor charging but with polarity opposite that in (*a*), current decreasing. (*e*) Capacitor with maximum charge having polarity opposite that in (*a*), no current. (*f*) Capacitor discharging, current increasing with direction opposite that in (*b*). (*g*) Capacitor fully discharged, current maximum. (*h*) Capacitor charging, current decreasing.

We now adopt the convention of representing *instantaneous values* of the electrical quantities of a sinusoidally oscillating circuit with small letters, such as *q*, and the *amplitudes* of those quantities with capital letters, such as *Q*. With this convention in mind, let us assume that initially the charge *q* on the capacitor in Fig. 33-1 is at its maximum value *Q* and that the current *i* through the inductor is zero. This initial state of the circuit is shown in Fig. 33-1*a*. The bar graphs for energy included there indicate that at this instant, with zero current through the inductor and maximum charge on the capacitor, the energy U_B of the magnetic field is zero and the energy U_E of the electric field is a maximum.

The capacitor now starts to discharge through the inductor, positive charge carriers moving counterclockwise, as shown in Fig. 33-1*b*. This means that a current *i*, given by *dq/dt* and pointing down in the inductor, is established. As the capacitor's charge decreases, the energy stored in the electric field within the capacitor also decreases. This energy is transferred to the magnetic field that appears around the inductor because of the current *i* that is building up there. Thus, the electric field decreases and the magnetic field builds up as energy is transferred from the electric field to the magnetic field.

The capacitor eventually loses all its charge (Fig. 33-1*c*) and thus also loses its electric field and the energy stored in that field. The energy has then been fully transferred to the magnetic field of the inductor. The magnetic field is then at its maximum magnitude, and the current through the inductor is then at its maximum value *I*.

Although the charge on the capacitor is now zero, the counterclockwise current must continue because the inductor does not allow it to change suddenly to zero. The current continues to transfer positive charge from the top plate to the bottom plate through the circuit (Fig. 33-1*d*). Energy now flows from the inductor back to the capacitor as the electric field within the capacitor builds up again. The current

Fig. 33-2 (*a*) The potential difference across the capacitor of the circuit of Fig. 33-1 as a function of time. This quantity is proportional to the charge on the capacitor. (*b*) A potential proportional to the current in the circuit of Fig. 33-1. The letters refer to the correspondingly labeled oscillation stages in Fig. 33-1.

gradually decreases during this energy transfer. When, eventually, the energy has been transferred completely back to the capacitor (Fig. 33-1*e*), the current has decreased to zero (momentarily). The situation of Fig. 33-1*e* is like the initial situation, except that the capacitor is now charged oppositely.

The capacitor then starts to discharge again but now with a clockwise current (Fig. 33-1*f*). Reasoning as before, we see that the clockwise current builds to a maximum (Fig. 33-1*g*) and then decreases (Fig. 33-1*h*), until the circuit eventually returns to its initial situation (Fig. 33-1*a*). The process then repeats at some frequency *f* and thus at an angular frequency $\omega = 2\pi f$. In the ideal *LC* circuit with no resistance, there are no energy transfers other than that between the electric field of the capacitor and the magnetic field of the inductor. Owing to the conservation of energy, the oscillations continue indefinitely. The oscillations need not begin with the energy all in the electric field; the initial situation could be any other stage of the oscillation.

To determine the charge *q* on the capacitor as a function of time, we can put in a voltmeter to measure the time-varying potential difference (or *voltage*) v_C that exists across the capacitor *C*. From Eq. 26-1 we can write

$$v_C = \left(\frac{1}{C}\right)q,$$

which allows us to find *q*. To measure the current, we can connect a small resistance *R* in series with the capacitor and inductor and measure the time-varying potential difference v_R across it; v_R is proportional to *i* through the relation

$$v_R = iR.$$

We assume here that *R* is so small that its effect on the behavior of the circuit is negligible. The variations in time of v_C and v_R, and thus of *q* and *i*, are shown in Fig. 33-2. All four quantities vary sinusoidally.

In an actual *LC* circuit, the oscillations will not continue indefinitely because there is always some resistance present that will drain energy from the electric and magnetic fields and dissipate it as thermal energy (the circuit may become warmer). The oscillations, once started, will die away as Fig. 33-3 suggests. Compare this figure with Fig. 16-16, which shows the decay of mechanical oscillations caused by frictional damping in a block–spring system.

✔**CHECKPOINT 1:**   A charged capacitor and an inductor are connected in series at time *t* = 0. In terms of the period *T* of the resulting oscillations, determine how much later the following reach their maximums: (a) the charge on the capacitor; (b) the voltage across the capacitor, with its original polarity; (c) the energy stored in the electric field; and (d) the current.

Fig. 33-3 An oscilloscope trace showing how the oscillations in an *RLC* circuit actually die away because energy is dissipated in the resistor as thermal energy.

Sample Problem 33-1

A 1.5 μF capacitor is charged to 57 V. The charging battery is then disconnected, and a 12 mH coil is connected in series with the capacitor so that *LC* oscillations occur. What is the maximum current in the coil? Assume that the circuit contains no resistance.

SOLUTION: The Key Ideas here are these:

1. Because the circuit contains no resistance, the electromagnetic energy of the circuit is conserved as the energy is transferred back and forth between the electric field of the capacitor and the magnetic field of the coil (inductor).

2. At any time t, the energy $U_B(t)$ of the magnetic field is related to the current $i(t)$ through the coil by Eq. 33-2 ($U_B = Li^2/2$). When all the energy is stored as magnetic energy, the current is at its maximum value I and that energy is $U_{B,\text{max}} = LI^2/2$.

3. At any time t, the energy $U_E(t)$ of the electric field is related to the charge $q(t)$ on the capacitor by Eq. 33-1 ($U_E = q^2/2C$). When all the energy is stored as electric energy, the charge is at its maximum value Q and that energy is $U_{E,\text{max}} = Q^2/2C$.

With these ideas, we can now write the conservation of energy as

$$U_{B,\text{max}} = U_{E,\text{max}}$$

or

$$\frac{LI^2}{2} = \frac{Q^2}{2C}.$$

Solving for I gives us

$$I = \sqrt{\frac{Q^2}{LC}}.$$

We know L and C, but not Q. However, with Eq. 26-1 ($q = CV$) we can relate Q to the maximum potential difference V across the capacitor, which is the initial potential difference of 57 V. Thus, substituting $Q = CV$ leads to

$$I = V\sqrt{\frac{C}{L}} = (57 \text{ V})\sqrt{\frac{1.5 \times 10^{-6} \text{ F}}{12 \times 10^{-3} \text{ H}}}$$

$$= 0.637 \text{ A} \approx 640 \text{ mA}. \qquad \text{(Answer)}$$

33-3 The Electrical–Mechanical Analogy

Let us look a little closer at the analogy between the oscillating *LC* system of Fig. 33-1 and an oscillating block–spring system. Two kinds of energy are involved in the block–spring system. One is potential energy of the compressed or extended spring; the other is kinetic energy of the moving block. These two energies are given by the familiar formulas in the left energy column in Table 33-1.

The table also shows, in the right energy column, the two kinds of energy involved in *LC* oscillations. By looking across the table, we can see an analogy between the forms of the two pairs of energies—the mechanical energies of the block–spring system and the electromagnetic energies of the *LC* oscillator. The equations for v and i at the bottom of the table help us see the details of the analogy. They tell us that q corresponds to x, and i corresponds to v (in both equations, the former is differentiated to obtain the latter). These correspondences then suggest that, in the energy expressions, $1/C$ corresponds to k and L corresponds to m. Thus,

$$q \text{ corresponds to } x, \qquad 1/C \text{ corresponds to } k,$$
$$i \text{ corresponds to } v, \quad \text{and} \quad L \text{ corresponds to } m.$$

These correspondences suggest that in an *LC* oscillator, the capacitor is mathematically like the spring in a block–spring system, and the inductor is like the block.

In Section 16-3 we saw that the angular frequency of oscillation of a (friction-

TABLE 33-1   The Energy in Two Oscillating Systems Compared

Block–Spring System		LC Oscillator	
Element	Energy	Element	Energy
Spring	Potential, $\frac{1}{2}kx^2$	Capacitor	Electric, $\frac{1}{2}(1/C)q^2$
Block	Kinetic, $\frac{1}{2}mv^2$	Inductor	Magnetic, $\frac{1}{2}Li^2$
	$v = dx/dt$		$i = dq/dt$

less) block–spring system is

$$\omega = \sqrt{\frac{k}{m}} \qquad \text{(block–spring system)}. \qquad (33\text{-}3)$$

The correspondences listed above suggest that to find the angular frequency of oscillation for a (resistanceless) *LC* circuit, *k* should be replaced by $1/C$ and *m* by *L*, yielding

$$\omega = \frac{1}{\sqrt{LC}} \qquad \text{(LC circuit)}. \qquad (33\text{-}4)$$

We derive this result in the next section.

33-4 *LC* Oscillations, Quantitatively

Here we want to show explicitly that Eq. 33-4 for the angular frequency of *LC* oscillations is correct. At the same time, we want to examine even more closely the analogy between *LC* oscillations and block–spring oscillations. We start by extending somewhat our earlier treatment of the mechanical block–spring oscillator.

The Block–Spring Oscillator

We analyzed block–spring oscillations in Chapter 16 in terms of energy transfers and did not—at that early stage—derive the fundamental differential equation that governs those oscillations. We do so now.

We can write, for the total energy *U* of a block–spring oscillator at any instant,

$$U = U_b + U_s = \tfrac{1}{2}mv^2 + \tfrac{1}{2}kx^2, \qquad (33\text{-}5)$$

where U_b and U_s are, respectively, the kinetic energy of the moving block and the potential energy of the stretched or compressed spring. If there is no friction—which we assume—the total energy *U* remains constant with time, even though *v* and *x* vary. In more formal language, $dU/dt = 0$. This leads to

$$\frac{dU}{dt} = \frac{d}{dt}\left(\tfrac{1}{2}mv^2 + \tfrac{1}{2}kx^2\right) = mv\frac{dv}{dt} + kx\frac{dx}{dt} = 0. \qquad (33\text{-}6)$$

However, $v = dx/dt$ and $dv/dt = d^2x/dt^2$. With these substitutions, Eq. 33-6 becomes

$$m\frac{d^2x}{dt^2} + kx = 0 \qquad \text{(block–spring oscillations)}. \qquad (33\text{-}7)$$

Equation 33-7 is the fundamental *differential equation* that governs the frictionless block–spring oscillations.

The general solution to Eq. 33-7—that is, the function $x(t)$ that describes the block–spring oscillations—is (as we saw in Eq. 16-3)

$$x = X\cos(\omega t + \phi) \qquad \text{(displacement)}, \qquad (33\text{-}8)$$

in which *X* is the amplitude of the mechanical oscillations (represented by x_m in Chapter 16), ω is the angular frequency of the oscillations, and ϕ is a phase constant.

The *LC* Oscillator

Now let us analyze the oscillations of a resistanceless *LC* circuit, proceeding exactly as we just did for the block–spring oscillator. The total energy *U* present at any instant in an oscillating *LC* circuit is given by

$$U = U_B + U_E = \frac{Li^2}{2} + \frac{q^2}{2C},$$ (33-9)

in which U_B is the energy stored in the magnetic field of the inductor and U_E is the energy stored in the electric field of the capacitor. Since we have assumed the circuit resistance to be zero, no energy is transferred to thermal energy and U remains constant with time. In more formal language, dU/dt must be zero. This leads to

$$\frac{dU}{dt} = \frac{d}{dt}\left(\frac{Li^2}{2} + \frac{q^2}{2C}\right) = Li\frac{di}{dt} + \frac{q}{C}\frac{dq}{dt} = 0.$$ (33-10)

However, $i = dq/dt$ and $di/dt = d^2q/dt^2$. With these substitutions, Eq. 33-10 becomes

$$L\frac{d^2q}{dt^2} + \frac{1}{C}q = 0 \qquad \text{(\textit{LC} oscillations).}$$ (33-11)

This is the *differential equation* that describes the oscillations of a resistanceless *LC* circuit. Equations 33-11 and 33-7 are exactly of the same mathematical form.

Charge and Current Oscillations

Since the differential equations are mathematically identical, their solutions must also be mathematically identical. Because q corresponds to x, we can write the general solution of Eq. 33-11, by analogy to Eq. 33-8, as

$$q = Q\cos(\omega t + \phi) \qquad \text{(charge),}$$ (33-12)

where Q is the amplitude of the charge variations, ω is the angular frequency of the electromagnetic oscillations, and ϕ is the phase constant.

Taking the first derivative of Eq. 33-12 with respect to time gives us the current of the *LC* oscillator:

$$i = \frac{dq}{dt} = -\omega Q\sin(\omega t + \phi) \qquad \text{(current).}$$ (33-13)

The amplitude I of this sinusoidally varying current is

$$I = \omega Q,$$ (33-14)

so we can rewrite Eq. 33-13 as

$$i = -I\sin(\omega t + \phi).$$ (33-15)

Angular Frequencies

We can test whether Eq. 33-12 is a solution of Eq. 33-11 by substituting it and its second derivative with respect to time into Eq. 33-11. The first derivative of Eq. 33-12 is Eq. 33-13. The second derivative is then

$$\frac{d^2q}{dt^2} = -\omega^2 Q\cos(\omega t + \phi).$$

Substituting for q and d^2q/dt^2 into Eq. 33-11, we obtain

$$-L\omega^2 Q\cos(\omega t + \phi) + \frac{1}{C}Q\cos(\omega t + \phi) = 0.$$

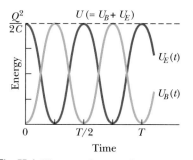

Fig. 33-4 The stored magnetic energy and electric energy in the circuit of Fig. 33-1 as a function of time. Note that their sum remains constant. T is the period of oscillation.

Canceling $Q \cos(\omega t + \phi)$ and rearranging lead to

$$\omega = \frac{1}{\sqrt{LC}}.$$

Thus, Eq. 33-12 is indeed a solution of Eq. 33-11 if ω has the constant value $1/\sqrt{LC}$. Note that this expression for ω is exactly that given by Eq. 33-4, which we arrived at by examining correspondences.

The phase constant ϕ in Eq. 33-12 is determined by the conditions that exist at any certain time, say, $t = 0$. If the conditions yield $\phi = 0$ at $t = 0$, Eq. 33-12 requires that $q = Q$ and Eq. 33-13 requires that $i = 0$; these are the initial conditions represented by Fig. 33-1a.

Electric and Magnetic Energy Oscillations

The electric energy stored in the *LC* circuit at any time t is, from Eqs. 33-1 and 33-12,

$$U_E = \frac{q^2}{2C} = \frac{Q^2}{2C} \cos^2(\omega t + \phi). \tag{33-16}$$

The magnetic energy is, from Eqs. 33-2 and 33-13,

$$U_B = \tfrac{1}{2}Li^2 = \tfrac{1}{2}L\omega^2 Q^2 \sin^2(\omega t + \phi).$$

Substituting for ω from Eq. 33-4 then gives us

$$U_B = \frac{Q^2}{2C} \sin^2(\omega t + \phi). \tag{33-17}$$

Figure 33-4 shows plots of $U_E(t)$ and $U_B(t)$ for the case of $\phi = 0$. Note that

1. The maximum values of U_E and U_B are both $Q^2/2C$.
2. At any instant the sum of U_E and U_B is equal to $Q^2/2C$, a constant.
3. When U_E is maximum, U_B is zero, and conversely.

✔**CHECKPOINT 2:**   A capacitor in an *LC* oscillator has a maximum potential difference of 17 V and a maximum energy of 160 μJ. When the capacitor has a potential difference of 5 V and an energy of 10 μJ, what are (a) the emf across the inductor and (b) the energy stored in the magnetic field?

Sample Problem 33-2

For the situation described in Sample Problem 33-1, let the coil (inductor) be connected to the charged capacitor at time $t = 0$. The result is an *LC* circuit like that in Fig. 33-1.

(a) What is the potential difference $v_L(t)$ across the inductor as a function of time?

SOLUTION: One **Key Idea** here is that the current and potential differences of the circuit undergo sinusoidal oscillations. Another **Key Idea** is that we can still apply the loop rule to this oscillating circuit—just as we did for the nonoscillating circuits of Chapter 28. At any time t during the oscillations, the loop rule and Fig. 33-1

give us

$$v_L(t) = v_C(t); \tag{33-18}$$

that is, the potential difference v_L across the inductor must always be equal to the potential difference v_C across the capacitor, so that the net potential difference around the circuit is zero. Thus, we will find $v_L(t)$ if we can find $v_C(t)$, and we can find $v_C(t)$ from $q(t)$ with Eq. 26-1 ($q = CV$).

Because the potential difference $v_C(t)$ is maximum when the oscillations begin at time $t = 0$, the charge q on the capacitor must also be maximum then. Thus, phase constant ϕ must be zero, so

that Eq. 33-12 gives us

$$q = Q \cos \omega t. \qquad (33\text{-}19)$$

(Note that this cosine function does indeed yield maximum $q \ (= Q)$ when $t = 0$.) To get the potential difference $v_C(t)$, we divide both sides of Eq. 33-19 by C to write

$$\frac{q}{C} = \frac{Q}{C} \cos \omega t,$$

and then use Eq. 26-1 to write

$$v_C = V_C \cos \omega t. \qquad (33\text{-}20)$$

Here, V_C is the amplitude of the oscillations in the potential difference v_C across the capacitor.

Next, substituting $v_C = v_L$ from Eq. 33-18, we find

$$v_L = V_C \cos \omega t. \qquad (33\text{-}21)$$

We can evaluate the right side of this equation by first noting that the amplitude V_C is equal to the initial (maximum) potential difference of 57 V across the capacitor. Then, using the values of L and C from Sample Problem 33-1, we find ω with Eq. 33-4:

$$\omega = \frac{1}{\sqrt{LC}} = \frac{1}{[(0.012 \text{ H})(1.5 \times 10^{-6} \text{ F})]^{0.5}}$$

$$= 7454 \text{ rad/s} \approx 7500 \text{ rad/s}.$$

Thus, Eq. 33-21 becomes

$$v_L = (57 \text{ V}) \cos(7500 \text{ rad/s})t. \qquad \text{(Answer)}$$

(b) What is the maximum rate $(di/dt)_{max}$ at which the current i changes in the circuit?

SOLUTION: The Key Idea here is that, with the charge on the capacitor oscillating as in Eq. 33-12, the current is in the form of Eq. 33-13. Because $\phi = 0$, that equation gives us

$$i = -\omega Q \sin \omega t.$$

Then

$$\frac{di}{dt} = \frac{d}{dt}(-\omega Q \sin \omega t) = -\omega^2 Q \cos \omega t.$$

We can simplify this equation by substituting CV_C for Q (because we know C and V_C but not Q) and $1/\sqrt{LC}$ for ω according to Eq. 33-4. We get

$$\frac{di}{dt} = -\frac{1}{LC} CV_C \cos \omega t = -\frac{V_C}{L} \cos \omega t.$$

This tells us that the current changes at a varying (sinusoidal) rate, with its maximum rate of change being

$$\frac{V_C}{L} = \frac{57 \text{ V}}{0.012 \text{ H}} = 4750 \text{ A/s} \approx 4800 \text{ A/s}. \qquad \text{(Answer)}$$

33-5   Damped Oscillations in an *RLC* Circuit

A circuit containing resistance, inductance, and capacitance is called an *RLC circuit.* We shall here discuss only *series RLC circuits* like that shown in Fig. 33-5. With a resistance R present, the total *electromagnetic energy* U of the circuit (the sum of the electric energy and magnetic energy) is no longer constant; instead, it decreases with time as energy is transferred to thermal energy in the resistance. Because of this loss of energy, the oscillations of charge, current, and potential difference continuously decrease in amplitude, and the oscillations are said to be *damped.* As you will see, they are damped in exactly the same way as those of the damped block–spring oscillator of Section 16-8.

To analyze the oscillations of this circuit, we write an equation for the total electromagnetic energy U in the circuit at any instant. Because the resistance does not store electromagnetic energy, we can use Eq. 33-9:

$$U = U_B + U_E = \frac{Li^2}{2} + \frac{q^2}{2C}. \qquad (33\text{-}22)$$

Now, however, this total energy decreases as energy is transferred to thermal energy. The rate of that transfer is, from Eq. 27-22,

$$\frac{dU}{dt} = -i^2R, \qquad (33\text{-}23)$$

where the minus sign indicates that U decreases. By differentiating Eq. 33-22 with respect to time and then substituting the result in Eq. 33-23, we obtain

$$\frac{dU}{dt} = Li\frac{di}{dt} + \frac{q}{C}\frac{dq}{dt} = -i^2R.$$

Fig. 33-5 A series *RLC* circuit. As the charge contained in the circuit oscillates back and forth through the resistance, electromagnetic energy is dissipated as thermal energy, damping (decreasing the amplitude of) the oscillations.

Substituting dq/dt for i and d^2q/dt^2 for di/dt, we obtain

$$L\frac{d^2q}{dt^2} + R\frac{dq}{dt} + \frac{1}{C}q = 0 \qquad (RLC \text{ circuit}), \qquad (33\text{-}24)$$

which is the differential equation for damped oscillations in an *RLC* circuit.

The solution to Eq. 33-24 is

$$q = Qe^{-Rt/2L}\cos(\omega't + \phi), \qquad (33\text{-}25)$$

in which

$$\omega' = \sqrt{\omega^2 - (R/2L)^2}, \qquad (33\text{-}26)$$

where $\omega = 1/\sqrt{LC}$, as with an undamped oscillator. Equation 33-25 tells us how the charge on the capacitor oscillates in a damped *RLC* circuit; that equation is the electromagnetic counterpart of Eq. 16-40, which gives the displacement of a damped block–spring oscillator.

Equation 33-25 describes a sinusoidal oscillation (the cosine function) with an *exponentially decaying amplitude* $Qe^{-Rt/2L}$ (the factor that multiplies the cosine). The angular frequency ω' of the damped oscillations is always less than the angular frequency ω of the undamped oscillations; however, we shall here consider only situations in which R is small enough for us to replace ω' with ω.

Let us next find an expression for the total electromagnetic energy U of the circuit as a function of time. One way to do so is to monitor the energy of the electric field in the capacitor, which is given by Eq. 33-1 ($U_E = q^2/2C$). By substituting Eq. 33-25 into Eq. 33-1, we obtain

$$U_E = \frac{q^2}{2C} = \frac{[Qe^{-Rt/2L}\cos(\omega't + \phi)]^2}{2C} = \frac{Q^2}{2C}e^{-Rt/L}\cos^2(\omega't + \phi). \qquad (33\text{-}27)$$

Thus, the energy of the electric field oscillates according to a cosine-squared term and the amplitude of that oscillation decreases exponentially with time.

Sample Problem 33-3

A series *RLC* circuit has inductance $L = 12$ mH, capacitance $C = 1.6\ \mu\text{F}$, and resistance $R = 1.5\ \Omega$.

(a) At what time t will the amplitude of the charge oscillations in the circuit be 50% of its initial value?

SOLUTION: The Key Idea here is that the amplitude of the charge oscillations decreases exponentially with time t: According to Eq. 33-25, the charge amplitude at any time t is $Qe^{-Rt/2L}$, in which Q is the amplitude at time $t = 0$. We want the time when the charge amplitude has decreased to $0.500Q$, that is, when

$$Qe^{-Rt/2L} = 0.50Q.$$

Canceling Q and taking the natural logarithms of both sides, we have

$$-\frac{Rt}{2L} = \ln 0.50.$$

Solving for t and then substituting given data yield

$$t = -\frac{2L}{R}\ln 0.50 = -\frac{(2)(12 \times 10^{-3}\ \text{H})(\ln 0.50)}{1.5\ \Omega}$$

$$= 0.0111\ \text{s} \approx 11\ \text{ms}. \qquad \text{(Answer)}$$

(b) How many oscillations are completed within this time?

SOLUTION: The Key Idea here is that the time for one complete oscillation is the period $T = 2\pi/\omega$, where the angular frequency for *LC* oscillations is given by Eq. 33-4 ($\omega = 1/\sqrt{LC}$). Thus, in the time interval $\Delta t = 0.0111$ s, the number of complete oscillations is

$$\frac{\Delta t}{T} = \frac{\Delta t}{2\pi\sqrt{LC}}$$

$$= \frac{0.0111\ \text{s}}{2\pi[(12 \times 10^{-3}\ \text{H})(1.6 \times 10^{-6}\ \text{F})]^{1/2}} \approx 13. \qquad \text{(Answer)}$$

Thus, the amplitude decays by 50% in about 13 complete oscillations. This damping is less severe than that shown in Fig. 33-3, where the amplitude decays by a little more than 50% in one oscillation.

33-6 Alternating Current

The oscillations in an *RLC* circuit will not damp out if an external emf device supplies enough energy to make up for the energy dissipated as thermal energy in the resistance *R*. Circuits in homes, offices, and factories, including countless *RLC* circuits, receive such energy from local power companies. In most countries the energy is supplied via oscillating emfs and currents—the current is said to be an **alternating current,** or **ac** for short. (The nonoscillating current from a battery is said to be a **direct current,** or **dc.**) These oscillating emfs and currents vary sinusoidally with time, reversing direction (in North America) 120 times per second and thus having frequency $f = 60$ Hz.

At first sight this may seem to be a strange arrangement. We have seen that the drift speed of the conduction electrons in household wiring may typically be 4×10^{-5} m/s. If we now reverse their direction every $\frac{1}{120}$ s, such electrons can move only about 3×10^{-7} m in a half-cycle. At this rate, a typical electron can drift past no more than about 10 atoms in the wiring before it is required to reverse its direction. How, you may wonder, can the electron ever get anywhere?

Although this question may be worrisome, it is a needless concern. The conduction electrons do not have to "get anywhere." When we say that the current in a wire is one ampere, we mean that charge passes through any plane cutting across that wire at the rate of one coulomb per second. The speed at which the charge carriers cross that plane does not matter directly; one ampere may correspond to many charge carriers moving very slowly or to a few moving very rapidly. Furthermore, the signal to the electrons to reverse directions—which originates in the alternating emf provided by the power company's generator—is propagated along the conductor at a speed close to that of light. All electrons, no matter where they are located, get their reversal instructions at about the same instant. Finally, we note that for many devices, such as lightbulbs and toasters, the direction of motion is unimportant as long as the electrons do move so as to transfer energy to the device via collisions with atoms in the device.

The basic advantage of alternating current is this: *As the current alternates, so does the magnetic field that surrounds the conductor.* This makes possible the use of Faraday's law of induction, which, among other things, means that we can step up (increase) or step down (decrease) the magnitude of an alternating potential difference at will, using a device called a transformer, as you will see later in this chapter. Moreover, alternating current is more readily adaptable to rotating machinery such as generators and motors than is (nonalternating) direct current.

Figure 33-6 shows a simple model of an ac generator. As the conducting loop is forced to rotate through the external magnetic field $\vec{B}$, a sinusoidally oscillating emf $\mathscr{E}$ is induced in the loop:

$$\mathscr{E} = \mathscr{E}_m \sin \omega_d t. \tag{33-28}$$

Fig. 33-6 The basic mechanism of an alternating-current generator is a conducting loop rotated in an external magnetic field. In practice, the alternating emf induced in a coil of many turns of wire is made accessible by means of slip rings attached to the rotating loop. Each ring is connected to one end of the loop wire and is electrically connected to the rest of the generator circuit by a conducting brush against which it slips as the loop (and it) rotates.

The *angular frequency* ω_d of the emf is equal to the angular speed with which the loop rotates in the magnetic field; the *phase* of the emf is $\omega_d t$; and the *amplitude* of the emf is $\mathscr{E}_m$ (where the subscript stands for maximum). When the rotating loop is part of a closed conducting path, this emf produces (*drives*) a sinusoidal (alternating) current along the path with the same angular frequency ω_d, which then is called the **driving angular frequency.** We can write the current as

$$i = I \sin(\omega_d t - \phi), \tag{33-29}$$

in which *I* is the amplitude of the driven current. (The phase $\omega_d t - \phi$ of the current is traditionally written with a minus sign instead of as $\omega_d t + \phi$.) We include a phase

Fig. 33-7 A single-loop circuit containing a resistor, a capacitor, and an inductor. A generator, represented by a sine wave in a circle, produces an alternating emf that establishes an alternating current; the directions of the emf and current are indicated here at only one instant.

(a)

(b)

(c)

Fig. 33-8 (a) A resistor is connected across an alternating-current generator. (b) The current i_R and the potential difference v_R across the resistor are plotted on the same graph, both versus time t. They are in phase and complete one cycle in one period T. (c) A phasor diagram shows the same thing as (b).

constant ϕ in Eq. 33-29 because the current i may not be in phase with the emf $\mathcal{E}$. (As you will see, the phase constant depends on the circuit to which the generator is connected.) We can also write the current i in terms of the **driving frequency** f_d of the emf, by substituting $2\pi f_d$ for ω_d in Eq. 33-29.

33-7 Forced Oscillations

We have seen that once started, the charge, potential difference, and current in both undamped LC circuits and damped RLC circuits (with small enough R) oscillate at angular frequency $\omega = 1/\sqrt{LC}$. Such oscillations are said to be *free oscillations* (free of any external emf), and the angular frequency ω is said to be the circuit's **natural angular frequency.**

When the external alternating emf of Eq. 33-28 is connected to an RLC circuit, the oscillations of charge, potential difference, and current are said to be *driven oscillations* or *forced oscillations*. These oscillations always occur at the driving angular frequency ω_d:

> Whatever the natural angular frequency ω of a circuit may be, forced oscillations of charge, current, and potential difference in the circuit always occur at the driving angular frequency ω_d.

However, as you will see in Section 33-9, the amplitudes of the oscillations very much depend on how close ω_d is to ω. When the two angular frequencies match—a condition known as **resonance**—the amplitude I of the current in the circuit is maximum.

33-8 Three Simple Circuits

Later in this chapter, we shall connect an external alternating emf device to a series RLC circuit as in Fig. 33-7. We shall then find expressions for the amplitude I and phase constant ϕ of the sinusoidally oscillating current in terms of the amplitude $\mathcal{E}_m$ and angular frequency ω_d of the external emf. First, however, let us consider three simpler circuits, each having an external emf and only one other circuit element: R, C, or L. We start with a resistive element (a purely *resistive load*).

A Resistive Load

Figure 33-8a shows a circuit containing a resistance element of value R and an ac generator with the alternating emf of Eq. 33-28. By the loop rule, we have

$$\mathcal{E} - v_R = 0.$$

With Eq. 33-28, this gives us

$$v_R = \mathcal{E}_m \sin \omega_d t.$$

Because the amplitude V_R of the alternating potential difference (or voltage) across the resistance is equal to the amplitude $\mathcal{E}_m$ of the alternating emf, we can write this as

$$v_R = V_R \sin \omega_d t. \qquad (33\text{-}30)$$

From the definition of resistance ($R = V/i$), we can now write the current i_R in the resistance as

$$i_R = \frac{v_R}{R} = \frac{V_R}{R} \sin \omega_d t. \qquad (33\text{-}31)$$

From Eq. 33-29, we can also write this current as

$$i_R = I_R \sin(\omega_d t - \phi), \qquad (33\text{-}32)$$

where I_R is the amplitude of the current i_R in the resistance. Comparing Eqs. 33-31 and 33-32, we see that for a purely resistive load the phase constant $\phi = 0°$. We also see that the voltage amplitude and current amplitude are related by

$$V_R = I_R R \qquad \text{(resistor)}. \qquad (33\text{-}33)$$

Although we found this relation for the circuit of Fig. 33-8a, it applies to any resistance in any ac circuit.

By comparing Eqs. 33-30 and 33-31, we see that the time-varying quantities v_R and i_R are both functions of $\sin \omega_d t$ with $\phi = 0°$. Thus, these two quantities are *in phase*, which means that their corresponding maxima (and minima) occur at the same times. Figure 33-8b, which is a plot of $v_R(t)$ and $i_R(t)$, illustrates this fact. Note that v_R and i_R do not decay here, because the generator supplies energy to the circuit to make up for the energy dissipated in R.

The time-varying quantities v_R and i_R can also be represented geometrically by *phasors*. Recall from Section 17-10 that phasors are vectors that rotate around an origin. Those that represent the voltage across and current in the resistor of Fig. 33-8a are shown in Fig. 33-8c at an arbitrary time t. Such phasors have the following properties:

Angular speed: Both phasors rotate counterclockwise about the origin with an angular speed equal to the angular frequency ω_d of v_R and i_R.

Length: The length of each phasor represents the amplitude of the alternating quantity: V_R for the voltage and I_R for the current.

Projection: The projection of each phasor on the *vertical* axis represents the value of the alternating quantity at time t: v_R for the voltage and i_R for the current.

Rotation angle: The rotation angle of each phasor is equal to the phase of the alternating quantity at time t. In Fig. 33-8c, the voltage and current are in phase, so their phasors always have the same phase $\omega_d t$ and the same rotation angle, and thus they rotate together.

Mentally follow the rotation. Can you see that when the phasors have rotated so that $\omega_d t = 90°$ (they point vertically upward), they indicate that just then $v_R = V_R$ and $i_R = I_R$? Equations 33-30 and 33-32 give the same results.

Sample Problem 33-4

Purely resistive load. In Fig. 33-8a, resistance R is 200 Ω and the sinusoidal alternating emf device operates at amplitude $\mathcal{E}_m = 36.0$ V and frequency $f_d = 60.0$ Hz.

(a) What is the potential difference $v_R(t)$ across the resistance as a function of time t, and what is the amplitude V_R of $v_R(t)$?

SOLUTION: When we applied the loop rule to the circuit of Fig. 33-8a, we found this Key Idea: In a circuit with a purely resistive load, the potential difference $v_R(t)$ across the resistance is always equal to the potential difference $\mathcal{E}(t)$ across the emf device. Thus, $v_R(t) = \mathcal{E}(t)$ and $V_R = \mathcal{E}_m$. Since $\mathcal{E}_m$ is given, we can write

$$V_R = \mathcal{E}_m = 36.0 \text{ V}. \qquad \text{(Answer)}$$

To find $v_R(t)$, we use Eq. 33-28 to write

$$v_R(t) = \mathcal{E}(t) = \mathcal{E}_m \sin \omega_d t, \qquad (33\text{-}34)$$

and then substitute $\mathcal{E}_m = 36.0$ V and

$$\omega_d = 2\pi f_d = 2\pi(60 \text{ Hz}) = 120\pi$$

to obtain

$$v_R = (36.0 \text{ V}) \sin(120\pi t). \qquad \text{(Answer)}$$

We can leave the argument of the sine in this form for convenience, or we can write it as $(377 \text{ rad/s})t$ or as $(377 \text{ s}^{-1})t$.

(b) What are the current $i_R(t)$ in the resistance and the amplitude I_R of $i_R(t)$?

SOLUTION: The Key Idea here is that in an ac circuit with a purely resistive load, the alternating current $i_R(t)$ in the resistance is *in phase* with the alternating potential difference $v_R(t)$ across the resistance; that is, the phase constant ϕ for the current is zero. Thus, we can write Eq. 33-29 as

$$i_R = I_R \sin(\omega_d t - \phi) = I_R \sin \omega_d t.\qquad(33\text{-}35)$$

From Eq. 33-33, the amplitude I_R is

$$I_R = \frac{V_R}{R} = \frac{36.0 \text{ V}}{200 \text{ }\Omega} = 0.180 \text{ A}.\qquad\text{(Answer)}$$

Substituting this and $\omega_d = 2\pi f_d = 120\pi$ into Eq. 33-35, we have

$$i_R = (0.180 \text{ A}) \sin(120\pi t).\qquad\text{(Answer)}$$

✓CHECKPOINT 3: If we increase the driving frequency in a circuit with a purely resistive load, do (a) amplitude V_R and (b) amplitude I_R increase, decrease, or remain the same?

A Capacitive Load

Figure 33-9*a* shows a circuit containing a capacitance and a generator with the alternating emf of Eq. 33-28. Using the loop rule and proceeding as we did when we obtained Eq. 33-30, we find that the potential difference across the capacitor is

$$v_C = V_C \sin \omega_d t,\qquad(33\text{-}36)$$

where V_C is the amplitude of the alternating voltage across the capacitor. From the definition of capacitance we can also write

$$q_C = C v_C = C V_C \sin \omega_d t.\qquad(33\text{-}37)$$

Our concern, however, is with the current rather than the charge. Thus, we differentiate Eq. 33-37 to find

$$i_C = \frac{dq_C}{dt} = \omega_d C V_C \cos \omega_d t.\qquad(33\text{-}38)$$

We now modify Eq. 33-38 in two ways. First, for reasons of symmetry of notation, we introduce the quantity X_C, called the **capacitive reactance** of a capacitor, defined as

$$X_C = \frac{1}{\omega_d C} \qquad\text{(capacitive reactance).}\qquad(33\text{-}39)$$

Its value depends not only on the capacitance but also on the driving angular frequency ω_d. We know from the definition of the capacitive time constant ($\tau = RC$) that the SI unit for C can be expressed as seconds per ohm. Applying this to Eq. 33-39 shows that the SI unit of X_C is the *ohm*, just as for resistance R.

Second, we replace $\cos \omega_d t$ in Eq. 33-38 with a phase-shifted sine:

$$\cos \omega_d t = \sin(\omega_d t + 90°).$$

You can verify this identity by shifting a sine curve in the negative direction by 90°. With these two modifications, Eq. 33-38 becomes

$$i_C = \left(\frac{V_C}{X_C}\right) \sin(\omega_d t + 90°).\qquad(33\text{-}40)$$

From Eq. 33-29, we can also write the current i_C in C as

$$i_C = I_C \sin(\omega_d t - \phi),\qquad(33\text{-}41)$$

where I_C is the amplitude of i_C. Comparing Eqs. 33-40 and 33-41, we see that for a purely capacitive load the phase constant ϕ for the current is $-90°$. We also see that the voltage amplitude and current amplitude are related by

$$V_C = I_C X_C \qquad\text{(capacitor).}\qquad(33\text{-}42)$$

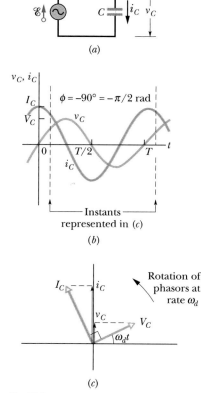

Fig. 33-9 (*a*) A capacitor is connected across an alternating-current generator. (*b*) The current in the capacitor leads the voltage by 90° (= $\pi/2$ rad). (*c*) A phasor diagram shows the same thing.

Although we found this relation for the circuit of Fig. 33-9a, it applies to any capacitance in any ac circuit.

Comparison of Eqs. 33-36 and 33-40, or inspection of Fig. 33-9b, shows that the quantities v_C and i_C are 90°, or one-quarter cycle, out of phase. Furthermore, we see that i_C *leads* v_C, which means that, if you monitored the current i_C and the potential difference v_C in the circuit of Fig. 33-9a, you would find that i_C reaches its maximum *before* v_C does, by one-quarter cycle.

This relation between i_C and v_C is illustrated by the phasor diagram of Fig. 33-9c. As the phasors representing these two quantities rotate counterclockwise together, the phasor labeled I_C does indeed lead that labeled V_C, and by an angle of 90°; that is, the phasor I_C coincides with the vertical axis one-quarter cycle before the phasor V_C does. Be sure to convince yourself that the phasor diagram of Fig. 33-9c is consistent with Eqs. 33-36 and 33-40.

✔**CHECKPOINT 4:**   The figure shows, in (a), a sine curve $S(t) = \sin(\omega_d t)$ and three other sinusoidal curves $A(t)$, $B(t)$, and $C(t)$, each of the form $\sin(\omega_d t - \phi)$. (a) Rank the three other curves according to the value of ϕ, most positive first and most negative last. (b) Which curve corresponds to which phasor in (b) of the figure? (c) Which curve leads the others?

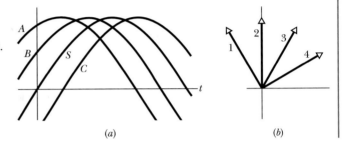

(a)   (b)

Sample Problem 33-5

Purely capacitive load. In Fig. 33-9a, capacitance C is 15.0 μF and the sinusoidal alternating emf device operates at amplitude $\mathcal{E}_m = 36.0$ V and frequency $f_d = 60.0$ Hz.

(a) What are the potential difference $v_C(t)$ across the capacitance and the amplitude V_C of $v_C(t)$?

SOLUTION: If we apply the loop rule to the circuit of Fig. 33-9a, we find this **Key Idea**: In a circuit with a purely capacitive load, the potential difference $v_C(t)$ across the capacitance is always equal to the potential difference $\mathcal{E}(t)$ across the emf device. Thus, $v_C(t) = \mathcal{E}(t)$ and $V_C = \mathcal{E}_m$. Since $\mathcal{E}_m$ is given, we have

$$V_C = \mathcal{E}_m = 36.0 \text{ V.} \qquad \text{(Answer)}$$

To find $v_C(t)$, we use Eq. 33-28 to write

$$v_C(t) = \mathcal{E}(t) = \mathcal{E}_m \sin \omega_d t. \qquad (33\text{-}43)$$

Then, substituting $\mathcal{E}_m = 36.0$ V and $\omega_d = 2\pi f_d = 120\pi$ into Eq. 33-43, we have

$$v_C = (36.0 \text{ V}) \sin(120\pi t). \qquad \text{(Answer)}$$

(b) What are the current $i_C(t)$ in the circuit as a function of time and the amplitude I_C of $i_C(t)$?

SOLUTION: The **Key Idea** here is that in an ac circuit with a purely capacitive load, the alternating current $i_C(t)$ in the capacitance leads the alternating potential difference $v_C(t)$ by 90°; that is, the phase constant ϕ for the current is −90° or −$\pi/2$ rad. Thus, we can write Eq. 33-29 as

$$i_C = I_C \sin(\omega_d t - \phi) = I_C \sin(\omega_d t + \pi/2). \qquad (33\text{-}44)$$

A second **Key Idea** is that we can find the amplitude I_C from Eq. 33-42 ($V_C = I_C X_C$) if we first find the capacitive reactance X_C. From Eq. 33-39 ($X_C = 1/\omega_d C$), with $\omega_d = 2\pi f_d$, we can write

$$X_C = \frac{1}{2\pi f_d C} = \frac{1}{(2\pi)(60.0 \text{ Hz})(15.0 \times 10^{-6} \text{ F})}$$
$$= 177 \; \Omega.$$

Then Eq. 33-42 tells us that the current amplitude is

$$I_C = \frac{V_C}{X_C} = \frac{36.0 \text{ V}}{177 \; \Omega} = 0.203 \text{ A.} \qquad \text{(Answer)}$$

Substituting this and $\omega_d = 2\pi f_d = 120\pi$ into Eq. 33-44, we have

$$i_C = (0.203 \text{ A}) \sin(120\pi t + \pi/2). \qquad \text{(Answer)}$$

✔**CHECKPOINT 5:**   If we increase the driving frequency in a circuit with a purely capacitive load, do (a) amplitude V_C and (b) amplitude I_C increase, decrease, or remain the same?

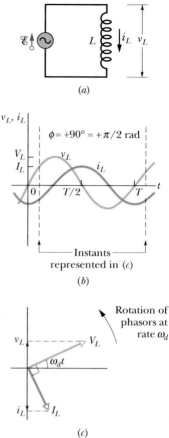

Fig. 33-10 (a) An inductor is connected across an alternating-current generator. (b) The current in the inductor lags the voltage by 90° (= π/2 rad). (c) A phasor diagram shows the same thing.

An Inductive Load

Figure 33-10a shows a circuit containing an inductance and a generator with the alternating emf of Eq. 33-28. Using the loop rule and proceeding as we did to obtain Eq. 33-30, we find that the potential difference across the inductance is

$$v_L = V_L \sin \omega_d t, \tag{33-45}$$

where V_L is the amplitude of v_L. From Eq. 31-37, we can write the potential difference across an inductance L, in which the current is changing at the rate di_L/dt, as

$$v_L = L\frac{di_L}{dt}. \tag{33-46}$$

If we combine Eqs. 33-45 and 33-46, we have

$$\frac{di_L}{dt} = \frac{V_L}{L}\sin \omega_d t. \tag{33-47}$$

Our concern, however, is with the current rather than with its time derivative. We find the former by integrating Eq. 33-47, obtaining

$$i_L = \int di_L = \frac{V_L}{L}\int \sin \omega_d t \, dt = -\left(\frac{V_L}{\omega_d L}\right)\cos \omega_d t. \tag{33-48}$$

We now modify this equation in two ways. First, for reasons of symmetry of notation, we introduce the quantity X_L, called the **inductive reactance** of an inductor, which is defined as

$$X_L = \omega_d L \quad \text{(inductive reactance).} \tag{33-49}$$

The value of X_L depends on the driving angular frequency ω_d. The unit of the inductive time constant τ_L indicates that the SI unit of X_L is the *ohm*, just as it is for X_C and for R.

Second, we replace $-\cos \omega_d t$ in Eq. 33-48 with a phase-shifted sine:

$$-\cos \omega_d t = \sin(\omega_d t - 90°).$$

You can verify this identity by shifting a sine curve in the positive direction by 90°.

With these two changes, Eq. 33-48 becomes

$$i_L = \left(\frac{V_L}{X_L}\right)\sin(\omega_d t - 90°). \tag{33-50}$$

From Eq. 33-29, we can also write this current in the inductance as

$$i_L = I_L \sin(\omega_d t - \phi), \tag{33-51}$$

where I_L is the amplitude of the current i_L. Comparing Eqs. 33-50 and 33-51, we see that for a purely inductive load the phase constant ϕ for the current is +90°. We also see that the voltage amplitude and current amplitude are related by

$$V_L = I_L X_L \quad \text{(inductor).} \tag{33-52}$$

Although we found this relation for the circuit of Fig. 33-10a, it applies to any inductance in any ac circuit.

Comparison of Eqs. 33-45 and 33-50, or inspection of Fig. 33-10b, shows that the quantities i_L and v_L are 90° out of phase. In this case, however, i_L *lags* v_L; that is, if you monitored the current i_L and the potential difference v_L in the circuit of Fig. 33-10a, you would find that i_L reaches its maximum value *after* v_L does, by one-quarter cycle.

The phasor diagram of Fig. 33-10c also contains this information. As the phasors rotate counterclockwise in the figure, the phasor labeled I_L does indeed lag that labeled V_L, and by an angle of 90°. Be sure to convince yourself that Fig. 33-10c represents Eqs. 33-45 and 33-50.

PROBLEM-SOLVING TACTICS

Tactic 1: *Leading and Lagging in AC Circuits*
Table 33-2 summarizes the relations between the current i and the voltage v for each of the three kinds of circuit elements we have considered. When an applied alternating voltage produces an alternating current in them, the current is in phase with the voltage across a resistor, leads the voltage across a capacitor, and lags the voltage across an inductor.

Many students remember these results with the mnemonic "*ELI* the *ICE* man." *ELI* contains the letter *L* (for inductor), and in

it the letter *I* (for current) comes after the letter *E* (for emf or voltage). Thus, for an inductor, the current *lags* the voltage. Similarly *ICE* (which contains a *C* for capacitor) means that the current *leads* the voltage. You might also use the modified mnemonic "*ELI positively* is the *ICE* man" to remember that the phase constant ϕ is positive for an inductor.

If you have difficulty in remembering whether X_C is equal to $\omega_d C$ (wrong) or $1/\omega_d C$ (right), try remembering that C is in the "cellar"—that is, in the denominator.

TABLE 33-2   **Phase and Amplitude Relations for Alternating Currents and Voltages**

Circuit Element	Symbol	Resistance or Reactance	Phase of the Current	Phase Constant (or Angle) ϕ	Amplitude Relation
Resistor	R	R	In phase with v_R	0° (= 0 rad)	$V_R = I_R R$
Capacitor	C	$X_C = 1/\omega_d C$	Leads v_C by 90° (= $\pi/2$ rad)	−90° (= −$\pi/2$ rad)	$V_C = I_C X_C$
Inductor	L	$X_L = \omega_d L$	Lags v_L by 90° (= $\pi/2$ rad)	+90° (= +$\pi/2$ rad)	$V_L = I_L X_L$

Sample Problem 33-6

Purely inductive load. In Fig. 33-10a, inductance L is 230 mH and the sinusoidal alternating emf device operates at amplitude $\mathcal{E}_m$ = 36.0 V and frequency f_d = 60.0 Hz.

(a) What are the potential difference $v_L(t)$ across the inductance and the amplitude V_L of $v_L(t)$?

SOLUTION: If we apply the loop rule to the circuit in Fig. 33-10a, we find this **Key Idea**: In a circuit with a purely inductive load, the potential difference $v_L(t)$ across the inductance is always equal to the potential difference $\mathcal{E}(t)$ across the emf device. Thus, $v_L(t) = \mathcal{E}(t)$ and $V_L = \mathcal{E}_m$. Since $\mathcal{E}_m$ is given, we have

$$V_L = \mathcal{E}_m = 36.0 \text{ V}. \qquad \text{(Answer)}$$

To find $v_L(t)$, we use Eq. 33-28 to write

$$v_L(t) = \mathcal{E}(t) = \mathcal{E}_m \sin \omega_d t. \qquad (33\text{-}53)$$

Then, substituting $\mathcal{E}_m$ = 36.0 V and $\omega_d = 2\pi f_d = 120\pi$ into Eq. 33-53, we have

$$v_L = (36.0 \text{ V}) \sin(120\pi t). \qquad \text{(Answer)}$$

(b) What are the current $i_L(t)$ in the circuit as a function of time and the amplitude I_L of $i_L(t)$?

SOLUTION: The **Key Idea** here is that in an ac circuit with a purely inductive load, the alternating current $i_L(t)$ in the inductance lags

the alternating potential difference $v_L(t)$ by 90°. (In the mnemonic of Problem-Solving Tactic 1, this circuit is "positively an *ELI* circuit," which tells us that the emf E leads the current I and that ϕ is *positive*.) Thus, the phase constant ϕ for the current is +90° or +$\pi/2$ rad, and we can write Eq. 33-29 as

$$i_L = I_L \sin(\omega_d t - \phi) = I_L \sin(\omega_d t - \pi/2). \qquad (33\text{-}54)$$

A second **Key Idea** is that we can find the amplitude I_L from Eq. 33-52 ($V_L = I_L X_L$) if we first find the inductive reactance X_L. From Eq. 33-49 ($X_L = \omega_d L$), with $\omega_d = 2\pi f_d$, we can write

$$X_L = 2\pi f_d L = (2\pi)(60.0 \text{ Hz})(230 \times 10^{-3} \text{ H})$$
$$= 86.7 \ \Omega.$$

Then Eq. 33-52 tells us that the current amplitude is

$$I_L = \frac{V_L}{X_L} = \frac{36.0 \text{ V}}{86.7 \ \Omega} = 0.415 \text{ A}. \qquad \text{(Answer)}$$

Substituting this and $\omega_d = 2\pi f_d = 120\pi$ into Eq. 33-54, we have

$$i_L = (0.415 \text{ A}) \sin(120\pi t - \pi/2). \qquad \text{(Answer)}$$

✔**CHECKPOINT 6:**   If we increase the driving frequency in a circuit with a purely inductive load, do (a) amplitude V_L and (b) amplitude I_L increase, decrease, or remain the same?

33-9   The Series *RLC* Circuit

We are now ready to apply the alternating emf of Eq. 33-28,

$$\mathscr{E} = \mathscr{E}_m \sin \omega_d t \qquad \text{(applied emf)}, \qquad (33\text{-}55)$$

to the full *RLC* circuit of Fig. 33-7. Because *R*, *L*, and *C* are in series, the same current

$$i = I \sin(\omega_d t - \phi) \qquad (33\text{-}56)$$

is driven in all three of them. We wish to find the current amplitude *I* and the phase constant ϕ. The solution is simplified by the use of phasor diagrams.

The Current Amplitude

We start with Fig. 33-11*a*, which shows the phasor representing the current of Eq. 33-56 at an arbitrary time *t*. The length of the phasor is the current amplitude *I*, the projection of the phasor on the vertical axis is the current *i* at time *t*, and the angle of rotation of the phasor is the phase $\omega_d t - \phi$ of the current at time *t*.

Figure 33-11*b* shows the phasors representing the voltages across *R*, *L*, and *C* at the same time *t*. Each phasor is oriented relative to the angle of rotation of current phasor *I* in Fig. 33-11*a*, based on the information in Table 33-2:

Resistor: Here current and voltage are in phase, so the angle of rotation of voltage phasor V_R is the same as that of phasor *I*.

Capacitor: Here current leads voltage by 90°, so the angle of rotation of voltage phasor V_C is 90° less than that of phasor *I*.

Inductor: Here current lags voltage by 90°, so the angle of rotation of voltage phasor v_L is 90° greater than that of phasor *I*.

Figure 33-11*b* also shows the instantaneous voltages v_R, v_C, and v_L across *R*, *C*, and *L* at time *t*; those voltages are the projections of the corresponding phasors on the vertical axis of the figure.

Figure 33-11*c* shows the phasor representing the applied emf of Eq. 33-55. The length of the phasor is the emf amplitude $\mathscr{E}_m$, the projection of the phasor on the vertical axis is the emf $\mathscr{E}$ at time *t*, and the angle of rotation of the phasor is the phase $\omega_d t$ of the emf at time *t*.

From the loop rule we know that at any instant the sum of the voltages v_R, v_C, and v_L is equal to the applied emf $\mathscr{E}$:

$$\mathscr{E} = v_R + v_C + v_L. \qquad (33\text{-}57)$$

Thus, at time *t* the projection $\mathscr{E}$ in Fig. 33-11*c* is equal to the algebraic sum of the projections v_R, v_C, and v_L in Fig. 33-11*b*. In fact, as the phasors rotate together, this equality always holds. This means that phasor $\mathscr{E}_m$ in Fig. 33-11*c* must be equal to the vector sum of the three voltage phasors V_R, V_C, and V_L in Fig. 33-11*b*.

Fig. 33-11 (*a*) A phasor representing the alternating current in the driven *RLC* circuit of Fig. 33-7 at time *t*. The amplitude *I*, the instantaneous value *i*, and the phase ($\omega_d t - \phi$) are shown. (*b*) Phasors representing the voltages across the inductor, resistor, and capacitor, oriented with respect to the current phasor in (*a*). (*c*) A phasor representing the alternating emf that drives the current of (*a*). (*d*) The emf phasor is equal to the vector sum of the three voltage phasors of (*b*). Here, voltage phasors V_L and V_C have been added to yield their net phasor ($V_L - V_C$).

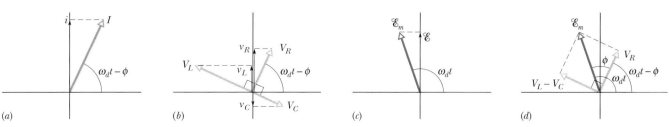

That requirement is indicated in Fig. 33-11d, where phasor $\mathcal{E}_m$ is drawn as the sum of phasors V_R, V_L, and V_C. Because phasors V_L and V_C have opposite directions in the figure, we simplify the vector sum by first combining V_L and V_C to form the single phasor $V_L - V_C$. Then we combine that single phasor with V_R to find the net phasor. Again, the net phasor must coincide with phasor $\mathcal{E}_m$, as shown.

Both triangles in Fig. 33-11d are right triangles. Applying the Pythagorean theorem to either one yields

$$\mathcal{E}_m^2 = V_R^2 + (V_L - V_C)^2. \tag{33-58}$$

From the amplitude information displayed in Table 33-2 we can rewrite this as

$$\mathcal{E}_m^2 = (IR)^2 + (IX_L - IX_C)^2, \tag{33-59}$$

and then rearrange it to the form

$$I = \frac{\mathcal{E}_m}{\sqrt{R^2 + (X_L - X_C)^2}}. \tag{33-60}$$

The denominator in Eq. 33-60 is called the **impedance** Z of the circuit for the driving angular frequency ω_d:

$$Z = \sqrt{R^2 + (X_L - X_C)^2} \qquad \text{(impedance defined)}. \tag{33-61}$$

We can then write Eq. 33-60 as

$$I = \frac{\mathcal{E}_m}{Z}. \tag{33-62}$$

If we substitute for X_C and X_L from Eqs. 33-39 and 33-49, we can write Eq. 33-60 more explicitly as

$$I = \frac{\mathcal{E}_m}{\sqrt{R^2 + (\omega_d L - 1/\omega_d C)^2}} \qquad \text{(current amplitude)}. \tag{33-63}$$

We have now accomplished half our goal: We have obtained an expression for the current amplitude I in terms of the sinusoidal driving emf and the circuit elements in a series RLC circuit.

The value of I depends on the difference between $\omega_d L$ and $1/\omega_d C$ in Eq. 33-63 or, equivalently, the difference between X_L and X_C in Eq. 33-60. In either equation, it does not matter which of the two quantities is greater because the difference is always squared.

The current that we have been describing in this section is the *steady-state current* that occurs after the alternating emf has been applied for some time. When the emf is first applied to a circuit, a brief *transient current* occurs. Its duration (before settling down into the steady-state current) is determined by the time constants $\tau_L = L/R$ and $\tau_C = RC$ as the inductive and capacitive elements "turn on." This transient current can be large and can, for example, destroy a motor on start-up if it is not properly taken into account in the motor's circuit design.

The Phase Constant

From the right-hand phasor triangle in Fig. 33-11d and from Table 33-2 we can write

$$\tan \phi = \frac{V_L - V_C}{V_R} = \frac{IX_L - IX_C}{IR}, \tag{33-64}$$

which gives us

$$\tan \phi = \frac{X_L - X_C}{R} \qquad \text{(phase constant)}. \qquad (33\text{-}65)$$

This is the other half of our goal: an equation for the phase constant ϕ in a sinusoidally driven series *RLC* circuit. In essence, it gives us three different results for the phase constant, depending on the relative values of X_L and X_C:

$X_L > X_C$: The circuit is said to be *more inductive than capacitive.* Equation 33-65 tells us that ϕ is positive for such a circuit, which means that phasor I rotates behind phasor $\mathscr{E}_m$ (Fig. 33-12*a*). A plot of $\mathscr{E}$ and i versus time is like that in Fig. 33-12*b*. (Figures 33-11*c* and *d* were drawn assuming $X_L > X_C$.)

$X_C > X_L$: The circuit is said to be *more capacitive than inductive.* Equation 33-65 tells us that ϕ is negative for such a circuit, which means that phasor I rotates ahead of phasor $\mathscr{E}_m$ (Fig. 33-12*c*). A plot of $\mathscr{E}$ and i versus time is like that in Fig. 33-12*d*.

$X_C = X_L$: The circuit is said to be in *resonance,* a state that is discussed next. Equation 33-65 tells us that $\phi = 0°$ for such a circuit, which means that phasors $\mathscr{E}_m$ and I rotate together (Fig. 33-12*e*). A plot of $\mathscr{E}$ and i versus time is like that in Fig. 33-12*f*.

As illustration, let us reconsider two extreme circuits: In the *purely inductive circuit* of Fig. 33-10*a*, where X_L is nonzero and $X_C = R = 0$, Eq. 33-65 tells us that $\phi = +90°$ (the greatest value of ϕ), consistent with Fig. 33-10*c*. In the *purely capacitive circuit* of Fig. 33-9*a*, where X_C is nonzero and $X_L = R = 0$, Eq. 33-65 tells us that $\phi = -90°$ (the least value of ϕ), consistent with Fig. 33-9*c*.

Resonance

Equation 33-63 gives the current amplitude I in an *RLC* circuit as a function of the driving angular frequency ω_d of the external alternating emf. For a given resistance R, that amplitude is a maximum when the quantity $\omega_d L - 1/\omega_d C$ in the denominator

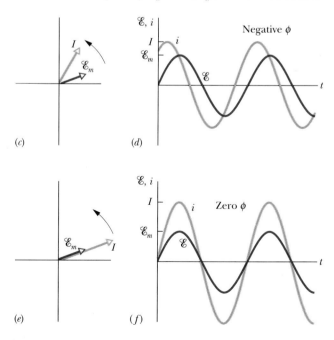

Fig. 33-12 Phasor diagrams and graphs of the alternating emf $\mathscr{E}$ and current i for the driven *RLC* circuit of Fig. 33-7. In the phasor diagram of (*a*) and the graph of (*b*), the current i lags the driving emf $\mathscr{E}$ and the current's phase constant ϕ is positive. In (*c*) and (*d*), the current i leads the driving emf $\mathscr{E}$ and its phase constant ϕ is negative. In (*e*) and (*f*), the current i is in phase with the driving emf $\mathscr{E}$ and its phase constant ϕ is zero.

Fig. 33-13 *Resonance curves* for the driven *RLC* circuit of Fig. 33-7 with $L = 100\ \mu\text{H}$, $C = 100$ pF, and three values of R. The current amplitude I of the alternating current depends on how close the driving angular frequency ω_d is to the natural angular frequency ω. The horizontal arrow on each curve measures the curve's width at the half-maximum level, a measure of the sharpness of the resonance. To the left of $\omega_d/\omega = 1.00$, the circuit is mainly capacitive, with $X_C > X_L$; to the right, it is mainly inductive, with $X_L > X_C$.

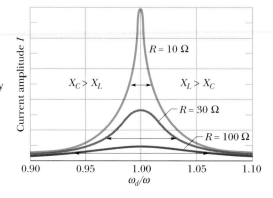

is zero—that is; when

$$\omega_d L = \frac{1}{\omega_d C}$$

or $$\omega_d = \frac{1}{\sqrt{LC}} \qquad \text{(maximum } I\text{)}. \tag{33-66}$$

Because the natural angular frequency ω of the *RLC* circuit is also equal to $1/\sqrt{LC}$, the maximum value of I occurs when the driving angular frequency matches the natural angular frequency—that is, at resonance. Thus, in an *RLC* circuit, resonance and maximum current amplitude I occur when

$$\omega_d = \omega = \frac{1}{\sqrt{LC}} \qquad \text{(resonance)}. \tag{33-67}$$

Figure 33-13 shows three *resonance curves* for sinusoidally driven oscillations in three series *RLC* circuits differing only in R. Each curve peaks at its maximum current amplitude I when the ratio ω_d/ω is 1.00, but the maximum value of I decreases with increasing R. (The maximum I is always $\mathscr{E}_m/R$; to see why, combine Eqs. 33-61 and 33-62.) In addition, the curves increase in width (measured in Fig. 33-13 at half the maximum value of I) with increasing R.

To make physical sense of Fig. 33-13, consider how the reactances X_L and X_C change as we increase the driving angular frequency ω_d, starting with a value much less than the natural frequency ω. For small ω_d, reactance X_L ($= \omega_d L$) is small and reactance X_C ($= 1/\omega_d C$) is large. Thus, the circuit is mainly capacitive and the impedance is dominated by the large X_C, which keeps the current low.

As we increase ω_d, reactance X_C remains dominant but decreases while reactance X_L increases. The decrease in X_C decreases the impedance, allowing the current to increase, as we see on the left side of any resonance curve in Fig. 33-13. When the increasing X_L and the decreasing X_C reach equal values, the current is greatest and the circuit is in resonance, with $\omega_d = \omega$.

As we continue to increase ω_d, the increasing reactance X_L becomes progressively more dominant over the decreasing reactance X_C. The impedance increases because of X_L and the current decreases, as on the right side of any resonance curve in Fig. 33-13. In summary, then: The low-angular-frequency side of a resonance curve is dominated by the capacitor's reactance, the high-angular-frequency side is dominated by the inductor's reactance, and resonance occurs in the middle.

✔**CHECKPOINT 7:**   Here are the capacitive reactance and inductive reactance, respectively, for three sinusoidally driven series *RLC* circuits: (1) 50 Ω, 100 Ω; (2) 100 Ω, 50 Ω; (3) 50 Ω, 50 Ω. (a) For each, does the current lead or lag the applied emf, or are the two in phase? (b) Which circuit is in resonance?

Sample Problem 33-7

In Fig. 33-7 let $R = 200\ \Omega$, $C = 15.0\ \mu F$, $L = 230$ mH, $f_d = 60.0$ Hz, and $\mathscr{E}_m = 36.0$ V. (These parameters are those used in Sample Problems 33-4, 33-5, and 33-6.)

(a) What is the current amplitude I?

SOLUTION: The Key Idea here is that current amplitude I depends on the amplitude $\mathscr{E}_m$ of the driving emf and on the impedance Z of the circuit, according to Eq. 33-62 ($I = \mathscr{E}_m/Z$). Thus, we need to find Z, which depends on the circuit's resistance R, capacitive reactance X_C, and inductive reactance X_L.

The circuit's only resistance is the given resistance R. Its only capacitive reactance is due to the given capacitance and, from Sample Problem 33-5, $X_C = 177\ \Omega$. Its only inductive reactance is due to the given inductance and, from Sample Problem 33-6, $X_L = 86.7\ \Omega$. Thus, the circuit's impedance is

$$Z = \sqrt{R^2 + (X_L - X_C)^2}$$
$$= \sqrt{(200\ \Omega)^2 + (86.7\ \Omega - 177\ \Omega)^2}$$
$$= 219\ \Omega.$$

We then find

$$I = \frac{\mathscr{E}_m}{Z} = \frac{36.0\ \text{V}}{219\ \Omega} = 0.164\ \text{A}. \qquad \text{(Answer)}$$

(b) What is the phase constant ϕ of the current in the circuit relative to the driving emf?

SOLUTION: The Key Idea here is that the phase constant depends on the inductive reactance, the capacitive reactance, and the resistance of the circuit, according to Eq. 33-65. Solving that equation for ϕ leads to

$$\phi = \tan^{-1}\frac{X_L - X_C}{R} = \tan^{-1}\frac{86.7\ \Omega - 177\ \Omega}{200\ \Omega}$$
$$= -24.3° = -0.424\ \text{rad}. \qquad \text{(Answer)}$$

The negative phase constant is consistent with the fact that the load is mainly capacitive; that is, $X_C > X_L$. In the mnemonic of Problem-Solving Tactic 1, this circuit is an *ICE* circuit—the current *leads* the driving emf.

33-10 Power in Alternating-Current Circuits

In the *RLC* circuit of Fig. 33-7, the source of energy is the alternating-current generator. Some of the energy that it provides is stored in the electric field in the capacitor, some is stored in the magnetic field in the inductor, and some is dissipated as thermal energy in the resistor. In steady-state operation—which we assume—the average energy stored in the capacitor and inductor together remains constant. The net transfer of energy is thus from the generator to the resistor, where electromagnetic energy is dissipated as thermal energy.

The instantaneous rate at which energy is dissipated in the resistor can be written, with the help of Eqs. 27-22 and 33-29, as

$$P = i^2 R = [I \sin(\omega_d t - \phi)]^2 R = I^2 R \sin^2(\omega_d t - \phi). \qquad (33\text{-}68)$$

The *average* rate at which energy is dissipated in the resistor, however, is the average of Eq. 33-68 over time. Over one complete cycle, the average value of $\sin \theta$, where θ is any variable, is zero (Fig. 33-14a) but the average value of $\sin^2 \theta$ is $\frac{1}{2}$ (Fig. 33-14b). (Note in Fig. 33-14b how the shaded areas under the curve but above the horizontal line marked $+\frac{1}{2}$ exactly fill in the unshaded spaces below that line.) Thus, we can write, from Eq. 33-68,

$$P_{\text{avg}} = \frac{I^2 R}{2} = \left(\frac{I}{\sqrt{2}}\right)^2 R. \qquad (33\text{-}69)$$

The quantity $I/\sqrt{2}$ is called the **root-mean-square**, or **rms**, value of the current i:

$$I_{\text{rms}} = \frac{I}{\sqrt{2}} \qquad \text{(rms current)}. \qquad (33\text{-}70)$$

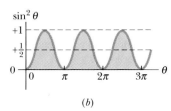

Fig. 33-14 (a) A plot of $\sin \theta$ versus θ. The average value over one cycle is zero. (b) A plot of $\sin^2 \theta$ versus θ. The average value over one cycle is $\frac{1}{2}$.

We can now rewrite Eq. 33-69 as

$$P_{\text{avg}} = I_{\text{rms}}^2 R \qquad \text{(average power)}. \qquad (33\text{-}71)$$

Equation 33-71 looks much like Eq. 27-22 ($P = i^2 R$); the message is that if we switch to the rms current, we can compute the average rate of energy dissipation for alternating-current circuits just as for direct-current circuits.

We can also define rms values of voltages and emfs for alternating-current circuits:

$$V_{rms} = \frac{V}{\sqrt{2}} \quad \text{and} \quad \mathcal{E}_{rms} = \frac{\mathcal{E}_m}{\sqrt{2}} \quad \text{(rms voltage; rms emf).} \quad (33\text{-}72)$$

Alternating-current instruments, such as ammeters and voltmeters, are usually calibrated to read I_{rms}, V_{rms}, and $\mathcal{E}_{rms}$. Thus, if you plug an alternating-current voltmeter into a household electric outlet and it reads 120 V, that is an rms voltage. The *maximum* value of the potential difference at the outlet is $\sqrt{2} \times (120 \text{ V})$, or 170 V.

Because the proportionality factor $1/\sqrt{2}$ in Eqs. 33-70 and 33-72 is the same for all three variables, we can write Eqs. 33-62 and 33-60 as

$$I_{rms} = \frac{\mathcal{E}_{rms}}{Z} = \frac{\mathcal{E}_{rms}}{\sqrt{R^2 + (X_L - X_C)^2}}, \quad (33\text{-}73)$$

and, indeed, this is the form that we almost always use.

We can use the relationship $I_{rms} = \mathcal{E}_{rms}/Z$ to recast Eq. 33-71 in a useful equivalent way. We write

$$P_{avg} = \frac{\mathcal{E}_{rms}}{Z} I_{rms} R = \mathcal{E}_{rms} I_{rms} \frac{R}{Z}. \quad (33\text{-}74)$$

From Fig. 33-11*d*, Table 33-2, and Eq. 33-62, however, we see that R/Z is just the cosine of the phase constant ϕ:

$$\cos \phi = \frac{V_R}{\mathcal{E}_m} = \frac{IR}{IZ} = \frac{R}{Z}. \quad (33\text{-}75)$$

Equation 33-74 then becomes

$$P_{avg} = \mathcal{E}_{rms} I_{rms} \cos \phi \quad \text{(average power),} \quad (33\text{-}76)$$

in which the term $\cos \phi$ is called the **power factor.** Because $\cos \phi = \cos(-\phi)$, Eq. 33-76 is independent of the sign of the phase constant ϕ.

To maximize the rate at which energy is supplied to a resistive load in an *RLC* circuit, we should keep the power factor $\cos \phi$ as close to unity as possible. This is equivalent to keeping the phase constant ϕ in Eq. 33-29 as close to zero as possible. If, for example, the circuit is highly inductive, it can be made less so by putting more capacitance in the circuit, connected in series. (Recall that putting an additional capacitance into a series of capacitances decreases the equivalent capacitance C_{eq} of the series.) Thus, the resulting decrease in C_{eq} in the circuit reduces the phase constant and increases the power factor in Eq. 33-76. Power companies place series-connected capacitors throughout their transmission systems to get these results.

✓CHECKPOINT 8: (a) If the current in a sinusoidally driven series *RLC* circuit leads the emf, would we increase or decrease the capacitance to increase the rate at which energy is supplied to the resistance? (b) Will this change bring the resonant angular frequency of the circuit closer to the angular frequency of the emf or put it farther away?

Sample Problem 33-8

A series *RLC* circuit, driven with $\mathcal{E}_{rms}$ = 120 V at frequency f_d = 60.0 Hz, contains a resistance R = 200 Ω, an inductance with X_L = 80.0 Ω, and a capacitance with X_C = 150 Ω.

(a) What are the power factor $\cos \phi$ and phase constant ϕ of the circuit?

SOLUTION: The Key Idea here is that the power factor $\cos \phi$ can be found from the resistance R and impedance Z via Eq. 33-75

($\cos \phi = R/Z$). To calculate Z, we use Eq. 33-61:

$$Z = \sqrt{R^2 + (X_L - X_C)^2}$$
$$= \sqrt{(200\ \Omega)^2 + (80.0\ \Omega - 150\ \Omega)^2} = 211.90\ \Omega.$$

Equation 33-75 then gives us

$$\cos \phi = \frac{R}{Z} = \frac{200\ \Omega}{211.90\ \Omega} = 0.9438 \approx 0.944. \quad \text{(Answer)}$$

Taking the inverse cosine then yields

$$\phi = \cos^{-1} 0.944 = \pm 19.3°.$$

Both $+19.3°$ and $-19.3°$ have a cosine of 0.944. To determine which sign is correct, we must consider whether the current leads or lags the driving emf. Because $X_C > X_L$, this circuit is mainly capacitive, with the current leading the emf. Thus, ϕ must be negative:

$$\phi = -19.3°. \quad \text{(Answer)}$$

We could, instead, have found ϕ with Eq. 33-65. A calculator would then have given us the complete answer, with the minus sign.

(b) What is the average rate P_{avg} at which energy is dissipated in the resistance?

SOLUTION: One way to answer this question is to use this Key Idea: Because the circuit is assumed to be in steady-state operation, the rate at which energy is dissipated in the resistance is equal to the rate at which energy is supplied to the circuit, as given by Eq. 33-76 ($P_{avg} = \mathscr{E}_{rms} I_{rms} \cos \phi$).

We are given the rms driving emf $\mathscr{E}_{rms}$ and we know $\cos \phi$ from part (a). To find I_{rms} we use the Key Idea that the rms current is determined by the rms value of the driving emf and the circuit's impedance Z (which we know), according to Eq. 33-73:

$$I_{rms} = \frac{\mathscr{E}_{rms}}{Z}.$$

Substituting this into Eq. 33-76 then leads to

$$P_{avg} = \mathscr{E}_{rms} I_{rms} \cos \phi = \frac{\mathscr{E}_{rms}^2}{Z} \cos \phi$$
$$= \frac{(120\ \text{V})^2}{211.90\ \Omega} (0.9438) = 64.1\ \text{W}. \quad \text{(Answer)}$$

A second way to answer the question is to use the Key Idea that the rate at which energy is dissipated in a resistance R depends on the square of the rms current I_{rms} through it, according to Eq. 33-71. We then find

$$P_{avg} = I_{rms}^2 R = \frac{\mathscr{E}_{rms}^2}{Z^2} R$$
$$= \frac{(120\ \text{V})^2}{(211.90\ \Omega)^2} (200\ \Omega) = 64.1\ \text{W}. \quad \text{(Answer)}$$

(c) What new capacitance C_{new} is needed to maximize P_{avg} if the other parameters of the circuit are not changed?

SOLUTION: One Key Idea here is that the average rate P_{avg} at which energy is supplied and dissipated is maximized if the circuit is brought into resonance with the driving emf. A second Key Idea is that resonance occurs when $X_C = X_L$. From the given data, we have $X_C > X_L$. Thus, we must decrease X_C to reach resonance. From Eq. 33-39 ($X_C = 1/\omega_d C$), we see that this means we must increase C to the new value C_{new}.

Using Eq. 33-39, we can write the condition $X_C = X_L$ as

$$\frac{1}{\omega_d C_{new}} = X_L.$$

Substituting $2\pi f_d$ for ω_d (because we are given f_d and not ω_d) and then solving for C_{new}, we find

$$C_{new} = \frac{1}{2\pi f_d X_L} = \frac{1}{(2\pi)(60\ \text{Hz})(80.0\ \Omega)}$$
$$= 3.32 \times 10^{-5}\ \text{F} = 33.2\ \mu\text{F}. \quad \text{(Answer)}$$

Following the procedure of part (b), you can show that with C_{new}, P_{avg} would then be at its maximum value of 72.0 W.

33-11 Transformers

Energy Transmission Requirements

When an ac circuit has only a resistive load, the power factor in Eq. 33-76 is $\cos 0° = 1$ and the applied rms emf $\mathscr{E}_{rms}$ is equal to the rms voltage V_{rms} across the load. Thus, with an rms current I_{rms} in the load, energy is supplied and dissipated at the average rate of

$$P_{avg} = \mathscr{E}I = IV. \tag{33-77}$$

(In Eq. 33-77 and the rest of this section, we follow conventional practice and drop the subscripts identifying rms quantities. Engineers and scientists assume that all time-varying currents and voltages are reported as rms values; that is what the meters read.) Equation 33-77 tells us that, to satisfy a given power requirement, we have a range of choices, from a relatively large current I and a relatively small voltage V to just the reverse, provided only that the product IV is as required.

At 5:17 p.m. on November 9, 1965, a faulty relay in the power system near Niagara Falls opened a circuit breaker on a transmission line, automatically causing the current to switch to other lines, which overloaded those lines and made other circuit breakers open. Within minutes a runaway shutdown had blacked out much of New York, New England, and Ontario.

In electric power distribution systems it is desirable for reasons of safety and for efficient equipment design to deal with relatively low voltages at both the generating end (the electric power plant) and the receiving end (the home or factory). Nobody wants an electric toaster or a child's electric train to operate at, say, 10 kV. On the other hand, in the transmission of electric energy from the generating plant to the consumer, we want the lowest practical current (hence the largest practical voltage) to minimize I^2R losses (often called *ohmic losses*) in the transmission line.

As an example, consider the 735 kV line used to transmit electric energy from the La Grande 2 hydroelectric plant in Quebec to Montreal, 1000 km away. Suppose that the current is 500 A and the power factor is close to unity. Then from Eq. 33-77, energy is supplied at the average rate

$$P_{avg} = \mathscr{E}I = (7.35 \times 10^5 \text{ V})(500 \text{ A}) = 368 \text{ MW}.$$

The resistance of the transmission line is about 0.220 Ω/km; thus, there is a total resistance of about 220 Ω for the 1000 km stretch. Energy is dissipated owing to that resistance at a rate of about

$$P_{avg} = I^2R = (500 \text{ A})^2(220 \text{ }\Omega) = 55.0 \text{ MW},$$

which is nearly 15% of the supply rate.

Imagine what would happen if we doubled the current and halved the voltage. Energy would be supplied by the plant at the same average rate of 368 MW as previously, but now energy would be dissipated at the rate of about

$$P_{avg} = I^2R = (1000 \text{ A})^2(220 \text{ }\Omega) = 220 \text{ MW},$$

which is *almost 60% of the supply rate.* Hence the general energy transmission rule: Transmit at the highest possible voltage and the lowest possible current.

The Ideal Transformer

The transmission rule leads to a fundamental mismatch between the requirement for efficient high-voltage transmission and the need for safe low-voltage generation and consumption. We need a device with which we can raise (for transmission) and lower (for use) the ac voltage in a circuit, keeping the product current × voltage essentially constant. The **transformer** is such a device. It has no moving parts, operates by Faraday's law of induction, and has no simple direct-current counterpart.

The *ideal transformer* in Fig. 33-15 consists of two coils, with different numbers of turns, wound around an iron core. (The coils are insulated from the core.) In use, the primary winding, of N_p turns, is connected to an alternating-current generator whose emf $\mathscr{E}$ at any time t is given by

$$\mathscr{E} = \mathscr{E}_m \sin \omega t. \qquad (33\text{-}78)$$

The secondary winding, of N_s turns, is connected to load resistance R, but its circuit is an open circuit as long as switch S is open (which we assume for the present). Thus, there can be no current through the secondary coil. We assume further for this ideal transformer that the resistances of the primary and secondary windings are negligible, as are energy losses due to magnetic hysteresis in the iron core. Well-designed, high-capacity transformers can have energy losses as low as 1%, so our assumptions are reasonable.

For the assumed conditions, the primary winding (or *primary*) is a pure inductance, and the primary circuit is like that in Fig. 33-10a. Thus, the (very small) primary current, also called the *magnetizing current* I_{mag}, lags the primary voltage V_p by 90°; the primary's power factor (= cos ϕ in Eq. 33-76) is zero, so no power is delivered from the generator to the transformer.

Fig. 33-15 An ideal transformer (two coils wound on an iron core) in a basic transformer circuit. An ac generator produces current in the coil at the left (the *primary*). The coil at the right (the *secondary*) is connected to the resistive load R when switch S is closed.

However, the small alternating primary current I_{mag} induces an alternating magnetic flux Φ_B in the iron core. Because the core extends through the secondary winding (or *secondary*), this induced flux also extends through the turns of the secondary. From Faraday's law of induction (Eq. 31-6), the induced emf per turn $\mathscr{E}_{turn}$ is the same for both the primary and the secondary. Also, the voltage V_p across the primary is equal to the emf induced in the primary, and the voltage V_s across the secondary is equal to the emf induced in the secondary. Thus, we can write

and thus,

$$\mathscr{E}_{turn} = \frac{d\Phi_B}{dt} = \frac{V_p}{N_p} = \frac{V_s}{N_s}$$

$$V_s = V_p \frac{N_s}{N_p} \qquad \text{(transformation of voltage).} \qquad (33\text{-}79)$$

If $N_s > N_p$, the transformer is called a *step-up transformer* because it steps the primary's voltage V_p *up* to a higher voltage V_s. Similarly, if $N_s < N_p$, the device is a *step-down transformer*.

So far, with switch S open, no energy is transferred from the generator to the rest of the circuit. Now let us close S to connect the secondary to the resistive load R. (In general, the load would also contain inductive and capacitive elements, but here we consider just resistance R.) We find that now energy *is* transferred from the generator. Let us see why.

Several things happen when we close switch S.

1. An alternating current I_s appears in the secondary circuit, with corresponding energy dissipation rate $I_s^2 R$ ($= V_s^2/R$) in the resistive load.

2. This current produces its own alternating magnetic flux in the iron core, and this flux induces (from Faraday's law and Lenz's law) an opposing emf in the primary windings.

3. The voltage V_p of the primary, however, cannot change in response to this opposing emf because it must always be equal to the emf $\mathscr{E}$ that is provided by the generator; closing switch S cannot change this fact.

4. To maintain V_p, the generator now produces (in addition to I_{mag}) an alternating current I_p in the primary circuit; the magnitude and phase constant of I_p are just those required for the emf induced by I_p in the primary to exactly cancel the emf induced there by I_s. Because the phase constant of I_p is not 90° like that of I_{mag}, this current I_p can transfer energy to the primary.

We want to relate I_s to I_p. However, rather than analyze the foregoing complex process in detail, let us just apply the principle of conservation of energy. The rate at which the generator transfers energy to the primary is equal to I_pV_p. The rate at which the primary then transfers energy to the secondary (via the alternating magnetic field linking the two coils) is I_sV_s. Because we assume that no energy is lost along the way, conservation of energy requires that

$$I_pV_p = I_sV_s.$$

Substituting for V_s from Eq. 33-79, we find that

$$I_s = I_p \frac{N_p}{N_s} \qquad \text{(transformation of currents).} \qquad (33\text{-}80)$$

This equation tells us that the current I_s in the secondary can differ from the current I_p in the primary, depending on the *turns ratio N_p/N_s*.

Current I_p appears in the primary circuit because of the resistive load R in the secondary circuit. To find I_p, we substitute $I_s = V_s/R$ into Eq. 33-80 and then we substitute for V_s from Eq. 33-79. We find

$$I_p = \frac{1}{R}\left(\frac{N_s}{N_p}\right)^2 V_p. \tag{33-81}$$

This equation has the form $I_p = V_p/R_{eq}$, where equivalent resistance R_{eq} is

$$R_{eq} = \left(\frac{N_p}{N_s}\right)^2 R. \tag{33-82}$$

This R_{eq} is the value of the load resistance as "seen" by the generator; the generator produces the current I_p and voltage V_p as if it were connected to a resistance R_{eq}.

Impedance Matching

Equation 33-82 suggests still another function for the transformer. For maximum transfer of energy from an emf device to a resistive load, the resistance of the emf device and the resistance of the load must be equal. The same relation holds for ac circuits except that the *impedance* (rather than just the resistance) of the generator must be matched to that of the load. Often this condition is not met. For example, in a music-playing system, the amplifier has high impedance and the speaker set has low impedance. We can match the impedances of the two devices by coupling them through a transformer with a suitable turns ratio N_p/N_s.

✓**CHECKPOINT 9:**   An alternating-current emf device has a smaller resistance than that of the resistive load; to increase the transfer of energy from the device to the load, a transformer will be connected between the two. (a) Should N_s be greater than or less than N_p? (b) Will that make it a step-up or step-down transformer?

Sample Problem 33-9

A transformer on a utility pole operates at $V_p = 8.5$ kV on the primary side and supplies electric energy to a number of nearby houses at $V_s = 120$ V, both quantities being rms values. Assume an ideal step-down transformer, a purely resistive load, and a power factor of unity.

(a) What is the turns ratio N_p/N_s of the transformer?

SOLUTION: The Key Idea here is that the turns ratio N_p/N_s is related to the (given) rms primary and secondary voltages via Eq. 33-79, which we can write as

$$\frac{V_s}{V_p} = \frac{N_s}{N_p}.$$

(Note that the right side of this equation is the *inverse* of the turns ratio.) Inverting both sides then gives us

$$\frac{N_p}{N_s} = \frac{V_p}{V_s} = \frac{8.5 \times 10^3 \text{ V}}{120 \text{ V}} = 70.83 \approx 71. \quad \text{(Answer)}$$

(b) The average rate of energy consumption (or dissipation) in the houses served by the transformer is 78 kW. What are the rms currents in the primary and secondary of the transformer?

SOLUTION: The Key Idea here is that for a purely resistive load, the power factor $\cos\phi$ is unity; thus, the average rate at which energy is supplied and dissipated is given by Eq. 33-77. In the primary circuit, with $V_p = 8.5$ kV, Eq. 33-77 yields

$$I_p = \frac{P_{avg}}{V_p} = \frac{78 \times 10^3 \text{ W}}{8.5 \times 10^3 \text{ V}} = 9.176 \text{ A} \approx 9.2 \text{ A}. \quad \text{(Answer)}$$

Similarly, in the secondary circuit,

$$I_s = \frac{P_{avg}}{V_s} = \frac{78 \times 10^3 \text{ W}}{120 \text{ V}} = 650 \text{ A}. \quad \text{(Answer)}$$

You can check that $I_s = I_p(N_p/N_s)$ as required by Eq. 33-80.

(c) What is the resistive load R_s in the secondary circuit? What is the corresponding resistive load R_p in the primary circuit?

SOLUTION: For both circuits, the Key Idea here is that we can relate the resistive load to the rms voltage and current with $V = IR$. For the secondary circuit, we find

$$R_s = \frac{V_s}{I_s} = \frac{120 \text{ V}}{650 \text{ A}} = 0.1846 \ \Omega \approx 0.18 \ \Omega. \quad \text{(Answer)}$$

Similarly, for the primary circuit we find

$$R_p = \frac{V_p}{I_p} = \frac{8.5 \times 10^3 \text{ V}}{9.176 \text{ A}} = 926 \ \Omega \approx 930 \ \Omega. \quad \text{(Answer)}$$

Another Key Idea that we can use to find R_p is that R_p is the equivalent resistive load "seen" from the primary side of the trans-

former, as given by Eq. 33-82. If we substitute R_p for R_{eq} and R_s for R, that equation yields

$$R_p = \left(\frac{N_p}{N_s}\right)^2 R_s = (70.83)^2 (0.1846 \ \Omega)$$

$$= 926 \ \Omega \approx 930 \ \Omega. \quad \text{(Answer)}$$

REVIEW & SUMMARY

LC Energy Transfers   In an oscillating LC circuit, energy is shuttled periodically between the electric field of the capacitor and the magnetic field of the inductor; instantaneous values of the two forms of energy are

$$U_E = \frac{q^2}{2C} \quad \text{and} \quad U_B = \frac{Li^2}{2}, \quad (33\text{-}1, 33\text{-}2)$$

where q is the instantaneous charge on the capacitor and i is the instantaneous current through the inductor. The total energy $U \ (= U_E + U_B)$ remains constant.

LC Charge and Current Oscillations   The principle of conservation of energy leads to

$$L \frac{d^2q}{dt^2} + \frac{1}{C} q = 0 \quad (LC \text{ oscillations}) \quad (33\text{-}11)$$

as the differential equation of LC oscillations (with no resistance). The solution of Eq. 33-11 is

$$q = Q \cos(\omega t + \phi) \quad \text{(charge),} \quad (33\text{-}12)$$

in which Q is the *charge amplitude* (maximum charge on the capacitor) and the angular frequency ω of the oscillations is

$$\omega = \frac{1}{\sqrt{LC}}. \quad (33\text{-}4)$$

The phase constant ϕ in Eq. 33-12 is determined by the initial conditions (at $t = 0$) of the system.

The current i in the system at any time t is

$$i = -\omega Q \sin(\omega t + \phi) \quad \text{(current),} \quad (33\text{-}13)$$

in which ωQ is the *current amplitude I*.

Damped Oscillations   Oscillations in an LC circuit are damped when a dissipative element R is also present in the circuit. Then

$$L \frac{d^2q}{dt^2} + R \frac{dq}{dt} + \frac{1}{C} q = 0 \quad (RLC \text{ circuit}). \quad (33\text{-}24)$$

The solution of this differential equation is

$$q = Q e^{-Rt/2L} \cos(\omega' t + \phi), \quad (33\text{-}25)$$

where

$$\omega' = \sqrt{\omega^2 - (R/2L)^2}. \quad (33\text{-}26)$$

We consider only situations with small R and thus small damping; then $\omega' \approx \omega$.

Alternating Currents; Forced Oscillations   A series RLC circuit may be set into *forced oscillation* at a *driving angular fre-*

quency ω_d by an external alternating emf

$$\mathscr{E} = \mathscr{E}_m \sin \omega_d t. \quad (33\text{-}28)$$

The current driven in the circuit by the emf is

$$i = I \sin(\omega_d t - \phi), \quad (33\text{-}29)$$

where ϕ is the phase constant of the current.

Resonance   The current amplitude I in a series RLC circuit driven by a sinusoidal external emf is a maximum ($I = \mathscr{E}_m/R$) when the driving angular frequency ω_d equals the natural angular frequency ω of the circuit (that is, at *resonance*). Then $X_C = X_L$, $\phi = 0$, and the current is in phase with the emf.

Single Circuit Elements   The alternating potential difference across a resistor has amplitude $V_R = IR$; the current is in phase with the potential difference.

For a *capacitor*, $V_C = IX_C$, in which $X_C = 1/\omega_d C$ is the **capacitive reactance;** the current here leads the potential difference by 90° ($\phi = -90° = -\pi/2$ rad).

For an *inductor*, $V_L = IX_L$, in which $X_L = \omega_d L$ is the **inductive reactance;** the current here lags the potential difference by 90° ($\phi = +90° = +\pi/2$ rad).

Series RLC Circuits   For a series RLC circuit with external emf given by Eq. 33-28 and current given by Eq. 33-29,

$$I = \frac{\mathscr{E}_m}{\sqrt{R^2 + (X_L - X_C)^2}}$$

$$= \frac{\mathscr{E}_m}{\sqrt{R^2 + (\omega_d L - 1/\omega_d C)^2}} \quad \text{(current amplitude)} \quad (33\text{-}60, 33\text{-}63)$$

and

$$\tan \phi = \frac{X_L - X_C}{R} \quad \text{(phase constant).} \quad (33\text{-}65)$$

Defining the impedance Z of the circuit as

$$Z = \sqrt{R^2 + (X_L - X_C)^2} \quad \text{(impedance)} \quad (33\text{-}61)$$

allows us to write Eq. 33-60 as $I = \mathscr{E}_m/Z$.

Power   In a series RLC circuit, the **average power** P_{avg} of the generator is equal to the production rate of thermal energy in the resistor:

$$P_{avg} = I_{rms}^2 R = \mathscr{E}_{rms} I_{rms} \cos \phi. \quad (33\text{-}71, 33\text{-}76)$$

Here rms stands for **root-mean-square;** the rms quantities are related to the maximum quantities by $I_{rms} = I/\sqrt{2}$, $V_{rms} = V_m/\sqrt{2}$, and $\mathscr{E}_{rms} = \mathscr{E}_m/\sqrt{2}$. The term $\cos\phi$ is called the **power factor** of the circuit.

Transformers   A *transformer* (assumed to be ideal) is an iron core on which are wound a primary coil of N_p turns and a secondary coil of N_s turns. If the primary coil is connected across an alternating-current generator, the primary and secondary voltages are related by

$$V_s = V_p \frac{N_s}{N_p} \qquad \text{(transformation of voltage).} \qquad (33\text{-}79)$$

The currents through the coils are related by

$$I_s = I_p \frac{N_p}{N_s} \qquad \text{(transformation of currents),} \qquad (33\text{-}80)$$

and the equivalent resistance of the secondary circuit, as seen by the generator, is

$$R_{eq} = \left(\frac{N_p}{N_s}\right)^2 R, \qquad (33\text{-}82)$$

where R is the resistive load in the secondary circuit. The ratio N_p/N_s is called the transformer's *turns ratio*.

QUESTIONS

1. A charged capacitor and an inductor are connected at time $t = 0$. In terms of the period T of the resulting oscillations, what is the first later time at which the following reach a maximum: (a) U_B, (b) the magnetic flux through the inductor, (c) di/dt, and (d) the emf of the inductor?

2. What values of phase constant ϕ in Eq. 33-12 allow situations (a), (c), (e), and (g) of Fig. 33-1 to occur at $t = 0$?

3. Figure 33-16 shows three oscillating LC circuits with identical inductors and capacitors. Rank the circuits according to the time taken to fully discharge the capacitors during the oscillations, greatest first.

Fig. 33-16 Question 3.

(a)   (b)   (c)

4. Figure 33-17 shows graphs of capacitor voltage v_C for LC circuits 1 and 2, which contain identical capacitances and have the same maximum charge Q. Are (a) the inductance L and (b) the maximum current I in circuit 1 greater than, less than, or the same as those in circuit 2?

Fig. 33-17 Question 4.

5. Charges on the capacitors in three oscillating LC circuits vary as follows: (1) $q = 2\cos 4t$, (2) $q = 4\cos t$, (3) $q = 3\cos 4t$ (with q in coulombs and t in seconds). Rank the circuits according to (a) the current amplitude and (b) the period, greatest first.

6. If you increase the inductance L in an oscillating LC circuit having a given maximum charge Q, do (a) the current magnitude I and (b) the maximum magnetic energy U_B increase, decrease, or stay the same?

7. An alternating emf source with a certain emf amplitude is connected, in turn, to a resistor, a capacitor, and then an inductor. Once connected to one of the devices, the driving frequency f_d is varied and the amplitude I of the resulting current through the device is measured and plotted. Which of the three plots in Fig. 33-18 corresponds to which of the three devices?

8. The values of the phase constant ϕ for four sinusoidally driven series RLC circuits are (1) $-15°$, (2) $+35°$, (3) $\pi/3$ rad, and (4) $-\pi/6$ rad. (a) In which is the load primarily capacitive? (b) In which does the current lag the alternating emf?

9. Figure 33-19 shows the current i and driving emf $\mathscr{E}$ for a series RLC circuit. (a) Does the current lead or lag the emf? (b) Is the circuit's load mainly capacitive or mainly inductive? (c) Is the angular frequency ω_d of the emf greater than or less than the natural angular frequency ω?

Fig. 33-18 Question 7.

Fig. 33-19 Questions 9 and 11.

10. Figure 33-20 shows three situations like those of Fig. 33-12. For each situation, is the driving angular frequency greater than, less than, or equal to the resonant angular frequency of the circuit?

(a)   (b)   (c)

Fig. 33-20 Question 10.

11. Figure 33-19 shows the current i and driving emf $\mathscr{E}$ for a series RLC circuit. Relative to the emf curve, does the current curve shift leftward or rightward and does the amplitude of that curve increase or decrease if we slightly increase (a) L, (b) C, and (c) ω_d?

12. Figure 33-21 shows the current i and driving emf $\mathscr{E}$ for a series RLC circuit. (a) Is the phase constant positive or negative? (b) To increase the rate at which energy is transferred to the resistive load, should L be increased or decreased? (c) Should, instead, C be increased or decreased?

Fig. 33-21 Question 12.

EXERCISES & PROBLEMS

SEC. 33-2   *LC* Oscillations, Qualitatively

1E. What is the capacitance of an oscillating *LC* circuit if the maximum charge on the capacitor is 1.60 μC and the total energy is 140 μJ?

2E. In an oscillating *LC* circuit, $L = 1.10$ mH and $C = 4.00$ μF. The maximum charge on the capacitor is 3.00 μC. Find the maximum current.

3E. An oscillating *LC* circuit consists of a 75.0 mH inductor and a 3.60 μF capacitor. If the maximum charge on the capacitor is 2.90 μC, (a) what is the total energy in the circuit and (b) what is the maximum current?   ssm

4E. In a certain oscillating *LC* circuit the total energy is converted from electric energy in the capacitor to magnetic energy in the inductor in 1.50 μs. (a) What is the period of oscillation? (b) What is the frequency of oscillation? (c) How long after the magnetic energy is a maximum will it be a maximum again?

5P. The frequency of oscillation of a certain *LC* circuit is 200 kHz. At time $t = 0$, plate *A* of the capacitor has maximum positive charge. At what times $t > 0$ will (a) plate *A* again have maximum positive charge, (b) the other plate of the capacitor have maximum positive charge, and (c) the inductor have maximum magnetic field?

SEC. 33-3   The Electrical–Mechanical Analogy

6E. A 0.50 kg body oscillates in SHM on a spring that, when extended 2.0 mm from its equilibrium, has an 8.0 N restoring force. (a) What is the angular frequency of oscillation? (b) What is the period of oscillation? (c) What is the capacitance of an *LC* circuit with the same period if *L* is chosen to be 5.0 H?

7P. The energy in an oscillating *LC* circuit containing a 1.25 H inductor is 5.70 μJ. The maximum charge on the capacitor is 175 μC. Find (a) the mass, (b) the spring constant, (c) the maximum displacement, and (d) the maximum speed for a mechanical system with the same period.   ssm

SEC. 33-4   *LC* Oscillations, Quantitatively

8E. *LC* oscillators have been used in circuits connected to loudspeakers to create some of the sounds of electronic music. What inductance must be used with a 6.7 μF capacitor to produce a frequency of 10 kHz, which is near the middle of the audible range of frequencies?

9E. In an oscillating *LC* circuit with $L = 50$ mH and $C = 4.0$ μF, the current is initially a maximum. How long will it take before the capacitor is fully charged for the first time?   ssm   ilw

10E. A single loop consists of inductors (L_1, L_2, . . .), capacitors (C_1, C_2, . . .), and resistors (R_1, R_2, . . .) connected in series as shown, for example, in Fig. 33-22*a*. Show that regardless of the sequence of these circuit elements in the loop, the behavior of this circuit is identical to that of the simple *LC* circuit shown in Fig. 33-22*b*. (*Hint:* Consider the loop rule and see Problem 43 in Chapter 31.)

Fig. 33-22 Exercise 10.

11P. An oscillating *LC* circuit consisting of a 1.0 nF capacitor and a 3.0 mH coil has a maximum voltage of 3.0 V. (a) What is the maximum charge on the capacitor? (b) What is the maximum current through the circuit? (c) What is the maximum energy stored in the magnetic field of the coil?   ilw

12P. In an oscillating *LC* circuit in which $C = 4.00$ μF, the maximum potential difference across the capacitor during the oscillations is 1.50 V and the maximum current through the inductor is 50.0 mA. (a) What is the inductance *L*? (b) What is the frequency of the oscillations? (c) How much time is required for the charge on the capacitor to rise from zero to its maximum value?

13P. In the circuit shown in Fig. 33-23 the switch is kept in position *a* for a long time. It is then thrown to position *b*. (a) Calculate the frequency of the resulting oscillating current. (b) What is the amplitude of the current oscillations?   ssm   ilw

Fig. 33-23 Problem 13.

14P. You are given a 10 mH inductor and two capacitors, of 5.0 μF and 2.0 μF capacitance. List the oscillation frequencies that can be generated by connecting these elements in various combinations.

15P. A variable capacitor with a range from 10 to 365 pF is used with a coil to form a variable-frequency *LC* circuit to tune the input to a radio. (a) What ratio of maximum to minimum frequencies may be obtained with such a capacitor? (b) If this circuit is to obtain frequencies from 0.54 MHz to 1.60 MHz, the ratio computed in (a) is too large. By adding a capacitor in parallel to the variable capacitor, this range may be adjusted. What should be the capacitance of this added capacitor, and what inductance should be used to obtain the desired range of frequencies?   ssm   www

16P. In an oscillating *LC* circuit, 75.0% of the total energy is stored in the magnetic field of the inductor at a certain instant. (a) In terms of the maximum charge on the capacitor, what is the charge there at that instant? (b) In terms of the maximum current in the inductor, what is the current there at that instant?

17P. In an oscillating *LC* circuit, $L = 25.0$ mH and $C = 7.80$ μF. At time $t = 0$ the current is 9.20 mA, the charge on the capacitor is 3.80 μC, and the capacitor is charging. (a) What is the total energy in the circuit? (b) What is the maximum charge on the capacitor? (c) What is the maximum current? (d) If the charge on the capacitor is given by $q = Q \cos(\omega t + \phi)$, what is the phase angle ϕ? (e) Suppose the data are the same, except that the capacitor is discharging at $t = 0$. What then is ϕ? ssm

18P. An inductor is connected across a capacitor whose capacitance can be varied by turning a knob. We wish to make the frequency of oscillation of this LC circuit vary linearly with the angle of rotation of the knob, going from 2×10^5 to 4×10^5 Hz as the knob turns through 180°. If $L = 1.0$ mH, plot the required capacitance C as a function of the angle of rotation of the knob.

19P. In an oscillating LC circuit, $L = 3.00$ mH and $C = 2.70$ μF. At $t = 0$ the charge on the capacitor is zero and the current is 2.00 A. (a) What is the maximum charge that will appear on the capacitor? (b) In terms of the period T of oscillation, how much time will elapse after $t = 0$ until the energy stored in the capacitor will be increasing at its greatest rate? (c) What is this greatest rate at which energy is transferred to the capacitor? ssm

20P. A series circuit containing inductance L_1 and capacitance C_1 oscillates at angular frequency ω. A second series circuit, containing inductance L_2 and capacitance C_2, oscillates at the same angular frequency. In terms of ω, what is the angular frequency of oscillation of a series circuit containing all four of these elements? Neglect resistance. (*Hint:* Use the formulas for equivalent capacitance and equivalent inductance; see Section 26-4 and Problem 43 in Chapter 31.)

21P. In an oscillating LC circuit with $C = 64.0$ μF, the current as a function of time is given by $i = (1.60) \sin(2500t + 0.680)$, where t is in seconds, i in amperes, and the phase constant in radians. (a) How soon after $t = 0$ will the current reach its maximum value? What are (b) the inductance L and (c) the total energy? ilw

22P. Three identical inductors L and two identical capacitors C are connected in a two-loop circuit as shown in Fig. 33-24. (a) Suppose the currents are as shown in Fig. 33-24a. What is the current in the middle inductor? Write the loop equations and show that they are satisfied if the current oscillates with angular frequency $\omega = 1/\sqrt{LC}$. (b) Now suppose the currents are as shown in Fig. 33-24b. What is the current in the middle inductor? Write the loop equations and show that they are satisfied if the current oscillates with angular frequency $\omega = 1/\sqrt{3LC}$. Because the circuit can oscillate at two different frequencies, we cannot find an equivalent single-loop LC circuit to replace it.

(a)

(b)

Fig. 33-24 Problem 22.

23P*. In Fig. 33-25, capacitor 1 with $C_1 = 900$ μF is initially charged to 100 V and capacitor 2 with $C_2 = 100$ μF is uncharged. The inductor has an inductance of 10.0 H. Describe in detail how

one might charge capacitor 2 to 300 V by manipulating switches S_1 and S_2. ssm   www

Fig. 33-25 Problem 23.

SEC. 33-5 Damped Oscillations in an *RLC* Circuit

24E. Consider a damped LC circuit.
(a) Show that the damping term $e^{-Rt/2L}$ (which involves L but not C) can be rewritten in a more symmetric manner (involving L and C) as $e^{-\pi R(\sqrt{C/L})t/T}$. Here T is the period of oscillation (neglecting resistance). (b) Using (a), show that the SI unit of $\sqrt{L/C}$ is the ohm. (c) Using (a), show that the condition that the fractional energy loss per cycle be small is $R \ll \sqrt{L/C}$.

25E. What resistance R should be connected in series with an inductance $L = 220$ mH and capacitance $C = 12.0$ μF for the maximum charge on the capacitor to decay to 99.0% of its initial value in 50.0 cycles? (Assume $\omega' \approx \omega$.) ssm   ilw

26P. A single-loop circuit consists of a 7.20 Ω resistor, a 12.0 H inductor, and a 3.20 μF capacitor. Initially the capacitor has a charge of 6.20 μC and the current is zero. Calculate the charge on the capacitor N complete cycles later for $N = 5$, 10, and 100.

27P. In an oscillating series RLC circuit, find the time required for the maximum energy present in the capacitor during an oscillation to fall to half its initial value. Assume $q = Q$ at $t = 0$. ssm

28P. At time $t = 0$ there is no charge on the capacitor of a series RLC circuit but there is current I through the inductor. (a) Find the phase constant ϕ in Eq. 33-25 for the circuit. (b) Write an expression for the charge q on the capacitor as a function of time t and in terms of the current amplitude and angular frequency ω' of the oscillations.

29P*. In an oscillating series RLC circuit, show that the fraction of the energy lost per cycle of oscillation, $\Delta U/U$, is given to a close approximation by $2\pi R/\omega L$. The quantity $\omega L/R$ is often called the Q of the circuit (for *quality*). A high-Q circuit has low resistance and a low fractional energy loss ($= 2\pi/Q$) per cycle. ssm   www

SEC. 33-8 Three Simple Circuits

30E. A 1.50 μF capacitor is connected as in Fig. 33-9a to an ac generator with $\mathscr{E}_m = 30.0$ V. What is the amplitude of the resulting alternating current if the frequency of the emf is (a) 1.00 kHz and (b) 8.00 kHz?

31E. A 50.0 mH inductor is connected as in Fig. 33-10a to an ac generator with $\mathscr{E}_m = 30.0$ V. What is the amplitude of the resulting alternating current if the frequency of the emf is (a) 1.00 kHz and (b) 8.00 kHz? ssm   ilw

32E. A 50 Ω resistor is connected as in Fig. 33-8a to an ac generator with $\mathscr{E}_m = 30.0$ V. What is the amplitude of the resulting alternating current if the frequency of the emf is (a) 1.00 kHz and (b) 8.00 kHz?

33E. (a) At what frequency would a 6.0 mH inductor and a 10 μF capacitor have the same reactance? (b) What would the reactance be? (c) Show that this frequency would be the natural frequency of an oscillating circuit with the same L and C. ssm

34P. An ac generator has emf $\mathscr{E} = \mathscr{E}_m \sin \omega_d t$, with $\mathscr{E}_m = 25.0$ V and $\omega_d = 377$ rad/s. It is connected to a 12.7 H inductor. (a) What is the maximum value of the current? (b) When the current is a maximum, what is the emf of the generator? (c) When the emf of the generator is -12.5 V and increasing in magnitude, what is the current?

35P. An ac generator has emf $\mathcal{E} = \mathcal{E}_m \sin(\omega_d t - \pi/4)$, where $\mathcal{E}_m = 30.0$ V and $\omega_d = 350$ rad/s. The current produced in a connected circuit is $i(t) = I \sin(\omega_d t - 3\pi/4)$, where $I = 620$ mA. (a) At what time after $t = 0$ does the generator emf first reach a maximum? (b) At what time after $t = 0$ does the current first reach a maximum? (c) The circuit contains a single element other than the generator. Is it a capacitor, an inductor, or a resistor? Justify your answer. (d) What is the value of the capacitance, inductance, or resistance, as the case may be? **ssm**

36P. The ac generator of Problem 34 is connected to a 4.15 μF capacitor. (a) What is the maximum value of the current? (b) When the current is a maximum, what is the emf of the generator? (c) When the emf of the generator is -12.5 V and increasing in magnitude, what is the current?

SEC. 33-9   The Series *RLC* Circuit

37E. (a) Find Z, ϕ, and I for the situation of Sample Problem 33-7 with the capacitor removed from the circuit, all other parameters remaining unchanged. (b) Draw to scale a phasor diagram like that of Fig. 33-11d for this new situation.

38E. (a) Find Z, ϕ, and I for the situation of Sample Problem 33-7 with the inductor removed from the circuit, all other parameters remaining unchanged. (b) Draw to scale a phasor diagram like that of Fig. 33-11d for this new situation.

39E. (a) Find Z, ϕ, and I for the situation of Sample Problem 33-7 with $C = 70.0$ μF, the other parameters remaining unchanged. (b) Draw a phasor diagram like that of Fig. 33-11d for this new situation and compare the two diagrams closely. **ssm www**

40P. In Fig. 33-26, a generator with an adjustable frequency of oscillation is connected to a variable resistance R, a capacitor of $C = 5.50$ μF, and an inductor of inductance L. The amplitude of the current produced in the circuit by the generator is at half-maximum level when the generator's frequency is 1.30 or 1.50 kHz.

Fig. 33-26 Problem 40.

(a) What is L? (b) If R is increased, what happens to the frequencies at which the current amplitude is at half-maximum level?

41P. In an *RLC* circuit, can the amplitude of the voltage across an inductor be greater than the amplitude of the generator emf? Consider an *RLC* circuit with $\mathcal{E}_m = 10$ V, $R = 10$ Ω, $L = 1.0$ H, and $C = 1.0$ μF. Find the amplitude of the voltage across the inductor at resonance. **ssm ilw**

42P. When the generator emf in Sample Problem 33-7 is a maximum, what is the voltage across (a) the generator, (b) the resistance, (c) the capacitance, and (d) the inductance? (e) By summing these with appropriate signs, verify that the loop rule is satisfied.

43P. A coil of inductance 88 mH and unknown resistance and a 0.94 μF capacitor are connected in series with an alternating emf of frequency 930 Hz. If the phase constant between the applied

voltage and the current is 75°, what is the resistance of the coil? **ssm ilw**

44P. An ac generator with $\mathcal{E}_m = 220$ V and operating at 400 Hz causes oscillations in a series *RLC* circuit having $R = 220$ Ω, $L = 150$ mH, and $C = 24.0$ μF. Find (a) the capacitive reactance X_C, (b) the impedance Z, and (c) the current amplitude I. A second capacitor of the same capacitance is then connected in series with the other components. Determine whether the values of (d) X_C, (e) Z, and (f) I increase, decrease, or remain the same.

45P. An *RLC* circuit such as that of Fig. 33-7 has $R = 5.00$ Ω, $C = 20.0$ μF, $L = 1.00$ H, and $\mathcal{E}_m = 30.0$ V. (a) At what angular frequency ω_d will the current amplitude have its maximum value, as in the resonance curves of Fig. 33-13? (b) What is this maximum value? (c) At what two angular frequencies ω_{d1} and ω_{d2} will the current amplitude be half this maximum value? (d) What is the fractional half-width $[= (\omega_{d1} - \omega_{d2})/\omega]$ of the resonance curve for this circuit? **ssm**

46P. An ac generator is to be connected in series with an inductor of $L = 2.00$ mH and a capacitance C. You are to produce C by using capacitors of capacitances $C_1 = 4.00$ μF and $C_2 = 6.00$ μF, either singly or together. What resonant frequencies can the circuit have, depending on how you use C_1 and C_2?

47P. Show that the fractional half-width (see Problem 45) of a resonance curve is given by

$$\frac{\Delta \omega_d}{\omega} = \sqrt{\frac{3C}{L}} R,$$

in which ω is the angular frequency at resonance and $\Delta \omega_d$ is the width of the resonance curve at half-amplitude. Note that $\Delta \omega_d/\omega$ increases with R, as Fig. 33-13 shows. Use this formula to check the answer to Problem 45d. **ssm**

48P. In Fig. 33-27, a generator with an adjustable frequency of oscillation is connected to resistance $R = 100$ Ω, inductances $L_1 = 1.70$ mH and $L_2 = 2.30$ mH, and capacitances $C_1 = 4.00$ μF, $C_2 = 2.50$ μF, and $C_3 = 3.50$ μF. (a) What is the resonant frequency of the circuit? (*Hint:* See

Fig. 33-27 Problem 48.

Problem 43 in Chapter 31.) What happens to the resonant frequency if (b) the value of R is increased, (c) the value of L_1 is increased, and (d) capacitance C_3 is removed from the circuit?

SEC. 33-10   Power in Alternating-Current Circuits

49E. What direct current will produce the same amount of thermal energy, in a particular resistor, as an alternating current that has a maximum value of 2.60 A? **ssm**

50E. An ac voltmeter with large impedance is connected in turn across the inductor, the capacitor, and the resistor in a series circuit having an alternating emf of 100 V (rms); it gives the same reading in volts in each case. What is this reading?

51E. What is the maximum value of an ac voltage whose rms value is 100 V?

52E. (a) For the conditions in Problem 34c, is the generator supplying energy to or taking energy from the rest of the circuit? (b) Repeat for the conditions of Problem 36c.

53E. Calculate the average rate of energy dissipation in the circuits of Exercises 31, 32, 37, and 38.

54E. Show that the average rate at which energy is supplied to the circuit of Fig. 33-7 can also be written as $P_{avg} = \mathscr{E}_{rms}^2 R/Z^2$. Show that this expression for average power gives reasonable results for a purely resistive circuit, for an *RLC* circuit at resonance, for a purely capacitive circuit, and for a purely inductive circuit.

55E. An air conditioner connected to a 120 V rms ac line is equivalent to a 12.0 Ω resistance and a 1.30 Ω inductive reactance in series. (a) Calculate the impedance of the air conditioner. (b) Find the average rate at which energy is supplied to the appliance. ssm

56P. In a series oscillating *RLC* circuit, $R = 16.0$ Ω, $C = 31.2$ μF, $L = 9.20$ mH, and $\mathscr{E} = \mathscr{E}_m \sin \omega_d t$ with $\mathscr{E}_m = 45.0$ V and $\omega_d = 3000$ rad/s. For time $t = 0.442$ ms find (a) the rate at which energy is being supplied by the generator, (b) the rate at which the energy in the capacitor is changing, (c) the rate at which the energy in the inductor is changing, and (d) the rate at which energy is being dissipated in the resistor. (e) What is the meaning of a negative result for any of (a), (b), and (c)? (f) Show that the results of (b), (c), and (d) sum to the result of (a).

57P. Figure 33-28 shows an ac generator connected to a "black box" through a pair of terminals. The box contains an *RLC* circuit, possibly even a multiloop circuit, whose elements and connections we do not know. Measurements outside the box reveal that

Fig. 33-28 Problem 57.

$$\mathscr{E}(t) = (75.0 \text{ V}) \sin \omega_d t$$

and

$$i(t) = (1.20 \text{ A}) \sin(\omega_d t + 42.0°).$$

(a) What is the power factor? (b) Does the current lead or lag the emf? (c) Is the circuit in the box largely inductive or largely capacitive? (d) Is the circuit in the box in resonance? (e) Must there be a capacitor in the box? An inductor? A resistor? (f) At what average rate is energy delivered to the box by the generator? (g) Why don't you need to know the angular frequency ω_d to answer all these questions? ssm www

58P. In Fig. 33-29 show that the average rate at which energy is dissipated in resistance R is a maximum when R is equal to the internal resistance r of the ac generator. (In the text discussion we have tacitly assumed that $r = 0$.)

59P. In an *RLC* circuit such as that of Fig. 33-7 assume that $R = 5.00$ Ω, $L = 60.0$ mH, $f_d = 60.0$ Hz, and $\mathscr{E}_m = 30.0$ V. For what values of the capacitance would the average rate at which energy is dissipated in the resistance be (a) a maximum and (b) a minimum? (c) What are these maximum and minimum energy dissipation rates? What are (d) the corresponding phase angles and (e) the corresponding power factors? ssm

60P. A typical "light dimmer" used to dim the stage lights in a theater consists of a variable inductor L (whose inductance is adjustable between zero and L_{max}) connected in series with the lightbulb B as shown in Fig. 33-30. The electrical supply is 120 V (rms) at 60.0 Hz; the lightbulb is rated as "120 V, 1000 W." (a) What L_{max} is required if the rate of energy dissipation in the lightbulb is to be varied by a factor of 5 from its upper limit of 1000 W? Assume that the resistance of the lightbulb is independent of its temperature. (b) Could one use a variable resistor (adjustable between zero and R_{max}) instead of an inductor? If so, what R_{max} is required? Why isn't this done?

Fig. 33-30 Problem 60.

61P. In Fig. 33-31, $R = 15.0$ Ω, $C = 4.70$ μF, and $L = 25.0$ mH. The generator provides a sinusoidal voltage of 75.0 V (rms) and frequency $f = 550$ Hz. (a) Calculate the rms current. (b) Find the rms voltages V_{ab}, V_{bc}, V_{cd}, V_{bd}, V_{ad}. (c) At what average rate is energy dissipated by each of the three circuit elements? ilw

Fig. 33-31 Problem 61.

SEC. 33-11 Transformers

62E. A generator supplies 100 V to the primary coil of a transformer of 50 turns. If the secondary coil has 500 turns, what is the secondary voltage?

63E. A transformer has 500 primary turns and 10 secondary turns. (a) If V_p is 120 V (rms), what is V_s with an open circuit? (b) If the secondary now has a resistive load of 15 Ω, what are the currents in the primary and secondary? ssm ilw

64E. Figure 33-32 shows an "autotransformer." It consists of a single coil (with an iron core). Three taps T_i are provided. Between taps T_1 and T_2 there are 200 turns, and between taps T_2 and T_3 there are 800 turns. Any two taps can be considered the "primary terminals" and any two taps can be considered the "secondary terminals." List all the ratios by which the primary voltage may be changed to a secondary voltage.

Fig. 33-32 Exercise 64.

65P. In Fig. 33-29 let the rectangular box on the left represent the (high-impedance) output of an audio amplifier, with $r = 1000$ Ω. Let $R = 10$ Ω represent the (low-impedance) coil of a loudspeaker. For maximum transfer of energy to the load R we must have $R = r$, and that is not true in this case. However, a transformer can be used to "transform" resistances, making them behave electrically as if they were larger or smaller than they actually are. Sketch the primary and secondary coils of a transformer that can be introduced between the amplifier and the speaker in Fig. 33-29 to match the impedances. What must be the turns ratio? ssm

Fig. 33-29 Problems 58 and 65.

APPENDIX A
The International System of Units (SI)*

1. The SI Base Units

Quantity	Name	Symbol	Definition
length	meter	m	". . . the length of the path traveled by light in vacuum in 1/299,792,458 of a second." (1983)
mass	kilogram	kg	". . . this prototype [a certain platinum–iridium cylinder] shall henceforth be considered to be the unit of mass." (1889)
time	second	s	". . . the duration of 9,192,631,770 periods of the radiation corresponding to the transition between the two hyperfine levels of the ground state of the cesium-133 atom." (1967)
electric current	ampere	A	". . . that constant current which, if maintained in two straight parallel conductors of infinite length, of negligible circular cross section, and placed 1 meter apart in vacuum, would produce between these conductors a force equal to 2×10^{-7} newton per meter of length." (1946)
thermodynamic temperature	kelvin	K	". . . the fraction 1/273.16 of the thermodynamic temperature of the triple point of water." (1967)
amount of substance	mole	mol	". . . the amount of substance of a system which contains as many elementary entities as there are atoms in 0.012 kilogram of carbon-12." (1971)
luminous intensity	candela	cd	". . . the luminous intensity, in the perpendicular direction, of a surface of 1/600,000 square meter of a blackbody at the temperature of freezing platinum under a pressure of 101.325 newtons per square meter." (1967)

*Adapted from "The International System of Units (SI)," National Bureau of Standards Special Publication 330, 1972 edition. The definitions above were adopted by the General Conference of Weights and Measures, an international body, on the dates shown. In this book we do not use the candela.

2. Some SI Derived Units

Quantity	Name of Unit	Symbol	
area	square meter	m^2	
volume	cubic meter	m^3	
frequency	hertz	Hz	s^{-1}
mass density (density)	kilogram per cubic meter	kg/m^3	
speed, velocity	meter per second	m/s	
angular velocity	radian per second	rad/s	
acceleration	meter per second per second	m/s^2	
angular acceleration	radian per second per second	rad/s^2	
force	newton	N	$kg \cdot m/s^2$
pressure	pascal	Pa	N/m^2
work, energy, quantity of heat	joule	J	$N \cdot m$
power	watt	W	J/s
quantity of electric charge	coulomb	C	$A \cdot s$
potential difference, electromotive force	volt	V	W/A
electric field strength	volt per meter (or newton per coulomb)	V/m	N/C
electric resistance	ohm	Ω	V/A
capacitance	farad	F	$A \cdot s/V$
magnetic flux	weber	Wb	$V \cdot s$
inductance	henry	H	$V \cdot s/A$
magnetic flux density	tesla	T	Wb/m^2
magnetic field strength	ampere per meter	A/m	
entropy	joule per kelvin	J/K	
specific heat	joule per kilogram kelvin	$J/(kg \cdot K)$	
thermal conductivity	watt per meter kelvin	$W/(m \cdot K)$	
radiant intensity	watt per steradian	W/sr	

3. The SI Supplementary Units

Quantity	Name of Unit	Symbol
plane angle	radian	rad
solid angle	steradian	sr

APPENDIX B
Some Fundamental Constants of Physics*

Constant	Symbol	Computational Value	Best (1998) Value Value[a]	Best (1998) Value Uncertainty[b]
Speed of light in a vacuum	c	3.00×10^8 m/s	2.997 924 58	exact
Elementary charge	e	1.60×10^{-19} C	1.602 176 462	0.039
Gravitational constant	G	6.67×10^{-11} m³/s² · kg	6.673	1500
Universal gas constant	R	8.31 J/mol · K	8.314 472	1.7
Avogadro constant	N_A	6.02×10^{23} mol^{-1}	6.022 141 99	0.079
Boltzmann constant	k	1.38×10^{-23} J/K	1.380 650 3	1.7
Stefan–Boltzmann constant	σ	5.67×10^{-8} W/m² · K⁴	5.670 400	7.0
Molar volume of ideal gas at STP[d]	V_m	2.27×10^{-2} m³/mol	2.271 098 1	1.7
Permittivity constant	ϵ_0	8.85×10^{-12} F/m	8.854 187 817 62	exact
Permeability constant	μ_0	1.26×10^{-6} H/m	1.256 637 061 43	exact
Planck constant	h	6.63×10^{-34} J · s	6.626 068 76	0.078
Electron mass[c]	m_e	9.11×10^{-31} kg	9.109 381 88	0.079
		5.49×10^{-4} u	5.485 799 110	0.0021
Proton mass[c]	m_p	1.67×10^{-27} kg	1.672 621 58	0.079
		1.0073 u	1.007 276 466 88	1.3×10^{-4}
Ratio of proton mass to electron mass	m_p/m_e	1840	1836.152 667 5	0.0021
Electron charge-to-mass ratio	e/m_e	1.76×10^{11} C/kg	1.758 820 174	0.040
Neutron mass[c]	m_n	1.68×10^{-27} kg	1.674 927 16	0.079
		1.0087 u	1.008 664 915 78	5.4×10^{-4}
Hydrogen atom mass[c]	m_{1_H}	1.0078 u	1.007 825 031 6	0.0005
Deuterium atom mass[c]	m_{2_H}	2.0141 u	2.014 101 777 9	0.0005
Helium atom mass[c]	$m_{4_{He}}$	4.0026 u	4.002 603 2	0.067
Muon mass	m_μ	1.88×10^{-28} kg	1.883 531 09	0.084
Electron magnetic moment	μ_e	9.28×10^{-24} J/T	9.284 763 62	0.040
Proton magnetic moment	μ_p	1.41×10^{-26} J/T	1.410 606 663	0.041
Bohr magneton	μ_B	9.27×10^{-24} J/T	9.274 008 99	0.040
Nuclear magneton	μ_N	5.05×10^{-27} J/T	5.050 783 17	0.040
Bohr radius	r_B	5.29×10^{-11} m	5.291 772 083	0.0037
Rydberg constant	R	1.10×10^7 m^{-1}	1.097 373 156 854 8	7.6×10^{-6}
Electron Compton wavelength	λ_C	2.43×10^{-12} m	2.426 310 215	0.0073

[a]Values given in this column should be given the same unit and power of 10 as the computational value.

[b]Parts per million.

[c]Masses given in u are in unified atomic mass units, where 1 u = 1.660 538 73 $\times$ 10^{-27} kg.

[d]STP means standard temperature and pressure: 0°C and 1.0 atm (0.1 MPa).

*The values in this table were selected from the 1998 CODATA recommended values (www.physics.nist.gov).

Some Astronomical Data

Some Distances from Earth

To the Moon*	3.82×10^8 m	To the center of our galaxy	2.2×10^{20} m
To the Sun*	1.50×10^{11} m	To the Andromeda Galaxy	2.1×10^{22} m
To the nearest star (Proxima Centauri)	4.04×10^{16} m	To the edge of the observable universe	$\sim 10^{26}$ m

*Mean distance.

The Sun, Earth, and the Moon

Property	Unit	Sun		Earth	Moon
Mass	kg	1.99×10^{30}		5.98×10^{24}	7.36×10^{22}
Mean radius	m	6.96×10^8		6.37×10^6	1.74×10^6
Mean density	kg/m³	1410		5520	3340
Free-fall acceleration at the surface	m/s²	274		9.81	1.67
Escape velocity	km/s	618		11.2	2.38
Period of rotation[a]	—	37 d at poles[b]	26 d at equator[b]	23 h 56 min	27.3 d
Radiation power[c]	W	3.90×10^{26}			

[a]Measured with respect to the distant stars.
[b]The Sun, a ball of gas, does not rotate as a rigid body.
[c]Just outside Earth's atmosphere solar energy is received, assuming normal incidence, at the rate of 1340 W/m².

Some Properties of the Planets

	Mercury	Venus	Earth	Mars	Jupiter	Saturn	Uranus	Neptune	Pluto
Mean distance from Sun, 10^6 km	57.9	108	150	228	778	1430	2870	4500	5900
Period of revolution, y	0.241	0.615	1.00	1.88	11.9	29.5	84.0	165	248
Period of rotation,[a] d	58.7	-243^b	0.997	1.03	0.409	0.426	-0.451^b	0.658	6.39
Orbital speed, km/s	47.9	35.0	29.8	24.1	13.1	9.64	6.81	5.43	4.74
Inclination of axis to orbit	<28°	≈3°	23.4°	25.0°	3.08°	26.7°	97.9°	29.6°	57.5°
Inclination of orbit to Earth's orbit	7.00°	3.39°		1.85°	1.30°	2.49°	0.77°	1.77°	17.2°
Eccentricity of orbit	0.206	0.0068	0.0167	0.0934	0.0485	0.0556	0.0472	0.0086	0.250
Equatorial diameter, km	4880	12 100	12 800	6790	143 000	120 000	51 800	49 500	2300
Mass (Earth = 1)	0.0558	0.815	1.000	0.107	318	95.1	14.5	17.2	0.002
Density (water = 1)	5.60	5.20	5.52	3.95	1.31	0.704	1.21	1.67	2.03
Surface value of g,[c] m/s²	3.78	8.60	9.78	3.72	22.9	9.05	7.77	11.0	0.5
Escape velocity,[c] km/s	4.3	10.3	11.2	5.0	59.5	35.6	21.2	23.6	1.1
Known satellites	0	0	1	2	16 + ring	18 + rings	17 + rings	8 + rings	1

[a]Measured with respect to the distant stars.
[b]Venus and Uranus rotate opposite their orbital motion.
[c]Gravitational acceleration measured at the planet's equator.

APPENDIX D
Conversion Factors

Conversion factors may be read directly from these tables. For example, 1 degree = 2.778×10^{-3} revolutions, so $16.7° = 16.7 \times 2.778 \times 10^{-3}$ rev. The SI units are fully capitalized. Adapted in part from G. Shortley and D. Williams, *Elements of Physics,* 1971, Prentice-Hall, Englewood Cliffs, NJ.

Plane Angle

	°	′	″	RADIAN	rev
1 degree =	1	60	3600	1.745×10^{-2}	2.778×10^{-3}
1 minute =	1.667×10^{-2}	1	60	2.909×10^{-4}	4.630×10^{-5}
1 second =	2.778×10^{-4}	1.667×10^{-2}	1	4.848×10^{-6}	7.716×10^{-7}
1 RADIAN =	57.30	3438	2.063×10^{5}	1	0.1592
1 revolution =	360	2.16×10^{4}	1.296×10^{6}	6.283	1

Solid Angle

1 sphere = 4π steradians = 12.57 steradians

Length

	cm	METER	km	in.	ft	mi
1 centimeter =	1	10^{-2}	10^{-5}	0.3937	3.281×10^{-2}	6.214×10^{-6}
1 METER =	100	1	10^{-3}	39.37	3.281	6.214×10^{-4}
1 kilometer =	10^{5}	1000	1	3.937×10^{4}	3281	0.6214
1 inch =	2.540	2.540×10^{-2}	2.540×10^{-5}	1	8.333×10^{-2}	1.578×10^{-5}
1 foot =	30.48	0.3048	3.048×10^{-4}	12	1	1.894×10^{-4}
1 mile =	1.609×10^{5}	1609	1.609	6.336×10^{4}	5280	1

1 angström = 10^{-10} m   1 fermi = 10^{-15} m   1 fathom = 6 ft   1 rod = 16.5 ft
1 nautical mile = 1852 m   1 light-year = 9.460×10^{12} km   1 Bohr radius = 5.292×10^{-11} m   1 mil = 10^{-3} in.
  = 1.151 miles = 6076 ft   1 parsec = 3.084×10^{13} km   1 yard = 3 ft   1 nm = 10^{-9} m

Area

	METER^2	cm^2	ft^2	in.^{2}
1 SQUARE METER =	1	10^{4}	10.76	1550
1 square centimeter =	10^{-4}	1	1.076×10^{-3}	0.1550
1 square foot =	9.290×10^{-2}	929.0	1	144
1 square inch =	6.452×10^{-4}	6.452	6.944×10^{-3}	1

1 square mile = 2.788×10^{7} ft^2 = 640 acres   1 acre = 43 560 ft^2
1 barn = 10^{-28} m^2 1 hectare = 10^{4} m^2 = 2.471 acres

Volume

	METER^3	cm^3	L	ft^3	in.^{3}
1 CUBIC METER = 1		10^6	1000	35.31	6.102×10^4
1 cubic centimeter = 10^{-6}		1	1.000×10^{-3}	3.531×10^{-5}	6.102×10^{-2}
1 liter = 1.000×10^{-3}		1000	1	3.531×10^{-2}	61.02
1 cubic foot = 2.832×10^{-2}		2.832×10^4	28.32	1	1728
1 cubic inch = 1.639×10^{-5}		16.39	1.639×10^{-2}	5.787×10^{-4}	1

1 U.S. fluid gallon = 4 U.S. fluid quarts = 8 U.S. pints = 128 U.S. fluid ounces = 231 in.^{3}
1 British imperial gallon = 277.4 in.^{3} = 1.201 U.S. fluid gallons

Mass

Quantities in the colored areas are not mass units but are often used as such. When we write, for example, 1 kg ''='' 2.205 lb, this means that a kilogram is a *mass* that *weighs* 2.205 pounds at a location where g has the standard value of 9.80665 m/s^2.

	g	KILOGRAM	slug	u	oz	lb	ton
1 gram = 1		0.001	6.852×10^{-5}	6.022×10^{23}	3.527×10^{-2}	2.205×10^{-3}	1.102×10^{-6}
1 KILOGRAM = 1000		1	6.852×10^{-2}	6.022×10^{26}	35.27	2.205	1.102×10^{-3}
1 slug = 1.459×10^4		14.59	1	8.786×10^{27}	514.8	32.17	1.609×10^{-2}
1 atomic mass unit = 1.661×10^{-24}		1.661×10^{-27}	1.138×10^{-28}	1	5.857×10^{-26}	3.662×10^{-27}	1.830×10^{-30}
1 ounce = 28.35		2.835×10^{-2}	1.943×10^{-3}	1.718×10^{25}	1	6.250×10^{-2}	3.125×10^{-5}
1 pound = 453.6		0.4536	3.108×10^{-2}	2.732×10^{26}	16	1	0.0005
1 ton = 9.072×10^5		907.2	62.16	5.463×10^{29}	3.2×10^4	2000	1

1 metric ton = 1000 kg

Density

Quantities in the colored areas are weight densities and, as such, are dimensionally different from mass densities. See note for mass table.

	slug/ft^3	KILOGRAM/ METER^3	g/cm^3	lb/ft^3	lb/in.^{3}
1 slug per foot^3 = 1		515.4	0.5154	32.17	1.862×10^{-2}
1 KILOGRAM per METER^3 = 1.940×10^{-3}		1	0.001	6.243×10^{-2}	3.613×10^{-5}
1 gram per centimeter^3 = 1.940		1000	1	62.43	3.613×10^{-2}
1 pound per foot^3 = 3.108×10^{-2}		16.02	16.02×10^{-2}	1	5.787×10^{-4}
1 pound per inch^3 = 53.71		2.768×10^4	27.68	1728	1

Time

	y	d	h	min	SECOND
1 year = 1		365.25	8.766×10^3	5.259×10^5	3.156×10^7
1 day = 2.738×10^{-3}		1	24	1440	8.640×10^4
1 hour = 1.141×10^{-4}		4.167×10^{-2}	1	60	3600
1 minute = 1.901×10^{-6}		6.944×10^{-4}	1.667×10^{-2}	1	60
1 SECOND = 3.169×10^{-8}		1.157×10^{-5}	2.778×10^{-4}	1.667×10^{-2}	1

Speed

	ft/s	km/h	METER/SECOND	mi/h	cm/s
1 foot per second = 1	1.097	0.3048	0.6818	30.48
1 kilometer per hour = 0.9113	1	0.2778	0.6214	27.78
1 METER per SECOND = 3.281	3.6	1	2.237	100
1 mile per hour = 1.467	1.609	0.4470	1	44.70
1 centimeter per second = 3.281×10^{-2}	3.6×10^{-2}	0.01	2.237×10^{-2}	1

1 knot = 1 nautical mi/h = 1.688 ft/s   1 mi/min = 88.00 ft/s = 60.00 mi/h

Force

Force units in the colored areas are now little used. To clarify: 1 gram-force (= 1 gf) is the force of gravity that would act on an object whose mass is 1 gram at a location where g has the standard value of 9.80665 m/s^2.

	dyne	NEWTON	lb	pdl	gf	kgf
1 dyne = 1	10^{-5}	2.248×10^{-6}	7.233×10^{-5}	1.020×10^{-3}	1.020×10^{-6}
1 NEWTON = 10^5	1	0.2248	7.233	102.0	0.1020
1 pound = 4.448×10^5	4.448	1	32.17	453.6	0.4536
1 poundal = 1.383×10^4	0.1383	3.108×10^{-2}	1	14.10	1.410×10^2
1 gram-force = 980.7	9.807×10^{-3}	2.205×10^{-3}	7.093×10^{-2}	1	0.001
1 kilogram-force = 9.807×10^5	9.807	2.205	70.93	1000	1

1 ton = 2000 lb

Pressure

	atm	dyne/cm^2	inch of water	cm Hg	PASCAL	lb/in.^{2}	lb/ft^2
1 atmosphere = 1	1.013×10^6	406.8	76	1.013×10^5	14.70	2116
1 dyne per centimeter^2 = 9.869×10^{-7}	1	4.015×10^{-4}	7.501×10^{-5}	0.1	1.405×10^{-5}	2.089×10^{-3}
1 inch of water^a at 4°C = 2.458×10^{-3}	2491	1	0.1868	249.1	3.613×10^{-2}	5.202
1 centimeter of mercury^a at 0°C = 1.316×10^{-2}	1.333×10^4	5.353	1	1333	0.1934	27.85
1 PASCAL = 9.869×10^{-6}	10	4.015×10^{-3}	7.501×10^{-4}	1	1.450×10^{-4}	2.089×10^{-2}
1 pound per inch^2 = 6.805×10^{-2}	6.895×10^4	27.68	5.171	6.895×10^3	1	144
1 pound per foot^2 = 4.725×10^{-4}	478.8	0.1922	3.591×10^{-2}	47.88	6.944×10^{-3}	1

^{a}Where the acceleration of gravity has the standard value of 9.80665 m/s^2.

1 bar = 10^6 dyne/cm^2 = 0.1 MPa   1 millibar = 10^3 dyne/cm^2 = 10^2 Pa   1 torr = 1 mm Hg

Energy, Work, Heat

Quantities in the colored areas are not energy units but are included for convenience. They arise from the relativistic mass–energy equivalence formula $E = mc^2$ and represent the energy released if a kilogram or unified atomic mass unit (u) is completely converted to energy (bottom two rows) or the mass that would be completely converted to one unit of energy (rightmost two columns).

	Btu	erg	ft·lb	hp·h	JOULE	cal	kW·h	eV	MeV	kg	u
1 British thermal unit =	1	1.055×10^{10}	777.9	3.929×10^{-4}	1055	252.0	2.930×10^{-4}	6.585×10^{21}	6.585×10^{15}	1.174×10^{-14}	7.070×10^{12}
1 erg =	9.481×10^{-11}	1	7.376×10^{-8}	3.725×10^{-14}	10^{-7}	2.389×10^{-8}	2.778×10^{-14}	6.242×10^{11}	6.242×10^{5}	1.113×10^{-24}	670.2
1 foot-pound =	1.285×10^{-3}	1.356×10^{7}	1	5.051×10^{-7}	1.356	0.3238	3.766×10^{-7}	8.464×10^{18}	8.464×10^{12}	1.509×10^{-17}	9.037×10^{9}
1 horsepower-hour =	2545	2.685×10^{13}	1.980×10^{6}	1	2.685×10^{6}	6.413×10^{5}	0.7457	1.676×10^{25}	1.676×10^{19}	2.988×10^{-11}	1.799×10^{16}
1 JOULE =	9.481×10^{-4}	10^{7}	0.7376	3.725×10^{-7}	1	0.2389	2.778×10^{-7}	6.242×10^{18}	6.242×10^{12}	1.113×10^{-17}	6.702×10^{9}
1 calorie =	3.969×10^{-3}	4.186×10^{7}	3.088	1.560×10^{-6}	4.186	1	1.163×10^{-6}	2.613×10^{19}	2.613×10^{13}	4.660×10^{-17}	2.806×10^{10}
1 kilowatt-hour =	3413	3.600×10^{13}	2.655×10^{6}	1.341	3.600×10^{6}	8.600×10^{5}	1	2.247×10^{25}	2.247×10^{19}	4.007×10^{-11}	2.413×10^{16}
1 electron-volt =	1.519×10^{-22}	1.602×10^{-12}	1.182×10^{-19}	5.967×10^{-26}	1.602×10^{-19}	3.827×10^{-20}	4.450×10^{-26}	1	10^{-6}	1.783×10^{-36}	1.074×10^{-9}
1 million electron-volts =	1.519×10^{-16}	1.602×10^{-6}	1.182×10^{-13}	5.967×10^{-20}	1.602×10^{-13}	3.827×10^{-14}	4.450×10^{-20}	10^{-6}	1	1.783×10^{-30}	1.074×10^{-3}
1 kilogram =	8.521×10^{13}	8.987×10^{23}	6.629×10^{16}	3.348×10^{10}	8.987×10^{16}	2.146×10^{16}	2.497×10^{10}	5.610×10^{35}	5.610×10^{29}	1	6.022×10^{26}
1 unified atomic mass unit =	1.415×10^{-13}	1.492×10^{-3}	1.101×10^{-10}	5.559×10^{-17}	1.492×10^{-10}	3.564×10^{-11}	4.146×10^{-17}	9.320×10^{8}	932.0	1.661×10^{-27}	1

Power

	Btu/h	ft·lb/s	hp	cal/s	kW	WATT
1 British thermal unit per hour =	1	0.2161	3.929×10^{-4}	6.998×10^{-2}	2.930×10^{-4}	0.2930
1 foot-pound per second =	4.628	1	1.818×10^{-3}	0.3239	1.356×10^{-3}	1.356
1 horsepower =	2545	550	1	178.1	0.7457	745.7
1 calorie per second =	14.29	3.088	5.615×10^{-3}	1	4.186×10^{-3}	4.186
1 kilowatt =	3413	737.6	1.341	238.9	1	1000
1 WATT =	3.413	0.7376	1.341×10^{-3}	0.2389	0.001	1

Magnetic Field

	gauss	TESLA	milligauss
1 gauss =	1	10^{-4}	1000
1 TESLA =	10^{4}	1	10^{7}
1 milligauss =	0.001	10^{-7}	1

1 tesla = 1 weber/meter^2

Magnetic Flux

	maxwell	WEBER
1 maxwell =	1	10^{-8}
1 WEBER =	10^{8}	1

APPENDIX E
Mathematical Formulas

Geometry

Circle of radius r: circumference $= 2\pi r$; area $= \pi r^2$.

Sphere of radius r: area $= 4\pi r^2$; volume $= \frac{4}{3}\pi r^3$.

Right circular cylinder of radius r and height h:
area $= 2\pi r^2 + 2\pi rh$; volume $= \pi r^2 h$.

Triangle of base a and altitude h: area $= \frac{1}{2}ah$.

Quadratic Formula

If $ax^2 + bx + c = 0$, then $x = \dfrac{-b \pm \sqrt{b^2 - 4ac}}{2a}$.

Trigonometric Functions of Angle θ

$$\sin\theta = \frac{y}{r} \quad \cos\theta = \frac{x}{r}$$

$$\tan\theta = \frac{y}{x} \quad \cot\theta = \frac{x}{y}$$

$$\sec\theta = \frac{r}{x} \quad \csc\theta = \frac{r}{y}$$

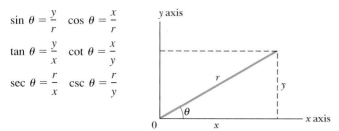

Pythagorean Theorem

In this right triangle,
$$a^2 + b^2 = c^2$$

Triangles

Angles are A, B, C

Opposite sides are a, b, c

Angles $A + B + C = 180°$

$$\frac{\sin A}{a} = \frac{\sin B}{b} = \frac{\sin C}{c}$$

$$c^2 = a^2 + b^2 - 2ab\cos C$$

Exterior angle $D = A + C$

Mathematical Signs and Symbols

$=$ equals

$\approx$ equals approximately

$\sim$ is the order of magnitude of

$\neq$ is not equal to

$\equiv$ is identical to, is defined as

$>$ is greater than ($\gg$ is much greater than)

$<$ is less than ($\ll$ is much less than)

$\geq$ is greater than or equal to (or, is no less than)

$\leq$ is less than or equal to (or, is no more than)

$\pm$ plus or minus

$\propto$ is proportional to

Σ the sum of

x_{avg} the average value of x

Trigonometric Identities

$$\sin(90° - \theta) = \cos\theta$$

$$\cos(90° - \theta) = \sin\theta$$

$$\sin\theta/\cos\theta = \tan\theta$$

$$\sin^2\theta + \cos^2\theta = 1$$

$$\sec^2\theta - \tan^2\theta = 1$$

$$\csc^2\theta - \cot^2\theta = 1$$

$$\sin 2\theta = 2\sin\theta\cos\theta$$

$$\cos 2\theta = \cos^2\theta - \sin^2\theta = 2\cos^2\theta - 1 = 1 - 2\sin^2\theta$$

$$\sin(\alpha \pm \beta) = \sin\alpha\cos\beta \pm \cos\alpha\sin\beta$$

$$\cos(\alpha \pm \beta) = \cos\alpha\cos\beta \mp \sin\alpha\sin\beta$$

$$\tan(\alpha \pm \beta) = \frac{\tan\alpha \pm \tan\beta}{1 \mp \tan\alpha\tan\beta}$$

$$\sin\alpha \pm \sin\beta = 2\sin\tfrac{1}{2}(\alpha \pm \beta)\cos\tfrac{1}{2}(\alpha \mp \beta)$$

$$\cos\alpha + \cos\beta = 2\cos\tfrac{1}{2}(\alpha + \beta)\cos\tfrac{1}{2}(\alpha - \beta)$$

$$\cos\alpha - \cos\beta = -2\sin\tfrac{1}{2}(\alpha + \beta)\sin\tfrac{1}{2}(\alpha - \beta)$$

Binomial Theorem

$$(1 + x)^n = 1 + \frac{nx}{1!} + \frac{n(n-1)x^2}{2!} + \cdots \qquad (x^2 < 1)$$

Exponential Expansion

$$e^x = 1 + x + \frac{x^2}{2!} + \frac{x^3}{3!} + \cdots$$

Logarithmic Expansion

$$\ln(1 + x) = x - \tfrac{1}{2}x^2 + \tfrac{1}{3}x^3 - \cdots \qquad (|x| < 1)$$

Trigonometric Expansions
(θ in radians)

$$\sin \theta = \theta - \frac{\theta^3}{3!} + \frac{\theta^5}{5!} - \cdots$$

$$\cos \theta = 1 - \frac{\theta^2}{2!} + \frac{\theta^4}{4!} - \cdots$$

$$\tan \theta = \theta + \frac{\theta^3}{3} + \frac{2\theta^5}{15} + \cdots$$

Cramer's Rule

Two simultaneous equations in unknowns x and y,

$$a_1 x + b_1 y = c_1 \quad \text{and} \quad a_2 x + b_2 y = c_2,$$

have the solutions

$$x = \frac{\begin{vmatrix} c_1 & b_1 \\ c_2 & b_2 \end{vmatrix}}{\begin{vmatrix} a_1 & b_1 \\ a_2 & b_2 \end{vmatrix}} = \frac{c_1 b_2 - c_2 b_1}{a_1 b_2 - a_2 b_1}$$

and

$$y = \frac{\begin{vmatrix} a_1 & c_1 \\ a_2 & c_2 \end{vmatrix}}{\begin{vmatrix} a_1 & b_1 \\ a_2 & b_2 \end{vmatrix}} = \frac{a_1 c_2 - a_2 c_1}{a_1 b_2 - a_2 b_1}.$$

Products of Vectors

Let $\hat{i}$, $\hat{j}$, and $\hat{k}$ be unit vectors in the x, y, and z directions. Then

$$\hat{i} \cdot \hat{i} = \hat{j} \cdot \hat{j} = \hat{k} \cdot \hat{k} = 1, \qquad \hat{i} \cdot \hat{j} = \hat{j} \cdot \hat{k} = \hat{k} \cdot \hat{i} = 0,$$

$$\hat{i} \times \hat{i} = \hat{j} \times \hat{j} = \hat{k} \times \hat{k} = 0,$$

$$\hat{i} \times \hat{j} = \hat{k}, \qquad \hat{j} \times \hat{k} = \hat{i}, \qquad \hat{k} \times \hat{i} = \hat{j}.$$

Any vector $\vec{a}$ with components a_x, a_y, and a_z along the x, y, and z axes can be written as

$$\vec{a} = a_x \hat{i} + a_y \hat{j} + a_z \hat{k}.$$

Let $\vec{a}$, $\vec{b}$, and $\vec{c}$ be arbitrary vectors with magnitudes a, b, and c. Then

$$\vec{a} \times (\vec{b} + \vec{c}) = (\vec{a} \times \vec{b}) + (\vec{a} \times \vec{c})$$

$$(s\vec{a}) \times \vec{b} = \vec{a} \times (s\vec{b}) = s(\vec{a} \times \vec{b}) \quad (s = \text{a scalar}).$$

Let θ be the smaller of the two angles between $\vec{a}$ and $\vec{b}$. Then

$$\vec{a} \cdot \vec{b} = \vec{b} \cdot \vec{a} = a_x b_x + a_y b_y + a_z b_z = ab \cos \theta$$

$$\vec{a} \times \vec{b} = -\vec{b} \times \vec{a} = \begin{vmatrix} \hat{i} & \hat{j} & \hat{k} \\ a_x & a_y & a_z \\ b_x & b_y & b_z \end{vmatrix}$$

$$= \hat{i} \begin{vmatrix} a_y & a_z \\ b_y & b_z \end{vmatrix} - \hat{j} \begin{vmatrix} a_x & a_z \\ b_x & b_z \end{vmatrix} + \hat{k} \begin{vmatrix} a_x & a_y \\ b_x & b_y \end{vmatrix}$$

$$= (a_y b_z - b_y a_z)\hat{i} + (a_z b_x - b_z a_x)\hat{j} + (a_x b_y - b_x a_y)\hat{k}$$

$$|\vec{a} \times \vec{b}| = ab \sin \theta$$

$$\vec{a} \cdot (\vec{b} \times \vec{c}) = \vec{b} \cdot (\vec{c} \times \vec{a}) = \vec{c} \cdot (\vec{a} \times \vec{b})$$

$$\vec{a} \times (\vec{b} \times \vec{c}) = (\vec{a} \cdot \vec{c})\vec{b} - (\vec{a} \cdot \vec{b})\vec{c}$$

Derivatives and Integrals

In what follows, the letters u and v stand for any functions of x, and a and m are constants. To each of the indefinite integrals should be added an arbitrary constant of integration. The *Handbook of Chemistry and Physics* (CRC Press Inc.) gives a more extensive tabulation.

1. $\dfrac{dx}{dx} = 1$

2. $\dfrac{d}{dx}(au) = a\dfrac{du}{dx}$

3. $\dfrac{d}{dx}(u + v) = \dfrac{du}{dx} + \dfrac{dv}{dx}$

4. $\dfrac{d}{dx}x^m = mx^{m-1}$

5. $\dfrac{d}{dx}\ln x = \dfrac{1}{x}$

6. $\dfrac{d}{dx}(uv) = u\dfrac{dv}{dx} + v\dfrac{du}{dx}$

7. $\dfrac{d}{dx}e^x = e^x$

8. $\dfrac{d}{dx}\sin x = \cos x$

9. $\dfrac{d}{dx}\cos x = -\sin x$

10. $\dfrac{d}{dx}\tan x = \sec^2 x$

11. $\dfrac{d}{dx}\cot x = -\csc^2 x$

12. $\dfrac{d}{dx}\sec x = \tan x \sec x$

13. $\dfrac{d}{dx}\csc x = -\cot x \csc x$

14. $\dfrac{d}{dx}e^u = e^u\dfrac{du}{dx}$

15. $\dfrac{d}{dx}\sin u = \cos u\dfrac{du}{dx}$

16. $\dfrac{d}{dx}\cos u = -\sin u\dfrac{du}{dx}$

1. $\displaystyle\int dx = x$

2. $\displaystyle\int au\,dx = a\int u\,dx$

3. $\displaystyle\int (u + v)\,dx = \int u\,dx + \int v\,dx$

4. $\displaystyle\int x^m\,dx = \dfrac{x^{m+1}}{m+1}\quad (m \neq -1)$

5. $\displaystyle\int \dfrac{dx}{x} = \ln |x|$

6. $\displaystyle\int u\dfrac{dv}{dx}\,dx = uv - \int v\dfrac{du}{dx}\,dx$

7. $\displaystyle\int e^x\,dx = e^x$

8. $\displaystyle\int \sin x\,dx = -\cos x$

9. $\displaystyle\int \cos x\,dx = \sin x$

10. $\displaystyle\int \tan x\,dx = \ln |\sec x|$

11. $\displaystyle\int \sin^2 x\,dx = \tfrac{1}{2}x - \tfrac{1}{4}\sin 2x$

12. $\displaystyle\int e^{-ax}\,dx = -\dfrac{1}{a}e^{-ax}$

13. $\displaystyle\int xe^{-ax}\,dx = -\dfrac{1}{a^2}(ax + 1)e^{-ax}$

14. $\displaystyle\int x^2 e^{-ax}\,dx = -\dfrac{1}{a^3}(a^2x^2 + 2ax + 2)e^{-ax}$

15. $\displaystyle\int_0^\infty x^n e^{-ax}\,dx = \dfrac{n!}{a^{n+1}}$

16. $\displaystyle\int_0^\infty x^{2n} e^{-ax^2}\,dx = \dfrac{1 \cdot 3 \cdot 5 \cdots (2n-1)}{2^{n+1}a^n}\sqrt{\dfrac{\pi}{a}}$

17. $\displaystyle\int \dfrac{dx}{\sqrt{x^2 + a^2}} = \ln(x + \sqrt{x^2 + a^2})$

18. $\displaystyle\int \dfrac{x\,dx}{(x^2 + a^2)^{3/2}} = -\dfrac{1}{(x^2 + a^2)^{1/2}}$

19. $\displaystyle\int \dfrac{dx}{(x^2 + a^2)^{3/2}} = \dfrac{x}{a^2(x^2 + a^2)^{1/2}}$

20. $\displaystyle\int_0^\infty x^{2n+1} e^{-ax^2}\,dx = \dfrac{n!}{2a^{n+1}}\quad (a > 0)$

21. $\displaystyle\int \dfrac{x\,dx}{x + d} = x - d\ln(x + d)$

Properties of the Elements

All physical properties are for a pressure of 1 atm unless otherwise specified.

Element	Symbol	Atomic Number Z	Molar Mass, g/mol	Density, g/cm^3 at 20°C	Melting Point, °C	Boiling Point, °C	Specific Heat, J/(g·°C) at 25°C
Actinium	Ac	89	(227)	10.06	1323	(3473)	0.092
Aluminum	Al	13	26.9815	2.699	660	2450	0.900
Americium	Am	95	(243)	13.67	1541	—	—
Antimony	Sb	51	121.75	6.691	630.5	1380	0.205
Argon	Ar	18	39.948	1.6626×10^{-3}	−189.4	−185.8	0.523
Arsenic	As	33	74.9216	5.78	817 (28 atm)	613	0.331
Astatine	At	85	(210)	—	(302)	—	—
Barium	Ba	56	137.34	3.594	729	1640	0.205
Berkelium	Bk	97	(247)	14.79	—	—	—
Beryllium	Be	4	9.0122	1.848	1287	2770	1.83
Bismuth	Bi	83	208.980	9.747	271.37	1560	0.122
Bohrium	Bh	107	262.12	—	—	—	—
Boron	B	5	10.811	2.34	2030	—	1.11
Bromine	Br	35	79.909	3.12 (liquid)	−7.2	58	0.293
Cadmium	Cd	48	112.40	8.65	321.03	765	0.226
Calcium	Ca	20	40.08	1.55	838	1440	0.624
Californium	Cf	98	(251)	—	—	—	—
Carbon	C	6	12.01115	2.26	3727	4830	0.691
Cerium	Ce	58	140.12	6.768	804	3470	0.188
Cesium	Cs	55	132.905	1.873	28.40	690	0.243
Chlorine	Cl	17	35.453	3.214×10^{-3} (0°C)	−101	−34.7	0.486
Chromium	Cr	24	51.996	7.19	1857	2665	0.448
Cobalt	Co	27	58.9332	8.85	1495	2900	0.423
Copper	Cu	29	63.54	8.96	1083.40	2595	0.385
Curium	Cm	96	(247)	13.3	—	—	—
Dubnium	Db	105	262.114	—	—	—	—
Dysprosium	Dy	66	162.50	8.55	1409	2330	0.172
Einsteinium	Es	99	(254)	—	—	—	—
Erbium	Er	68	167.26	9.15	1522	2630	0.167
Europium	Eu	63	151.96	5.243	817	1490	0.163
Fermium	Fm	100	(237)	—	—	—	—
Fluorine	F	9	18.9984	1.696×10^{-3} (0°C)	−219.6	−188.2	0.753
Francium	Fr	87	(223)	—	(27)	—	—
Gadolinium	Gd	64	157.25	7.90	1312	2730	0.234
Gallium	Ga	31	69.72	5.907	29.75	2237	0.377

Element	Symbol	Atomic Number Z	Molar Mass, g/mol	Density, g/cm³ at 20°C	Melting Point, °C	Boiling Point, °C	Specific Heat, J/(g · °C) at 25°C
Germanium	Ge	32	72.59	5.323	937.25	2830	0.322
Gold	Au	79	196.967	19.32	1064.43	2970	0.131
Hafnium	Hf	72	178.49	13.31	2227	5400	0.144
Hassium	Hs	108	(265)	—	—	—	—
Helium	He	2	4.0026	0.1664×10^{-3}	−269.7	−268.9	5.23
Holmium	Ho	67	164.930	8.79	1470	2330	0.165
Hydrogen	H	1	1.00797	0.08375×10^{-3}	−259.19	−252.7	14.4
Indium	In	49	114.82	7.31	156.634	2000	0.233
Iodine	I	53	126.9044	4.93	113.7	183	0.218
Iridium	Ir	77	192.2	22.5	2447	(5300)	0.130
Iron	Fe	26	55.847	7.874	1536.5	3000	0.447
Krypton	Kr	36	83.80	3.488×10^{-3}	−157.37	−152	0.247
Lanthanum	La	57	138.91	6.189	920	3470	0.195
Lawrencium	Lr	103	(257)	—	—	—	—
Lead	Pb	82	207.19	11.35	327.45	1725	0.129
Lithium	Li	3	6.939	0.534	180.55	1300	3.58
Lutetium	Lu	71	174.97	9.849	1663	1930	0.155
Magnesium	Mg	12	24.312	1.738	650	1107	1.03
Manganese	Mn	25	54.9380	7.44	1244	2150	0.481
Meitnerium	Mt	109	(266)	—	—	—	—
Mendelevium	Md	101	(256)	—	—	—	—
Mercury	Hg	80	200.59	13.55	−38.87	357	0.138
Molybdenum	Mo	42	95.94	10.22	2617	5560	0.251
Neodymium	Nd	60	144.24	7.007	1016	3180	0.188
Neon	Ne	10	20.183	0.8387×10^{-3}	−248.597	−246.0	1.03
Neptunium	Np	93	(237)	20.25	637	—	1.26
Nickel	Ni	28	58.71	8.902	1453	2730	0.444
Niobium	Nb	41	92.906	8.57	2468	4927	0.264
Nitrogen	N	7	14.0067	1.1649×10^{-3}	−210	−195.8	1.03
Nobelium	No	102	(255)	—	—	—	—
Osmium	Os	76	190.2	22.59	3027	5500	0.130
Oxygen	O	8	15.9994	1.3318×10^{-3}	−218.80	−183.0	0.913
Palladium	Pd	46	106.4	12.02	1552	3980	0.243
Phosphorus	P	15	30.9738	1.83	44.25	280	0.741
Platinum	Pt	78	195.09	21.45	1769	4530	0.134
Plutonium	Pu	94	(244)	19.8	640	3235	0.130
Polonium	Po	84	(210)	9.32	254	—	—
Potassium	K	19	39.102	0.862	63.20	760	0.758
Praseodymium	Pr	59	140.907	6.773	931	3020	0.197
Promethium	Pm	61	(145)	7.22	(1027)	—	—
Protactinium	Pa	91	(231)	15.37 (estimated)	(1230)	—	—
Radium	Ra	88	(226)	5.0	700	—	—
Radon	Rn	86	(222)	9.96×10^{-3} (0°C)	(−71)	−61.8	0.092
Rhenium	Re	75	186.2	21.02	3180	5900	0.134

Element	Symbol	Atomic Number Z	Molar Mass, g/mol	Density, g/cm³ at 20°C	Melting Point, °C	Boiling Point, °C	Specific Heat, J/(g · °C) at 25°C
Rhodium	Rh	45	102.905	12.41	1963	4500	0.243
Rubidium	Rb	37	85.47	1.532	39.49	688	0.364
Ruthenium	Ru	44	101.107	12.37	2250	4900	0.239
Rutherfordium	Rf	104	261.11	—	—	—	—
Samarium	Sm	62	150.35	7.52	1072	1630	0.197
Scandium	Sc	21	44.956	2.99	1539	2730	0.569
Seaborgium	Sg	106	263.118	—	—	—	—
Selenium	Se	34	78.96	4.79	221	685	0.318
Silicon	Si	14	28.086	2.33	1412	2680	0.712
Silver	Ag	47	107.870	10.49	960.8	2210	0.234
Sodium	Na	11	22.9898	0.9712	97.85	892	1.23
Strontium	Sr	38	87.62	2.54	768	1380	0.737
Sulfur	S	16	32.064	2.07	119.0	444.6	0.707
Tantalum	Ta	73	180.948	16.6	3014	5425	0.138
Technetium	Tc	43	(99)	11.46	2200	—	0.209
Tellurium	Te	52	127.60	6.24	449.5	990	0.201
Terbium	Tb	65	158.924	8.229	1357	2530	0.180
Thallium	Tl	81	204.37	11.85	304	1457	0.130
Thorium	Th	90	(232)	11.72	1755	(3850)	0.117
Thulium	Tm	69	168.934	9.32	1545	1720	0.159
Tin	Sn	50	118.69	7.2984	231.868	2270	0.226
Titanium	Ti	22	47.90	4.54	1670	3260	0.523
Tungsten	W	74	183.85	19.3	3380	5930	0.134
Un-named	Uun	110	(269)	—	—	—	—
Un-named	Uuu	111	(272)	—	—	—	—
Un-named	Uub	112	(264)	—	—	—	—
Un-named	Uut	113	—	—	—	—	—
Un-named	Unq	114	(285)	—	—	—	—
Un-named	Uup	115	—	—	—	—	—
Un-named	Uuh	116	(289)	—	—	—	—
Un-named	Uus	117	—	—	—	—	—
Un-named	Uuo	118	(293)	—	—	—	—
Uranium	U	92	(238)	18.95	1132	3818	0.117
Vanadium	V	23	50.942	6.11	1902	3400	0.490
Xenon	Xe	54	131.30	5.495×10^{-3}	−111.79	−108	0.159
Ytterbium	Yb	70	173.04	6.965	824	1530	0.155
Yttrium	Y	39	88.905	4.469	1526	3030	0.297
Zinc	Zn	30	65.37	7.133	419.58	906	0.389
Zirconium	Zr	40	91.22	6.506	1852	3580	0.276

The values in parentheses in the column of molar masses are the mass numbers of the longest-lived isotopes of those elements that are radioactive. Melting points and boiling points in parentheses are uncertain.

The data for gases are valid only when these are in their usual molecular state, such as H_2, He, O_2, Ne, etc. The specific heats of the gases are the values at constant pressure.

Source: Adapted from J. Emsley, *The Elements,* 3rd ed., 1998, Clarendon Press, Oxford. See also www.webelements.com for the latest values and newest elements.

Periodic Table of the Elements

The names for elements 104 through 109 (Rutherfordium, Dubnium, Seaborgium, Bohrium, Hassium, and Meitnerium, respectively) were adopted by the International Union of Pure and Applied Chemistry (IUPAC) in 1997. Elements 110, 111, 112, 114, 116, and 118 have been discovered but, as of 2000, have not yet been named. See www.webelements.com for the latest information and newest elements.

ANSWERS

to Checkpoints and Odd-Numbered Questions, Exercises, and Problems

CHAPTER 22

CP **1.** C and D attract; B and D attract **2.** (a) leftward; (b) leftward; (c) leftward **3.** (a) a, c, b; (b) less than **4.** $-15e$ (net charge of $-30e$ is equally shared) **Q** **1.** no, only for charged particles, charged particle-like objects, and spherical shells (including solid spheres) of uniform charge **3.** a and b **5.** $2q^2/4\pi\varepsilon_0 r^2$, up the page **7.** (a) same; (b) less than; (c) cancel; (d) add; (e) the adding components; (f) positive direction of y; (g) negative direction of y; (h) positive direction of x; (i) negative direction of x **9.** (a) possibly; (b) definitely **11.** no (the person and the conductor share the charge) **EP** **1.** 1.38 m **3.** (a) 4.9×10^{-7} kg; (b) 7.1×10^{-11} C **5.** (a) 0.17 N; (b) -0.046 N **7.** either $-1.00\ \mu$C and $+3.00\ \mu$C or $+1.00\ \mu$C and $-3.00\ \mu$C **9.** (a) charge $-4q/9$ must be located on the line joining the two positive charges, a distance $L/3$ from charge $+q$. **11.** (a) 5.7×10^{13} C, no; (b) 6.0×10^5 kg **13.** $q = Q/2$ **15.** (b) $\pm 2.4 \times 10^{-8}$ C **17.** (a) $\dfrac{L}{2}\left(1 + \dfrac{1}{4\pi\varepsilon_0}\dfrac{qQ}{Wh^2}\right)$; (b) $\sqrt{3qQ/4\pi\varepsilon_0 W}$ **19.** -1.32×10^{13} C **21.** (a) 3.2×10^{-19} C; (b) two **23.** 6.3×10^{11} **25.** 122 mA **27.** (a) 0; (b) 1.9×10^{-9} N **29.** (a) ^{9}B; (b) ^{13}N; (c) ^{12}C

CHAPTER 23

CP **1.** (a) rightward; (b) leftward; (c) leftward; (d) rightward (p and e have same charge magnitude, and p is farther) **2.** all tie **3.** (a) toward positive y; (b) toward positive x; (c) toward negative y **4.** (a) leftward; (b) leftward; (c) decrease **5.** (a) all tie; (b) 1 and 3 tie, then 2 and 4 tie **Q** **1.** (a) toward positive x; (b) downward and to the right; (c) A **3.** two points: one to the left of the particles, the other between the protons **5.** (a) yes; (b) toward; (c) no (the field vectors are not along the same line); (d) cancel; (e) add; (f) adding components; (g) toward negative y **7.** e, b, then a and c tie, then d (zero) **9.** (a) toward the bottom; (b) 2 and 4 toward the bottom, 3 toward the top **11.** (a) 4, 3, 1, 2; (b) 3, then 1 and 4 tie, then 2 **EP** **1.** (a) 6.4×10^{-18} N; (b) 20 N/C **5.** 56 pC **7.** 3.07×10^{21} N/C, radially outward **9.** 50 cm from q_1 and 100 cm from q_2 **11.** 0 **13.** 1.02×10^5 N/C, upward **15.** 6.88×10^{-28} C·m **21.** $q/\pi^2\varepsilon_0 r^2$, vertically downward **23.** (a) $-q/L$; (b) $q/4\pi\varepsilon_0 a(L + a)$ **27.** $R/\sqrt{3}$ **29.** 3.51×10^{15} m/s^2 **31.** 6.6×10^{-15} N **33.** (a) 1.5×10^3 N/C; (b) 2.4×10^{-16} N, up; (c) 1.6×10^{-26} N; (d) 1.5×10^{10} **35.** (a) 1.92×10^{12} m/s^2; (b) 1.96×10^5 m/s **37.** $-5e$ **39.** (a) 2.7×10^6 m/s; (b) 1000 N/C **41.** 27 μm **43.** (a) yes; (b) upper plate, 2.73 cm **45.** (a) 0; (b) 8.5×10^{-22} N·m; (c) 0 **47.** $(1/2\pi)\sqrt{pE/I}$

CHAPTER 24

CP **1.** (a) $+E\Lambda$; (b) $-EA$; (c) 0; (d) 0 **2.** (a) 2; (b) 3; (c) 1 **3.** (a) equal; (b) equal; (c) equal **4.** (a) $+50e$; (b) $-150e$ **5.** 3 and 4 tie, then 2, 1 **Q** **1.** (a) 8 N·m^2/C; (b) 0 **3.** (a) all four; (b) neither (they are equal) **5.** (a) S_3, S_2, S_1; (b) all tie; (c) S_3, S_2, S_1; (d) all tie (zero) **7.** 2σ, σ, 3σ; or 3σ, σ, 2σ **9.** (a) all tie ($E = 0$); (b) all tie **EP** **1.** (a) 693 kg/s; (b) 693 kg/s; (c) 347 kg/s; (d) 347 kg/s; (e) 575 kg/s **3.** (a) 0; (b) -3.92 N·m^2/C; (c) 0; (d) 0 for each field **5.** 2.0×10^5 N·m^2/C **7.** (a) 8.23 N·m^2/C; (b) 8.23 N·m^2/C; (c) 72.8 pC in each case **9.** 3.54 μC **11.** 0 through each of the three faces meeting at q, $q/24\varepsilon_0$ through each of the other faces **13.** (a) 37 μC; (b) 4.1×10^6 N·m^2/C **15.** (a) -3.0×10^{-6} C; (b) $+1.3 \times 10^{-5}$ C **17.** 5.0 μC/m **19.** (a) $E = q/2\pi\varepsilon_0 LR$, radially inward; (b) $-q$ on both inner and outer surfaces; (c) $E = q/2\pi\varepsilon_0 Lr$, radially outward **21.** (a) 2.3×10^6 N/C, radially out; (b) 4.5×10^5 N/C, radially in **23.** 3.6 nC **25.** (b) $\rho R^2/2\varepsilon_0 r$ **27.** (a) 5.3×10^7 N/C; (b) 60 N/C **29.** 5.0 nC/m^2 **31.** 0.44 mm **33.** (a) $\rho x/\varepsilon_0$; (b) $\rho d/2\varepsilon_0$ **35.** -7.5 nC **39.** -1.04 nC **43.** (a) $E = (q/4\pi\varepsilon_0 a^3)r$; (b) $E = q/4\pi\varepsilon_0 r^2$; (c) 0; (d) 0; (e) inner, $-q$; outer, 0 **45.** $q/2\pi a^2$ **47.** $6K\varepsilon_0 r^3$

CHAPTER 25

CP **1.** (a) negative; (b) increase **2.** (a) positive; (b) higher **3.** (a) rightward; (b) 1, 2, 3, 5: positive; 4, negative; (c) 3, then 1, 2, and 5 tie, then 4 **4.** all tie **5.** a, c (zero), b **6.** (a) 2, then 1 and 3 tie; (b) 3; (c) accelerate leftward **Q** **1.** (a) higher; (b) positive; (c) negative; (d) all tie **3.** $-4q/4\pi\varepsilon_0 d$ **5.** (a)–(c) $Q/4\pi\varepsilon_0 R$; (d) a, b, c **7.** (a) 2, 4, and then a tie of 1, 3, and 5 (where $E = 0$); (b) negative x direction; (c) positive x direction **9.** (a)–(d) zero **EP** **1.** (a) 3.0×10^5 C; (b) 3.6×10^6 J **3.** (a) 3.0×10^{10} J; (b) 7.7 km/s; (c) 9.0×10^4 kg **5.** 8.8 mm **7.** (a) 136 MV/m; (b) 8.82 kV/m **9.** (b) because $V = 0$ point is chosen differently; (c) $q/(8\pi\varepsilon_0 R)$; (d) potential differences are independent of the choice of $V = 0$ point **11.** (a) $Q/4\pi\varepsilon_0 r$; (b) $\dfrac{\rho}{3\varepsilon_0}\left(\dfrac{3}{2}r_2^2 - \dfrac{1}{2}r^2 - \dfrac{r_1^3}{r}\right)$, $\rho = \dfrac{Q}{\dfrac{4\pi}{3}(r_2^3 - r_1^3)}$; (c) $\dfrac{\rho}{2\varepsilon_0}(r_2^2 - r_1^2)$, with ρ as in (b); (d) yes **13.** (a) -4.5 kV; (b) -4.5 kV **15.** $x = d/4$ and $x = -d/2$ **17.** (a) 0.54 mm; (b) 790 V **19.** 6.4×10^8 V **21.** $2.5q/4\pi\varepsilon_0 d$ **25.** (a) $-5Q/4\pi\varepsilon_0 R$; (b) $-5Q/4\pi\varepsilon_0(z^2 + R^2)^{1/2}$ **27.** $(\sigma/8\varepsilon_0)[(z^2 + R^2)^{1/2} - z]$ **29.** $(c/4\pi\varepsilon_0)[L - d\ln(1 + L/d)]$ **31.** 17 V/m at $135°$ counterclockwise from $+x$

35. (a) $\dfrac{Q}{4\pi\varepsilon_0 d(d+L)}$, leftward; (b) 0   **37.** $-0.21q^2/\varepsilon_0 a$

39. (a) $+6.0\times10^4$ V; (b) -7.8×10^5 V; (c) 2.5 J; (d) increase;

(e) same; (f) same   **41.** $W-\dfrac{qQ}{8\pi\varepsilon_0}\left(\dfrac{1}{r_1}-\dfrac{1}{r_2}\right)$ **43.** 2.5 km/s

45. (a) 0.225 J; (b) A, 45.0 m/s²; B, 22.5 m/s²; (c) A, 7.75 m/s;
B, 3.87 m/s   **47.** 0.32 km/s   **49.** 1.6×10^{-9} m
51. 2.5×10^{-8} C   **53.** (a) -180 V; (b) 2700 V, -8900 V
55. (a) -0.12 V; (b) 1.8×10^{-8} N/C, radially inward

CHAPTER 26

CP   **1.** (a) same; (b) same   **2.** (a) decreases; (b) increases;
(c) decreases   **3.** (a) V, $q/2$; (b) $V/2$, q **4.** (a) $q_0=q_1+q_{34}$;
(b) equal (C_3 and C_4 are in series)   **5.** (a) same; (b)–(d) increase; (e) same (same potential difference across same plate separation)   **6.** (a) same; (b) decrease; (c) increase   **Q**   **1.** a, 2;
b, 1; c, 3   **3.** a, series; b, parallel; c, parallel   **5.** (a) $C/3$;
(b) $3C$; (c) parallel   **7.** (a) same; (b) same; (c) more; (d) more
9. (a) 2; (b) 3; (c) 1   **11.** (a) increases; (b) increases; (c) decreases; (d) decreases; (e) same, increases, increases, increases
EP   **1.** 7.5 pC   **3.** 3.0 mC   **5.** (a) 140 pF; (b) 17 nC
7. $5.04\pi\varepsilon_0 R$ **11.** 9090   **13.** 3.16 μF   **17.** 43 pF
19. (a) 50 V; (b) 5.0×10^{-5} C; (c) 1.5×10^{-4} C

21. $q_1=\dfrac{C_1C_2+C_1C_3}{C_1C_2+C_1C_3+C_2C_3}\,C_1V_0$,

$q_2=q_3=\dfrac{C_2C_3}{C_1C_2+C_1C_3+C_2C_3}\,C_1V_0$

23. 72 F   **25.** 0.27 J   **27.** (a) 2.0 J   **29.** (a) $2V$;
(b) $U_i=\varepsilon_0 AV^2/2d$, $U_f=2U_i$; (c) $\varepsilon_0 AV^2/2d$ **35.** Pyrex
37. 81 pF/m   **39.** 0.63 m²   **43.** (a) 10 kV/m; (b) 5.0 nC;
(c) 4.1 nC

45. (a) $C=4\pi\varepsilon_0\kappa\left(\dfrac{ab}{b-a}\right)$; (b) $q=4\pi\varepsilon_0\kappa V\left(\dfrac{ab}{b-a}\right)$;
(c) $q'=q(1-1/\kappa)$

CHAPTER 27

CP   **1.** 8 A, rightward   **2.** (a)–(c) rightward   **3.** a and c tie,
then b **4.** Device 2   **5.** (a) and (b) tie, then (d), then (c)
Q   **1.** a, b, and c tie, then d (zero)   **3.** b, a, c **5.** tie of A, B,
and C, then tie of $A+B$ and $B+C$, then $A+B+C$
7. (a)–(c) 1 and 2 tie, then 3   **9.** C, A, B **EP**   **1.** (a) 1200 C;
(b) 7.5×10^{21} **3.** 5.6 ms   **5.** (a) 6.4 A/m², north; (b) no, cross-sectional area   **7.** 0.38 mm   **9.** (a) 2×10^{12}; (b) 5000;
(c) 10 MV   **11.** 13 min   **13.** 2.0×10^{-8} $\Omega\cdot$m   **15.** 100 V
17. 2.4 Ω **19.** 54 Ω **21.** 3.0   **23.** 8.2×10^{-4} $\Omega\cdot$m
25. 2000 K   **27.** (a) 0.43%, 0.0017%, 0.0034%
29. (a) $R=\rho L/\pi ab$ **31.** 560 W   **33.** (a) 1.0 kW; (b) 25¢
35. 0.135 W   **37.** (a) 10.9 A; (b) 10.6 Ω; (c) 4.5 MJ
39. 660 W   **41.** (a) 3.1×10^{11}; (b) 25 μA; (c) 1300 W, 25 MW
43. (a) 17 mV/m; (b) 243 J   **45.** (a) $J=I/2\pi r^2$;
(b) $E=\rho I/2\pi r^2$; (c) $\Delta V=\rho I(1/r-1/b)/2\pi$; (d) 0.16 A/m²;
(e) 16 V/m; (f) 0.16 MV

CHAPTER 28

CP   **1.** (a) rightward; (b) all tie; (c) b, then a and c tie;
(d) b, then a and c tie   **2.** (a) all tie; (b) R_1, R_2, R_3 **3.** (a) less;

(b) greater; (c) equal   **4.** (a) $V/2$, i; (b) V, $i/2$ **5.** (a) 1, 2, 4, 3;
(b) 4, tie of 1 and 2; then 3   **Q**   **1.** 3, 4, 1, 2   **3.** (a) no;
(b) yes; (c) all tie   **5.** parallel, R_2, R_1, series   **7.** (a) same;
(b) same; (c) less; (d) more   **9.** (a) less; (b) less; (c) more
11. c, b, a **EP**   **1.** (a) \$320; (b) 4.8 cents   **3.** 14 h 24 min
5. (a) 0.50 A; (b) $P_1=1.0$ W, $P_2=2.0$ W; (c) $P_1=6.0$ W
supplied, $P_2=3.0$ W absorbed   **7.** (a) 14 V; (b) 100 W;
(c) 600 W; (d) 10 V, 100 W   **9.** (a) 50 V; (b) 48 V; (c) B is
connected to the negative terminal   **11.** 2.5 V   **13.** 8.0 Ω
15. (a) r_1-r_2; (b) battery with r_1 **19.** 5.56 A   **21.** $i_1=$
50 mA, $i_2=60$ mA, $V_{ab}=9.0$ V   **23.** (a) bulb 2; (b) bulb 1
25. $3d$ **27.** nine   **29.** (a) $R=r/2$; (b) $P_{\max}=\mathcal{E}^2/2r$
31. (a) 0.346 W; (b) 0.050 W; (c) 0.709 W; (d) 1.26 W;
(e) -0.158 W   **33.** (a) battery 1, 0.67 A down; battery 2, 0.33 A
up; battery 3, 0.33 A up; (b) 3.3 V   **35.** (a) Cu: 1.11 A, Al:
0.893 A; (b) 126 m   **37.** 0.45 A   **39.** -3.0% **45.** 4.6
47. (a) 2.41 μs; (b) 161 pF   **49.** (a) 0.955 μC/s; (b) 1.08 μW;
(c) 2.74 μW; (d) 3.82 μW   **51.** (a) 2.17 s; (b) 39.6 mV
53. (a) 1.0×10^{-3} C; (b) 10^{-3} A; (c) $V_C=10^3e^{-t}$ V,
$V_R=10^3e^{-t}$ V; (d) $P=e^{-2t}$ W   **55.** (a) at $t=0$, $i_1=1.1$ mA,
$i_2=i_3=0.55$ mA; at $t=\infty$, $i_1=i_2=0.82$ mA, $i_3=0$;
(c) at $t=0$, $V_2=400$ V; at $t=\infty$, $V_2=600$ V; (d) after several
time constants ($\tau=7.1$ s) have elapsed

CHAPTER 29

CP   **1.** a, $+z$; b, $-x$; c, $\vec{F}_B=0$ **2.** (a) 2, then tie of 1 and 3
(zero); (b) 4   **3.** (a) $+z$ and $-z$ tie, then $+y$ and $-y$ tie, then
$+x$ and $-x$ tie (zero); (b) $+y$ **4.** (a) electron; (b) clockwise
5. $-y$ **6.** (a) all tie; (b) 1 and 4 tie, then 2 and 3 tie
Q   **1.** (a) no, $\vec{v}$ and $\vec{F}_B$ must be perpendicular; (b) yes; (c) no, $\vec{B}$
and $\vec{F}_B$ must be perpendicular   **3.** (a) $\vec{F}_E$; (b) $\vec{F}_B$ **5.** (a) negative; (b) equal; (c) equal; (d) half-circle   **7.** (a) $\vec{B}_1$; (b) B_1 into
page, B_2 out of page; (c) less
9. (a) 1, 180°; 2, 270°; 3, 90°; 4, 0°; 5, 315°; 6, 225°; 7, 135°;
8, 45°; (b) 1 and 2 tie, then 3 and 4 tie;
(c) 8, then 5 and 6 tie, then 7   **EP**   **1.** (a) 6.2×10^{-18} N;
(b) 9.5×10^8 m/s²; (c) remains equal to 550 m/s
3. (a) 400 km/s; (b) 835 eV   **5.** (a) east;
(b) 6.28×10^{14} m/s²; (c) 2.98 mm   **7.** (a) 3.4×10^{-4} T, horizontal and to the left as viewed along $\vec{v}_0$; (b) yes, if its velocity is
the same as the electron's velocity   **9.** 0.27 mT   **11.** 680 kV/m
13. (b) 2.84×10^{-3} **15.** 21 μT   **17.** (a) 2.05×10^7 m/s;
(b) 467 μT; (c) 13.1 MHz; (d) 76.3 ns   **19.** (a) 0.978 MHz;
(b) 96.4 cm   **23.** (a) 1.0 MeV; (b) 0.5 MeV   **25.** (a) 495 mT;
(b) 22.7 mA; (c) 8.17 MJ   **27.** (a) 0.36 ns; (b) 0.17 mm;
(c) 1.5 mm   **29.** (a) $-q$; (b) $\pi m/qB$ **31.** 240 m   **33.** 28.2 N,
horizontally west   **35.** 467 mA, from left to right   **37.** 0.10 T,
at 31° from the vertical   **39.** 4.3×10^{-3} N$\cdot$m, negative y
43. $2\pi aiB\sin\theta$, normal to the plane of the loop (up)
45. (a) 540 Ω, connected in series with the galvanometer;
(b) 2.52 Ω, connected in parallel   **47.** 2.45 A
49. (a) 12.7 A; (b) 0.0805 N$\cdot$m   **51.** (a) 0.30 J/T;
(b) 0.024 N$\cdot$m   **53.** (a) 2.86 A$\cdot$m²; (b) 1.10 A$\cdot$m²
55. (a) $(8.0\times10^{-4}$ N$\cdot$m$)(-1.2\hat{i}-0.90\hat{j}+1.0\hat{k})$; (b) $-6.0\times$
10^{-4} J   **57.** $-(0.10$ V/m$)\hat{k}$ **59.** -2.0 T

CHAPTER 30

CP **1.** a, c, b **2.** b, c, a **3.** d, tie of a and c, then b
4. d, a, tie of b and c (zero)   **Q** **1.** c, d, then a and b tie
3. c, a, b **5.** (a) 1, 3, 2; (b) less **7.** c and d tie, then b, a
9. d, then tie of a and e, then b, c **EP** **1.** (a) 3.3 μT; (b) yes
3. (a) 16 A; (b) west to east **5.** (a) $\mu_0 qvi/2\pi d$, antiparallel to i;
(b) same magnitude, parallel to i **7.** 2 rad **9.** $\dfrac{\mu_0 i\theta}{4\pi}\left(\dfrac{1}{b} - \dfrac{1}{a}\right)$,
out of page **19.** $(\mu_0 i/2\pi w)\ln(1 + w/d)$, up **21.** (a) it is im-
possible to have other than $B = 0$ midway between them;
(b) 30 A **23.** 4.3 A, out of page **25.** 80 μT, up the page
27. $0.791\mu_0 i^2/\pi a$, 162° counterclockwise from the horizontal
29. 3.2 mN, toward the wire **31.** (a) $(-2.0\ \text{A})\mu_0$; (b) 0
35. $\mu_0 J_0 r^2/3a$ **41.** 0.30 mT **43.** (a) 533 μT; (b) 400 μT
47. (a) 4.77 cm; (b) 35.5 μT **49.** 0.47 A $\cdot$ m^2
51. (a) 2.4 A $\cdot$ m^2; (b) 46 cm **57.** (a) 79 μT;
(b) 1.1×10^{-6} N $\cdot$ m

CHAPTER 31

CP **1.** b, then d and e tie, and then a and c tie (zero) **2.** a and
b tie, then c (zero) **3.** c and d tie, then a and b tie **4.** b, out; c,
out; d, into; e, into **5.** d and e **6.** (a) 2, 3, 1 (zero); (b) 2, 3, 1
7. a and b tie, then c **Q** **1.** (a) all tie (zero); (b) 2, then tie of
1 and 3 (zero) **3.** (a) into; (b) counterclockwise; (c) larger
5. c, a, b **7.** c, b, a **9.** (a) more; (b) same; (c) same;
(d) same (zero) **EP** **1.** 1.5 mV **3.** (a) 31 mV; (b) right to
left **5.** (a) 1.1×10^{-3} Ω; (b) 1.4 T/s **7.** 30 mA
9. (a) $\mu_0 iR^2\pi r^2/2x^3$; (b) $3\mu_0 i\pi R^2 r^2 v/2x^4$; (c) in the same direction as the current in the large loop **11.** (b) no **13.** 29.5 mC
15. (a) 21.7 V; (b) counterclockwise **17.** (b) design it so that
$Nab = (5/2\pi)$ m^2 **19.** 5.50 kV **21.** 80 μV, clockwise
23. (a) 13 μWb/m; (b) 17%; (c) 0 **25.** 3.66 μW
27. (a) 48.1 mV; (b) 2.67 mA; (c) 0.128 mW **29.** (a) 600 mV,
up the page; (b) 1.5 A, clockwise; (c) 0.90 W; (d) 0.18 N;
(e) same as (c) **31.** (a) 240 μV; (b) 0.600 mA; (c) 0.144 μW;
(d) 2.88×10^{-8} N; (e) same as (c) **33.** (a) 71.5 μV/m;
(b) 143 μV/m **37.** 0.10 μWb **41.** let the current change at
5.0 A/s **43.** (b) so that the changing magnetic field of one does
not induce current in the other; (c) $L_{\text{eq}} = \displaystyle\sum_{j=1}^{N} L_j$ **45.** $6.91\tau_L$
47. 46 Ω **49.** (a) 8.45 ns; (b) 7.37 mA **51.** 12.0 A/s
53. (a) $i_1 = i_2 = 3.33$ A; (b) $i_1 = 4.55$ A, $i_2 = 2.73$ A; (c) $i_1 = 0$, $i_2 = 1.82$ A (reversed); (d) $i_1 = i_2 = 0$ **55.** (a) $i(1 - e^{-Rt/L})$
57. 25.6 ms **59.** (a) 97.9 H; (b) 0.196 mJ **63.** (a) 34.2 J/m^3;
(b) 49.4 mJ **65.** 1.5×10^8 V/m **67.** (a) 1.0 J/m^3; (b) 4.8 $\times$
10^{-15} J/m^3 **69.** (a) 1.67 mH; (b) 6.00 mWb **71.** (b) have the
turns of the two solenoids wrapped in opposite directions
73. magnetic field exists only within the cross section of
solenoid 1 **75.** (a) $\dfrac{\mu_0 Nl}{2\pi} \ln\left(1 + \dfrac{b}{a}\right)$; (b) 13 μH

CHAPTER 32

CP **1.** d, b, c, a (zero) **2.** (a) 2; (b) 1 **3.** (a) away; (b) away;
(c) less **4.** (a) toward; (b) toward; (c) less **5.** a, c, b, d (zero)
6. tie of b, c, and d, then a **Q** **1.** supplied **3.** (a) all down;

(b) 1 up, 2 down, 3 zero **5.** (a) 1 up, 2 up, 3 down; (b) 1 down,
2 up, 3 zero **7.** (a) 1, up; 2, up; 3, down; (b) and (c) 2, then 1
and 3 tie **9.** (a) rightward; (b) leftward; (c) into **11.** 1, a; 2, b;
3, c and d **EP** **1.** (b) sign is minus; (c) no, there is com-
pensating positive flux through open end near magnet
3. 47.4 μWb, inward **5.** 55 μT **7.** (a) 31.0 μT, 0°;
(b) 55.9 μT, 73.9°; (c) 62.0 μT, 90° **9.** (a) $-9.3 \times$
10^{-24} J/T; (b) 1.9×10^{-23} J/T **11.** (a) 0; (b) 0; (c) 0;
(d) $\pm 3.2 \times 10^{-25}$ J; (e) -3.2×10^{-34} J $\cdot$ s, 2.8×10^{-23} J/T,
$+9.7 \times 10^{-25}$ J, $\pm 3.2 \times 10^{-25}$ J **13.** $\Delta\mu = e^2 r^2 B/4m$
15. 20.8 mJ/T **17.** yes **19.** (b) K_i/B, opposite to the field;
(c) 310 A/m **21.** (a) 3.0 μT; (b) 5.6×10^{-10} eV
23. 5.15×10^{-24} A $\cdot$ m^2 **25.** (a) 180 km; (b) 2.3×10^{-5}
27. 2.4×10^{13} V/m $\cdot$ s **33.** (a) 0.63 μT; (b) 2.3×10^{12}
V/m $\cdot$ s **35.** (a) 710 mA; (b) 0; (c) 1.1 A **37.** (a) 2.0 A;
(b) 2.3×10^{11} V/m $\cdot$ s; (c) 0.50 A; (d) 0.63 μT $\cdot$ m

CHAPTER 33

CP **1.** (a) $T/2$, (b) T, (c) $T/2$, (d) $T/4$ **2.** (a) 5 V; (b) 150 μJ
3. (a) remains the same; (b) remains the same **4.** (a) C, B, A;
(b) 1, A; 2, B; 3, S; 4, C; (c) A **5.** (a) remains the same;
(b) increases **6.** (a) remains the same; (b) decreases **7.** (a) 1,
lags; 2, leads; 3, in phase; (b) 3 ($\omega_d = \omega$ when $X_L = X_C$)
8. (a) increase (circuit is mainly capacitive; increase C to decrease X_C to be closer to resonance for maximum P_{avg}); (b) closer
9. (a) greater; (b) step-up **Q** **1.** (a) $T/4$; (b) $T/4$; (c) $T/2$ (see
Fig. 33-2); (d) $T/2$ (see Eq. 31-37) **3.** b, a, c **5.** (a) 3, 1, 2;
(b) 2, tie of 1 and 3 **7.** a, inductor; b, resistor; c, capacitor
9. (a) leads; (b) capacitive; (c) less **11.** (a) rightward, increase
(X_L increases, closer to resonance); (b) rightward, increase (X_C
decreases, closer to resonance); (c) rightward, increase (ω_d/ω increases, closer to resonance)   **EP** **1.** 9.14 nF **3.** (a) 1.17 μJ;
(b) 5.58 mA **5.** with n a positive integer: (a) $t = n(5.00\ \mu\text{s})$;
(b) $t = (2n - 1)(2.50\ \mu\text{s})$; (c) $t = (2n - 1)(1.25\ \mu\text{s})$
7. (a) 1.25 kg; (b) 372 N/m; (c) 1.75×10^{-4} m; (d) 3.02 mm/s
9. 7.0×10^{-4} s **11.** (a) 3.0 nC; (b) 1.7 mA; (c) 4.5 nJ
13. (a) 275 Hz; (b) 364 mA **15.** (a) 6.0:1; (b) 36 pF, 0.22 mH
17. (a) 1.98 μJ; (b) 5.56 μC; (c) 12.6 mA; (d) $-46.9°$; (e) $+46.9°$
19. (a) 0.180 mC; (b) $T/8$; (c) 66.7 W **21.** (a) 356 μs;
(b) 2.50 mH; (c) 3.20 mJ **23.** Let T_2 ($= 0.596$ s) be the period
of the inductor plus the 900 μF capacitor and let T_1
($= 0.199$ s) be the period of the inductor plus the 100 μF capacitor. Close S_2, wait $T_2/4$; quickly close S_1, then open S_2; wait
$T_1/4$ and then open S_1. **25.** 8.66 mΩ **27.** $(L/R)\ln 2$
31. (a) 0.0955 A; (b) 0.0119 A **33.** (a) 0.65 kHz; (b) 24 Ω
35. (a) 6.73 ms; (b) 11.2 ms; (c) inductor; (d) 138 mH
37. (a) $X_C = 0$, $X_L = 86.7\ \Omega$, $Z = 182\ \Omega$, $I = 198$ mA,
$\phi = 28.5°$ **39.** (a) $X_C = 37.9\ \Omega$, $X_L = 86.7\ \Omega$, $Z = 167\ \Omega$,
$I = 216$ mA, $\phi = 17.1°$ **41.** 1000 V **43.** 89 Ω
45. (a) 224 rad/s; (b) 6.00 A; (c) 228 rad/s, 219 rad/s; (d) 0.040
49. 1.84 A **51.** 141 V **53.** 0, 9.00 W, 3.14 W, 1.82 W
55. (a) 12.1 Ω; (b) 1.19 kW **57.** (a) 0.743; (b) leads;
(c) capacitive; (d) no; (e) yes, no, yes; (f) 33.4 W
59. (a) 117 μF; (b) 0; (c) 90.0 W, 0; (d) 0°, 90°; (e) 1, 0
61. (a) 2.59 A; (b) 38.8 V, 159 V, 224 V, 64.2 V, 75.0 V;
(c) 100 W for R, 0 for L and C **63.** (a) 2.4 V; (b) 3.2 mA,
0.16 A **65.** 10

PHOTO CREDITS

INDEX

Figures are noted by page numbers in *italics;* tables are indicated by t following the page number.

Proper frequency, 936
Proper length, 930
Proper time interval, 926
Proper wavelength, 937–938
Proton(s), 507–508. *See also* Nucleus (nuclei), atomic
  cosmic ray, 519 (problem 25)
  magnetic resonance of, 1015–1017
Proton–antiproton annihilation, 1121
Proton number, 1065
Proton-proton (p-p) cycle, 1106–1108
Pulley, massless, frictionless, 83, *83*
Pulsar, 9 (problem 18)

Q

QED (quantum electrodynamics), 749, 1132, 1133
Quadrupole field, 1142 (problem 41)
Quadrupole, electric, 540 (problem 17)
Quantization, and wave confinement, 980
Quantized electric charge, 514–515
Quantum, 954
Quantum chromodynamics (QCD), 1133
Quantum corrals, 979, *979*, 991
Quantum dots, 991
Quantum electrodynamics (QED), 749, 1132, 1133
Quantum jump, 983
Quantum mechanics. *See* Quantum physics
Quantum number(s), 981
  for hydrogen, 996
  orbital, 996
  orbital magnetic, 996
  principal, 996
Quantum physics, 954
  and atomic structure, 980
  and gas molecules, 472–473
  matter waves and, 980. *See also* Matter waves
Quantum states. *See also* Occupancy probability
  density of, 1042–1043
Quantum theory. *See* Quantum physics
Quark(s), 1129
  and baryons, 1129–1130
  and beta decay, 1130–1131
  charm, top and bottom, 1131
  forces between, 1133
  and mesons, 1130
Quark flavors, 1129
Quartz, crystalline, 972
Quasar(s), 1134

R

R-value, 443
Rad (unit), 1080
Radar, ground penetrating, 832 (problem 64)
Radial motion, 937
Radial probability density, 997
Radians, 37
  and definition of angles, 216
Radiation
  absorption of, 812

cosmic background, 1135–1136
  of electromagnetic waves, 803–804
  reflection of, 812
  thermal, 444–445
Radiation dosage, measuring, 1080
Radiation pressure, of electromagnetic waves, 812–814
Radio waves, 802–803
Radioactive dating, 1079
Radioactive decay, 515–516, 1070–1074
Radioactive elements, 1063
Radioactivity, and nuclidic chart, 1077–1078
Radionuclide(s), 1065
  activity of, 1072
  half-life of, 1072
  mass excess of, 1077–1078
Radionuclide decay, 1065
Radius of gyration, 242 (problem 43)
Rail gun, 693–694
Rainbow
  formation of, 821
  Maxwell's, 802, 827
  primary, 827
  secondary, 827
Rayleigh's criterion, for resolvability, 899
Rays, 399
  light. *See* Light waves
  x. *See* X-rays
RBE (relative biological effectiveness), 1080
RC circuits, 648–651
Real images, 834
  reflected, 839
  refracted, 842
Rectangular box, 1017
Rectangular corral, 1017
Red giant star, 1107
Red-shift, cosmological, 1143 (problem 43)
Reference circle, 359
Reference configuration, 144
Reference frame(s). *See also* System(s)
  inertial, 74–75, 920
  of motion, 62–63
  noninertial, 75
  in relativity, 920–921
Reference point/position, 144
Reflecting planes, 910
Reflection
  of electromagnetic waves, 818–819
  image formation by, 834
    location of, by ray drawing, 840
    mirror lateral magnification and, 839, 840–841
    plane mirror, 835–836
    spherical mirror, 835–841, 852–853
  law of, 819
  polarization of light by, 824–825
  of radiation, 812
  total internal, 823
Reflection phase shifts, 875–876
Refracting surface formula, 853
Refracting telescope lenses, 851–852
Refraction
  of electromagnetic waves, 818–819

image formation by, 841–843
    formula proof of, 853
    lens lateral magnification and, 840–841, 846
    lenses and, 843–849
    thin-lens formula proof of, 853–855
  indexes of, 819t
  law of, 819
Refrigerator, 494–495
Reines, F., 1077
Relative motion, 62–64
  acceleration and, 62–63
  one-dimensional, 62–63
  two-dimensional, 63–64
Relativity
  and Doppler effect for light, 936–940
  and energy, 941–946
  general theory of, 312, 920
  of length, 929–931, 934
  Lorentz transformation in, 932–935
  and measuring an event, 922–923
  and momentum, 940–941
  postulates of, 921–922
  of simultaneity, 923–925, 934
  special theory of, 920
  of time, 925–929, 934
  of velocities, 936
Relaxed state, of spring force, 126, *126*
Rem (unit), 1080
Resistance, 617–621
  calculating, 618–621
  difference between resistivity and, 618
  internal, 638
  in parallel, 643–644
  in series, 638–639
Resistance rule, 637
Resistive dissipation, 625
Resistive load, 779–781
Resistivity, 618
  calculating, 618–621
  difference between resistance and, 618
  and *Hindenburg* explosion, 620
  of metals, 622–624
  of semiconductors, 1047
  of solids, 1038
  temperature coefficient of, 1038
  and variation with temperature, 620
Resistor, 617–618, 780
Resolvability, 899–900
  of electron microscope, 900
  of pointillistic painting, 899–900
Resolving power, 852, 907
  dispersion compared to, 908–909
  of refraction grating, 907
Resonance, 362–363
  defined, 390
  of electric circuits, 779
  standing waves and, 389–391
Resonance condition, of protons, 672
Resonant frequencies, 390
  of string, 410
Rest energy, 941–942
Rest frame, 927

sound, 399. *See also* Sound waves
speed of, 378–380
standing, 387–389. *See also* Standing waves
string, 378–382. *See also* String waves
superposition of, 382–383
transverse, 372–373
traveling, 372–373
  speed of, 375–376
traveling string, 378–382
types of, 371
wavelength of, 373–375
Wave forms, 372–373
Wave function, 967
  of electrons, 985–988
  normalizing, 987
Wave speed, 375–376
  derivation of, from Newton's second law, 379–380
  dimensional analysis of, 378–379
  on a stretched string, 378–380
Wavefronts, 399, 805
Wavelength
  and angular wave number, 374
  Compton, 961
  cutoff, 956, 1022–1023
  de Broglie, 964
  and frequency, 373–375
  and index of refraction, 864–866
  proper, 937–938
Weak force, 1119, 1132–1133
Weakly interacting massive particles, 1136
Weight, 80–81
  and mass, 81
Well depth, 988–990

Wheeler, John, 1096
White dwarf, 305t
Wilson, Robert, 1135
WIMPS (weakly interacting massive particles), 1136
Work, 118
  and applied forces, 568
  applied forces and, 127–128
  Carnot engine and energy conversion, 489–492
  conservation of energy and, 155–158
  conservative forces and, 141–143
  constant force and, 119
  defined, 118, 152
  emf devices and, 634–635
  expression of, 119
  external forces and, 152–155
    frictional, 153–154
    frictionless, 152
  gravitational force and, 122–125, 140
  heat, volume change and, 438–440
  isothermal volume change of gas and, 457
  kinetic energy and, 119–120. *See also* Work–kinetic energy theorem
    rotational, 233–236
  lifting and lowering an object and, 123–125
  meaning of term, 493
  mechanical energy and, 146–149
  net, 120
  nonconservative forces and, 141. *See also* Frictional force
  positive and negative, 120, 127, 140, 152
  potential energy and, 140–141. *See also* Potential energy; Potential energy curve

power and, 131–133
rotational kinetic energy and, 233–236
spring force and, 126–128
Stirling engine and energy conversion, 492
units of, 120
variable forces and, 119, 126, 129–131
  one-dimensional analysis, 129, *129*
  three-dimensional analysis, 129–130
Work function, 957
Work–kinetic energy theorem, 120–122
  with variable force, 130–131

X

X-rays
  characteristic spectrum of, 1023
  continuous spectrum of, 1022–1023
  diffraction of, 909–911
  and numbering elements, 1022–1026
  scattered, 910

Y

Yield strength, 284
Yo-yo, physics of, 250–251
Young, Thomas, 866–867, 962
Young's interference experiment, 866–870
Young's modulus, 285

Z

Zero angular position, 216
Zero-point energy, 987
Zeroth law, of thermodynamics, 426–427
Zweig, George, 1129